Chapter 8: Ordinary Differential Equations (8.1/8.7/8.22)

SUBROUTINE SDRIV2(N,T,Y,F,TOUT,MSTATE,NROOT,EPS,EWT,MINT,W,LW,IW,LIW,G)
Solves N first order ODEs $Y' = F$ from T to TOUT to accuracy EPS with right hand sides YP calculated by SUBROUTINE F(N,T,Y,YP). Needs EXTERNAL statement for user-selected subroutine name. Y contains initial values and returns solution. Set MINT = 2, EWT = 0.0 for stiff solver using relative error control. Set MSTATE = 1 on first call. Place in loop, update TOUT each call. Needs W(N*N+10*N+2*NROOT+204), IW(N+21). For routine use, set NROOT = 0 and G as same name as selected for F.

Chapter 9: Optimization and Nonlinear Least Squares (9.1/9.3,9.5/9.10)

REAL FUNCTION FMIN(AX,BX,F,EPS)
Finds minimum of REAL FUNCTION F(X) on interval [AX,BX] to accuracy EPS. Needs EXTERNAL statement for user selected function name. Returns FMIN as location of minimum with AX,BX bracketing this value.

SUBROUTINE UNCMIN(N,X0,FCN,X,F,INFO,W,LW)
Finds minimum of N-variable function given initial values X0(N). Returns solution in X(N) and minimum function value in F. Needs EXTERNAL statement for user selected SUBROUTINE FCN(N,X,F) returning value of F at X(N). Needs W(N*N+10*N).

Chapter 10: Simulation and Random Numbers (10.1/10.6,10.8/10.11)

REAL FUNCTION UNI()
Uniform random-number generator on [0,1). Initialize with Z = USTART(ISEED), ISEED $\geq$ 0, then evaluate by repeating Z = UNI() as needed.

REAL FUNCTION RNOR()
Standard normal generator, with zero mean and unit standard deviation. Initialize with Z = RSTART(ISEED) as above, and evaluate via Z = RNOR().

Chapter 11: Trigonometric Approximation and the FFT (11.1/11.6,11.11,11.12 11.16)

SUBROUTINE EZFFTI(N,W)
Initialize for forward/reverse real FFT. Needs W(3*N+15).

SUBROUTINE EZFFTF(N,R,AZERO,A,B,W)
Forward FFT of N data points R(N). Returns A(N/2) and B(N/2) as cosine and sine coefficients, AZERO as mean. Needs W(3*N+15).

SUBROUTINE EZFFTB(N,R,AZERO,A,B,W)
Backward FFT of N data points in AZERO, A(N/2), B(N/2). Returns data in R(N). Needs W(3*N+15).

SUBROUTINE CFFTI(N,W)
Initialize for forward/back complex FFT. Ne

SUBROUTINE CFFTF(N,C,W)
Forward complex FFT of complex C(N). R   l5).

SUBROUTINE CFFTB(N,C,W)
Reverse complex FFT of complex C(N). R   5).

SUBROUTINE CFFT2D(N,F,LDF,W,FORWD)
Two-dimensional complex FFT of N by N complex array F, dimensioned F(LDF,N). Forward or reverse as FORWD is .TRUE. or .FALSE. Needs real W(6*N+15).

Numerical Methods and Software

Programming in the 1940's

Numerical Methods and Software

David Kahaner
National Institute of Standards and Technology
(National Bureau of Standards)

Cleve Moler
Ardent Corporation

Stephen Nash
George Mason University

Prentice Hall, *Englewood Cliffs, New Jersey 07632*

Library of Congress Cataloging-in-Publication Data

KAHANER, DAVID.
 Numerical methods and software / David Kahaner, Cleve Moler,
Stephen Nash.

 Bibliography: p.
 Includes index.
 ISBN 0-13-627258-4
 1. Engineering mathematics—Data processing. I. Moler, Cleve B.
 II. Nash, Stephen. III. Title.
 TA345.K34 1989
 620'.0042—dc19   88-15540

Editorial/production supervision and
   interior design: BARBARA MARTTINE
Cover design: EDSAL ENTERPRISES
Manufacturing buyer: MARY ANN GLORIANDE

 © 1989 by Prentice-Hall, Inc.
A Division of Simon & Schuster
Englewood Cliffs, New Jersey 07632

Previously published as Forsythe, Malcolm and Moler's
COMPUTER METHODS FOR MATHEMATICAL COMPUTATIONS
© 1977 by Prentice Hall

If your diskette is defective or damaged in transit, return it directly to Prentice Hall at the address below for a no-charge replacement within 90 days of the date of purchase. Mail the defective diskette together with your name and address.

Prentice Hall
Attention: College Software
Englewood Cliffs, New Jersey 07632

Printed in the United States of America

10 9 8 7 6 5 4 3 2 1

ISBN   0-13-627258-4

Prentice-Hall International (UK) Limited, *London*
Prentice-Hall of Australia Pty. Limited, *Sydney*
Prentice-Hall Canada Inc., *Toronto*
Prentice-Hall Hispanoamericana, S.A., *Mexico*
Prentice-Hall of India Private Limited, *New Delhi*
Prentice-Hall of Japan, Inc., *Tokyo*
Simon & Schuster Asia Pte. Ltd., *Singapore*
Editora Prentice-Hall do Brasil, Ltda., *Rio de Janeiro*

CONTENTS

3   *Linear Systems of Equations*   *41*

4   *Interpolation*   *81*

5   *Numerical Quadrature*   *138*

ACKNOWLEDGMENTS

The numerical analysis and mathematical software communities are not large. Nevertheless they are blessed with men and women who understand the need to provide the highest quality algorithms and software to other parts of the scientific world. During the writing of this book we have had enthusiastic help and advice from many of these gifted people. The chapters have been scrutinized and the programs tested by specialists. Specifically we would like to note our special appreciation and thanks to Javier Bernal, James Blue, Tom Booth, Ralph Byers, George Byrne, Paul Calamai, Alfred Carasso, Janet Donaldson, Graeme Fairweather, Fred Fritsch, Michael Greene, Mark Johnson, W. Kahan, Bruce Kellogg, Randall Leveque, Daniel Lozier, George Marsaglia, John Nash, Roger Pinkham, Philip Rabinowitz, Bert Rust, Robert Schnabel, C. Dexter Sutherland, Paul Swarztrauber, Skip Thompson, Layne Watson, and H. Joseph Weaver for their comments and ideas. We have tried to incorporate as many of their suggestions as was possible, but of course, any errors, omissions or inconsistencies are entirely our own responsibility. A special note of thanks is due to Gerald Candela who helped to generate many of our figures. In addition we would like to thank the management of the Center for Computing and Applied Mathematics, National Institute of Standards and Technology (National Bureau of Standards), particularly, Paul Boggs, Burton Colvin, Glenn Ingram, and Francis Sullivan for providing the kind of nurturing environment in which mathematical software is a legitimate discipline.

A major motivation for writing this text was to update and expand the material in the book by G. Forsythe, et al., (1977). We liked its philosophy and informal style as well as the inclusion of programs. Our original intent was to produce a revision of that book but with Forsythe deceased and Malcolm involved in other activities the current authors decided to begin afresh. The publisher granted us permission to extract those portions of the original book that seemed appropriate. In practice this meant that several of the problems have been retained. The text is almost all new, although we were certainly inspired by the original. We would also like to acknowledge our debt to all those authors who have written on this subject before us; perhaps we may have made a few matters a little clearer or a little sounder.

Finally, to our long suffering families, thanks for the support and patience.

DAVID KAHANER
CLEVE MOLER
STEPHEN NASH

1

Introduction

1.1 WHY A NEW BOOK?

Computers are changing rapidly. The mainframe computer with its traditional one scalar operation at a time is giving way to a plethora of different architectures. Vector computers such as the Cray 1, CDC Cyber 205, or ETA 10 can now manipulate arrays, take dot products, and do several operations in parallel with a single machine instruction. Multiprocessors provide up to a few thousand highly connected computers that can work more or less independently on different aspects of a single problem. Many researchers are trying to take advantage of these new computer architectures.

The way scientists and engineers communicate with computers is changing too. The slow link between user and remote computer is being replaced by the personal computer or workstation with its high data transmission rate. This in turn is fueling the development of highly interactive and graphically oriented applications requiring little programming effort.

Together with these developments better numerical methods are also appearing. Because of the rapid changes there is a gap between the computational techniques taught in physics, chemistry, and engineering courses and the software that will confront a practising scientist. What is needed is a text in tune with current numerical methods that recognizes the new developments in hardware and communications, but still suitable as a first course. This book fills this role. We give four simple examples. If the terms used below are unfamiliar be assured that they will be explained in the subsequent chapters.

(1) "Splines" are often recommended for fitting a smooth curve through accurate data, but the best current software uses "Hermite cubics." For interactive design, such

as in Computer Aided Design and Computer Aided Manufacturing (CAD/CAM), "Bézier" curves are also a good choice.

(2) Programs for matrix manipulation, such as linear system solvers, can be organized to run very efficiently on both vector and scalar computers by implementing the algorithms in terms of lower level "Basic Linear Algebra Subroutines" (BLAS).

(3) "Adaptive" methods are recommended for the evaluation of definite integrals, but a heavily used software package gets much better results by incorporating "Kronrod" formulas into them.

(4) "Runge Kutta" methods have largely been replaced by "multi-value" methods in modern software for solving differential equations.

The book discusses the topics which we feel are among the most important for a scientist who will be doing numerical computation. This includes many standard items such as solution of systems of linear equations, interpolation, fast Fourier transform, evaluation of integrals and differential equations, optimization, and simulation. In each case we have attempted to bring the discussion into conformance with what people are actually doing in the field. This is illustrated by the examples above. Every chapter also includes as many up-to-date ideas as is practical to present. For example, the linear equations chapter describes column-oriented algorithms, the optimization chapter explains the use of quasi-Newton methods, the simulation chapter introduces fractals, etc.

The book can be used for a one-semester course in numerical methods by following the plan given below. Each chapter has been organized to permit students to get to the software as rapidly as possible. The fundamental ideas on each major topic have been presented early in the chapter so that the section describing the software can provide a convenient stopping point. Later sections cover more advanced material and may be used selectively as time allows.

PLAN FOR A ONE-SEMESTER COURSE

Chapter 1

Chapter 2: Sections 1–4   Computer arithmetic and computational errors.

Chapter 3: Sections 1–3   Solving stored linear systems of equations.

Chapter 4: Sections 1–9   Polynomial and piecewise polynomial interpolation.

Chapter 5: Sections 1–7   Quadrature and adaptive quadrature.

Chapter 6: Sections 1–5   Solving linear least squares problems.

Chapter 7: Sections 1–3   Solving a single nonlinear equation.

Chapter 10: Sections 1–6   Random numbers (time permitting).

The chapters need not all be taught, and need not be taught in order. In particular, once the first seven chapters have been covered, the final four chapters may be taught in

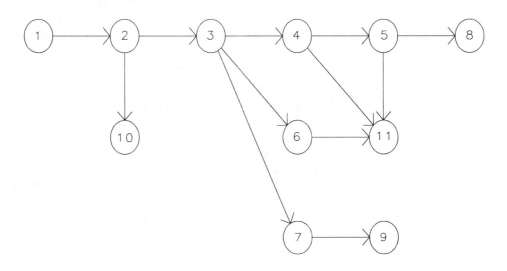

Note: 1. Chapter 9 Section 6 depends on Chapter 6

2. Chapter 8 Section 12 depends on Chapter 7

Figure 1.1 Chapter Dependencies

any order. A more detailed picture of the dependencies among the chapters is given in Figure 1.1.

The one-semester course plan covers only the most important topics in Chapters 1–7. In a two-semester course an instructor may choose to cover the basic material in the remaining four chapters, or to teach fewer topics in more detail.

Some sections have been marked with a ∗ because of the somewhat higher level of mathematical sophistication required. These sections may be skipped without interrupting the flow of material. Even with these omissions, there is enough basic material for a two-semester course. We have deliberately omitted some important topics either because they are too advanced to treat in a single chapter (partial differential equations) or seem of more specialized interest (eigenvalues).

This book is written for students of science and engineering. It is intermediate between a cookbook and a numerical analysis text. The reader is assumed to have completed two years of university mathematics including differential and integral calculus as well as a little matrix theory and differential equations. The reader is also assumed to have access to a computer and have rudimentary knowledge of the Fortran language. At the conclusion of a course organized around this text, a student will

(1) understand the capabilities and limitations of the techniques used in numerical scientific computing,

(2) have used more than a dozen of the most highly regarded subroutines implementing these algorithms,

(3) have sharpened Fortran programming skills, ample for "real" applications.

1.1.1 COMMENTS ON THE ENCLOSED FLOPPY DISK

A 5.25 inch floppy disk has been included with this text. It contains the source of all the Fortran routines that are discussed, as well as subsidiary routines and data files associated with the problems. The disk contains a READ-ME.NMS file which explains in detail how the files are organized. The disk has a double-sided 360KB format readable by any personal computer with an MS-DOS 2 or 3 operating system, such as the IBM PC/XT or PC/AT. Questions regarding the software may be directed to

David Kahaner   or   Stephen Nash
NBS, Center for Applied Mathematics   George Mason University
Technology Building, Room A151   Operations Research Department
Gaithersburg, MD 20899   Fairfax, VA 22030

The same software is also available for a fee in other formats and precisions. For information contact one of the authors.

1.2 THE SUBROUTINES

A significant part of this book is a set of Fortran subroutines. In fact, the book might well be regarded as an extensive "user's guide" for them. These are not simple illustrative programs, but rather are representative of the state of the art in current mathematical software. Together they form a useful software library. For the areas of computation discussed here there exist well written and documented routines which are in worldwide use because of their robustness and ease of application. They appear in collections such as Linpack, Quadpack, Minpack, SLATEC, and others, which are known and used throughout the scientific community. From these we have selected what we feel are the most useful for inclusion. Further, *we have resisted the temptation to write our own versions.* Thus the student derives the double benefit of having access to some of the best routines available and having the satisfaction of knowing that they will appear again in later professional activities. The only exceptions are (a) we have standardized the documentation of each routine to agree with the conventions of the Department of Energy SLATEC Library, widely used within many U.S. national laboratories, and (b) in those cases when experience suggests that a particularly popular subroutine will substantially benefit from a simple user interface, this has been written on top of the original. The main reasons for designing a text around these routines are to (a) avoid "reinventing the wheel" in writing software for well understood mathematical problems, and (b) to promote quality software that has been carefully engineered, documented, and tested (Gaffney, 1987).

We provide here a list of the routines which have been included. More complete information about their origins can be found in the Bibliography.

Linear Systems of Equations

SGEFS:

>   For solving $Ax = b$ with accuracy estimation, from the SLATEC library. This is a driver for the Linpack subroutines SGECO and SGESL.

Interpolation

PCHEZ

PCHEV:

>   For spline and visually pleasing piecewise polynomial manipulation, including interpolation. These are easy to use drivers for the PCHIP package, which is in the SLATEC library.

Quadrature

Q1DA:

>   For evaluation of one dimensional integrals adaptively, including singular integrands. This is an easy to use driver for Quadpack subroutine QK15.

QK15:

>   For quick estimates of one dimensional integrals, without adaption. Uses a fifteen point Gauss Kronrod quadrature rule. From Quadpack.

PCHQA:

>   For integration of data in one dimension. This is an easy to use driver for the PCHIP package, which is in the SLATEC library.

Linear Least-Squares Data Fitting

SQRLS:

>   For finding the least squares solution to a system of equations. This is an easy to use driver for routines from the SLATEC library. The latter in turn call routines from Linpack.

Solution of Nonlinear Equations

FZERO:

>   For solving $f(x) = 0$ in one dimension (a single equation). This is from the SLATEC library.

SNSQE:

>   For solving $f(x) = 0$ in n dimensions (a system of equations). This is an easy to use driver for Minpack routines. It is from the SLATEC library.

Ordinary Differential Equations

SDRIV2:

> For solving $y' = f(t, y)$, including stiff equations. This is an easy to use driver from the SDRIV package from the program library at Los Alamos National Laboratory.

Optimization and Nonlinear Least Squares

FMIN:

> For finding min $F(x)$ in one dimension, by Brent (1973).

UNCMIN:

> For finding min $F(x)$ in n dimensions. This is an easy to use driver for a routine from the paper by Schnabel, Koontz and Weiss (1982).

Random Numbers and Simulation

UNI

RNOR:

> For generation of uniform and normal random numbers, from the SLATEC library.

Trigonometric Approximation and the Fast Fourier Transform

EZFFTF

EZFFTB

CFFTF

CFFTB

CFFT2D:

> For easy to use real and complex fast Fourier transforms (one and two dimensions), from the SLATEC library.

Many of these routines are intricate, and to describe in detail how and why they work would require much longer than most students can spend on this type of material. (See, for example the following section.) Hence it is necessary to treat them as black, or perhaps grey, boxes. This is consistent with the trend to think in terms of modules, systems, or transforms which convert input into output according to well defined rules. Most programmers already treat the standard functions (SIN, EXP, etc.) in exactly this way without understanding them in detail. These are accepted as primitives and are expected to work. As the underlying algorithms become more reliable they (1) become primitives of the language rather than subroutines which must be added, (2) are implemented in hardware for more efficient operation, and (3) disappear from elementary texts. This is exactly the situation with the algorithms for the arithmetic of floating point numbers (i.e., "real" as opposed to integer) and the computation of the standard functions. It is beginning to occur with algorithms for random numbers, fast Fourier transforms, and linear systems of equations.

For the more complicated algorithms it is not sufficient to simply explain how to call each subroutine and say they "usually work." There are many pitfalls in numerical computation and students must be properly warned of these, so that they can correctly diagnose symptoms of numerical ill health. This requires a certain level of understanding of the methods employed; see Chapter 10 for an example of the effects of a poorly designed random number algorithm.

We often consult with scientists who are anxious to study the source text of a subroutine so that they can modify it for some special purpose. In our experience this desire is common, but the need is not. Usually, a well designed subroutine is already capable of dealing with most situations and it is unnecessary and unwise to modify it frivolously. A better approach is to consult with a local expert who might suggest alternatives. For this reason, and also because of the complex nature of the routines, we do not provide listings in the book, but only the subroutine documentation. We hope this will encourage a cautious attitude toward modification. The complete programs as well as the sample problems are available on the disk that is included in the insert at the end of the text, and we expect that readers will copy these onto their computer in order to be able to work the problems in the chapters.

Note that mistakes in programs can take a long time to be uncovered, but because the software in this book is widely used all the known errors have been corrected. The authors will try to fix any new errors that are reported to them. Finally, we emphasize that the development of algorithms continues at a rapid pace and readers are encouraged to check with the program library at their institution for the newest routines.

1.3 MATHEMATICAL SOFTWARE—AN EXAMPLE: SQUARE ROOT OF SUM OF SQUARES

To illustrate some of the issues which must be dealt with in designing a piece of software suitable for inclusion in a heavily used library, or reliable enough to put on a space vehicle, we consider the problem of computing the **Euclidean norm** (length) of an array $a = (a_1, \ldots, a_n)$, defined by

$$\|a\|_2 = \sqrt{\sum_{i=1}^{n} a_i^2}.$$

Such a calculation is required by many of the subroutines in this book. It appears easy, and a Fortran Function to compute it might look like

```
      REAL FUNCTION SNORM (N,A)
C
C Computes the square root of sum of squares of array A,
C of length N.   (VERSION 1)
C
      REAL   A(N)
      INTEGER N, I
C
      SNORM = 0.0
C
```

```
      DO 20 I = 1, N
   20    SNORM = SNORM + A(I)**2
      SNORM = SQRT(SNORM)
      RETURN
      END
```

SNORM is a straightforward implementation of the formula defining $\|a\|_2$, but it doesn't always work. Even if a_i is not too large, its square a_i^2 may be too large to store in the computer and the program will abort: an **overflow** occurs. We discuss this topic more carefully in the next chapter. On the other hand if a_i is near zero, its square may be too small to represent and **underflow** will occur. Underflows are often set to zero automatically. As long as some of the other a_i's are large enough, then $\|a\|_2$ will be correct, but if all the components of a are small then SNORM will return zero.

One way to overcome this problem is to find the largest component of a and divide every component of the vector by it. This scaled vector will have its largest component equal to 1 and the computation of the norm can then be carried out safely. This is illustrated below. Notice that the program also works when $a = 0$ or when $n = 0$.

```
      REAL FUNCTION SNORM (N,A)
C
C Computes the square root of sum of squares of array A,
C of length N.   (VERSION 2)
C Assumes that underflows are set to zero.
C
      REAL    A(*), AMAX
      INTEGER N, I
C
      SNORM = 0.0
      IF (N .LE. 0) RETURN
      AMAX = 0.0
      DO 10 I = 1, N
   10    AMAX = MAX (AMAX, ABS(A(I)))
C
      IF (AMAX .LE. 0.0) RETURN
      DO 20 I = 1, N
   20    SNORM = SNORM + (A(I)/AMAX)**2
      SNORM = AMAX*SQRT(SNORM)
      RETURN
      END
```

This function is more reliable, but it can still be improved. If the vector a has components that are neither too big nor too small, the scaling is not necessary, and the above program does twice as many calculations as it should. A more complex algorithm which deals with these and other problems was given by Lawson (1978) and is implemented as routine SNRM2 in the Linpack collection mentioned above. SNRM2 is included on the disk at the end of the text. It is not necessary to study this routine in detail in order to understand the major point: building efficient reliable programs, even for simple tasks, requires careful engineering combined with computer understanding.

1.4 PORTABILITY

A portable program is one that runs on a variety of machines without any alterations, and produces identical results. This is a goal, but it is rarely possible to achieve. The Fortran program you write for one computer might not work when you try to move it to another.

(1) The program may not compile. Even though Fortran originated in the 1950's, a precise definition of the language didn't exist until recently. Each computer manufacturer provided a compiler with different features, or sometimes equivalent features with different syntax. All the programs in this book have been written in Fortran-77, a modern and carefully defined dialect. Every computer vendor catering to scientists provides a Fortran-77 compiler. Although extensions and extra features are permitted, such a compiler is required to flag all non-standard usages. In this way we can hope that our programs will move easily. However, even a program written using Fortran-77 syntax is not guaranteed to compile. For example, the span of computable numbers is not part of the standard. Thus the statement X=1.E50 will compile on a CDC Cyber but not on a VAX 11/780; VAX Fortran does not normally permit values greater than about 10^{38}. Similarly, maximum array lengths are not standardized but are determined by hardware restrictions and software organization. On IBM personal computers some Fortran compilers require arrays to be shorter than $64K = 64 \cdot 1024 = 65536$ bytes (1 byte = 8 binary digits).[1] This normally means that a matrix A(I,J) must be smaller than 128×128, if each element occupies 4 bytes. Sometimes a program will fail to compile for more idiosyncratic reasons; the compiler has a defect or the language definition is not precise enough. These are not trivial problems. One commercial library vendor has estimated that almost 40% of their routines, written in "standard" Fortran, fail to compile correctly on one or more of the machines they support. The subroutines provided in this book have been tested on a variety of computers, including IBM personal computers, VAX, CDC Cyber, and Sun computers.

(2) The program may compile but abort during execution, or execute but produce different answers. Variability in computer hardware is reflected in fundamental differences in accuracy. Thus Cray 1 hardware produces about 12 decimal digits accuracy on X=SQRT(7.) but Sperry Univac hardware only gives about 7 digits. Of course, arbitrary accuracy is possible on any computer, but only by using sophisticated software that will run slowly. The job of writing portable programs is further complicated by the fact that many programs make use of machine-dependent constants. The commonest

---

[1] The notation $64K$ for $64 \cdot 1024$ was a computer-era idea inspired by the use of binary numbers for addressing memory. However, the approximation $K = 1024 \approx 1000 =$ kilo has at least one earlier usage—in cooking. Puff pastry can be made by rolling out dough, spreading butter on it, folding it in half and repeating the procedure, typically 8–10 times. This produces 2^8–2^{10} layers. In France, puff pastry is the basis for the dessert "mille-feuille"="1000 leaf."

machine-dependent parameter found in mathematical software is the machine epsilon. This number, which will be defined in Chapter 2, is related to the accuracy of floating point arithmetic.

1.5 SOFTWARE DESIGN: ERROR HANDLING

Errors will occur when using software. Asking for the square root of a negative number or the tangent of $\pi/2$ are obvious examples. A robust software package should check its input for validity. For example, it will not be possible to find the intersection of two parallel lines, and a routine that is asked to do that should generate an error message. Other errors might occur during the computation; for example, a subroutine might not be able to solve a problem to the requested accuracy. Designing a mechanism for reporting errors to users is both boring and challenging: boring because most users do not expect to make errors, and algorithm designers don't want to think about all the combinations of unusual situations which can occur; challenging because the design should be simple, efficient, and flexible.

The easiest way to deal with errors is to ignore them; a routine which develops an error will either abort, or continue with whatever bad data are in place. Some errors can be processed by making some decision and continuing, such as returning a fixed large number for $\tan(\pi/2)$, or taking the absolute value of all arguments to SQRT. Sometimes a routine will write a message such as "ERROR: Square root of negative number," and either continue or stop. More sophisticated programs might also return an **error code**, which is keyed to a list of error messages given in the documentation. But this won't work for every routine: would you expect to compute a square root by calling SQRT(X, IERROR)?

The environment you are working in determines your attitude toward errors. Also, there are several types of errors: serious (where the program cannot determine how to proceed, and which may indicate a fundamental flaw in the problem), warnings (such as that the problem has multiple solutions), while others are merely informational (for example, that the solution is valid but trivial). For simple programs with one or two subroutines, printed error messages are adequate. But for large programs with thousands of pieces, printed messages might go unnoticed, especially if they are brief and overwhelmed by other output. Similarly you probably would not want to see a warning message in the middle of an otherwise elegantly formatted tabulation, or in the center of a plot.

It is frustrating to run a large program and get a message such as "Bad arguments," but not know where it came from or how to turn it off. Among professionals, there is general agreement that error codes are convenient but may not be flexible enough, and that error messages generated by WRITE or PRINT statements should never appear in a subroutine unless an input parameter allows them to be skipped.

Most large collections of programs try to adopt a consistent approach to error processing. A typical example of this is XERROR (Jones et al. (1983)), the set of subroutines developed for the SLATEC library. Some of them are used in this text. The philosophy of the package is that all error output goes through one routine, which is the only place where there are WRITE statements. Thus the SLATEC library with more than 200,000 lines of Fortran has only two WRITE statements, both in XERROR. The disadvantage of such an error package is the need for extra routines.

Errors are classified by the library software writer, for example, fatal, recoverable, warning, etc., and some flexibility is provided to the general user. Thus the user can decide that a recoverable error such as "Too much accuracy requested" is fatal for a specific case, or that warning messages should not be printed. There is also the provision for routing the error messages to selected files so that they all appear in a convenient place. Since the library writer builds the error message, specific problem-dependent information can be included which is difficult to provide with error codes. Some writers will use both XERROR and error codes.

1.6 SOFTWARE DESIGN: SCRATCH STORAGE

Many of the programs in this book require extra storage for intermediate calculations. This is often called **work** or **scratch** storage. If the amount of storage is fixed then internal declarations suffice. But if the amount is dependent on input parameters, for example on the number of data points, or the size of a matrix, then a mechanism must be available for providing the required space. There are three common approaches.

(1) Have the called subroutine declare and dimension an array of fixed size. A routine MAT that manipulates matrices might have an internal array declared REAL ATEMP(500), restricting the maximum size of the problem which can be solved. This is effective as long as a call to MAT with a larger problem results in an error indication. The disadvantages are that all users pay the storage costs for the largest allowable problem, and that MAT must be altered internally for larger cases. In your own software this is easy, but in most computing facilities the software libraries are precompiled and made available for all users. To discourage multiple copies, the computer center usually does not make source text available. For proprietary libraries it may not be allowed to do so.

(2) Some coordinated software collections such as the PORT library[2] simulate **dynamic storage allocation**. One large array, W, is declared in a general utility routine. When any subroutine in the library such as MAT needs work storage, it calls a storage allocator with a request for a specific amount of space. The allocator returns to MAT an address inside W of the beginning of a block of unused space of the correct length. Before MAT returns to the user it calls a deallocator to turn back the now unneeded storage for other routines to use. The main disadvantages are that (a) everyone pays the storage costs for the large storage array, (b) extra routines are needed to manage the storage allocation, and (c) there is a complex interrelationship among the library routines making it more difficult to extract individual programs. On the other hand, because the scratch array is large it is unlikely that problems will exceed available space. Also, in a complicated multi-routine software project, managing all the scratch storage in one place can be more efficient than having each routine deal with it separately.

---

[2] The PORT Mathematical Subroutine Library, AT&T Bell Laboratories, 600 Mountain Ave., Murray Hill, NJ 07974

(3) A subroutine like MAT needing work storage can require that the *calling* program declare and dimension an appropriate sized array W, and pass it to MAT as a parameter. Inside MAT the array has an unspecified or "dummy" declaration, such as REAL A(*), allowing MAT to be written for problems of arbitrary size. With this approach MAT may require LENW in the call sequence, the exact number that the user has specified in the DIMENSION for W. MAT can check LENW against the input problem size to verify that the dimensioned array is large enough. (Recall that in Fortran there is no automatic way for the called routine to know the array's length.) The disadvantages are the need for extra arguments in the call sequence, and the possibility that MAT will write past the end of W destroying other data if the user *both* under-dimensions W and incorrectly specifies LENW. Advantages are generality and independence of different routines. Most mathematical software is written using this technique and you will see several examples in this text.

1.7 HISTORICAL PERSPECTIVE: BACKUS AND THE FORTRAN LANGUAGE

In university computing science curricula today it is fashionable to teach programming via Pascal, C, APL or some other "structured" language. Little or no time is spent on Fortran which is usually described as old-fashioned, and unstructured. Fortran was the first high level language though, developed in the mid 1950s by an IBM team led by John Backus and is still *the* scientific computing language; all others taking a minor back seat.

From World War II through 1954 almost all programming was done (mostly by women) in machine language or assembly language. Programmers thought of themselves and their work as a creative process akin to art, and much clever effort was devoted to overcoming the difficulties caused by computers of that era. There were already a few semi-automatic programming languages such as Autocode, Speedcoding, and Dual, but they were not very efficient. For the most part this didn't matter because inefficiencies in coding logic for looping/testing and computing addresses were masked by the large amount of time spent simulating floating point operations by software. The introduction of the IBM 704 with built-in hardware for floating point and indexing, radically altered the situation; inefficient code now had nowhere to hide. For this reason, programmers were skeptical of techniques to automate programming, and Backus was convinced that the only way for such a system to be accepted was for it to translate reasonable scientific source programs into object programs which ran no more than half as slowly as a hand coded counterpart. To this day he feels that modern languages suffer from too much emphasis on language design and not enough on the generation of efficient programs.

Fortran was never really designed. In fact Backus states that "We did not regard language design as a difficult problem, merely a simple prelude to the real problem; designing a compiler which would produce efficient programs ... We knew nothing of many issues which were later thought to be important, e.g., block structure, conditional expressions, type declarations—it seemed to us that once one had the notions of the assignment statement, the subscripted variable and the DO statement in hand then the remaining problems of language design were trivial." (Backus (1979)) Fortran ignored blanks; it still does. This was criticised but was included because keypunchers had difficulty recognizing and counting blanks and this caused many problems. The team thought that Fortran "should virtually eliminate coding and debugging" and not much thought was given to catching syntax errors.

Originally, the reception to Fortran was only lukewarm, but by 1958 over half the IBM 704 installations were using it for more than 50% of their work, and several for over 80%. Today all large scientific computer centers support Fortran, much improved but still resembling the original, as their main language for scientific computing. For example, the US National Center for Atmospheric Research (NCAR) in Boulder, Colorado is "exclusively a Fortran shop,"[3] and at Lawrence Livermore National Laboratory (LLNL) in California, "almost all computation done on our mainframes is done in Fortran, it's still the best language for physics."[4]   Backus believes that structured programming "can be viewed as a modest effort to introduce a small amount of order into the chaotic world of statements ... New programming styles one day will offer far greater intellectual and computational power. Fortran is a good language for some applications. As a language of the future, however, we ought to do a lot better. The problem is, we haven't."

The current investment in Fortran programs guarantees that the language will be around perhaps even until the next century. But changes are being made. The Fortran 8X committee is developing a standard for a new revision, which will probably appear around 1990. Extensions to the language are planned in order to make it more suitable for array processing and graphics, two areas where Fortran-77 is lacking. Other extensions are designed to strengthen its numerical computations, add derived data types and give its syntax a more modern look. Cynics claim that thirty years from now the language of scientific computing will look very different from anything we have today, but it will still be called Fortran.

1.8 OTHER USEFUL SOURCES OF INFORMATION

In any text of this type we are only able to cover a small portion of the material which is available. In some cases you may discover that another introductory book explains some topic better, or that a more advanced book goes into more depth. Ultimately, journals and monographs provide the most up-to-date presentation. This section lists several specific references along with brief descriptions to place them in context. Complete citations can be found in the bibliography.

---

[3] J. Adams, chair of the ANSI Fortran committee and consultant for NCAR.
[4] C. Hendrickson, assistant associate director for defense systems computing at LLNL.

CHAPTER 2—COMPUTER ARITHMETIC

1. The IEEE Floating-Point Standard. A standard for computer arithmetic, specifying precision, rules for rounding, evaluation of special functions, and many other topics.

2. *Rounding Errors in Algebraic Processes* by J.H. Wilkinson. A discussion of rounding-error analysis applied to various computation problems.

3. *Handbook of Mathematical Functions* by M. Abramowitz and I.A. Stegun. A comprehensive collection of mathematical tables, identities, and other information on special functions and constants.

4. *A History of Computing Technology* by M.R. Roy.

CHAPTER 3—LINEAR EQUATIONS AND
CHAPTER 6—LINEAR LEAST SQUARES DATA FITTING

1. The *Linpack Users' Guide* by J.J. Dongarra et al. The manual for the Linpack collection of linear algebra subroutines.

2. *Matrix Computations* by G.H. Golub and C. Van Loan. A comprehensive discussion of numerical linear algebra.

3. *Computer Solution of Large Sparse Positive Definite Systems* by A. George and J.W. Liu. A discussion of direct methods for sparse matrices.

4. *Matrix Iterative Analysis* by R.S. Varga. A discussion of iterative methods, especially useful for those systems that arise when solving differential equations.

5. *Solving Least Squares Problems* by C. Lawson and R. Hanson.

CHAPTER 4—INTERPOLATION

1. *A Practical Guide to Splines* by C. de Boor. This contains a careful but readable theoretical development as well as many examples and programs.

2. *Interpolation and Approximation* by P. Davis. This is not really a text on numerical methods, but it provides the theoretical background for most forms of interpolation.

CHAPTER 5—QUADRATURE

1. *Methods of Numerical Integration* by P.J. Davis and P. Rabinowitz. This is the most comprehensive survey of the field and also has the best bibliography. It contains many numerical examples.

2. *Quadpack: A Subroutine Package for Automatic Integration* by R. Piessens et al. The manual and explanation for the one-dimensional Quadpack subroutines.

CHAPTER 7—NONLINEAR EQUATIONS

1. *Iterative Solution of Nonlinear Equations in Several Variables* by J.M. Ortega and W.C. Rheinboldt. An extensive discussion of methods and theory.

2. *User Guide for Minpack–1* by J.J. Moré, B.S. Garbow, and K.E. Hillstrom. Manual for the subroutine SNSQE.

CHAPTER 8—ORDINARY DIFFERENTIAL EQUATIONS

1. *Numerical Solution of Ordinary Differential Equations* by L. Lapidus and J. Seinfeld.

2. *Computational Methods in Ordinary Differential Equations* by J. Lambert. Both these references contain an abundance of mathematical and computational details.

CHAPTER 9—OPTIMIZATION AND NONLINEAR DATA FITTING

1. *Algorithms for Minimization Without Derivatives* by R.P. Brent.

2. *Practical Optimization* by P.E. Gill, W. Murray, and M.H. Wright.

CHAPTER 10—RANDOM NUMBERS AND SIMULATION

1. *The Art of Computer Programming: Volume 2, Seminumerical Algorithms* by D.E. Knuth. An extensive discussion of algorithms for generating random numbers.

2. *Simulation and Monte Carlo Method* by R. Rubinstein. Contains algorithms and a good bibliography.

CHAPTER 11—TRIGONOMETRIC APPROXIMATION AND THE FAST FOURIER TRANSFORM

1. *Applications of Discrete and Continuous Fourier Analysis* by H.J. Weaver.

2. *Digital Picture Processing* by A. Rosenfeld and A. Kak.

3. *Fundamentals of Digital Signal Processing* by L. Ludeman.

4. *Fourier Analysis of Time Series: An Introduction* by P. Bloomfield.

1.9 PROBLEMS

P1–1.–The **error function** is an important function in many branches of applied mathematics. It is defined by an integral,

$$\mathrm{erf}(x) = \frac{2}{\sqrt{\pi}} \int_0^x \exp(-t^2)\, dt.$$

The integral cannot be expressed in terms of more elementary functions. Each chapter of this book will include a problem relating the material in the chapter to some aspect of the error function. We begin with

(a) Find a table of $\mathrm{erf}(x)$ and look up the value of $\mathrm{erf}(1.0)$.

(b) Is there a function or subroutine available on your computer for evaluating $\mathrm{erf}(x)$? If so, find out how to use it, print out the value of $\mathrm{erf}(1.0)$, and compare it with the value obtained from the table. (A program to evaluate $\mathrm{erf}(x)$ has been provided on the software disk that accompanies the book.)

(c) If the formula for $\mathrm{erf}(x)$ cannot be expressed in terms of more elementary functions, how is it possible for a subroutine to compute its values?

P1–2.–In 250 B.C.E., the Greek mathematician Archimedes estimated the number π as follows. He looked at a circle with diameter 1, and hence circumference π. Inside the circle he inscribed a square; see Figure P1.2. The perimeter of the square is smaller than the circumference of the circle, and so it is a lower bound for π. Archimedes then considered an inscribed octagon, 16-gon, etc., each time doubling the number of sides of the inscribed polygon, and producing ever better estimates for π. Using 96-sided inscribed and circumscribed polygons, he was able to show that $223/71 < \pi < 22/7$. There is a recursive formula for these estimates. Let p_n be the perimeter of the inscribed polygon with 2^n sides. Then p_2 is the perimeter of the inscribed square, $p_2 = 2\sqrt{2}$. In general

$$p_{n+1} = 2^n \sqrt{2(1 - \sqrt{1 - (p_n/2^n)^2})}$$

Compute p_n for $n = 3, 4, \ldots, 60$. Try to explain your results. (This problem was suggested by Alan Cline.)

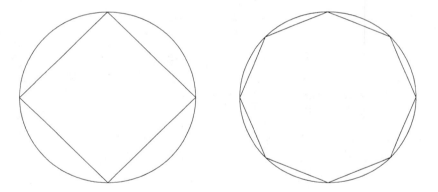

Figure P1.2

2

Computer Arithmetic
and Computational Errors

2.1 INTRODUCTION

Humans have been calculating for thousands of years. The Pythagorean formula, an early landmark of mathematics, is a computational formula. In ancient Greece, Archimedes and others approximated π. Hundreds of years ago mathematical tables were used in warfare and navigation. And yet the field of Numerical Analysis only came into being about forty years ago, just after World War II. How did the human race avert computational disaster for all those centuries?

Even though Numerical Analysis as a separate topic is relatively new, the underlying ideas and goals are not. It is only with the invention of the electronic computer in the 1940's that large-scale automated calculations became an important tool for science and technology. This invention has two implications for us.

(1) Computer arithmetic is not the same as "pencil and paper" arithmetic. In hand calculations, it is possible to monitor the intermediate results and adjust the accuracy of the calculation as required. With computer arithmetic, each number has a fixed number of digits which in some cases may be inadequate for a calculation.

(2) A hand calculation will usually be short, whereas a computer calculation can involve millions of steps. Tiny errors that would be negligible in a short calculation can be devastating when accumulated over a long calculation. Also, methods that are perfectly adequate for a small problem may be hopelessly inefficient when scaled to a large problem (see the discussion of Cramer's rule in Section 1 of Chapter 3).

In this chapter we will be mainly concerned with (1): the properties of computer arithmetic. The differences between computer arithmetic and "real" arithmetic may seem subtle, but their effects are not. To illustrate this, we begin with some simple examples.

Example 2.1 Taylor Series for e^x.

We know the Taylor series for e^x converges for all finite values of x

$$e^x = 1 + x + \frac{x^2}{2!} + \frac{x^3}{3!} + \cdots .$$

Below we list a Fortran program for summing this series.

```
        WRITE (*,*) 'ENTER X'
        READ  (*,*) X
        SUM  = 1.0
        TERM = 1.0
        I    = 1
1       TERM = TERM * (X/I)
        IF (SUM+TERM .EQ. SUM) THEN
            WRITE (*,*) 'E(X) = ', SUM, EXP(X)
            STOP

C STOP SUMMING WHEN THERE IS NO FURTHER CONTRIBUTION TO THE SERIES.
        ELSE
            SUM = SUM + TERM
            I   = I + 1
            GO TO 1
        ENDIF
        END
```

Notice the test which terminates the sum. This relies on the fact that computer arithmetic is only approximate. The expression SUM+TERM will have the same value as SUM if TERM is small enough. (We discuss these ideas more fully in Section 3.) If we run this program on a VAX computer for various values of x we get the numbers listed below. For $x > 0$ these results are what we would expect, but for $x < 0$ they are not.

x	$E(X)$	e^x
1	2.718282	2.718282
5	148.4132	148.4132
10	22026.47	22026.46
15	3269017.	3269017.
20	4.8516531×10^8	4.8516520×10^8
-1	.3678794	.3678795
-5	6.7377836×10^{-3}	6.7379470×10^{-3}
-10	$-1.6408609 \times 10^{-4}$	4.5399930×10^{-5}
-15	$-2.2377001 \times 10^{-2}$	3.0590232×10^{-7}
-20	1.202966	2.0611537×10^{-9}

For $x > 0$ these results are what we would expect, but for $x < 0$ they are not. Even the signs of the answers are incorrect in some cases. ■

Example 2.2   Difference Quotient for a Derivative.

The derivative of f at x is defined by

$$f'(x) = \lim_{h \to 0} \frac{f(x+h) - f(x)}{h},$$

so if h is "small,"

$$f'(x) \approx \frac{f(x+h) - f(x)}{h} \equiv \Delta_h f(x).$$

The table below shows the results of this calculation with $f(x) = e^x$ to compute $f'(x)|_{x=1} = e = 2.71828175$ on the same VAX.

h	$\Delta_h f(1)$	e	error
10^0	4.67077446	2.71828175	1.95×10^0
10^{-1}	2.85884380	2.71828175	1.41×10^{-1}
10^{-2}	2.73191929	2.71828175	1.36×10^{-2}
10^{-3}	2.71987939	2.71828175	1.60×10^{-3}
10^{-4}	2.72035623	2.71828175	2.07×10^{-3}
10^{-5}	2.71797204	2.71828175	3.10×10^{-4}
10^{-6}	2.62260461	2.71828175	9.57×10^{-2}
10^{-7}	4.76837206	2.71828175	2.05×10^0
10^{-8}	0.00000000	2.71828175	2.72×10^0

We see that as h decreases, the difference quotient approximation improves but then begins to get progressively worse. The result in the final row indicates that we have reached the limiting accuracy of the arithmetic. ■

Example 2.3   Two Linear Equations.

Solve the system of two equations

$$.780x + .563y = .217,$$
$$.457x + .330y = .127.$$

Using the standard techniques (described in Chapter 3) and working in three-digit decimal arithmetic, we get the solution

$$x = \quad 1.71,$$
$$y = -1.98.$$

If we take these numbers and substitute them back into the equations, we get

$$.780 * (1.71) + .563 * (-1.98) - .217 = 0.00206,$$
$$.457 * (1.71) + .330 * (-1.98) - .127 = 0.00107.$$

The difference between the right and left sides, called the residual, is satisfyingly small. Nevertheless the exact solution to this system is

$$x_{\text{exact}} = \quad 1.000,$$
$$y_{\text{exact}} = -1.000.$$

While this example is contrived, it does illustrate that the computed solution can be very different from the exact solution (the computed values do not look like the exact values) but can behave like the exact solution (the computed values almost "solve" the equations). ∎

Arithmetic on computers only approximates what we know as ordinary arithmetic. The differences account for the seemingly strange values illustrated in these examples. More complete explanations will be given later.

Good numerical methods must take into account the effects of computer arithmetic. Even so, it is not always possible to compute an accurate solution to a problem. In Example 2.3 above, the algorithm used to solve the problem is of high quality, yet the computed solution is inaccurate. This is not the fault of the algorithm. This particular problem is **ill-conditioned**, **badly posed**, or **sensitive**, meaning that tiny changes to the coefficients of the problem produce large changes in the solution. You should not expect an algorithm to perform well on an ill-conditioned problem.

In the case of Example 2.1, we will see below that there are effective ways of computing e^x for all values of x, but that using the Taylor series is not one of them. In this case the algorithm is at fault; it is **unstable**. Unstable algorithms should be avoided.

This leads to the question of what is meant by a "solution" to a numerical problem. As in Example 2.3 it may be unreasonable to expect that the computed solution be close to the true solution of the problem. However, for a stable algorithm, we will often be able to guarantee that the computed solution *exactly solves* a perturbed problem, that is, a problem where the coefficients have been slightly modified. If the problem is also well-conditioned this will mean that the computed and true solutions are similar (the computed solution is a good approximation of the true solution). For an ill-conditioned problem this will probably not be true, although it will often happen that the computed solution behaves like the true solution, as in Example 2.3.

2.2 REPRESENTATION OF NUMBERS

Computer calculations can involve two types of numbers: integer (whole numbers) and "floating point" (real numbers). When computers were first developed, only integer arithmetic was available. Fractions were represented by inserting an imaginary decimal point at some fixed position within the integer (four digits from the left, say). This was referred to as "fixed point" arithmetic. But in 1954 IBM began development of the 704 computer which implemented all the algorithms for floating-point arithmetic as machine instructions, greatly simplifying the use of non-integer arithmetic. Fixed-point arithmetic is not common anymore, except in certain special-purpose devices such as graphics terminals. Today *hardware* floating-point arithmetic is often available, at least as an option.

Most computers represent numbers internally in binary representation, i.e., as a sequence of 0's and 1's. This is convenient for the technology, but unnatural for humans who learned to count on ten fingers. In order to make the ideas about computer arithmetic clear, and to separate them from the details of computer hardware, most of our discussion will be based on decimal arithmetic rather than binary.

We sketch below the important properties of computer arithmetic, to the extent that they are required for better understanding of numerical computing. These will be illustrated primarily using decimal examples, but essential details about binary arithmetic are also given, based on the IEEE Standard for Binary Floating Point Arithmetic (1985) which is discussed below.

Other forms of arithmetic could also be used for computer calculations. A few computers have been designed with base 3 arithmetic. Even humans have used other bases, such as 20. Recall Abraham Lincoln's "four score and seven years ago" with 1 score = 20, a word derived from the beginnings of the stock market, when stocks (sticks) were scored to record financial transactions. Or notice the traces of a counting system based on 20, in such French words as *quatre-vingt* for 80, which is literally "four-twenty;" or even the way we count in English, with special words like "twelve" and "sixteen" instead of the more consistent "ten-two" and "ten-six" in analogy with "twenty-two" and "fifty-six."

2.2.1 Computer Representation of Integers

Computer integers are represented by a finite number of digits. To illustrate, suppose we have six decimal-digit integers. Each integer will also have a sign + or −. This means that there are only finitely many integers, in this case the smallest integer will be −999999 and the largest will be 999999. An integer outside this range does not exist as far as the computer is concerned.

If we calculate with integers well within the computer's range, integer arithmetic works as expected, so that $5 + 7 = 12$, $8 - 27 = -19$, and $27 \times 3 = 81$. Integer division produces an integer result, with the quotient retained as the answer and the remainder discarded. This means that $1/3 \rightarrow 0$, $4/2 \rightarrow 2$, $7/(-3) \rightarrow -2$, etc.

If the result of an integer calculation is too big or too small for the computer, then unpredictable things can happen. On some computers, an error message will be printed and the calculation will be terminated abruptly. On others, the calculation will "wrap" around without any error indication. For example, in our case $999999 + 1 = -999999$. It is not good practice to rely on any results that go outside a computer's range.

These comments apply equally to integer calculations performed in binary arithmetic, although the actual values of the largest and smallest integers will vary. On many common computers, an integer is stored in 32 binary digits, with one of the digits being allocated for the sign of the number. In this case, the largest integer is $2^{31} - 1 = 2,147,483,647$ and the smallest is -2^{31}.

2.2.2 Computer Representation of Real Numbers: Floating Point Arithmetic

A decimal floating point number is a number represented in the form $a \times 10^b$; a is called the **mantissa** and b is called the **exponent**. Usually b is an integer, and a is a number with one digit to the left of the decimal point. Both a and b only have finitely many digits. Because of this, there are only finitely many floating point numbers, and in particular there are largest and smallest floating point numbers.

To illustrate in more detail we consider a system of floating point arithmetic where a has four and b has two decimal digits. Some floating point numbers are:

$$4.678 \times 10^{00} = 4.678, \quad -3.355 \times 10^{03} = -3355., \quad 9.876 \times 10^{-12} = .000000000009876.$$

In this system, the smallest and largest numbers are -9.999×10^{99} and 9.999×10^{99}. There is also a gap between zero and the smallest positive number, $0.001 \times 10^{-99} = 10^{-102}$.

Numbers in this floating point system do not always have a unique representation, since

$$1.000 \times 10^{00} = 0.100 \times 10^{01}.$$

To avoid this ambiguity, it is usual to insist that the first digit in the mantissa a be non-zero, so that the second representation here would not be allowed; this is called **normalization**. One exception is the representation of zero: 0.000×10^{00}. In a normalized floating point number system there are fewer numbers. For example, in our system the smallest positive floating point number is now $1.000 \times 10^{-99} = 10^{-99}$.

We have already mentioned that there is a largest floating point number. If a calculation goes outside this range, an **overflow** occurs and on most computers the calculation will be terminated. If a calculation results in a number that is too close to zero (in our case smaller than 10^{-99}) then **underflow** occurs. This is generally less disastrous than overflow, and many computers will declare the answer to be zero without any indication that anything out of the ordinary has occurred. Nevertheless, there are computations where underflow is not benign. See for example, problem P2–6.

Floating point calculations typically do not produce an exact result. As an example consider in our system

$$1.234 \times 10^{00} + 5.678 \times 10^{-04} = 1.234 + .0005678 = 1.2345678$$

where the answer cannot be represented exactly as a floating point number, but must be reduced to four digits. If we **chop** the answer then the final digits are discarded giving 1.234×10^{00}. It is preferable to **round** to the nearest floating point number, giving the more accurate result 1.235×10^{00}. The difference between the true answer and the floating point answer is called the **rounding error** regardless of which rule is used.

On a computer which uses binary arithmetic, floating point numbers are typically represented in memory by 32 binary digits or bits. In the IEEE Standard, 24 bits are set aside for the mantissa and 8 bits for the exponent; these bits must also store the signs of the mantissa and the exponent. The number is assumed to have 1 as its leftmost bit and this is not stored. Also, for technical reasons, the exponent is stored as an integer in the range [0, 255]; to get the actual exponent, 127 is subtracted from the stored value. The exponent value 255 is reserved to represent infinity and improper results such as square roots of negative numbers; an improper result is called a "not a number." This means that the largest floating point number is

$$+1.1 \cdots 1_2 \times 2^{127} \approx 10^{38}.$$

The smallest positive floating point number is approximately 10^{-38}. A mantissa of 24 bits corresponds to about 7 decimal digits. The IEEE Standard also recommends that computers perform arithmetic in higher precision even though the final result will be stored in 32 bits. Computers supporting the Standard often have all floating point arithmetic implemented internally in 80 bit form.

All Fortran compilers also support a double precision floating point number which uses twice as many bits as floating point to represent each number. Unless this is supported in hardware, double precision calculations will be much slower than floating point. Most computers that conform to the IEEE Standard provide hardware double precision, at least as an option. Fortran also supports complex floating point and complex double precision arithmetic using a pair of floating point or double precision numbers, but this is rarely supported in hardware; complex arithmetic must be done by software.

Designing computer hardware to perform reliable floating point arithmetic has proven to be difficult. There are numerous examples of ordinary calculations which give incorrect results, with potentially serious consequences. W. Kahan, at the University of California, Berkeley, has taken a major role in pointing out difficulties with the arithmetic on most existing computers. He has also led in the establishment of a well defined standard, referred to as the *IEEE Floating Point Standard*. This document has enabled computer manufacturers to build hardware floating point chips which perform intelligently and reliably. The IEEE Standard states exactly how rounding should be done, and what to do if a calculation results in a potential over/underflow or negative argument to a square root. Several computer chips, including the Intel 8087/80287/80387 chips and the Motorola 6888X series, incorporate the Standard and are used in common personal computers and workstations.

2.3 MACHINE CONSTANTS

When floating point numbers of different magnitudes are added, the result of the calculation may be exactly equal to one of the summands. For example, in our four-digit floating point arithmetic

$$1.000 \times 10^{00} + 1.000 \times 10^{-04} = 1.000 + .0001000 = 1.0001 \rightarrow 1.000 \times 10^{00}.$$

The smaller number might as well have been zero. The smallest floating point number that we can add to 1.0 and obtain a floating point result larger than 1.0 is called **machine epsilon**, written ϵ_{mach}. The actual value of ϵ_{mach} depends upon whether arithmetic is rounded or chopped. On our four digit machine $\epsilon_{mach} = .001 = 1.000 \times 10^{-03}$ if the arithmetic is chopped, and $\epsilon_{mach} = .0005$ if it is rounded. In either case, the addition of a floating point number to 1.0 will only be accurate to the third decimal place.

Machine epsilon determines the relative accuracy of the computer arithmetic. If x, y are two positive floating point numbers $x > y$, then their sum can be written

$$x + y = x \left(1 + \frac{y}{x}\right).$$

We can see that unless $y/x \geq \epsilon_{mach}$ the floating point sum of x and y will be x. A more careful argument shows that the relative error in the floating-point sum can be as large as ϵ_{mach}.

For 32-bit floating point arithmetic, $\epsilon_{mach} = 2^{-22} \approx 1.2 \cdot 10^{-7}$ if rounding is used. Thus it is hopeless to expect more than seven decimal digits accuracy from any floating point calculation, nor can we resolve relative differences finer than that level. Several subroutines in this book have an input parameter ϵ, representing desired accuracy. It is unreasonable to choose $\epsilon < \epsilon_{mach}$.

The act of reading a decimal number, such as 0.3, into the computer causes an error, because 0.3 does not have an exact floating point representation. If the input/output routines provided by the Fortran compiler work correctly the error is in the last bit. We sometimes express this as

$$x_{stored} = x(1 + \delta_x), \quad \text{or} \quad x_{stored} - x = x\delta_x, \quad |\delta_x| \leq \epsilon_{mach},$$

meaning that the stored floating point approximation to x can have a relative error of up to ϵ_{mach}.

We can use this to understand the concept of **catastrophic cancellation**, the error growth due to subtraction of nearly equal numbers. Consider our model computer with $\epsilon_{mach} = 0.0005$, and let $x_{stored} = 1.001$, $y_{stored} = 1.002$. The difference $y_{stored} - x_{stored} = .001$, even in floating-point arithmetic. However, in general we cannot assume that x_{stored} and y_{stored} are exact. They will normally be the stored versions of other numbers x, y and have relative errors δ_x and δ_y of up to ϵ_{mach}. The relative error in the true difference $y_{\text{stored}} - x_{\text{stored}}$ can be estimated by

$$\frac{(y_{\text{stored}} - x_{\text{stored}}) - (y - x)}{y - x} = \frac{\delta_y - \delta_x}{.001}.$$

The magnitude of the right hand side can be as large as

$$\frac{2\epsilon_{\text{mach}}}{.001} \approx 2000\epsilon_{\text{mach}} = 1.0.$$

Thus the difference can be wrong in every digit.

Example 2.4   Further Analysis of the Taylor Series for e^x.

Consider Example 2.1. In computing e^{-15} by Taylor series we have

$$e^{-15} = 1 - 15 + \cdots - 312540.3 + 334864.6 - \cdots.$$

This is much easier to understand if we write out the values of these terms to all the digits that are computed on the VAX.

n	n-th term in Taylor Series
0	1.000000
1	-15.00000
2	112.5000
3	-562.5000
$\vdots$	
13	-312540.3
14	334864.6
15	-334864.6
16	313935.5
17	-277001.9
18	230834.9
19	-182238.1
20	136678.6
21	-97627.55
$\vdots$	
51	-0.00000061660813
52	0.00000017786773
53	-0.000000050339924
54	0.000000013983313
55	-0.0000000038136312
56	0.0000000010215083
57	-0.00000000026881800

Note that some of these terms and hence intermediate sums are much larger than the final answer ($\approx 10^{-7}$). When we subtract one of these large terms from the evolving sum the difference is small and has a large relative error. Ultimately the differences have no digits of accuracy. This is another example of the catastrophic cancellation described above. One solution to this problem is to use a computer with a smaller ϵ_{mach}. The table below gives the results of the same program as in Example 2.1, but run on a CDC Cyber 855, which has $\epsilon_{mach} \approx 10^{-14}$.

x	$E(X)$	e^x
1	2.718282	2.718282
5	148.4132	148.4132
10	22026.47	22026.46
15	3269017.	3269017.
20	4.8516520×10^8	4.8516520×10^8
−1	.3678794	.3678795
−5	6.7379470×10^{-3}	6.7379470×10^{-3}
−10	4.5399952×10^{-5}	4.5399930×10^{-5}
−15	3.0508183×10^{-7}	3.0590232×10^{-7}
−20	3.865358×10^{-7}	2.0611537×10^{-9}

Note that the answers are better than before, but there are still difficulties when $x < -10$. In Example 2.8 we will see another solution to this problem. ∎

The most important points about floating point computation are summarized below. The specific numbers listed are for the IEEE binary 32-bit floating point Standard. Also, problem P2–4 will help you gain familiarity with floating point numbers.

(1) There are finitely many floating point numbers, about 2^{31} of them.

(2) There is a largest floating point number, OFL = overflow level $\approx 10^{38}$.

(3) There is a smallest positive floating point number, UFL = underflow level $\approx 10^{-38}$.

(4) The floating point numbers between 0 and OFL are not uniformly distributed. There are 2^{22} floating point numbers between each power of 2, e.g. 2^{22} between 2^{-128} and 2^{-127} as well as between 2^{126} and 2^{127}. Thus, floating-point numbers are concentrated near zero.

(5) Arithmetic operations on floating point numbers cannot always be represented exactly, and must therefore be chopped or rounded to the nearest floating point number.

(6) ϵ_{mach} is the smallest floating point number such that $1.0 + \epsilon_{mach} > 1.0$ in floating point arithmetic; $\epsilon_{mach} \approx 10^{-7}$ on a 32-bit computer. It represents the relative accuracy of computer arithmetic.

(7) OFL and UFL are determined mostly by the number of bits in the exponent, whereas ϵ_{mach} is determined by the number of bits in the mantissa. Thus they measure different parts of the floating point representation. $0 < UFL < \epsilon_{mach} < OFL$.

The software in this book sometimes needs to know the value of machine constants like ϵ_{mach} and OFL. The programs obtain their values from the Machine Constants Package, originally developed at AT&T Bell Telephone Laboratories (Fox (1978)). This consists of three Fortran functions, I1MACH(K), R1MACH(K), and D1MACH(K) which have a single integer argument, K, and return an integer, a floating point or a double precision number, respectively. By definition, R1MACH(1) = UFL, R1MACH(2) = OFL, and R1MACH(4) = ϵ_{mach}. Other values are defined similarly. The advantage of this approach is that it localizes all machine dependent data in a standardized way rather than having them appear in individual subroutines. The disadvantage is the need for extra routines.

Sometimes users will insert specific constants into their programs, for example the values PI=3.14159, or SQ2=1.414. This is portable, but may not be as accurate as possible. A better way is to place the burden on the compiler whenever possible, PI=ATAN(1.0)*4, or SQ2=SQRT(2.0). The standard Fortran functions like SQRT or ATAN usually provide results that agree with "exact" values to within one bit because they have been written carefully.

It is possible for the computer not only to perform floating-point calculations, but also to keep track of the rounding errors that have been made, and then to provide bounds on the errors in a calculation. This is called **interval analysis**, and is described in more detail in the book by Moore (1979).

2.4 ERRORS IN SCIENTIFIC COMPUTING

If the results of our floating point computation differ from what we expect, we have an error,

$$\text{error} = \text{true value} - \text{approximate value}.$$

Another term we often use is **relative error**,

$$\text{relative error} = \frac{\text{error}}{\text{true value}},$$

which is defined as long as the denominator is nonzero. Relative error is often a more useful measure of accuracy since it is less dependent on the way the numbers are scaled. For example, if we decide to measure our data in grams instead of kilograms, the relative error remains the same, but the error is magnified by 1000.

Errors arise due to

(1) Machine hardware malfunctions. These are very rare today but were common in the early days of computing when the mean time between failures was only a few minutes.

(2) Blunders. Programming the wrong formula, etc.

(3) Experimental error. This occurs when the data are acquired by a means which has limited precision, for example via an instrument.

(4) Ignoring a significant feature of the problem. If we use the first 5 terms in the Taylor series as an approximation for e^x, then independent of what computer or precision we use, there is some irrevocable "truncation" error. See Example 2.5.

Example 2.5   Truncation Error for a Difference Quotient.

Consider Example 2.2. If we expand $f(x+h)$ in a Taylor series about x, the right hand side of the difference quotient becomes

$$\Delta_h f(x) \equiv \frac{f(x+h) - f(x)}{h} = \frac{f(x) + hf'(x) + \frac{h^2}{2}f''(\xi) - f(x)}{h}$$

$$= f'(x) + \frac{h}{2}f''(\xi), \quad x < \xi < x + h,$$

hence the truncation error in this approximation is $\frac{h}{2}f''(\xi)$.

Often in numerical analysis an error depends upon a parameter such as h above. In many cases it is enough to display the dependence of the error on that parameter without giving the expression in more detail. Then the "big oh" notation developed by Bachmann and Landau in 1927, will be used, $O(h)$, meaning only that the truncation error goes to zero no slower than h does. More precisely, we will write

$$p(t) = O(q(t)), \qquad \text{as} \qquad t \to t_0$$

to mean that $|p(t)/q(t)|$ is bounded for t sufficiently close to t_0. We will see more examples of this useful notation later. ■

(5) Numerical or rounding error. This is due to a combination of

(a) Ill conditioning or sensitivity of the problem.

(b) Stability of the algorithm.

It is sometimes said that truncation error arises from the terms we *exclude* and that rounding error from the terms we *include*. See the examples below.

Example 2.6   An Ill-conditioned Problem: Roots of a Quartic.

Compute all four roots of $x^4 - 4x^3 + 8x^2 - 16x + 15.99999999 = (x-2)^4 - 10^{-8} = 0$. The roots satisfy $(x-2)^2 = \pm 10^{-4}$, and hence $x - 2 = \pm\sqrt{\pm 10^{-4}} = \pm 10^{-2}$ and $\pm 10^{-2}i$. The roots are $x_1 = 2.01$, $x_2 = 1.99$, $x_3 = 2 + .01i$, $x_4 = 2 - .01i$. If we use a computer with $\epsilon_{\text{mach}} > 10^{-10}$ the quartic's constant will be rounded to 16.0. From the perspective of the computer we are now solving the problem

$$(x-2)^4 = 0.$$

This new problem has four equal roots at 2.0 which differ from the roots of the original quartic by 0.5%. In this problem small changes in the data (such as changing a coefficient by 10^{-8}) result in much larger changes in the solution *independent of the method used to compute the solution*. Such problems are termed **ill conditioned**. There is no computational trick we can use to reduce this sensitivity; it is associated with the problem, not the method. ■

Example 2.7   Taylor Series for e^x: A Better Algorithm.

Consider Example 2.1 for $x < 0$. Since the series approximation works well for positive x, we expect that

$$e^{-x} = \frac{1}{e^x} = \frac{1}{1 + x + \frac{x^2}{2!} + \cdots}$$

can be used for negative values. The program in Example 2.1 was modified to compute e^x for $x < 0$ by the formula above, and the table below lists the results, computed on a VAX.

x	$E(X)$	e^x
1	2.718282	2.718282
5	148.4132	148.4132
10	22026.47	22026.46
15	3269017.	3269017.
20	4.8516531×10^8	4.8516520×10^8
-1	.3678794	.3678795
-5	6.7379461×10^{-3}	6.7379470×10^{-3}
-10	4.5399924×10^{-5}	4.5399930×10^{-5}
-15	3.0590232×10^{-7}	3.0590232×10^{-7}
-20	2.0611530×10^{-9}	2.0611537×10^{-9}

Now the results are almost completely accurate. In this case the fault was not with the problem but rather with the algorithm we chose; direct summation of the series is an *unstable algorithm* for negative x. ■

Example 2.8   Roundoff Error for a Difference Quotient.

Consider again Example 2.2. We can analyze the effect of rounding on the computation of $f'(x)$. Assume no errors are made in forming x or $x + h$ and that the only error in evaluating f occurs when it is stored. Then it can be shown that the rounding error in the difference quotient $\Delta_h f(x)$ is bounded by $2|f(x)|\epsilon_{\text{mach}}/h$. Thus the rounding error in the difference quotient grows as h decreases. From this and Example 2.5, we see that the total error, truncation plus rounding is bounded by

$$|\text{Error}| \leq E_{\text{total}} = \frac{h}{2}|f''(\xi)| + \frac{2}{h}\epsilon_{\text{mach}}|f(x)| = O(h) + O(1/h).$$

This expression shows the same behavior as the numbers in Example 2.2, initial decrease and eventual increase. If we differentiate E_{total} with respect to h and set the result equal to zero, we see that the value of h which minimizes this expression is

$$h = 2\sqrt{\frac{|f(x)|\epsilon_{\text{mach}}}{|f''(\xi)|}}.$$

If $|f| \approx |f''|$, we get the rule of thumb

$$h \approx \sqrt{\epsilon_{mach}}$$

as a good choice to minimize E_{total}. ∎

To obtain more accurate estimates of the derivative we can work on either the rounding or the truncation error. If we can reduce the former, perhaps by computing on a machine with a smaller ϵ_{mach}, then we can carry the computation to smaller values of h (giving smaller truncation error) before rounding becomes dominant. This, in turn, will result in a smaller E_{total}. To illustrate, we repeat the calculation of Example 2.2, but on the Cyber used in Example 2.4. The results are

h	$\Delta_h f(1)$	e	error
10^0	4.67077427	2.71828183	1.95×10^0
10^{-1}	2.85884195	2.71828183	1.41×10^{-1}
10^{-2}	2.73191866	2.71828183	1.36×10^{-2}
10^{-3}	2.71964142	2.71828183	1.36×10^{-3}
10^{-4}	2.71841775	2.71828183	1.36×10^{-4}
10^{-5}	2.71829542	2.71828183	1.36×10^{-5}
10^{-6}	2.71828318	2.71828183	1.36×10^{-6}
10^{-7}	2.71828199	2.71828183	1.62×10^{-7}
10^{-8}	2.71828213	2.71828183	3.04×10^{-7}
10^{-9}	2.71826650	2.71828183	1.51×10^{-5}

For the VAX and Cyber approximate values of $\sqrt{\epsilon_{mach}}$ are $4 \cdot 10^{-4}$ and $8 \cdot 10^{-8}$ respectively. Note in Example 2.2 and in the table above that these values of h give accurate estimates of $f'(1)$.

An alternative approach to more accuracy is to reduce somehow the truncation error for the same values of h. One obvious way is to use a more accurate formula. For example, if we estimated the derivative by a "centered" difference quotient,

$$\delta_h f(x) \equiv \frac{f(x+h) - f(x-h)}{2h},$$

then exactly the same type of Taylor series manipulation as in Example 2.5 shows that

$$\delta_h f(x) = f'(x) + \frac{h^2}{6} f'''(x) + \frac{h^4}{120} f^{(5)}(x) + \cdots,$$

and the truncation error is then $O(h^2)$. As an example, we use the Cyber to approximate $e'(1)$ but this time with the centered difference approximation.

TABLE 2.1. CENTERED DIFFERENCE APPROXIMATION

h	$D_h f(1)$	e	error
10^0	3.194528049	2.718281828	4.76×10^{-1}
10^{-1}	2.722814564	2.718281828	4.53×10^{-3}
10^{-2}	2.718327133	2.718281828	4.53×10^{-5}
10^{-3}	2.718282285	2.718281828	4.53×10^{-7}
10^{-4}	2.718281833	2.718281828	4.62×10^{-9}
10^{-5}	2.718281829	2.718281828	8.58×10^{-10}
10^{-6}	2.718281820	2.718281828	8.38×10^{-9}
10^{-7}	2.718281849	2.718281828	2.00×10^{-8}
10^{-8}	2.718282133	2.718281828	3.04×10^{-7}
10^{-9}	2.718266501	2.718281828	1.53×10^{-5}

Example 2.1 illustrates how a badly conceived algorithm can get a poor answer to a perfectly well posed problem; the difficulty was corrected by changing the algorithm. Example 2.6 shows that for certain problems, "good" answers cannot be obtained by any algorithm because the problem is sensitive to small errors made in the data and the arithmetic. Remember, we are talking about the result of performing the calculation on a computer; in a theoretical analysis there are no rounding errors. It is important to distinguish between these two classes of pitfalls because there are unstable algorithms and ill-conditioned problems in nearly all branches of numerical mathematics. Once you become aware of their symptoms these problems are fairly easy to diagnose. With respect to Example 2.8, we mention that computing derivatives is not necessarily sensitive, but most algorithms involving difference quotients are unstable for small h.

*2.5 EXTRAPOLATION

The centered difference quotient illustrates one approach to reducing truncation error. A second, widely used technique is **extrapolation**. It is applicable to many diverse problems including evaluation of integrals and solution of differential equations. The key step in the use of extrapolation is to express the truncation error in a series expansion. The truncation error for a difference quotient is an ideal example. Looked at directly, the series for the centered difference tells us that if we compute the centered difference using h and $h/10$, the error for $h/10$ should be about $1/100$-th the error for h. This is exactly what Table 2.1 above shows, at least until roundoff error begins to dominate. Because our truncation error involves higher order Taylor series terms we can extract much more information from this table. Abbreviating $\delta_h f(x)$ by δ_h

$$\delta_h = f'(x) + \frac{h^2}{6} f'''(x) + \frac{h^4}{120} f^{(5)}(x) + \cdots,$$

$$\delta_{h/10} = f'(x) + \frac{1}{100}\frac{h^2}{6} f'''(x) + \frac{1}{10,000}\frac{h^4}{120} f^{(5)}(x) + \cdots.$$

If we combine these estimates

$$\frac{100\delta_{h/10} - \delta_h}{99} = f'(x) - \frac{h^4}{100 \cdot 120} f^{(5)}(x) + \cdots.$$

The expression on the left is an estimate of $f'(x)$. Its truncation error begins with a higher power of h, $O(h^4)$, and at least for small h this estimate ought to be better. We have taken two estimates with different values of h and combined them to get a third estimate which is more accurate than either. In practice, the left hand side above is usually calculated by rewriting it as

$$\delta_{h/10} + \frac{\delta_{h/10} - \delta_h}{99}$$

which is less susceptible to roundoff error. In numerical computing generally, it is usually advisable to organize formulas so that they express the desired quantity as a small correction to a good approximation.

To illustrate, we consider the centered difference Table 2.1. The table below is similar except that the E' (1) column contains the extrapolated estimates. An element in the extrapolation column is the result of combining two estimates from the original centered difference column according to the rule above. For example, the first entry in the E' (1) column is computed from the first two entries in the $\delta_h f(1)$ column by $2.722815 + (2.722815 - 3.194528)/99 = 2.718050$. The error column shows that the extrapolated estimates improve very rapidly, each by about a factor of 10,000, until roundoff appears. This is consistent with the $O(h^4)$ truncation error we have predicted.

h	E' (1)	e	error
10^{-1}	2.71804978127	2.71828182846	2.32×10^{-4}
10^{-2}	2.71828180580	2.71828182846	2.27×10^{-8}
10^{-3}	2.71828182845	2.71828182846	9.08×10^{-12}
10^{-4}	2.71828182855	2.71828182846	9.44×10^{-11}
10^{-5}	2.71828182928	2.71828182846	8.20×10^{-10}
10^{-6}	2.71828181999	2.71828182846	8.47×10^{-9}
10^{-7}	2.71828184879	2.71828182846	2.03×10^{-8}
10^{-8}	2.71828213559	2.71828182846	3.07×10^{-7}

The best extrapolated estimate, with error 9.08×10^{-12}, is several orders of magnitude better than any of the non-extrapolated estimates.

It is possible to generalize these ideas. For example, the extrapolated estimates themselves have a truncation error which is given by a series in powers of h, beginning with h^4. Thus we should be able to combine several of these estimates to get even better approximations. The literature on the subject of extrapolation is vast, see for example the article by Joyce (1971). You should always be alert to the possibility of such a technique whenever a sequence is being generated; individual elements of the sequence may be approaching a limit slowly, but by combining the elements in a careful way you can often generate a new sequence which approaches the same limit much more quickly.

2.6 HISTORICAL PERSPECTIVE: ECKERT AND MAUCHLY

In 1971 the *New York Times* stated that is was a "gross injustice" that the names of Eckert and Mauchly "were not likely ever to become household words on a par with the Wright Brothers or Thomas A. Edison, let alone the Beatles." John William Mauchly (1907–1980) and John Presper Eckert (1919–): who were these unknown men?

In 1940, Mauchly, a physics professor at Ursinus College in Pennsylvania, became interested in a better way to perform the complex calculations involved in solving meteorological equations. Mechanical calculators existed at that time, but were too slow for the applications he had in mind. In December of that year, he met John Atanasoff at a scientific meeting, and visited him at Iowa State College the following summer. Atanasoff, together with his graduate student Clifford Berry, were building an electronic calculator using vacuum tubes, which had the potential to be several orders of magnitude faster than anything then in existence. Mauchly returned from his visit immensely excited by what he had seen. At about the same time, he attended a summer course at the Moore School of Electrical Engineering at the University of Pennsylvania, to learn more about electronic devices. The staff was impressed with his abilities and asked him to stay on as an instructor. The Moore School had a contract with the Ballistics Research Laboratory (BRL) to produce range tables which were essential to aim and fire new artillery and Mauchly was naturally drawn into the computational problems.

At that time Eckert, a graduate student in electrical engineering, was working on improving the "differential analyzers" which were then used for these ballistic calculations. By incorporating several hundred vacuum tubes into the machine he was able to increase its speed and accuracy by a factor of ten, but it was clear that further improvements would not be successful. Independently, Eckert and Mauchly had built electronic counter circuits, and this along with the electronic experiences of the differential analyzer impressed them with the future potential of electronics in computing. They proposed to the U.S. Army Ordinance Department to build a general-purpose digital machine. In 1943, Mauchly as principal consultant, and Eckert as chief engineer, were empowered to design and build a device based on many of these ideas; this was the ENIAC (Electronic Numerical Integrator And Computer), the first electronic digital computer. ENIAC was completed, amidst front page publicity, in 1946.

It was too late to help in the war effort, but it was put into service at BRL, and remained in continuous use until it was shut down for the last time in 1955. It has been said that ENIAC, in the ten years it was in operation, did more calculations than had been done by the entire human race up until 1945.

The ENIAC was a breakthrough but it had many deficiencies—it was physically large (8 feet high, 3 feet wide, and 100 feet long, arranged in a horseshoe shape), had a small memory (20 ten-digit numbers), and was programmable only by manual external setting of switches. Eckert and Mauchly began thinking about a new computer, to be called EDVAC, in the winter of 1943–1944. It was to have a 1000-word memory, use mercury delay lines for storing data, but most significantly it would have its program stored in the computer memory. John von Neumann joined the project in September 1944 as a consultant, and assisted with the later design of the machine. Von Neumann described the work in "First Draft of a Report on the EDVAC" in June 1945. Unfortunately for Eckert and Mauchly, this "draft" became widely circulated shortly after it was written. Because only von Neumann's name appeared on the report, and because von Neumann was a famous mathematician, it was assumed that all the ideas in the report were his, even though the most important idea (the stored program) had been developed before he joined the project.

The report also ruined any chances for a patent. In 1946, before the EDVAC was finished, Eckert and Mauchly left the Moore School over a controversial school patent policy. The university felt that all patents should belong to the school, whereas Eckert and Mauchly thought that the inventors should get them. After leaving, Eckert and Mauchly applied for a patent for their computer design. The application was turned down in 1947: lawyers declared that the "First Draft ... " constituted publication of the ideas, and thus they were in the public domain and not patentable. In addition, there was controversy over which ideas could be attributed to Atanasoff.

Out in the private sector they formed a partnership for the purpose of designing a commercially successful computer. Mauchly was president and was responsible for the logic design; Eckert, preferring to remain behind the scenes, was vice president and chief engineer. (During the period 1948–1966 Eckert was granted 85 patents by the U.S. Patent Office.) The Census Bureau and the National Bureau of Standards were to be the first recipients of this new machine, dubbed the UNIVAC. Like high technology companies today, theirs was continually in need of additional capital and although the UNIVAC was a success, financial troubles forced them to sell the company to Remington Rand in 1950. Both Eckert and Mauchly continued to work there until nearly 1960. Remington Rand merged with Sperry in 1955 to form Sperry Rand. In 1986 Sperry Rand was purchased by Burroughs Corporation; the combined corporation was renamed Unisys and has marketed a computer line under the Sperry name.

Computers captured the public imagination, thanks in part to television. In 1952, CBS arranged to have a UNIVAC computer predict the outcome of the presidential election as part of the election coverage. Programs were written, with the runs to be made on the fifth UNIVAC, which was still sitting on the factory floor. In the studio in New York, a dummy control panel was shown, with the control lights hooked up to Christmas tree lights. After only a few polls had come in, UNIVAC predicted a landslide victory for Eisenhower over Stevenson. Since this was unexpected, it was assumed that there was a bug[1] in the programs, and the algorithms were manipulated several times to make them predict a narrow victory for Eisenhower. As more polls came in, though, it became obvious that UNIVAC had been right all along; a little after nine o'clock that evening, Walter Cronkite admitted the fakery to the viewers.

2.7 PROBLEMS

P2–1.–The Taylor series for the error function is

$$\text{erf}(x) = \frac{2}{\sqrt{\pi}} \sum_{n=0}^{\infty} \frac{(-1)^n x^{2n+1}}{n!(2n+1)}.$$

The series converges for all x. Write a program to evaluate $\text{erf}(x)$ using this series. Use as many terms in the series as are necessary so that the first neglected term does not alter the accumulated sum when it is added to it in floating point arithmetic. Since this is an alternating series, the error caused by truncating the infinite sum will then be less than the roundoff error. Investigate the effect of roundoff error by comparing the computed sum with the value obtained from a table of $\text{erf}(x)$ or the value obtained from a reliable subroutine (see Problem P1–1). Try $x = 0.5$, 1.0, 5.0 and 10.0. Explain your results. *Hint:* The inner loop of your program might look something like this:

```
10    OLDS = S
      EN = EN + 1.0
      T = -XSQ*T*(2.0*EN-1.0)/(EN*(2.0*EN+1.0))
      S = S + T
      IF (OLDS .NE. S) GO TO 10
```

What values should be assigned to T, S, EN, and XSQ before entering this loop? Do not forget the $2/\sqrt{\pi}$ factor.

P2–2.–Consider the following two Fortran programs:

```
      EPS = 1.0
10    EPS = EPS/2.0
      WRITE(*,*) EPS
      EPSP1 = EPS +1.0
      IF (EPSP1 .GT. 1.0) GO TO 10
      STOP
```

---

[1] The term originated in 1947 at Harvard during the days of the Mark computer. The project was proposed and run by strong-willed Howard Aiken (referred to as "Commander" by junior faculty and students) who continually emphasized reliability over speed. The electro-mechanical Mark II computer was experiencing inexplicable malfunctions. Finally, a researcher opened the cabinet and there, lodged between two relay contacts, was a dead moth. Grace Hopper, who later became instrumental in the development of COBOL, retrieved the moth and pasted it in her logbook, alongside a notation about the "bug" that caused the trouble.

```
      END


      EPS = 1.0
10    EPS = EPS/2.0
      WRITE(*,*) EPS
      IF (EPS .GT. 0.0) GO TO 10
      STOP
      END
```

Run the programs on your system, and explain the differences in the output. If you have access to an IBM personal computer with a math chip run the programs there too, and explain the results.

P2–3. Write a program to evaluate the infinite sum for $x = 0.0, 0.1, \ldots, 1.0$,

$$\phi(x) = \sum_{k=1}^{\infty} \frac{1}{k(k+x)},$$

with an error less than $0.5 \cdot 10^{-6}$. *Note:* this requires both human analysis and computer power, and neither is likely to succeed without the other. Above all, do not waste a years's computer budget trying to sum the series by brute force. *Hint:* Use the fact that

$$\frac{1}{k(k+1)} = \frac{1}{k} - \frac{1}{k+1}$$

to prove that $\phi(1) = 1$. Then express $\phi(x) - \phi(1)$ as an infinite series which converges faster than the one defining $\phi(x)$. You will have to repeat this trick before you get a series for computing $\phi(x)$ that converges fast enough. Reference: Hamming (1962), pp. 48–50.

P2–4. Consider a fictitious floating point number system composed of the following numbers

$$S = \left\{ \pm b_1.b_2b_3 \times 2^{\pm y} : b_2, b_3, y = 0 \text{ or } 1, \text{ and } b_1 = 1 \text{ unless } b_1 = b_2 = b_3 = y = 0 \right\}.$$

  (a) Draw a portion of the real axis and plot the elements of S.
  (b) Show that S contains twenty five elements. What are the values of OFL, UFL, and ϵ_{mach}?

P2–5. Consider trying to approximate the derivatives of the following functions

$$f_1(x) = \sin x, \text{ at } x = 1, \quad \text{small}$$
$$f_2(x) = 10000 \sin x, \text{ at } x = 1, \quad \text{large}$$
$$f_3(x) = \tan x, \text{ at } x = 1.59.$$

To estimate the first derivative we will use the central difference quotient, and for the second derivative we will use

$$f''(x) \approx \frac{f(x+h) - 2f(x) + f(x-h)}{h^2}.$$

(a) Estimate the first derivatives of the three functions as indicated. Typically we would choose $h \approx \sqrt[3]{\epsilon_{mach}}$ to obtain the most accurate estimate of the first derivative. Try the central difference formula for $h = 10^k \sqrt[3]{\epsilon_{mach}}$ with $k = -2, -1, 0, 1, 2$.

(b) Now estimate the second derivatives. In this case the traditional choice of h is $h \approx \sqrt[4]{\epsilon_{mach}}$. Try using $h = 10^k \sqrt[4]{\epsilon_{mach}}$ with $k = -2, -1, 0, 1, 2$ on the three functions.

In each case, compare the computed derivative estimate with the actual derivative. Can you explain how the choice of the best h is affected by the behavior of $f(x)$ near the point x?

(c) Use the techniques of extrapolation (Section 5) to improve your estimates in (a) and (b).

P2–6.–The task is to write a program to compute

$$X(p, k, N) = \binom{N}{k} p^k (1 - p)^{N-k}, \qquad 0 \leq p \leq 1, \ 0 \leq k \leq N \leq 2000$$

which occurs frequently in statistical studies. The symbol $\binom{N}{k}$ is the binomial coefficient

$$\binom{N}{k} = \frac{N!}{k!(N-k)!} = \frac{N(N-1) \cdots (N-k+1)}{k(k-1) \cdots 2 \cdot 1}.$$

Consider the following Fortran program fragment to compute $X(p, k, N)$

```
        Q = 1.0 - P
        X = 10.0
        DO 1 J = 1,K
1           X = X*(N+1-J)
        DO 2 J = 1,K
2           X = X/J
        DO 3 J = 1,K
3           X = X*P
        DO 4 J = K+1,N
4           X = X*Q
```

(a) Using this program try to compute $X(0.1, 200, 2000) \approx 0.03$. Comment on the difficulties you encounter.

(b) One approach is developed in the book by P. Sterbenz (1974). Let B be a large power of 2 chosen so that $2000B$ will not overflow and $B/2000$ will not underflow. For example $B = 2^{100}$. If X becomes larger than B we shall divide it by B. Then the true value we are computing is represented by X·B. If we use a counter I to count the number of times we divide by B the correct output is

$$X(p, k, N) = \text{X} \cdot (B)^I, \qquad I \geq 0.$$

Similarly set $S = 1/B$. If X becomes smaller than S, multiply X by B and subtract one from I. At the conclusion of the computation if $I < 0$ we may be able to divide X by B I times without underflow and get a meaningful answer. Otherwise the best answer we can get is zero. Modify your program and try it on the problem in (a).

P2–7. The formula derived in Problem P1–2 to estimate π,

$$p_{n+1} = 2^n \sqrt{2(1 - \sqrt{1 - (p_n/2^n)^2})},$$
$$p_2 = 2\sqrt{2},$$

fails due to a combination of underflow and catastrophic cancellation.

(a) Explain the failure in the formula.

(b) The formula can be improved so that the subtraction is eliminated. First write p_{n+1} as

$$p_{n+1} = 2^n \sqrt{r_{n+1}},$$

where

$$r_{n+1} = 2(1 - \sqrt{1 - (p_n/2^n)^2}), \qquad r_3 = 2/(2 + \sqrt{2}).$$

Show that

$$r_{n+1} = \frac{r_n}{2 + \sqrt{4 - r_n}}.$$

Use the last iteration to calculate r_n and p_n for $n = 3, 4, \ldots, 60$. (This revision was suggested by W. Kahan.)

(c) Eventually, $4 - r_n$ will round to 4, and so the formula derived in (b) is still affected by rounding errors for large values of n. Should this concern us?

P2–8. Archimedes' method for estimating π can be derived in a different way. Let $P(h)$ denote the perimeter of the n-sided inscribed polygon, where $h = 1/n$.

(a) Using geometry, show that

$$P(h) = (1/h) \times \sin(\pi h).$$

Write a program to evaluate this function for $n = 3, 6, 12, 24, 48$ and 96. Print your estimate and the error for each value of n.

(b) By expanding $\sin(x)$ in a series show that

$$P(h) = \pi + \sum_{j=1}^{\infty} \frac{(-1)^j \pi^{2j+1}}{(2j+1)!2j}.$$

In other words, $P(h) = \pi + O(h^2)$. Modify your program to see if the convergence is really $O(h^2)$. Do this by printing the ratio of the error on one step to the error on the preceding step. What limit does the ratio approach? Why?

(c) Since we know that the error in $P(h)$ is $O(h^2)$, extrapolation can be used to improve the estimates. Show that the new sequence

$$P_{(1)}(h) \equiv \frac{4P(h) - P(2h)}{3}$$

should converge to π with an error that is $O(h^4)$. Rewrite the formula in the form of a correction as described in Section 5, and modify your program to calculate these numbers. Are they more accurate than $P(h)$? Hint: Both your understanding and the computations are aided if you construct a table with $P(h)$ in the first column and $P_{(1)}(h)$ in the second. It will also help if you perform these computations in double precision.

(d) Show that the numbers

$$P_{(2)}(h) = \frac{2^4 P_{(1)}(h) - P_{(1)}(2h)}{2^4 - 1}, \qquad P_{(3)}(h) = \frac{2^6 P_{(2)}(h) - P_{(2)}(2h)}{2^6 - 1}$$

converge like $O(h^6)$ and $O(h^8)$ respectively. Rewrite the formulas as in part (c), compute these numbers, and add them as third and fourth columns in your table.

(e) Develop the formulas for columns five and six and compute these numbers. Where in the table is your best estimate and how much better is it than Archimedes' (see Problem P1–2)? Is there any evidence of roundoff error?

P2–9.–Fortran compilers are not required to supply a function to evaluate the cotangent function. Instead you might write your own using `COT(X)=COS(X)/SIN(X)`. There are many ways to check on the accuracy, but one simple test is to compare `COT(X)` against `(COT(X/2)-1.0/COT(X/2))/2.0`, which is mathematically equal to the cotangent. Write a program to evaluate both functions for 2000 points in the interval $(6\pi, 25\pi/4)$. Print the average error and the maximum error. How satisfactory are the results? How do you know whether the errors you observe are from the computation of the `COT(X)` or from the subtraction in the second expression? As a check use double precision to compute the latter.

P2–10.–This problem concerns some details of floating point arithmetic.

(a) If `I` and `J` are integers explain why the Fortran expression `(I+J)*(I-J)` is preferred to `I*I-J*J`. Do the same reasons apply if `I` and `J` are real?

(b) Write a program to estimate OFL in both single and double precision, and to determine the largest integer on your computer. What happens if you exceed any of these values?

P2–11.–The examples in Section 4 suggest that an unstable algorithm results because of cancellation. This need not be the case as the following problem (due to W. Kahan) illustrates. Let N be the number of bits in the floating point word that your computer/compiler uses. (On a DEC Vax, IBM PC, Sun, or Apple Macintosh, $N = 32$.) Consider the Fortran function

```
REAL FUNCTION F(X,N)
REAL X
F = ABS(X)
DO 10 I = 1,N
    F = SQRT(F)
10   CONTINUE
DO 20 I = 1,N
    F = F*F
20   CONTINUE
RETURN
END
```

If there were no rounding errors we should get F = ABS(X), but this is not what happens. Experiment with this function for several values of X, both $|X| < 1$ and $|X| > 1$. Explain your results. Set your compiler options so that underflows are set to zero.

3

Linear Systems of Equations

3.1 INTRODUCTION

One of the most frequent problems encountered in scientific computation is the solution of a system of simultaneous linear equations—usually with as many equations as unknowns. Such a system can be written in the form

$$Ax = b,$$

where A is a given square matrix of order n, b is a given column vector of n components, and x is an unknown column vector of n components.

Linear equations arise directly in many applications, such as in the bridge engineering problem at the end of the chapter. They also arise indirectly as a step in the solution of a more complex problem. Systems of *nonlinear* equations, such as those arising in the design of integrated circuits, are solved via a sequence of linear approximations, hence a sequence of linear equations; these are discussed in Chapter 7. Boundary value problems in differential equations are often handled by seeking the solution only at a finite set of points, leading in many important cases to the solution of linear equations. An example application of this is given in Section 9.2.

Data-fitting problems can also lead to systems of linear equations. Consider Example 3.1.

Example 3.1 Data Fitting via Linear Equations.

Suppose we wanted to fit a quadratic polynomial to the data

$$(1, 0), \ (2, -1), \ (3, 2).$$

We seek a polynomial of the form

$$p(x) = a + bx + cx^2$$

satisfying

$$p(1) = 0, \ p(2) = -1, \ p(3) = 2.$$

Writing out these three conditions

$$
\begin{aligned}
a + \ b + \ \ c &= \ \ 0, \\
a + 2b + 4c &= -1, \\
a + 3b + 9c &= \ \ 2,
\end{aligned}
$$

we obtain a system of three linear equations in three unknowns, which in matrix/vector notation has the form

$$
\begin{pmatrix} 1 & 1 & 1 \\ 1 & 2 & 4 \\ 1 & 3 & 9 \end{pmatrix}
\begin{pmatrix} a \\ b \\ c \end{pmatrix} =
\begin{pmatrix} 0 \\ -1 \\ 2 \end{pmatrix}.
$$

The solution is

$$a = 5, \ b = -7, \ c = 2.$$

It is easily verified that the polynomial $p(x) = 5 - 7x + 2x^2$ fits the data as required. This subject is discussed in much more detail in Chapter 4. ∎

Many methods have been proposed for solving linear equations. A famous one is Cramer's rule, where each component of the solution is determined as the ratio of two determinants. If you tried to solve a system of 30 equations using Cramer's rule, you would need to compute 31 determinants each of order 30. If computed in the straightforward way, solution of the linear system takes $31 \times 30! \times 29$ multiplications, plus a similar number of additions. On a very fast computer doing $1,000,000,000$ multiplications per second, near the limits of current technology, the multiplications alone would take about $7,556,414,967,271,268,000$ years, which is more time than we would really care to spend on this problem. (Of course, this example doesn't give Cramer's rule a fair chance. The determinants were computed in the worst possible way. There are much better ways to compute determinants; however, with a good method it is possible to solve a linear system in about the time it takes to compute one determinant. Cramer's rule has at least one attractive property: it computes every element of the solution independently. For this reason, it can be a practical method for some special linear systems on parallel computers.)

Another approach with mathematical appeal but computational pitfalls writes the solution to $Ax = b$ as $x = A^{-1}b$, where A^{-1} is the inverse of A. However, in virtually every application it is unnecessary and inadvisable to compute A^{-1} explicitly. As an extreme but illustrative example, consider just one equation

$$7x = 21.$$

The best way to solve this "system" is by division,

$$x = \tfrac{21}{7} = 3.$$

Using the matrix inverse gives

$$x = (7^{-1})(21)$$
$$= (.142857)(21) = 2.99997.$$

The inverse requires more arithmetic—a division and a multiplication instead of just a division—and produces a less accurate answer. However, the extra arithmetic is the main reason we avoid computing the inverse. Similar considerations apply to systems of more than one equation. This is even true in the common situation where there are several systems having the same matrix A but different right-hand sides b. Consequently, we shall concentrate on the direct solution of systems of equations rather than the computation of the inverse.

The methods developed in the next section have five advantages over computing the inverse: (1) they are cheaper by a factor of 3, (2) they generally produce more accurate answers, (3) they are more flexible, in that they put the matrix in a form such that Ab, $A^{-1}b$, A^Tb, and $A^{-T}b$ can all be easily computed for any vector b, (4) in general they are better at preserving the "structure" of the matrix A (even if A has many zero entries, A^{-1} can have none), (5) they are more informative, in that they provide estimates of the accuracy of the solution. Instead of forming the inverse, the methods factor the matrix A into the product of matrices of simpler form.

It is important to distinguish between two types of matrices:

1. A **stored matrix** is one for which all n^2 matrix elements a_{ij} are stored in the computer memory. This limits the order n to be a few hundred on medium-scale computers and about a thousand on larger computers.

2. A **sparse matrix** is one for which most of the matrix elements are zero, and the nonzero elements can be either stored in some special data structure or regenerated as needed. This type of matrix often results from finite-difference and finite-element methods for partial differential equations. The order n is frequently as large as several tens of thousands and occasionally even larger. An example of a sparse matrix for which the elements can be easily regenerated is

$$\begin{pmatrix} 4 & 1 & 0 & 0 & 0 & 0 & 0 \\ 1 & 4 & 1 & 0 & 0 & 0 & 0 \\ 0 & 1 & 4 & 1 & 0 & 0 & 0 \\ 0 & 0 & 1 & 4 & 1 & 0 & 0 \\ 0 & 0 & 0 & 1 & 4 & 1 & 0 \\ 0 & 0 & 0 & 0 & 1 & 4 & 1 \\ 0 & 0 & 0 & 0 & 0 & 1 & 4 \end{pmatrix}.$$

These two types overlap somewhat. A stored matrix may have many zero elements and hence also be sparse, but if the zero elements are stored explicitly in memory, the sparsity is not important. A large but non-sparse matrix may be stored in secondary memory such as disk or tape and thereby require more elaborate data-handling techniques. A **band matrix** is one for which all the nonzero elements are near the diagonal, specifically $a_{ij} = 0$ for all i, j with $|i - j| > m$, where $m \ll n$. The **band width** is said to be $2m + 1$, and the nonzero elements are situated on $2m + 1$ diagonals. The matrix shown above is a three-band matrix, called a **tridiagonal matrix**, and satisfies the above conditions with $m = 1$.

Some of the computational techniques appropriate for stored matrices are quite different from those appropriate for sparse matrices. The stored matrix methods are the most basic and will be emphasized here. The methods can be modified to handle matrices in secondary storage, band matrices, and other types of large or moderately sparse matrices.

Let x^* be the computed solution to the linear system $Ax = b$. There are two common measures of the discrepancy in x^*: The **error**

$$e = x - x^*$$

and the **residual**

$$r = b - Ax^* = A(x - x^*) = Ae.$$

The residual measures the amount by which the computed solution, when substituted into the equations, differs from the right-hand side. Matrix theory tells us that, since A is nonsingular, if one of these is zero the other must also be zero. But they are not necessarily both "small" at the same time, as we will discuss. The methods presented in this chapter will produce an accurate answer in the sense that the residual will be "small." It will not be possible to guarantee in advance that the error will also be small. Thus, the computed solution will "nearly solve" the equations, but it might not resemble the true solution. For many applications this will be adequate. When the matrix A is nearly singular (i.e., when a small change in the equations will make the system singular), tiny changes in the coefficients of the system can produce large changes in the solution, even ignoring rounding errors. Thus, for these systems of equations it is unreasonable to expect that x^* will be accurate. If it is known that the matrix is far from being singular, then both the residual and the error will be satisfyingly small. It is possible to estimate how close A is to being singular as a byproduct of solving $Ax = b$, and so the accuracy of x^* can be estimated if desired.

3.2 LINEAR SYSTEMS FOR STORED MATRICES

In this section we shall consider the solution of a linear system of algebraic equations

$$Ax = b$$

with a stored, n-by-n matrix A and n-vectors b and x. We assume that A is a nonsingular matrix. If A is singular, in principle this will be revealed during the computation, but in practice it may be difficult to decide. (By the way, a **singular** matrix A is one that does not have an inverse. *Equivalent* definitions are that the determinant of A is zero, that the rows of A are linearly dependent, and that there exists some non-zero vector z such that $Az = 0$. From this last definition we can see that if A is singular and $Ax = b$ for some vector x, then $A(x + \alpha z) = b$ for any α and hence the solution of the linear system is not unique. It is also a consequence of these definitions that $Ax = b$ will not have any solution for certain values of the right-hand side b if A is singular.)

The algorithm that is almost universally used is one of the oldest numerical methods—the systematic elimination method, generally named after C. F. Gauss. In the late 1940's, soon after the development of the electronic computer, it was one of the first algorithms to be analyzed with regard to its behavior when programmed in finite-precision arithmetic. The results that John von Neumann and others obtained were considered pessimistic, in part because of their technical complexity, but also because their point of view emphasized some of the worst aspects of the method. Gaussian elimination fell into disfavor. Around 1960, primarily through the work of J.H. Wilkinson, the method was revealed to be a nearly ideal algorithm, producing solutions that were as good as any conceivable algorithm for the problem. Gaussian elimination has had a central place in numerical analysis ever since.

Modern research on Gaussian elimination has revealed the importance of two ideas: the need to pivot, and the interpretation of rounding errors. Gaussian elimination and other aspects of matrix computation are studied in detail in the book by Golub and Van Loan (1983). The reader who wishes more information than we have in this chapter should consult this reference.

Gaussian elimination is based on the idea of reducing a general system of equations to one of simpler form, one that is easier to solve. Here, "simpler" is taken to mean "triangular," as discussed in Example 3.2.

Example 3.2   Solving a Triangular System of Equations.

Consider a system of the form

$$10x_1 - 7x_2 + 0x_3 = 7,$$
$$2.5x_2 + 5x_3 = 2.5,$$
$$6.2x_3 = 6.2.$$

In matrix form this is

$$\begin{pmatrix} 10 & -7 & 0 \\ 0 & 2.5 & 5 \\ 0 & 0 & 6.2 \end{pmatrix} \begin{pmatrix} x_1 \\ x_2 \\ x_3 \end{pmatrix} = \begin{pmatrix} 7 \\ 2.5 \\ 6.2 \end{pmatrix},$$

and the non-zero elements all occur in the upper-right "triangle" of the matrix. This system is easy to solve. From the last equation

$$6.2x_3 = 6.2,$$

we determine that $x_3 = 1$. This value is substituted into the second equation:

$$2.5x_2 + (5)(1) = 2.5.$$

Hence $x_2 = -1$. Finally the values of x_3 and x_2 are substituted into the first equation:

$$10x_1 + (-7)(-1) + (0)(1) = 7.$$

Hence $x_1 = 0$. This solution can be easily verified using the original equations. ∎

This technique of solving triangular systems of equations is called **back substitution**. For a general upper triangular system:

$$\begin{pmatrix} a_{11} & a_{12} & \cdot & \cdot & \cdot & a_{1n} \\ & a_{22} & \cdot & \cdot & \cdot & a_{2n} \\ & & & & & \cdot \\ & & & & & \cdot \\ & & & & & \cdot \\ & & & & & a_{nn} \end{pmatrix} \begin{pmatrix} x_1 \\ x_2 \\ \cdot \\ \cdot \\ \cdot \\ x_n \end{pmatrix} = \begin{pmatrix} b_1 \\ b_2 \\ \cdot \\ \cdot \\ \cdot \\ b_n \end{pmatrix},$$

the solution can be obtained from the following formulas

$$x_i = \begin{cases} b_n/a_{nn}, & \text{if } i = n; \\ (b_i - \sum_{j=i+1}^n a_{ij}x_j)/a_{ii}, & \text{otherwise.} \end{cases}$$

This is an efficient algorithm, requiring about $n^2/2$ multiplications (note that there are about $n^2/2$ non-zero elements in the matrix). Also, the algorithm can be applied whenever $a_{ii} \neq 0$ for all i. If some $a_{ii} = 0$, then the matrix is singular. (Why?).

Back substitution is the second half of Gaussian elimination. The first half, the **forward elimination**, takes a general non-singular matrix and reduces it to upper triangular form. This algorithm is illustrated in Example 3.3.

Example 3.3   Gaussian Elimination.

Consider the following example of order 3:

$$\begin{pmatrix} 10 & -7 & 0 \\ -3 & 2 & 6 \\ 5 & -1 & 5 \end{pmatrix} \begin{pmatrix} x_1 \\ x_2 \\ x_3 \end{pmatrix} = \begin{pmatrix} 7 \\ 4 \\ 6 \end{pmatrix}.$$

This represents the three simultaneous equations

$$10x_1 - 7x_2 \qquad = 7,$$
$$-3x_1 + 2x_2 + 6x_3 = 4,$$
$$5x_1 - \quad x_2 + 5x_3 = 6.$$

The first step uses the first equation to eliminate x_1 from the other equations. This is accomplished by subtracting -0.3 times the first equation to the second equation and subtracting 0.5 times the first equation to the third equation. The quantities -0.3 and 0.5 are called the **multipliers**.

$$\begin{pmatrix} 10 & -7 & 0 \\ 0 & -0.1 & 6 \\ 0 & 2.5 & 5 \end{pmatrix} \begin{pmatrix} x_1 \\ x_2 \\ x_3 \end{pmatrix} = \begin{pmatrix} 7 \\ 6.1 \\ 2.5 \end{pmatrix}.$$

The second step might involve using the second equation to eliminate x_2 from the third equation. However, the coefficient of x_2 in the second equation is a small number, -0.1. Consequently, the last two equations are interchanged. This is not actually necessary in this example because there are no roundoff errors, but it is crucial in general for reasons discussed below.

$$\begin{pmatrix} 10 & -7 & 0 \\ 0 & 2.5 & 5 \\ 0 & -0.1 & 6 \end{pmatrix} \begin{pmatrix} x_1 \\ x_2 \\ x_3 \end{pmatrix} = \begin{pmatrix} 7 \\ 2.5 \\ 6.1 \end{pmatrix}.$$

Now, the second equation can be used to eliminate x_2 from the third equation. This is accomplished by subtracting -0.04 times the second equation to the third equation. (What would the multiplier have been if the equations had not been interchanged?)

$$\begin{pmatrix} 10 & -7 & 0 \\ 0 & 2.5 & 5 \\ 0 & 0 & 6.2 \end{pmatrix} \begin{pmatrix} x_1 \\ x_2 \\ x_3 \end{pmatrix} = \begin{pmatrix} 7 \\ 2.5 \\ 6.2 \end{pmatrix}.$$

This is the same upper triangular system we examined earlier. ∎

In general, the forward elimination consists of $n - 1$ steps. At the k-th step, multiples of the k-th equation are subtracted from the remaining equations to eliminate the k-th variable. If the coefficient of x_k is "small," it is advisable to interchange equations before this is done. The back substitution consists of solving the last equation for x_n, then the next-to-last equation for x_{n-1}, and so on until x_1 is computed from the first equation.

At the k-th step of Gaussian elimination, the elements of the matrix A are modified according to the following formula

$$a_{ij} \leftarrow a_{ij} - (a_{ik}/a_{kk})a_{kj}.$$

This calculation is carried out for $i > k$ and $j > k$. The term (a_{ik}/a_{kk}) is referred to as the **multiplier**.

We have already noted that the back substitution requires about $n^2/2$ multiplications. A similar analysis can be performed on the forward elimination process. This shows that about $n^3/3$ multiplications are required for that step. For all practical values of n the forward elimination dominates the computation. In analogy with Example 2.5 of Chapter 2, we often write that Gaussian elimination requires $O(n^3)$ operations.

The entire algorithm can be compactly expressed in matrix notation. It corresponds to the factorization of A into simpler matrices: $A = PLU$, where L and U are triangular and P is a permutation matrix that records the row interchanges. For the above example,

$$\begin{pmatrix} 10 & -7 & 0 \\ -3 & 2 & 6 \\ 5 & -1 & 5 \end{pmatrix} = \begin{pmatrix} 1 & 0 & 0 \\ 0 & 0 & 1 \\ 0 & 1 & 0 \end{pmatrix} \begin{pmatrix} 1 & 0 & 0 \\ 0.5 & 1 & 0 \\ -0.3 & -0.04 & 1 \end{pmatrix} \begin{pmatrix} 10.0 & -7.0 & 0.0 \\ 0 & 2.5 & 5.0 \\ 0 & 0 & 6.2 \end{pmatrix}.$$

The matrix P is just a rearrangement of the rows of the identity matrix, and records that rows 2 and 3 of A were interchanged during the forward elimination. The matrix U

is the final upper triangular matrix obtained at the end of the forward elimination. The matrix L records the multipliers used during the forward elimination.

The factorization $A = PLU$ is called an LU **factorization** or **triangular factorization** of A. It should be emphasized that no new algorithm has been introduced. Triangular factorization is simply Gaussian elimination expressed in matrix notation.

This factorization can be formed without knowledge of the right-hand side b. The linear system $Ax = b$ can be solved by solving a sequence of simpler linear systems: first solve $Pz = b$ by reordering the elements of b; next solve $Ly = z$ using "forward" substitution; finally obtain x by solving $Ux = y$ which is also a triangular system, via back substitution. Mathematically,

$$x = U^{-1}y = U^{-1}L^{-1}z = U^{-1}L^{-1}P^{-1}b = (PLU)^{-1}b = A^{-1}b,$$

so this is equivalent to solving the original system. When we apply these ideas to the example above, we obtain

$$Pz = b: \quad z = \begin{pmatrix} 7 \\ 6 \\ 4 \end{pmatrix}$$

$$Ly = z: \quad y = \begin{pmatrix} 7.0 \\ 2.5 \\ 6.2 \end{pmatrix}$$

$$Ux = y: \quad x = \begin{pmatrix} 0.0 \\ -1.0 \\ 1.0 \end{pmatrix}$$

which is the same as before.

The diagonal elements of U are called **pivots**. The k-th pivot is the coefficient of the k-th variable in the k-th equation at the k-th step of the elimination. In the example, the pivots are 10, 2.5, and 6.2.

The subroutine SGEFS in this chapter takes advantage of the fact that elimination and back substitution can occur independently. The user's input matrix is overwritten. In place of the original matrix, U is stored in the upper triangle. Returned in the lower triangle is the "below the diagonal" portion of L. This information, along with one extra array that describes the sequence of pivots, is sufficient to allow the solution for any right hand side b as we have just illustrated.

Both the computation of the multipliers and the back substitution require divisions by the pivots. Consequently, the algorithm cannot be carried out if any of the pivots are zero. Intuition should tell us that it is a bad idea to complete the computation if any of the pivots are nearly zero. To see this, consider Example 3.4.

Example 3.4   Near-zero Pivots.

Let us change our example slightly to

$$\begin{pmatrix} 10 & -7 & 0 \\ -3 & 2.099 & 6 \\ 5 & -1 & 5 \end{pmatrix} \begin{pmatrix} x_1 \\ x_2 \\ x_3 \end{pmatrix} = \begin{pmatrix} 7 \\ 3.901 \\ 6 \end{pmatrix}.$$

The $(2, 2)$ element of the matrix has been changed from 2.000 to 2.099, and the right-hand side has also been changed so that the exact answer is still $(0, -1, 1)^T$. Let us assume that the solution is to be computed on a hypothetical machine which does decimal floating-point arithmetic with five significant digits.

The first step of the elimination produces

$$\begin{pmatrix} 10 & -7 & 0 \\ 0 & -1.0 \times 10^{-3} & 6 \\ 0 & 2.5 & 5 \end{pmatrix} \begin{pmatrix} x_1 \\ x_2 \\ x_3 \end{pmatrix} = \begin{pmatrix} 7 \\ 6.001 \\ 2.5 \end{pmatrix}.$$

The $(2, 2)$ element is now quite small compared with the other elements in the matrix. Nevertheless, let us complete the elimination without using any interchanges. The next step requires adding 2.5×10^3 times the second equation to the third. On the right-hand side, this involves multiplying 6.001 by 2.5×10^3, giving 1.50025×10^4, which cannot be exactly represented in our hypothetical floating-point number system. It must either be chopped to 1.5002×10^4 or rounded to 1.5003×10^4. The result is then added to 2.5. Let us assume chopped arithmetic is used. Then the last equation becomes

$$1.5005 \times 10^4 x_3 = 1.5004 \times 10^4,$$

and so

$$x_3 = \frac{1.5004 \times 10^4}{1.5005 \times 10^4} = 0.99993.$$

Since the exact answer is $x_3 = 1$, it does not appear that the error is too serious. Unfortunately, x_2 must be determined from the equation

$$-1.0 \times 10^{-3} x_2 + (6)(0.99993) = 6.001,$$

which gives

$$x_2 = \frac{1.5 \times 10^{-3}}{-1.0 \times 10^{-3}} = -1.5.$$

Finally x_1 is determined from the first equation,

$$10x_1 + (-7)(-1.5) = 7,$$

which gives

$$x_1 = -0.35.$$

Instead of $(0, -1, 1)^T$, we have obtained $(-0.35, -1.50, 0.99993)^T$. ■

Where did things go wrong? There was no "accumulation of rounding errors" caused by doing thousands of arithmetic operations. The matrix is not close to singular. The difficulty comes from choosing a small pivot at the second step of the elimination. As a result, the multiplier is 2.5×10^3, and the final equation involves coefficients which are 10^3 times as large as those in the original problem. Roundoff errors which are small when compared to these large coefficients are unacceptable in terms of the original matrix and the actual solution.

We leave it to the reader to verify that if the second and third equations are interchanged, then no large multipliers are necessary and the final result is satisfactory. This turns out to be true in general: If the multipliers are all less than or equal to 1 in magnitude, then Gaussian elimination is a stable algorithm, producing a computed solution likely to be as accurate as that produced by any algorithm for this problem.

Keeping the multipliers less than one in absolute value can be ensured by a process known as **partial pivoting**: At the k-th step of the forward elimination, the pivot is taken to be the largest (in absolute value) element in the unreduced part of the k-th column. The row containing this pivot is interchanged with the k-th row to bring the pivot element into the (k, k) position. The same interchanges must be done with the elements of the right-hand side b. The unknowns in x are not reordered because the columns of A are not interchanged.

The rounding errors introduced during the computation almost always cause the computed solution—which we now denote by x^*—to differ somewhat from the theoretical solution $x = A^{-1}b$. In fact, it *must* differ because the elements of x are usually not floating-point numbers.

The following example illustrates that the residual and the error, two common measures in the accuracy of our solution, need not both be small at the same time.

Example 3.5   Errors and Residuals.

We now examine Example 2.3 of Chapter 2 in more detail. Consider the problem

$$\begin{pmatrix} 0.780 & 0.563 \\ 0.457 & 0.330 \end{pmatrix} \begin{pmatrix} x_1 \\ x_2 \end{pmatrix} = \begin{pmatrix} 0.217 \\ 0.127 \end{pmatrix}.$$

What will happen if we carry out Gaussian elimination with partial pivoting on a hypothetical three-digit decimal computer which rounds? First, the multiplier

$$\frac{0.457}{0.780} = 0.586 \quad \text{(to three places)}$$

is computed. Next, 0.586 times the first row is subtracted from the second row to produce the system

$$\begin{pmatrix} 0.780 & 0.563 \\ 0 & 0.0000820 \end{pmatrix} \begin{pmatrix} x_1 \\ x_2 \end{pmatrix} = \begin{pmatrix} 0.254 \\ -0.000162 \end{pmatrix}.$$

Finally the back substitution is carried out:

$$x_2 = \frac{-0.00162}{0.0000820} = -1.98 \quad \text{(to three places)},$$

$$x_1 = \frac{[0.217 - 0.563 x_2]}{0.780}$$

$$= 1.71 \quad \text{(to three places)}.$$

Thus the computed solution is

$$x^* = \begin{pmatrix} 1.71 \\ -1.98 \end{pmatrix}.$$

To assess the accuracy without knowing the exact answer, we compute the residuals (exactly):

$$r = b - Ax^* = \begin{pmatrix} 0.217 - [(0.780)(1.71) + (0.563)(-1.98] \\ 0.127 - [(0.467)(1.71) + (0.330)(-1.98)] \end{pmatrix}$$
$$= \begin{pmatrix} -0.00206 \\ -0.00107 \end{pmatrix}.$$

The *residuals* are less than 10^{-2}. We could hardly expect better on a three-digit machine. However, it is easy to see that the exact solution to this system is

$$x = \begin{pmatrix} 1.000 \\ -1.000 \end{pmatrix}.$$

So the *error* is almost as large as the solution. ■

Were the small residuals just a lucky fluke? First, the reader should begin to realize by now that this example is highly contrived. The matrix is incredibly close to being singular and is not typical of most problems encountered in practice. Nevertheless, let us track down the reason for the small residuals.

If Gaussian elimination with partial pivoting is carried out for this example on a computer with six or more digits, the forward elimination will produce a system something like

$$\begin{pmatrix} 0.780000 & 0.563000 \\ 0 & 0.000140 \end{pmatrix} \begin{pmatrix} x_1 \\ x_2 \end{pmatrix} = \begin{pmatrix} 0.212000 \\ -0.000140 \end{pmatrix}.$$

Now the back substitution produces

$$x_2 = \frac{-0.000140}{0.000140} = -1.00000,$$
$$x_1 = \frac{0.217 - 0.563x_2}{0.780} = 1.00000,$$

the exact answer. On our three-digit machine, x_2 was computed by dividing two quantities both of which were on the order of rounding errors. Hence x_2 can turn out to be almost anything. In fact, if we use a machine with nine binary digits, we shall obtain a completely different value. Then this completely arbitrary value of x_2 was substituted into the first equation to obtain x_1. We can reasonably expect the residual from the first equation to be small—x_1 was computed in such a way as to make this certain. Now comes a subtle but crucial point. We can also expect the residual from the second equation to be small, *precisely because the matrix is so close to being singular*. The two equations are nearly multiples of one another, so any pair (x_1, x_2) which nearly satisfies the first equation will also nearly satisfy the second. If the matrix were known to be exactly singular, we would not need the second equation at all—any solution of the first would automatically satisfy the second.

Although this example is contrived and atypical, the conclusion we reached is not. It is probably the single most important fact which people concerned with matrix computations have learned in the past 40 years: *Gaussian elimination with partial pivoting is guaranteed to produce small residuals.*

Now that we have stated it so strongly, we must make a couple of qualifying remarks. By "guaranteed" we mean it is possible to prove a precise theorem which assumes certain technical details about how the floating-point arithmetic system works and which establishes certain inequalities which the components of the residual must satisfy. If the arithmetic units work some other way or if there is a bug in the particular program, then the "guarantee" is void. Furthermore, by "small" we mean on the order of roundoff error *relative to* three things: the elements of the original coefficient matrix, the elements of the coefficient matrix at intermediate steps of the elimination process, and the elements of the computed solution. If any of these are "large," then the residual will not necessarily be small in an absolute sense. We can say this roughly as

$$\text{Size of residuals} \propto \text{Size of solution} \times \text{Size of } A \times \epsilon_{\text{mach}}$$

However, even if the residual is small this does *not* imply that the error will be small. The relationship between the size of the residual and the size of the error is determined in part by a quantity known as the **condition number** of the matrix, cond(A), in roughly the following way,

$$\text{Size of error in solution} \propto \text{Size of solution} \times \text{cond}(A) \times \epsilon_{\text{mach}}$$

The condition number is an inherent property of a matrix and has nothing to do with how we solve a system of equations. Matrices with larger condition numbers result in larger errors when we solve $Ax = b$. One useful interpretation of the condition number is that its logarithm approximates the number of digits which will be lost while solving $Ax = b$. Thus if cond(A) = 10^5 and if machine epsilon is 10^{-8}, then the best we can expect is that the solution will be accurate to about three digits.

What is the condition number and why does it play such an important role? We show in Section 6 that cond(A) measures how close A is to being singular, and more to the point it measures how sensitive the solution of $Ax = b$ is to changes in A and b. The coefficients in the matrix and right hand side of a system of equations are rarely known exactly. Some systems arise from experiments, and so the coefficients are subject to observational errors. Other systems have coefficients given by formulas which involve roundoff error in their evaluation. Even if the system can be stored exactly in the computer, it is almost inevitable that roundoff errors will be introduced during its solution. Thus knowing how the solution is affected by changes in A and b is useful, but more importantly it can be shown that roundoff errors in Gaussian elimination have the same effect on the answer as roundoff errors in the original components. We discuss these ideas in detail in Section 6, but the two results above concerning the size of residuals and errors are the key points you need to remember to effectively use the software in this chapter.

The above results can be misleading when the solution is badly scaled, that is, when the components of x vary widely in size. An example of a badly-scaled vector would be

$$x = (10^5, 10^{-6})^T.$$

When x is the solution to a system of linear equations, the errors in x will usually be proportional to the largest component of x, in this case 10^5. If x_1 were accurate to four decimal digits, the error in x_2 would be approximately $10^5 \times 10^{-4} = 10^1$, and thus it would be unreasonable to expect that x_2 had any correct digits. This problem can sometimes be corrected when setting up the system by changing the units of measurement.

Some software for solving linear equations attempts to perform this rescaling automatically. Unfortunately, automatic methods are not guaranteed to be successful, and can sometimes increase the error in the computed solution. For this reason, the software in this chapter does not attempt to scale the equations.

3.2.1 Vector Norms

A discussion of Gaussian elimination is clarified and made more precise by introducing the idea of a **norm** of a vector or a matrix. This will allow us to compare the sizes of vectors and matrices. To compare numbers, we use the absolute-value function $| \cdot |$, so that

$$|7| > |5| \quad \text{and} \quad |-3| > |-1.5|.$$

A **vector norm** is a single number which measures the general size of the elements of a vector x. This is written $\|x\|$. There are many ways to do this. Theoretically, any function which satisfies the four conditions below is an acceptable norm. Most theorems which are couched in terms of norms are true regardless of the precise norm that is being used. Consequently, you can comfortably read these results by replacing the norm symbol $\|x\|$ by "size of x." However norms also occur within computations and then the particular norm being used matters a great deal. The most common vector norm is the Euclidean length, or 2-norm

$$\|x\|_2 = \left(\sum_{i=1}^n |x_i|^2 \right)^{\frac{1}{2}}.$$

This has the advantage that it corresponds to our intuitive notion of distance. It will be used extensively in Chapter 6. However, when it is used to study linear equations, the 2-norm leads to expensive computations. Instead, in this chapter, we define the norm of a vector with n components to be

$$\|x\|_1 = \sum_{i=1}^n |x_i|,$$

which is called the 1-norm. This is sometimes referred to as the "Manhattan distance" since $\|x - y\|$ measures the number of blocks you would walk to get between two locations x and y in the city. The Euclidean length is "as the crow flies." A third commonly used norm is the max-norm or ∞-norm,

$$\|x\|_\infty = \max |x_i|.$$

When we need to distinguish between these we will use subscript notation

$$\|x\|_2 , \qquad \|x\|_1 , \text{ or } \|x\|_\infty .$$

If we omit the subscript the 1-norm is being used.

To return to the earlier point, a norm is any function satisfying

$$\|x\| > 0 \text{ if } x \neq 0,$$
$$\|0\| = 0,$$
$$\|cx\| = |c| \cdot \|x\| \quad \text{for all scalars } c,$$
$$\|x + y\| \le \|x\| + \|y\| .$$

Thus a norm has many of the analytic properties of Euclidean length. Some of the geometrical properties of Euclidean length are lost, but they are not too important.

3.3 SUBROUTINE SGEFS

Almost any computer library has subroutines based on variants of Gaussian elimination with partial pivoting for solving systems of simultaneous linear equations. The implementation details of various subroutines are quite different. These details can have important effects on the execution time of a particular subroutine, but if the subroutine is properly written, they should have little effect on its accuracy.

In this section, we shall describe one such subroutine, SGEFS that is built upon routines from Linpack library developed at the Argonne National Laboratory. Linpack contains routines for solving linear systems of equations and related matrix computations. The algorithms are based upon Gaussian elimination, sometimes adapted to systems having special form, such as banded or symmetric systems. Each routine has four variants, for real single, real double, complex single, and complex double precision arithmetic. Linpack has a set of core routines, called the Basic Linear Algebra Subroutines (BLAS), which perform routine tasks such as computing the norm of a vector. When Linpack is installed on a special computer, such as a vector or parallel machine, a version of the BLAS specially designed for the machine can be used, and the programs will run at near-optimal efficiency. Without this modular construction, Linpack would have to be rewritten for the special computer at great cost. The BLAS have been so useful that an extended set of basic subroutines is now being developed to perform more powerful matrix operations.

SGEFS handles both the factorization of the matrix, and the back-substitution; if it is necessary to solve additional systems of equations where only the right-hand side changes, then the factorization need only be done once. SGEFS also returns an estimate of the number of accurate digits in the solution, derived from an estimate for the condition number of the matrix. If the matrix is computationally singular, an error code is returned.

The subroutine parameter IND returns an estimate of the number of accurate digits in X. IND is computed from RCOND, an estimate for the reciprocal of the condition number. 1/RCOND is a lower bound for the actual condition, but it is computed in such a way that it is almost always within a factor of n of the actual condition, and it is usually much closer. In other words, for almost all matrices the estimate RCOND satisfies

$$\frac{\text{cond}(A)}{n} \leq \frac{1}{\text{RCOND}} \leq \text{cond}(A).$$

In those situations where $1/\text{RCOND} < \text{cond}(A)/n$, it still measures the sensitivity of solutions for most right-hand sides.

Roundoff error usually prevents SGEFS, or any other Gaussian elimination subroutine, from determining whether or not the input matrix is singular. If a computed pivot is exactly zero during the elimination, SGEFS sets an error code and does not attempt the back substitution. However, the occurrence of a zero pivot does not necessarily mean that the matrix is singular, nor does a singular matrix necessarily produce a zero pivot. In fact, the most common source of zero pivots is some kind of bug in the calling program!

It should be realized that with partial pivoting, *any* matrix has a triangular factorization. SGEFS actually works faster when zero pivots occur because they mean that the corresponding column is already in triangular form. The only difficulty with a zero pivot is that back substitution will fail.

To comment upon some details in SGEFS, we need to examine how Fortran systems store matrices. If a program contains the dimension statement

DIMENSION A(3,5)

then $3 \times 5 = 15$ locations will be reserved in memory for the elements of A. They will be stored in the following order:

A(1,1) A(2,1) A(3,1) A(1,2) A(2,2) A(3,2) A(1,3) ....

In other words, the elements of each *column* are stored together. The elements of each *row* are separated from each other by a number of locations equal to the first subscript in the dimension statement. This convention has been written into the American National Standards Institute specifications for Fortran.

Most, but not all, Fortran dialects have provision for *variable dimensions* in arrays which are subroutine parameters. In a main program, one may specify

DIMENSION A(50,70)

but intend to actually work with an $N \times N$ matrix where N may vary from problem to problem. Subroutines such as SGEFS need both N, the actual working order, and the quantity 50 used in the dimension statement because that is the memory increment between successive elements of a row. This dimension information is called LDA in SGEFS, and is an acronym for Leading Dimension of A.

SGEFS can be used to compute determinants. This is possible because of three basic properties of determinants. Subtracting a multiple of one row from another row does not change the determinant. Interchanging two rows changes the sign of the determinant. The determinant of a *triangular* matrix is simply the product of its diagonal elements. Thus, except for the sign, the determinant is the product of the diagonal elements of the output matrix. One annoying feature of computing determinants is that the intermediate products, and often the determinant itself, tend to be very large or very small numbers and consequently may easily cause floating-point overflow or underflow. One easy remedy is to compute the logarithm of the absolute value of the determinant as the sum of logarithms of its factors. Linpack contains routines for computing the determinant and the inverse of a matrix; it is recommended that these be used when determinants are required.

The complete documentation for SGEFS can be found in Section 11. The following main program illustrates the use of SGEFS. Note that LDA = 10, the declared dimension of the array A, while N = 3, the actual order of the matrix. In our experience, improper setting of LDA is a frequent source of error. We have set up the matrix and right-hand side using assignment statements simply to avoid worrying about the format of the data.

The example is the one used in Section 2. The output is

```
COEFFICIENT MATRIX =
        10.000000    -7.000000     0.000000
        -3.000000     2.000000     6.000000
         5.000000    -1.000000     5.000000
RIGHT-HAND SIDE =
         7.000000     4.000000     6.000000
NUMBER OF ACCURATE DIGITS =      6
SOLUTION =
         0.000000    -1.000000     1.000000
```

Here is the program:

```
C SAMPLE PROGRAM FOR SGEFS
C
      PARAMETER (LDA=10)
      REAL       A(LDA,LDA), B(LDA), WORK(LDA), RCOND
      INTEGER    IWORK(LDA), I, J, N, ITASK, IND
C
C SET UP PROBLEM
C
      N     = 3
      ITASK = 1
      A(1,1) = 10.0
      A(2,1) = -3.0
```

```
          A(3,1)  =   5.0
          A(1,2)  =  -7.0
          A(2,2)  =   2.0
          A(3,2)  =  -1.0
          A(1,3)  =   0.0
          A(2,3)  =   6.0
          A(3,3)  =   5.0
          B(1)    =   7.0
          B(2)    =   4.0
          B(3)    =   6.0
C
C PRINT PROBLEM INFORMATION
C
          WRITE (*,*) ' COEFFICIENT MATRIX ='
          DO 10 I = 1,N
             WRITE (*,800) (A(I,J), J = 1,N)
10        CONTINUE
          WRITE (*,*) ' RIGHT-HAND SIDE ='
          WRITE (*,800) (B(J), J = 1,N)
C
C SOLVE LINEAR SYSTEM
C
          CALL SGEFS (A, LDA, N, B, ITASK, IND, WORK, IWORK, RCOND)
C
C PRINT RESULTS
C
          IF (IND .EQ. -10) THEN
             WRITE (*,*) ' ERROR CODE =', IND
          ELSE IF (IND .LT.0) THEN
             WRITE (*,*) ' ERROR CODE =', IND
             STOP
          ELSE
             WRITE (*,*) ' NUMBER OF ACCURATE DIGITS =', IND
          END IF
C
          WRITE (*,*) ' SOLUTION ='
          WRITE (*,800) (B(J), J = 1,N)
C
          STOP
800       FORMAT (4X, 3F12.6)
          END
```

3.4 HISTORICAL PERSPECTIVE: J.H. WILKINSON

The work of James H. Wilkinson (1919–1986) has completely altered our perception
of computer algorithms. Up until the 1950's the performance of a numerical algorithm

would be determined by the accuracy of its solution; that is, we would assess an algorithm by asking, "How big is $x_{computed} - x_{true}$?" This seems like a reasonable question, in fact it seems like the only reasonable question to ask of an algorithm.

Before the invention of the electronic computer, algorithms were rarely judged in terms of rounding errors. When working with pencil and paper, the accuracy of calculations can be adjusted to suit the circumstances, potential difficulties are handled almost instinctively, and the total number of calculations is small, so rounding errors are not an issue. However, when calculations are automated, arithmetic operations are done using a fixed and finite number of digits, and the number of calculations can stretch into the thousands and millions. The early inventors of the computer realized this, and began analyzing the rounding-error properties of algorithms in the 1940's, as soon as the machines were available.

The first problem examined was the solution of $Ax = b$ using Gaussian elimination. Alan Turing in Britain and John von Neumann in the United States among others discovered that the computed solution might be quite different from the true solution for general systems of equations. The 2×2 matrix in Example 3.5 illustrates this. The results were complicated because they had to take into account the technicalities of the computer arithmetic, and were mysterious because of a strange factor that arose in the results. This "strange factor" was the condition number, which was not well understood at that time. The results seemed to imply that for many systems of linear equations, rounding errors would overwhelm the calculation and the computed solution would be of no value.

James Wilkinson had been involved in ballistics calculations during World War II. The computer was not available in Britain until the late 1940's, and so calculations were done by people (often women since men were taken for military service) with desk calculators. At various times, Wilkinson had had to solve systems of linear equations this way, with as many as 12 variables. This is well over 1000 hand calculations, since each calculation would have to be checked for correctness. As one final check on the solution, he would compute the residual $b - Ax$, which would always be gratifyingly small.

This experience made Wilkinson suspicious of the pessimistic rounding-error analysis of Gaussian elimination. It was not that the theoretical results were wrong. As our example illustrated, the computed solution need bear no relation to the true solution. This will happen whenever the matrix is nearly singular. Instead, it was that the wrong question was being asked. It is not fair to expect that an algorithm will produce an accurate solution when the solution is not well determined by the data for the problem, i.e., when the problem is nearly the same as one with multiple solutions.

Wilkinson had been educated at Trinity College, Cambridge, as a mathematician; after the war, he could have returned there, but instead in 1946 went to work at the National Physical Laboratory. Half of his time was assigned to desk machine calculations, but the other half was spent assisting Alan Turing in the design of the ACE (Automatic Computing Engine), one of the first British computers. Turing was already at work at version V of his design, an outgrowth of his wartime work in deciphering German military codes. Turing was brilliant, and quickly moved on to versions VI, VII, and

VIII. However, he was also eccentric and sometimes difficult to deal with, and he left the Laboratory to work at Cambridge before a machine could be built. The Laboratory continued to develop a computer under different management, but for practical reasons the project was scaled back. Version V of the ACE was simplified, and was built as the Pilot ACE.

Wilkinson had stayed on the project, and gained his first computing experience with this machine. To test the arithmetic of the Pilot ACE, he wrote a program to find the roots of polynomials, and tried it out on some examples where he knew the answers. One of his first tests was finding the roots of

$$p(x) = (x - 1)(x - 2) \cdots (x - 20) = x^{20} - 210x^{19} + \cdots = 0.$$

To his surprise, the computed roots weren't even close to the true roots. Since the machine was experimental, he assumed the problem must be in his program, or in the hardware, but a week's search left him back where he started.

The only thing left to consider was the problem itself. He discovered that the roots of this polynomial are incredibly sensitive to changes in the coefficients. In other words, even though the computed roots seem inaccurate, they are still the roots of a polynomial that is close to the original one. This throws the difficulty back on the problem, and allows us to consider the method more objectively.

Wilkinson later applied this point of view to the solution of $Ax = b$. As he reasoned, the computed solution from Gaussian elimination may not be accurate, but it is the *exact* solution to a *nearby* problem. What more could you want? Almost any problem, just by storing it in the finite precision arithmetic of the computer, is changed to some nearby *storable* problem. So in general, no algorithm can be expected to exactly solve a set of linear equations. But a good algorithm will always exactly solve a slightly altered problem. Another way of putting this is that the computed solution will "behave like" the true solution. In our case, this means that the residuals will be small, that the computed solution will nearly solve the equations.

This point of view has pervaded numerical analysis. The techniques Wilkinson used are now referred to as "backward" error analysis. A good algorithm is one that is good in this "backward" sense.

3.5   COLUMN-ORIENTED ALGORITHMS

Many of the common matrix operations are most naturally described in terms of rows. For example, in Gaussian elimination, a multiple of one row is subtracted from another row. When implemented in Fortran, such operations typically have the innermost loops varying the second index of arrays. Since Fortran stores matrices by columns, this has two potentially adverse effects on program efficiency: (1) Subscript calculations may be more costly. (2) Operating systems which automatically move data between *high-speed* and *secondary* memory units during computation may have to do an excessive amount of work. For these reasons, in SGEFS Gaussian elimination has been implemented in a somewhat unconventional manner with all the inner loops varying the first index. Such an implementation can be significantly more efficient with some operating systems. This is especially true on certain "supercomputers" such as the Cray computers that have special hardware instructions to move consecutive cells in memory quickly. It is also true when the matrix is so large that it will not fit in the high-speed memory of the computer and must be moved in pieces to and from other devices.

We will illustrate the ideas on a simpler algorithm, the multiplication of a vector by a matrix. Consider the following 3×3 problem

$$\text{Compute } b = Ax = \begin{pmatrix} 1 & 2 & 3 \\ 4 & 5 & 6 \\ 7 & 8 & 9 \end{pmatrix} \begin{pmatrix} 10 \\ 11 \\ 12 \end{pmatrix}.$$

The traditional formulas for this calculation would compute b as follows

$$b = \begin{pmatrix} 1 \cdot 10 & + & 2 \cdot 11 & + & 3 \cdot 12 \\ 4 \cdot 10 & + & 5 \cdot 11 & + & 6 \cdot 12 \\ 7 \cdot 10 & + & 8 \cdot 11 & + & 9 \cdot 12 \end{pmatrix} = \begin{pmatrix} 68 \\ 167 \\ 266 \end{pmatrix}.$$

When carried out in this way, the calculation uses the matrix A by rows; that is, the elements of A are used in the following order: 1, 2, 3, 4, 5, 6, 7, 8, 9. To use A by columns (that is, in the order 1, 4, 7, 2, 5, 8, 3, 6, 9) we rearrange the computation in the form

$$b = 10 \begin{pmatrix} 1 \\ 4 \\ 7 \end{pmatrix} + 11 \begin{pmatrix} 2 \\ 5 \\ 8 \end{pmatrix} + 12 \begin{pmatrix} 3 \\ 6 \\ 9 \end{pmatrix}.$$

This gives the same solution as before. In fact, it gives exactly the same rounding errors as before, so that even on a computer the two answers will be the same.

More generally, if we are trying to compute the product $b = Ax$ for arbitrary A and x, the traditional formulas are

$$b_i = \sum_{j=1}^{n} a_{ij} x_j.$$

To obtain a column-oriented method, let a_j be the j-th column of the matrix A, so that

$$A = (a_1 \ a_2 \ \cdots \ a_n).$$

The revised but equivalent formula is then

$$b = \sum_{j=1}^{n} x_j a_j.$$

Converting from a row-oriented computer program to a column-oriented one is often easy. In this case, the row-oriented program is

```
DO 20 I = 1,N
```

```
            B(I) = 0.0
            DO 10 J = 1,N
               B(I) = B(I) + A(I,J)*X(J)
   10       CONTINUE
   20    CONTINUE
```

The column-oriented program is

```
         DO 5 I = 1,N
            B(I) = 0.0
   5     CONTINUE
         DO 20 J = 1,N
            DO 10 I = 1,N
               B(I) = B(I) + A(I,J)*X(J)
   10       CONTINUE
   20    CONTINUE
```

The only significant change is the reordering of the DO loops; the arithmetic in the program remains the same.

*3.6 MORE ABOUT CONDITION NUMBERS

To understand cond(A) we have to make the idea of "nearly singular" precise. If A is a singular matrix, then for some b's a solution x will not exist, while for others it will not be unique. If this is not familiar, review the definition of a singular matrix at the beginning of Section 2. If A is nearly singular, we can expect small changes in A and b to cause large changes in x. On the other hand, if A is the identity matrix, then b and x are the same vector. So if A is nearly the identity, small changes in A and b should result in correspondingly small changes in x.

At first glance, it might appear that there is some connection between the size of the pivots encountered in Gaussian elimination with partial pivoting and nearness to singularity, because if the arithmetic could be done exactly, all the pivots would be nonzero if and only if the matrix is nonsingular. To some extent, it is also true that if the pivots are small, then the matrix is close to singular. However, when roundoff errors are encountered, the converse is no longer true—a matrix might be close to singular even though none of the pivots are small.

To get a more precise and reliable measure of nearness to singularity than the size of the pivots, we need to utilize the concept of a vector norm introduced in Section 2.1. Multiplication of a vector x by a matrix A results in a new vector Ax which may have a different norm from x. This change in norm is directly related to the sensitivity we wish to measure. The range of the possible change can be expressed by two numbers,

$$M = \max_x \frac{\|Ax\|}{\|x\|},$$

$$m = \min_x \frac{\|Ax\|}{\|x\|}.$$

The max and min are taken over all nonzero vectors. Note that if A is singular, then $m = 0$. The ratio M/m is called the **condition number** of A,

$$\text{cond}(A) = \frac{\max_x \frac{\|Ax\|}{\|x\|}}{\min_x \frac{\|Ax\|}{\|x\|}}.$$

Consider a system of equations

$$Ax = b$$

and a second system obtained by altering the right-hand side:

$$A(x + \Delta x) = b + \Delta b.$$

We think of Δb as being the error in b and Δx as being the resulting error in x, although we need not make any assumptions that the errors are small. Since $A(\Delta x) = \Delta b$, the definitions of M and m immediately lead to

$$\|b\| \leq M \|x\|$$

and

$$\|\Delta b\| \geq m \|\Delta x\|.$$

Consequently, if $m \neq 0$,

$$\frac{\|\Delta x\|}{\|x\|} \leq \text{cond}(A) \frac{\|\Delta b\|}{\|b\|}.$$

The quantity $\|\Delta b\| / \|b\|$ is the *relative* change in the right-hand side, and the quantity $\|\Delta x\| / \|x\|$ is the *relative* error caused by this change. The advantage of using relative changes is that they are dimensionless.

This shows that the condition number is a relative error magnification factor. Changes in the right-hand side may cause changes $\text{cond}(A)$ times as large in the solution. It turns out that the same is true of changes in the coefficient matrix.

The condition number is also a measure of nearness to singularity. Although we will not develop the mathematical tools necessary to make the idea precise, the condition number can be thought of as the reciprocal of the relative distance from the matrix to the set of singular matrices. So, if $\text{cond}(A)$ is large, A is close to singular.

Some of the basic properties of the condition number are easily derived. Clearly $M \geq m$, and so

$$\text{cond}(A) \geq 1.$$

If P is a permutation matrix, then the components of Px are simply a rearrangement of the components of x. It follows that $\|Px\| = \|x\|$ for all x, and so

$$\text{cond}(P) = 1.$$

In particular, $\text{cond}(I) = 1$. If A is multiplied by a scalar c, then M and m are both multiplied by the same scalar, and so

$$\text{cond}(cA) = \text{cond}(A).$$

If D is a diagonal matrix, then

$$\text{cond}(D) = \frac{\max |d_{ii}|}{\min |d_{ii}|}.$$

Thus for the matrix

$$D = \begin{pmatrix} 1 & & & & \\ & 2 & & & \\ & & 3 & & \\ & & & 4 & \\ & & & & 5 \end{pmatrix}$$

we have $M = 5$ and $m = 1$ so that $\text{cond}(D) = 5/1 = 5$.

The last two properties are two of the reasons that $\text{cond}(A)$ is a better measure of nearness to singularity than the determinant of A. As an extreme example, consider a 100×100 diagonal matrix with 0.1 on the diagonal. Then $\det(A) = 10^{-100}$, which is usually regarded as a small number. But $\text{cond}(A) = 1$, and the components of Ax are simply 0.1 times the corresponding components of x. For linear systems of equations, such an A behaves more like the identity than like a singular matrix.

The following example illustrates the condition number.

Example 3.6   The Condition Number.

Consider the linear system with

$$A = \begin{pmatrix} 9.7 & 6.6 \\ 4.1 & 2.8 \end{pmatrix}$$

$$b = \begin{pmatrix} 9.7 \\ 4.1 \end{pmatrix}, \quad x = \begin{pmatrix} 1 \\ 0 \end{pmatrix}.$$

Clearly $Ax = b$, and

$$\|b\| = 13.8, \quad \|x\| = 1.$$

For this example it is possible to compute the condition number, although we will not work out the details here,

$$\text{cond}(A) = 2249.4.$$

(Normally this can only be estimated.) Recall that this means, roughly, that a relative change in the right-hand side propagates into a relative change in the solution about four orders of magnitude larger. To test this, we change the right-hand side to

$$b' = \begin{pmatrix} 9.70 \\ 4.11 \end{pmatrix},$$

the solution becomes

$$x' = \begin{pmatrix} 0.34 \\ 0.97 \end{pmatrix}.$$

Let $\Delta b = b - b'$ and $\Delta x = x - x'$. Then

$$\|\Delta b\| = 0.01, \quad \|\Delta x\| = 1.63.$$

We have made a small perturbation in b which completely changes x. In fact, the relative changes are

$$\frac{\|\Delta b\|}{\|b\|} = 0.0007246, \quad \frac{\|\Delta x\|}{\|x\|} = 1.63.$$

The ratio of these is $1.63/0.0007246 = 2249.4$, i.e., exactly equal to cond(A). Of course, this doesn't usually happen. We know for sure that

$$\frac{\|\Delta x\|}{\|x\|} \leq \text{cond}(A)\frac{\|\Delta b\|}{\|b\|},$$

but for this example, we have carefully chosen b and Δb to illustrate the worst-case behavior. However, if b and Δb had been chosen randomly, similar behavior would have been observed. ∎

It is important to realize that this example is concerned with the *exact* solutions to two slightly different systems of equations and that the method used to obtain the solutions is irrelevant. The example is constructed to have a fairly large condition number so that the effect of changes in b is pronounced, but similar behavior can be expected in any problem with a large condition number. This is an ill-conditioned problem.

Suppose we wish to solve a problem in which $a_{1,1} = 0.1$, all the other elements of A and b are integers, and cond(A) = 10^5. Suppose further that we have a binary computer with 24 bits in the fraction and that we can somehow compute the exact solution to the system actually stored in the computer. Then the only error is caused by representing 0.1 in binary, but we can expect

$$\frac{\|\Delta x\|}{\|x\|} \approx \text{cond}(A) \times 2^{-24} \approx 6 \times 10^{-3}.$$

In other words, the simple act of storing the coefficient matrix in the machine might cause changes in the third significant figures of the true solution. We may summarize all these ideas in a practical rule of thumb; *in solving a linear system of equations, relative accuracy in the solution is proportional to relative accuracy in the coefficient matrix or right-hand side, with constant of proportionality as large as the condition number.*

The condition number also plays a fundamental role in the analysis of the roundoff errors introduced during the solution by Gaussian elimination. Let us assume that A and b have elements which are exact floating-point numbers and let x^* be the vector of floating-point numbers obtained from a linear equation solver such as the SGEFS. We also assume that exact singularity is not detected and that there are no underflows or overflows. Then it is possible to establish the following inequalities:

$$\frac{\|b - Ax^*\|}{\|A\| \cdot \|x^*\|} \le C\epsilon_{\text{mach}},$$

$$\frac{\|x - x^*\|}{\|x^*\|} \le C \operatorname{cond}(A)\epsilon_{\text{mach}}.$$

Here ϵ_{mach} is the machine epsilon discussed in Section 5 of Chapter 2. The constant C is discussed further below, but it usually is not much larger than 1.

The first inequality says that the relative residual will usually be about the size of roundoff error, no matter how badly conditioned the matrix is. This was illustrated by Example 3.5. The second inequality requires that A be nonsingular and involves the exact solution x. It follows directly from the first inequality and the definition of $\operatorname{cond}(A)$ and says that the relative error will also be small if $\operatorname{cond}(A)$ is small but might be quite large if the matrix is nearly singular. In the extreme case where A is singular but the singularity is not detected, the first inequality still holds, but the second has no meaning.

The basic result in the study of roundoff error in Gaussian elimination is due to J. H. Wilkinson. He proved that the computed solution x^* exactly satisfies

$$(A + E)x^* = b,$$

where E is a matrix whose elements are about the size of roundoff errors in the elements of A. Thus all the rounding errors can be lumped together and considered as a single perturbation made when the matrix is stored in the computer, with no errors thereafter. Since storage of almost any matrix in the computer will lead to perturbations the size of E, this is the best that can be said about *any* algorithm for solving linear equations. Thus in this sense Gaussian elimination is an ideal algorithm for solving $Ax = b$.

To be more precise about the constant C and the perturbation matrix E, it is necessary to introduce the idea of a matrix norm and establish some further inequalities. For those readers who are interested, this is discussed in the next section.

*3.7 NORMS AND ERROR ANALYSIS

In this section, we examine more carefully the Gaussian elimination algorithm, especially the effect of rounding errors on the accuracy of the computed solution. As above with vectors, we require some measure of the size of a matrix, that is, we need to define a **matrix norm**. We could define a matrix norm $\|A\|$ in the same way we defined a vector norm, for example

$$\|A\| = \sum_{i=1}^{n} \sum_{j=1}^{n} |a_{ij}|.$$

It would have all the properties of a vector norm and would allow us to compare the sizes of matrices. However, since we are working with linear equations, we will be interested in measuring $\|Ax\|$, and so it would be desirable if

$$\|Ax\| \le \|A\| \cdot \|x\|,$$

and that $\|A\bar{x}\| = \|A\| \cdot \|\bar{x}\|$ for *some* particular $\bar{x}$. To permit this, a different norm will be used. We will define the norm of a matrix to be the quantity M defined earlier. Thus

$$\|A\| = M = \max_{x \neq 0} \frac{\|Ax\|}{\|x\|}.$$

This has all the properties of a vector norm as well as the extra properties mentioned above. Because of our particular definition of $\|x\|$, it is not hard to show that if A has columns a_j, then

$$\|A\| = \max_{j} \|a_j\|.$$

If we had chosen to use the Euclidean length of a vector, then $\|A\|$ would be more expensive to compute.

We now return to Wilkinson's result that the computed solution x^* exactly satisfies $(A + E)x^* = b$, where E has elements at roundoff level. There are some rare situations where the intermediate matrices obtained during Gaussian elimination have elements which are larger than those of A, leading to especially large rounding errors, but it can be expected that if C is defined by

$$\frac{\|E\|}{\|A\|} = C\epsilon_{\text{mach}},$$

then C will rarely be much bigger than n.

From this basic result, we can immediately derive inequalities involving the residual and the error in the computed solution. The residual is given by

$$b - Ax^* = Ex^*,$$

and hence

$$\|b - Ax^*\| = \|Ex^*\| \leq \|E\| \|x^*\| .$$

The residual involves the product Ax^*, so it is appropriate to consider the relative residual which compares the norm of $b - Ax^*$ with the norms of A and x^*. It follows directly from the above inequalities that

$$\frac{\|b - Ax^*\|}{\|A\| \|x^*\|} \leq C\epsilon_{\text{mach}}.$$

When A is nonsingular, the error can be expressed using the inverse of A by

$$x - x^* = A^{-1}(b - Ax^*),$$

and so

$$\|x - x^*\| \leq \|A^{-1}\| \|E\| \|x^*\| .$$

It is simplest to compare the norm of the error with the norm of the computed solution. Thus the relative error satisfies

$$\frac{\|x - x^*\|}{\|x^*\|} \leq C \|A\| \|A^{-1}\| \epsilon_{\text{mach}}.$$

It turns out that $\|A^{-1}\| = 1/m$, and so

$$\text{cond}(A) = \|A\| \|A^{-1}\| .$$

Thus

$$\frac{\|x - x^*\|}{\|x^*\|} \leq C \, \text{cond}(A)\epsilon_{\text{mach}}.$$

The actual computation of $\text{cond}(A)$ requires knowing A^{-1}. If a_j are the columns of A and $\tilde{a}_j$ are the columns of A^{-1}, then in terms of the vector norm we are using

$$\text{cond}(A) = \max_j \|a_j\| \cdot \max_j \|\tilde{a}_j\| .$$

It is easy to compute $\|A\|$, but finding $\|A^{-1}\|$ would roughly triple the time required for Gaussian elimination. Fortunately, the exact value of $\text{cond}(A)$ is rarely required. Any reasonably good estimate of it is satisfactory.

The subroutine SGEFS described in Section 3 estimates the condition of a matrix by solving two auxiliary systems of linear equations. With a factorization of A already available, this only requires $O(n^2)$ arithmetic operations, and thus is inexpensive when compared with the cost of solving the original system of equations. Some details of how the estimate is computed are given in Section 8.

*3.8 ESTIMATING THE CONDITION NUMBER

After we have solved a system of linear equations, we would like to be able to estimate how accurate the computed solution is. Our earlier analysis shows that the error in the solution can be bounded in terms of the condition number of the matrix in the linear system. Recall that the condition number is defined by

$$\text{cond}(A) = \frac{\max \frac{\|Ax\|}{\|x\|}}{\min \frac{\|Ax\|}{\|x\|}}.$$

This is mathematically equivalent to the following

$$\text{cond}(A) = \|A\| \cdot \|A^{-1}\|.$$

For convenience, we have used the 1-norm. If a_j is the j-th column of the matrix A, then the norm is defined by

$$\|A\| = \max_j \|a_j\|.$$

This is fine for computing $\|A\|$, in fact we will use this formula in our estimate of the condition number. However, if we were to use it to compute $\|A^{-1}\|$ then we would have to know A^{-1} explicitly. Computing A^{-1} would be more expensive than solving the system of linear equations in the first place, and so this is undesirable.

Instead, we will attempt to estimate $\|A^{-1}\|$ using far fewer calculations. Our estimate will not be accurate for every matrix A, but in virtually all cases it will produce a condition number estimate that has the correct order of magnitude. Since we are primarily interested in the condition number to estimate the accuracy of our solution, this will allow us to estimate the number of accurate digits in our solution, which is adequate for most applications. The cost of the estimate will be $O(n^2)$ arithmetic operations, about the cost of a couple of back-substitutions, and much lower than the total cost of $O(n^3)$ operations required to factor the matrix.

To estimate $\|A^{-1}\|$ we return to the first definition above

$$\|A^{-1}\| = 1/\min_x \frac{\|Ax\|}{\|x\|} = \max_x \frac{\|x\|}{\|Ax\|} = \max_y \frac{\|A^{-1}y\|}{\|y\|}$$

where we have made the change of variables $y = Ax$. To estimate this quantity, we will carefully pick a vector y, solve $Az = y$ via back-substitution, and use $\|z\| / \|y\| = \|A^{-1}y\| / \|y\|$ as our estimate of $\|A^{-1}\|$.

If the vector y were chosen at random, then there is a small chance that we might severely underestimate $\|A^{-1}\|$. To reduce the possibility of failure, the vector y is selected in a more elaborate way. One approach is to first solve

$$A^T y = c$$

where c is a vector with entries $c_j = \pm 1$. The signs of the components of c are chosen to make y large (recall that we are trying to guess the solution to a *maximization* problem). This approach is illustrated in Example 3.7.

Example 3.7   Estimating the Condition Number.

Consider the following 2×2 example (see Section 6 above)

$$A = \begin{pmatrix} 9.7 & 6.6 \\ 4.1 & 2.8 \end{pmatrix} = \begin{pmatrix} 1 & 0 \\ .4227 & 1 \end{pmatrix} \begin{pmatrix} 9.7000 & 6.6000 \\ 0 & 0.0103 \end{pmatrix}.$$

To get the vector y we solve $A^T y = c$ for the special c mentioned above. This is done via $U^T(L^T y) = c$. The first component of c is arbitrary and we choose $c_1 = 1$. We will choose $c_2 = \pm 1$ so that $L^T y$ is as large as possible:

$$(L^T y)_1 = c_1/U_{11} = 1/9.7 = .1031,$$

$$(L^T y)_2 = (c_2 - U_{21}(L^T y)_1)/U_{22} = (\pm 1 - (6.6)(.1031))/.0103.$$

This is larger when $c_2 = -1$. With this choice of c_2, $L^T y = (.1031, -163)^T$ and thus $y = (-163, 69)^T$.

If we now solve $Az = y$ we obtain $z = (12690, -18640)^T$, and thus our estimate is

$$\left\| A^{-1} \right\| \approx \frac{\|z\|}{\|y\|} = \frac{|12690| + |-18640|}{|-163| + |69|} = \frac{31330}{232} = 163.$$

As before $\|A\| = 13.8$, so our condition number estimate is

$$\text{cond}(A) \approx 13.8 \times 163 = 1863.$$

The actual value is $\text{cond}(A) = 2249.4$, and so our estimate is accurate to within 17%, which is easily within one order of magnitude. ∎

Routine SGEFS uses a more elaborate technique than this to estimate the condition number, but the underlying principle is the same. The estimate produced will always be a lower bound for the actual condition number, but there is some theoretical basis for expecting it to be an accurate estimate in all but very rare cases.

One subtle point remains. Since the condition number estimate is based on the factors from Gaussian elimination, we are in fact estimating the condition number of a perturbed matrix $A + E$ and not the original matrix A. This is not a serious difficulty. Let $\epsilon = \|E\| / \|A\|$; for many problems $\epsilon \approx \epsilon_{\text{mach}}$. As long as $\text{cond}(A)\epsilon \leq 0.1$, roughly speaking if even one digit of our solution is correct, then the following result is valid

$$\frac{8}{9}(1 - \epsilon) \leq \frac{\text{cond}(A + E)}{\text{cond}(A)} \leq \frac{10}{9}(1 + \epsilon).$$

This means that the condition number of the perturbed matrix is almost the same as the condition number of the original matrix. Thus, even though we are working in rounded arithmetic, we are still estimating the condition number of the original exact problem.

3.9 FURTHER IDEAS

In this section are mentioned a few more advanced ideas with references to more detailed discussions in other sources.

*3.9.1 Updating Solutions

Suppose that a particular set of linear equations $Ax = b$ has been solved, but that it is discovered that some of the entries in the matrix are incorrect, or that new data are available. This is especially common in data-fitting problems, where new data may be generated over time and a model is continually re-evaluated to reflect the new information. Of course, it is possible to re-solve the system of equations each time a new piece of data appears, but this is unnecessarily expensive. It is possible to modify the solution in much less time.

The basic formula used is the **Sherman-Morrison formula**. Suppose that we have computed A^{-1} for some matrix A. As always, we do not recommend that A^{-1} be explicitly computed; here we use it temporarily to simplify the derivation. Suppose that A is changed to

$$\tilde{A} = A - uv^T,$$

where u and v are n-vectors. Then $\tilde{A}^{-1}$ can be computed from

$$\tilde{A}^{-1} = A^{-1} + \alpha(A^{-1}u)(v^T A^{-1}),$$

where $\alpha = 1/(1 - v^T A^{-1}u)$. This would cost $O(n^2)$ arithmetic operations, as opposed to $O(n^3)$ operations to compute the new inverse from scratch.

To show the use of this formula, consider the following example.

Example 3.8   The Sherman-Morrison Formula.

Suppose that

$$A = \begin{pmatrix} 1 & 2 & 3 \\ 4 & 5 & 6 \\ 7 & 8 & 10 \end{pmatrix} \quad A^{-1} = \begin{pmatrix} -0.6667 & -1.3333 & 1.0000 \\ -0.6667 & 3.6667 & -2.0000 \\ 1.0000 & -2.0000 & 1.0000 \end{pmatrix},$$

and that the modified matrix is

$$\tilde{A} = \begin{pmatrix} 1 & 2 & 3 \\ 4 & 5 & 6 \\ 7 & 8 & 12 \end{pmatrix}, \quad u = \begin{pmatrix} 0 \\ 0 \\ -2 \end{pmatrix}, \quad v = \begin{pmatrix} 0 \\ 0 \\ 1 \end{pmatrix}.$$

Then

$$A^{-1}u = \begin{pmatrix} -2 \\ 4 \\ -2 \end{pmatrix}, \quad v^T A^{-1} = (1 \quad -2 \quad 1), \quad \alpha = 1/(1 - (-2)) = 1/3 \approx .3333,$$

and the Sherman-Morrison formula gives

$$\tilde{A}^{-1} = \begin{pmatrix} -0.6667 & -1.3333 & 1.0000 \\ -0.6667 & 3.6667 & -2.0000 \\ 1.0000 & -2.0000 & 1.0000 \end{pmatrix} + (0.3333) \begin{pmatrix} -2 \\ 4 \\ -2 \end{pmatrix} (1 \quad -2 \quad 1)$$

$$= \begin{pmatrix} -1.3333 & 0.0000 & 0.3333 \\ 0.6667 & 1.0000 & -0.6667 \\ 0.3333 & -0.6667 & 0.3333 \end{pmatrix}. \quad \blacksquare$$

If linear equations are being solved, so that the solution of $Ax = b$ must be converted to the solution of $\tilde{A}\tilde{x} = b$, then the Sherman-Morrison formula can also be used. Suppose that we have a routine like SGEFS that can solve linear systems involving A or A^T. From above,

$$\tilde{A}^{-1}b = A^{-1}b + \alpha A^{-1}uv^T A^{-1}b.$$

The solution of the modified system can then be obtained from the following algorithm:

1. Solve $Ax = b$ for x, so that $x = A^{-1}b$.

2. Solve $Ay = u$ for y, so that $y = A^{-1}u$.

3. Solve $A^T z = v$ for z, so that $z^T = v^T A^{-1}$.

4. Form $\alpha = 1/(1 - v^T y)$.

5. Form $\beta = z^T b$.

6. Form $\tilde{x} = x + \alpha\beta y$, the solution of $\tilde{A}\tilde{x} = b$.

This only requires back-substitutions and inner-products, so that the cost is only $O(n^2)$ operations; it also avoids the explicit computation of the matrix inverse, an undesirable operation.

It is also possible to update the $A = PLU$ triangular factorization when A is modified. For further information on this and related topics, see the paper by Gill, Golub, Murray, and Saunders (1974).

3.9.2 Sparse Systems—Elimination Methods

When a linear system has a large number n of equations and variables it may be impossible to store a full square matrix of n^2 elements. Typically such problems arise in the discretization of differential equations or from problems involving network structures.

Example 3.9   A Simple Two Point Boundary Value Problem.

Consider the solution of the boundary-value problem

$$u''(x) = f(x), \quad u(0) = 1, \quad u(1) = 2,$$

on the interval $[0, 1]$. Split the interval into $n + 1$ equal pieces each of length $h = 1/(n + 1)$ and define $u_i = u(ih)$, $i = 1, \ldots, n$, the value of the solution at the end of the i-th subinterval. Instead of solving for the function $u(x)$ we will approximate the solution at n discrete points. Note that the values of $u_0 = u(0)$ and $u_{n+1} = u(1)$ are determined by the boundary conditions.

One approximation to $u''(x)$ is

$$u''(x) \approx \frac{u(x-h) - 2u(x) + u(x+h)}{h^2}.$$

If we let $x = ih$ then

$$u''(ih) \approx \frac{u_{i-1} - 2u_i + u_{i+1}}{h^2}.$$

Substituting this into the original differential equation gives the system of linear equations

$$u_{i-1} - 2u_i + u_{i+1} = h^2 f_i, \quad i = 1, \ldots, n$$

where $f_i = f(ih)$. From the boundary conditions, $u_0 = 1$ and $u_{n+1} = 2$. Substituting these values in the first and last equations, and putting the system in matrix form gives

$$\begin{pmatrix} -2 & 1 & & & & & \\ 1 & -2 & 1 & & & & \\ & 1 & -2 & 1 & & & \\ & & \cdot & \cdot & \cdot & & \\ & & & \cdot & \cdot & \cdot & \\ & & & 1 & -2 & 1 & \\ & & & & 1 & -2 & 1 \\ & & & & & 1 & -2 \end{pmatrix} \begin{pmatrix} u_1 \\ u_2 \\ \cdot \\ \cdot \\ \cdot \\ u_{n-2} \\ u_{n-1} \\ u_n \end{pmatrix} = \begin{pmatrix} h^2 f_1 - 1 \\ h^2 f_2 \\ \cdot \\ \cdot \\ \cdot \\ h^2 f_{n-2} \\ h^2 f_{n-1} \\ h^2 f_n - 2 \end{pmatrix}.$$

At most three entries in each equation are non-zero. ∎

Frequently, as in the preceding example, the matrices of these problems are so sparse that there is plenty of high-speed storage for all the nonzero elements, together with some coding which represents the location of each element stored. How shall the associated linear system be solved?

When it is possible, Gaussian elimination remains an economical, accurate, and useful algorithm. Elimination is possible as long as there is space to store all the nonzero elements of the triangular matrices associated with the elimination and when the coding necessary to locate these elements can be programmed. Let LU represent the array whose lower triangle is the matrix of multipliers and whose upper triangle is the triangularized matrix. Then LU is usually more dense than A, although it is still a sparse matrix. The elements of LU that are nonzero in positions where those of A are zero are said to be **filled in** by Gaussian elimination.

For certain matrices, the amount of fill-in is easy to bound. One example is a band matrix A. Let us ignore any zeros within the band. If Gaussian elimination can be carried out without pivoting, which is safe for certain types of matrices known as **positive definite** (see Forsythe and Moler (1967)) then there is no fill-in at all: LU has the same bandwidth as A. If pivoting is necessary, say for an arbitrary nonsingular matrix A, then the fill-in is limited to a band three-halves as wide as A.

It is easy to store the band matrix A in a rectangular array of length n and width equal to the bandwidth, and the wider band array LU can also be stored and handled easily. As a result, linear systems with band matrices can easily be solved by elimination, provided that the band array LU can be stored in the high-speed storage. Programs exist for dealing with band matrices, for example those in Linpack; these are comparatively simple modifications of such elimination programs as SGEFS.

For more general sparse matrices, it is more difficult to apply Gaussian elimination. In this case, algorithms examine the pattern of zero and non-zero elements in the matrix before beginning the elimination. By re-ordering the equations and rearranging the variables, it is often possible to reduce the amount of fill-in that will occur during the LU factorization. Finding the perfect rearrangement, the one that leads to the least fill-in, is expensive for general matrices, so algorithms use *heuristic* techniques to find a *good* rearrangement. Software that incorporates these ideas can be found in Sparspack, the Yale sparse matrix package, and the Harwell software library. For more information, see the book by George and Liu (1981).

Sometimes a full matrix, or even a band matrix, is too large to keep in the high-speed storage. If so, it is necessary for part of it to be in secondary storage—on the disks or magnetic tape. If a problem is that large, Gaussian elimination is still possible, but the sheer volume of computation makes it rather expensive. The execution time required for the arithmetic operations is usually substantially larger than the time required for transferring parts of the matrix to and from the secondary store. It is therefore important for economy's sake to organize the computation or the operating system environment in such a way that the processor is never waiting for input/output. This can be done in various ways. Programs will not be found on the shelf, although most large installations have had experience with the solution of such large systems.

A good deal of systems effort has gone into giving the programmer the illusion of having a very large (so-called **virtual**) high-speed memory available for data, although in fact the data are grouped into **pages** or **segments** which are constantly being swapped in and out of secondary storage. The presence of the virtual memory keeps the programmer from having to worry about input/output of data. However, this freedom from worry may come at a large price depending on the paging strategy: If the program is forced to wait while the swapping mechanism retrieves each new row of the matrix, then the execution time can go up prohibitively for a large matrix. However, the virtual memory is usually coupled with multiprogramming, and the processor will usually take up another program during the page swap. With many operating systems, one is not charged for this interrupted time, even when another program is not ready. Hence the "cost" of executing a matrix program remains approximately the same, whether or not there is page swapping. However, whether one is charged or not, swapping prolongs the *elapsed* time until the program is completed.

*3.9.3 Sparse Matrices—Iterative Methods

There is a substantial class of linear equation systems for which the elements of A are known by some simple formula and so can be generated as needed. This is true for example 3.10. The elements never need be stored but instead can be generated as needed. Moreover, often the orders n are so large that it would be impossible to store the filled-in array LU.

It is desirable to solve such linear systems $Ax = b$ by methods that do not factor the matrix A and never require storing more than a few vectors of length n. (Note that b must usually be stored, as well as x.)

Methods for this purpose exist and are called **iterative**. One starts with a trial solution vector $x^{(0)}$ and carries out some process using A, b, and $x^{(0)}$ to get a new vector $x^{(1)}$. Then one repeats. At the k-th stage, one uses the iterative process to get $x^{(k)}$ from A, b, and $x^{(k-1)}$. Under appropriate hypotheses, the vectors $x^{(k)}$ converge to a limit as $k \to \infty$. There is a wide variety of such iterative processes. The most successful of them are closely coupled with the actual problem being solved. Even though the iterative process may be mathematically simple, the structure of the matrix A is likely to be intricate and special to the problem. It is not usual to find library subroutines for iterative solution methods. However, a collection of such routines can be found in Itpack, described in the paper by Kincaid et al. (1982).

In the first days of computing iterative methods were of much more interest than Gaussian elimination, or so-called "direct" methods. Early computers had limited storage capacity (a few dozen words for data was common) and iterative methods are very storage efficient.

One simple iterative method is discussed in Section 24 of Forsythe and Moler (1967): the method is known as the **Gauss-Seidel** or **successive displacements method**. In it, the basic iterative step is to solve the i-th equation for the i-th component x_i of the new vector x, using for each other component of x its most recently computed value. It can be proved to converge for various types of matrices, including any symmetric positive-definite matrix A. However, convergence is ordinarily slow.

With many iterative processes in numerical analysis, convergence is so slow that the most important problem is to find a way of *accelerating* the convergence—e.g., of $x^{(k)}$ to the solution. Indeed, algorithms for accelerating the convergence of sequences form an important part of numerical analysis. The method of **successive over-relaxation** (SOR)[1] is one type of acceleration of the Gauss-Seidel process. It can speed up the convergence to the point where the SOR method is widely used in solving finite-difference equations that model elliptic boundary-value problems in two dimensions. Software for these methods is less frequently found in program libraries.

There is a family of iterative methods known as the methods of **conjugate gradients** or **conjugate directions**. A good explanation of the algorithms can be found in the book by Golub and Van Loan (1983). These methods are applicable to symmetric positive-definite matrices and involve no assumptions on the structure of the matrix A. In exact arithmetic, they converge in a finite number of iterations, but on a computer they must be considered as general iterative algorithms because of rounding errors. There are a number of published algorithms for the conjugate-gradient method.

In many cases, a given set of linear equations can be approximated by another system that is much easier to solve. For example, if the linear system arose when solving Laplace's equation on an irregular region the problem could be approximated by Laplace's equation on a square, for which there are special algorithms. It would

---

[1] The term comes from its first application, in structural engineering.

be useful to take advantage of this approximation when solving the problem. Many iterative methods can do this, and the effect on the performance of the algorithms can be dramatic. This idea is called **preconditioning**. For a discussion of this idea applied to the conjugate-gradient method, see the paper by Concus, Golub, and O'Leary (1976).

3.10 PROBLEMS

P3–1.–Use SGEFS to solve the 3-by-3 system

$$\begin{pmatrix} 1.00 & 0.80 & 0.64 \\ 1.00 & 0.90 & 0.81 \\ 1.00 & 1.10 & 1.21 \end{pmatrix} \begin{pmatrix} x_1 \\ x_2 \\ x_3 \end{pmatrix} = \begin{pmatrix} \text{erf}(0.80) \\ \text{erf}(0.90) \\ \text{erf}(1.10) \end{pmatrix}.$$

See Problem P1–1 for the definition of erf. Print out the estimated accuracy of the solution (IND) and the solution x_1, x_2, x_3. Also print out the sum $x_1 + x_2 + x_3$, and compare it with erf(1.00). Why are the two close to each other? If you cannot answer this last question, see Section 1.

P3–2.–The inverse of a matrix A can be defined as the matrix X whose columns x_j satisfy

$$Ax_j = e_j,$$

where e_j is the j-th column of the identity matrix.

(a) Write a subroutine with the heading

```
SUBROUTINE INVERT (A, LDA, N, X, IND, WORK, IWORK, RCOND)
```

which accepts a matrix of order N as input and which returns a matrix X, an approximation to the inverse of A, as well as the condition estimate and the pivot information. Your subroutine should only factor the matrix once, on the first call to SGEFS, and on the remaining $N - 1$ calls should only do back substitution, once for each column of X. Leave X undefined if SGEFS detects singularity. Test your subroutine on some matrices whose elements can be exactly represented as floating-point numbers and for which you know A^{-1}.

(b) There are several measures of the accuracy of the results:

$$\|AX - I\|,$$
$$\|XA - I\|,$$
$$\left\|X - A^{-1}\right\|.$$

You might also use INVERT twice, once to invert A and a second time to invert X. The result is a matrix Z which would be equal to A if there were no roundoff error. So, another measure of accuracy would be

$$\|Z - A\|.$$

Can you derive an inequality involving C, cond(A), and ϵ_{mach} which predicts how large $\|Z - A\|$ might be? (See Section 6.)

P3–3.–Let

$$A = \begin{pmatrix} 0.1 & 0.2 & 0.3 \\ 0.4 & 0.5 & 0.6 \\ 0.7 & 0.8 & 0.9 \end{pmatrix}, \quad b = \begin{pmatrix} 0.1 \\ 0.3 \\ 0.5 \end{pmatrix}.$$

(a) Show that the set of linear equations $Ax = b$ has many solutions. Describe the set of possible solutions.

(b) Suppose SGEFS were used to solve $Ax = b$ on a hypothetical computer which does exact arithmetic. Since there are many solutions, it is unreasonable to expect one particular solution to be computed. What does happen?

(c) Use SGEFS to compute a solution on a computer that uses binary arithmetic. Since some of the elements of A are not exact floating-point numbers on such a computer, the matrix which is given to SGEFS is not exactly singular. What solution is obtained? Why? In what sense is it a "good" solution? In what sense is it a "bad" solution?

P3–4.–The following tridiagonal matrix occurs in the interpolation of data by cubic splines, as will be discussed in the next chapter

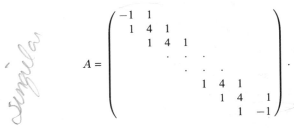

$$A = \begin{pmatrix} -1 & 1 & & & & & \\ 1 & 4 & 1 & & & & \\ & 1 & 4 & 1 & & & \\ & & \cdot & \cdot & \cdot & & \\ & & & \cdot & \cdot & \cdot & \\ & & & 1 & 4 & 1 & \\ & & & & 1 & 4 & 1 \\ & & & & & 1 & -1 \end{pmatrix}.$$

What special properties does the factored array returned by SGEFS have? How might the Gaussian elimination algorithm be simplified for this special case? How would you solve a large linear system involving this matrix? Use SGEFS to solve a linear system with this matrix and right-hand side $b = (1, \ldots, 1)^T$. How does RCOND change as the number of equations n increases?

P3–5.–Consider the linear system $Ax = b$ with

$$A = \begin{pmatrix} 7 & 9 & 2 \\ 6 & 4 & 1 \\ 3 & 9 & 3 \end{pmatrix}, \quad b = \begin{pmatrix} 6 \\ 4 \\ 2 \end{pmatrix}.$$

(a) Solve this linear system using SGEFS.

(b) Suppose that the right-hand side is changed to

$$b = \begin{pmatrix} 5 \\ 2 \\ 3 \end{pmatrix}.$$

Use SGEFS to solve the modified system, without refactoring the matrix A.

(c) Suppose that it was discovered that a mistake had been made when collecting data for the matrix A, and that the $(3, 3)$ entry should have been $A_{33} = 4$. Using the Sherman-Morrison formula together with routine SGEFS, determine the solution of the modified system without refactoring the matrix A.

P3–6.–It is required to determine the member forces in the 17-member plane truss in Figure P3.6. The members of the truss are assumed to be joined at the joints by frictionless pins. A theorem of elementary mechanics tells us that since the number of joints j is related to the number of members m by $2j - 3 = m$, the truss is statically determinant. This means that the member forces are determined entirely by the conditions of static equilibrium at the nodes. Let F_x denote horizontal force components and F_y denote vertical force components. If we let $\alpha = \sin 45° = \cos 45°$ and assume small displacements, then these equilibrium conditions are

$$\text{joint 2} \begin{cases} \sum F_x = -\alpha f_1 + f_4 + \alpha f_5 = 0, \\ \sum F_y = -\alpha f_1 - f_3 - \alpha f_5 = 0; \end{cases}$$

$$\text{joint 3} \begin{cases} \sum F_x = -f_2 + f_6 = 0, \\ \sum F_y = f_3 - 10 = 0; \end{cases}$$

$$\text{joint 4} \begin{cases} \sum F_x = -f_4 + f_8 = 0, \\ \sum F_y = -f_7 = 0; \end{cases}$$

$$\text{joint 5} \begin{cases} \sum F_x = -\alpha f_5 - f_6 + \alpha f_9 + f_{10} = 0, \\ \sum F_y = \alpha f_5 + f_7 + \alpha f_9 - 15 = 0; \end{cases}$$

$$\text{joint 6} \begin{cases} \sum F_x = -f_8 - \alpha f_9 + f_{12} + \alpha f_{13} = 0, \\ \sum F_y = -\alpha f_9 - f_{11} - \alpha f_{13} = 0; \end{cases}$$

$$\text{joint 7} \begin{cases} \sum F_x = -f_{10} + f_{14} = 0, \\ \sum F_y = f_{11} = 0; \end{cases}$$

$$\text{joint 8} \begin{cases} \sum F_x = -f_{12} + \alpha f_{16} = 0, \\ \sum F_y = -f_{15} - \alpha f_{16} = 0; \end{cases}$$

$$\text{joint 9} \begin{cases} \sum F_x = -\alpha f_{13} - f_{14} + f_{17} = 0, \\ \sum F_y = \alpha f_{13} + f_{15} - 10 = 0; \end{cases}$$

$$\text{joint 10} \left\{ \sum F_x = -\alpha f_{16} - f_{17} = 0. \right.$$

$h = 10^0$

Write a Fortran program which uses SGEFS to solve this linear system of equations for the member forces. Is the matrix of the linear system well conditioned?

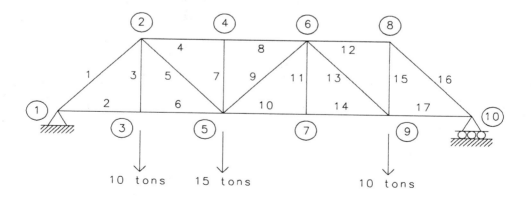

Figure P3.6

P3–7.–A paint company is trying to use up excess quantities of four shades of green paint by mixing them to form a more popular shade. One gallon of the new paint will be made of x_1 gallons of paint 1, x_2 gallons of paint 2, etc. Each of the paints is made up of four pigments, and they are related by the following system of linear equations

$$\begin{pmatrix} 80 & 0 & 30 & 10 \\ 0 & 80 & 10 & 10 \\ 16 & 20 & 60 & 72 \\ 4 & 0 & 0 & 8 \end{pmatrix} \begin{pmatrix} x_1 \\ x_2 \\ x_3 \\ x_4 \end{pmatrix} = \begin{pmatrix} 40 \\ 27 \\ 31 \\ 2 \end{pmatrix}.$$

Each number represents a percentage; for example, paint 4 contains 72% of pigment 3, and the more popular shade should contain 27% of pigment 2. Solve this system using SGEFS.

P3–8.–As was illustrated in Example 3.5, the error in the computed solution x can be large even though the residual is small. Sometimes it is necessary to accurately determine x, and in such cases a technique known as **iterative improvement** can be used. For this to be effective, you must be able to do a portion of the calculation in double precision. Consider the system of linear equations $Ax = b$ with

$$A = \begin{pmatrix} 21.0 & 67.0 & 88.0 & 73.0 \\ 76.0 & 63.0 & 7.0 & 20.0 \\ 0.0 & 85.0 & 56.0 & 54.0 \\ 19.3 & 43.0 & 30.2 & 29.4 \end{pmatrix} \text{ and } b = \begin{pmatrix} 141.0 \\ 109.0 \\ 218.0 \\ 93.7 \end{pmatrix}.$$

The solution of this system is $x = (1, 1, 1, 1)^T$.

(a) Use SGEFS to solve the linear system in single precision.

(b) Use double precision arithmetic to form the residual $r = b - Ax$. Remember that SGEFS destroys the original coefficients in A.

(c) Use SGEFS to solve the linear system $Ae = r$ for e. Since the matrix is the same as in (a), there is no need to refactor A. The vector e will be an estimate of the error in x, and so an improved estimate of the solution can be obtained by setting

$$x \leftarrow x + e.$$

The process can then be repeated starting with step (b) if more accuracy is desired.

3.11 PROLOGUE: SGEFS

```
      SUBROUTINE SGEFS (A, LDA, N, V, ITASK, IND, WORK, IWORK, RCOND)
C***BEGIN PROLOGUE   SGEFS
C***DATE WRITTEN     800317    (YYMMDD)
C***REVISION DATE    870916    (YYMMDD)
C***CATEGORY NO.   D2A1
C***KEYWORDS   GENERAL SYSTEM OF LINEAR EQUATIONS, LINEAR EQUATIONS
C***AUTHOR  VOORHEES, E., (LOS ALAMOS NATIONAL LABORATORY)
C***PURPOSE   SGEFS solves a GENERAL single precision real
C             NXN system of linear equations.
```

```
C***DESCRIPTION
C
C    From the book "Numerical Methods and Software"
C        by D. Kahaner, C. Moler, S. Nash
C            Prentice Hall 1988
C
C    Subroutine SGEFS solves a general NxN system of single
C    precision linear equations using LINPACK subroutines SGECO
C    and SGESL.  That is, if A is an NxN real matrix and if X
C    and B are real N-vectors, then SGEFS solves the equation
C
C                    A*X=B.
C
C    The matrix A is first factored into upper and lower tri-
C    angular matrices U and L using partial pivoting.  These
C    factors and the pivoting information are used to find the
C    solution vector X.  An approximate condition number is
C    calculated to provide a rough estimate of the number of
C    digits of accuracy in the computed solution.
C
C    If the equation A*X=B is to be solved for more than one vector
C    B, the factoring of A does not need to be performed again and
C    the option to only solve (ITASK .EQ. 2) will be faster for
C    the succeeding solutions.  In this case, the contents of A,
C    LDA, N and IWORK must not have been altered by the user follow-
C    ing factorization (ITASK=1).  IND will not be changed by SGEFS
C    in this case.  Other settings of ITASK are used to solve linear
C    systems involving the transpose of A.
C
C  Argument Description ***
C
C    A       REAL(LDA,N)
C                on entry, the doubly subscripted array with dimension
C                    (LDA,N) which contains the coefficient matrix.
C                on return, an upper triangular matrix U and the
C                    multipliers necessary to construct a matrix L
C                    so that A=L*U.
C    LDA     INTEGER
C                the leading dimension of the array A.  LDA must be great-
C                er than or equal to N.  (terminal error message IND=-1)
C    N       INTEGER
C                the order of the matrix A.  The first N elements of
C                the array A are the elements of the first column of
C                the  matrix A.  N must be greater than or equal to 1.
C                (terminal error message IND=-2)
C    V       REAL(N)
C                on entry, the singly subscripted array(vector) of di-
C                mension N which contains the right hand side B of a
```

```
C                     system of simultaneous linear equations A*X=B.
C                     on return, V contains the solution vector, X .
C     ITASK   INTEGER
C                     If ITASK=1, the matrix A is factored and then the
C                        linear equation is solved.
C                     If ITASK=2, the equation is solved using the existing
C                        factored matrix A and IWORK.
C                     If ITASK=3, the matrix is factored and A'x=b is solved
C                     If ITASK=4, the transposed equation is solved using the
C                        existing factored matrix A and IWORK.
C                     If ITASK .LT. 1 or ITASK .GT. 4, then the terminal error
C                        message IND=-3 is printed.
C     IND     INTEGER
C                     GT. 0  IND is a rough estimate of the number of digits
C                               of accuracy in the solution, X.
C                     LT. 0  see error message corresponding to IND below.
C     WORK    REAL(N)
C                     a singly subscripted array of dimension at least N.
C     IWORK   INTEGER(N)
C                     a singly subscripted array of dimension at least N.
C     RCOND   REAL
C                     estimate of 1.0/cond(A)
C
C  Error Messages Printed ***
C
C     IND=-1  fatal     N is greater than LDA.
C     IND=-2  fatal     N is less than 1.
C     IND=-3  fatal     ITASK is less than 1 or greater than 4.
C     IND=-4  fatal     The matrix A is computationally singular.
C                          A solution has not been computed.
C     IND=-10 warning    The solution has no apparent significance.
C                          The solution may be inaccurate or the matrix
C                          A may be poorly scaled.
C
C***REFERENCES   SUBROUTINE SGEFS WAS DEVELOPED BY GROUP C-3, LOS ALAMOS
C                  SCIENTIFIC LABORATORY, LOS ALAMOS, NM 87545.
C                  THE LINPACK SUBROUTINES USED BY SGEFS ARE DESCRIBED IN
C                  DETAIL IN THE *LINPACK USERS GUIDE* PUBLISHED BY
C                  THE SOCIETY FOR INDUSTRIAL AND APPLIED MATHEMATICS
C                  (SIAM) DATED 1979.
C***ROUTINES CALLED  R1MACH,SGECO,SGESL,XERROR
C***END PROLOGUE  SGEFS
```

4

Interpolation

4.1 INTRODUCTION

In the years 1913–1923, long before the development of computers, Sir Edmund Whittaker lectured to undergraduate and graduate students at the University of Edinburgh on Numerical Mathematics. Many of his students and disciples have gone on to play important roles in this field. One of them, G. Robinson, transcribed his notes and with Whittaker published what we now consider the first modern numerical analysis text in 1924. Their explanation of **interpolation** is as cogent today as it was then.

"If a function y of an argument x is defined by an equation $y = g(x)$, where $g(x)$ is an algebraical expression involving only arithmetical operations such as squaring, dividing, etc., then by performing these operations we can find accurately the value of y, which corresponds to any value of x. But if $y = \log_{10} x$ (say), it is not possible to calculate y by performing simple arithmetical operations on x (at any rate it is not possible to calculate y accurately by performing a finite number of such operations), and we are compelled to have recourse to a *table*, which gives the values of y corresponding to certain selected values of x. The question then arises as to how we can find the values of the function $\log_{10} x$ for values of the argument x which are intermediate between the tabulated values. The answer to this question is furnished by the theory of interpolation, which in its most elementary aspect may be described as the science of "reading between the lines of a mathematical table.""

Interpolation is a common operation both in everyday life and on computers. For example, if the speedometer in our automobile is between the marked lines we mentally interpolate to estimate our speed. If we have computational data that has been obtained at great expense at a few points we may want to determine values between the data points. Census data which are only available every ten years is an example of this.

The easiest case is that of **univariate** interpolation. We are given data points (x_i, y_i), $i = 1, \ldots, n$ and are required to find a function $f(x)$ which passes through the data, i.e.,

$$f(x_i) = y_i, \qquad i = 1, \ldots, n.$$

The function f is said to **interpolate the data** and f is called an **interpolant** or **interpolating function**. As Whittaker explained, we usually perform the interpolation because we need values which are not among the (x_i, y_i), for example for plotting, or for another calculation which requires a continuous function. Although the definition does not require it, the interpolant $f(x)$ is usually a function which can be computed at any x of interest. Thus $\log_{10} x$ certainly interpolates in a table of logarithms, but Whittaker would have said that it was not a very suitable interpolant, and should be replaced by another that is easier to compute.

In the years since Whittaker, mathematicians have extended their use of the term interpolation to include any process which determines a function that agrees with certain given "data." To return to the example of $y = g(x) = \log_{10} x$ we recall that $g'(x) = (\log_{10} e)/x$. Since division is a simple arithmetic operation we might augment our table by including not only values of x and $\log_{10} x$, but $0.4342944819/x \equiv d$ as well. Then for each x_i our "data" are a pair of numbers (y_i, d_i) and we want to find an easily computable function $f(x)$ which satisfies

$$f(x_i) = y_i, \qquad \text{and} \qquad f'(x_i) = d_i.$$

In that case f is called an **osculatory** or **Hermite** interpolant. ("Osculari" means "to kiss" in Latin.) Intuition and analysis tell us that since this function agrees not only with the values of $\log_{10} x$ but also with its derivatives it should do a better job approximating values between those in the table.

You are already familiar with an even more general case of this kind of interpolation. The Taylor expansion for $g(x)$ is

$$g(x) = \sum_{i=0}^{n} a_i x^i + \frac{g^{(n+1)}(\xi) x^{n+1}}{n!}, \qquad a_i = \frac{g^{(i)}(0)}{i!}.$$

The finite sum, $\sum a_i x^i$, is a polynomial of degree n which interpolates in the sense that it and its first n derivatives agree with those of g at zero, i.e., the "data" are $g^{(i)}(0), i = 0, \ldots, n$. Such a polynomial interpolant is an excellent approximation to $g(x)$ near $x = 0$, but deviates more and more as x moves away from the origin.

Interpolation can also be performed in more than one dimension. If our data were the temperatures T of a gas at accurately measured values of pressure P and volume V, we could look for a **bivariate** interpolant $f(P, V)$ so that

$$T_i = f(P_i, V_i).$$

There are many other possibilities, but the essential point is to find a function which matches some given information.

This chapter is entirely concerned with univariate interpolation. Sections 2–7 deal with polynomials, Sections 8–14 discuss piecewise polynomials, cubic splines, and Bézier curves. The latter are not always interpolants, but play an important role in computer aided design, and fit nicely into the general discussion of this chapter.

In Figure 4.1 we show several data values and three different interpolants. If the data represent some physical process each of the three can be viewed as an approximation to the underlying process. From this we see that

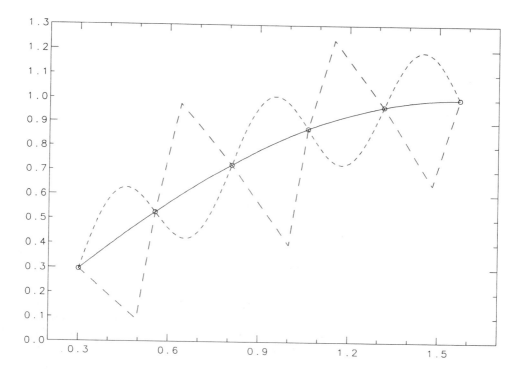

Figure 4.1   Three Different Interpolants to the Same Data

a) The data alone cannot determine the interpolant. There are arbitrarily many interpolants to a data set.

b) Interpolation can be useful only if the data are free of error. Experimental data, contaminated with errors, need to be approximated in some other way. Figure 4.2 shows experimental data and a function which "represents" the data better than any interpolant. In this chapter we assume that (x_i, y_i) are exact values of the function we want to approximate. **Least squares** approximation is the most common technique to use with data containing errors, and is considered in Chapter 6.

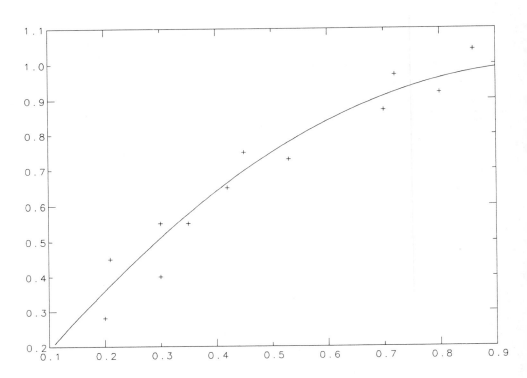

Figure 4.2   Non-Interpolant Represents Inaccurate Data

   Assume that $x_1 < x_2 < \cdots < x_n$. Our task is to find an interpolant f which
provides reasonable values when $x \neq x_i$. This can never be made perfectly rigorous,
since it depends upon the process which generated the data, our expectation of "good-
ness," etc. The standard approach is to decide on a set of **basis** functions in advance,
$b_1(x), b_2(x), \ldots, b_n(x)$. These can be selected by experience, recommendation, or mathematical or physical insight, but are thought of as known. The basis functions are
combined to form a **model**

$$f(x) = \sum_{j=1}^{n} \alpha_j b_j(x)$$

where the numbers α_j are *unknown* and are determined to make f an interpolant. Thus
the model must satisfy the interpolation properties

$$f(x_i) = y_i \qquad \text{or} \qquad \sum_{j=1}^{n} \alpha_j b_j(x_i) = y_i, \qquad i = 1, \ldots, n.$$

This represents a system of n linear equations for the α's with coefficient matrix

$$B = [b_{ij}], \qquad b_{ij} = b_j(x_i), \qquad i, j = 1, \ldots, n.$$

If we have chosen the b's judiciously we will be able to solve the system for the unknown α's and hence find the interpolant f. Since Chapter 3 has discussed solution of linear equations you might wonder if the observation that interpolation leads to such a system is all there is to say about the subject. This chapter considers two different sets of basis functions—polynomials and piecewise polynomials. For each of these the matrix B is so special that by taking advantage of it we can find the interpolant much more efficiently. Furthermore, just knowing that we can compute the unknown α's doesn't give us any information about whether the interpolant does a good job. The essence of the chapter then is to develop methods for the efficient computation of interpolants, and insights into selection of the most appropriate tool for each particular problem.

Before leaving this introductory section we wish to repeat that interpolation is only one way to approximate data. In (b) above, we mentioned that for data with significant errors the least squares approach is preferred. In other situations different methods are required. For example, consider the problem of writing a subroutine for a computer manufacturer that will provide approximations to $\sin(x)$. We can probably assume that x is in some fixed interval, say $0 \leq x \leq \pi/2$, because other values can be obtained from these. But within this interval all x's are equally likely to be arguments. There is no reason to select a particular set of x_i and force our approximate function to interpolate $\sin(x_i)$ at these points. What we would prefer is that the approximate function never have an error of more than one or two in the last place, for *all* x in the interval, although it need not be *exactly* correct i.e., it need not interpolate, at *any* specific x. This type of approximation requires more sophisticated mathematics and will not be discussed here. Interested readers are referred to the book by Davis (1963).

Finally readers should appreciate that depending upon the problem, there are two distinct goals for interpolation or any data fitting process.

(1) To determine the unknown coefficients α_i and make inferences about them. The methods we will discuss solve a system of equations for the coefficients. If the matrix is ill conditioned these will be inaccurate.

(2) To evaluate the interpolant $f(x)$ for plotting or for other reasons. Recall from Chapter 3 that solving a set of linear equations with an ill conditioned matrix results in inaccurate solution components, but small residuals. Inaccurate solution components mean that the interpolation coefficients α_i will have little or no accuracy; small residuals mean that $B\alpha \approx y$. The effect of this on the value of the interpolant $\sum \alpha_j b_j(x)$ at an arbitrary x can be inferred by considering evaluating it at one of the x_j. But this is exactly equivalent to asking for the residuals in the equations, and we have observed that these will be small. Thus if values of the interpolant are of main interest, even fairly large condition numbers are acceptable. Of course, it is always a good idea to go for the lowest condition numbers possible, and as we will see in Section 3, there are often several different basis functions that lead to the identical interpolant. For a concrete example see problem P4–2.

4.1.1 Historical Perspective: Table Making

As Whittaker has observed, the most common form of interpolation occurs when we seek data from a table which does not have the exact values we want. Methods for interpolation were often tested on tables and developed simultaneously with them. For several centuries one of the most important applications of numerical analysis was making tables of special functions, such as trigonometric functions like $\sin x$. *Mathematics of Computation*, the first modern journal devoted to numerical methods, was originally entitled *Mathematical Tables and Other Aids to Computation*. The name was changed in 1960. It is only in the last forty years, and especially since the invention of the pocket calculator, that these tables have gone out of use. Their primary application was in ocean navigation, where they were used to determine latitude and longitude. A scientist of the late eighteenth century might well own over a hundred volumes of tables to aid in calculations.

Creating a comprehensive set of tables was a monumental task. After the French Revolution, the French government produced such a set, made necessary by the newly invented metric system, but also motivated by the glory attached to such a project. It was supervised by the country's best mathematicians, but much of the tedious work was done by unemployed hairdressers. Many of the fashionable heads that wore elaborate powdered wigs had been lost to the guillotine.

More often, a set of tables was produced by plagiarism, copying numbers wholesale from earlier work. This meant that most tables were unreliable. Not only were the original calculation errors reproduced, but new errors were introduced with every new edition. It was an attempt to solve this problem that led Charles Babbage (1792–1871) to invent his "difference engine." The Cambridge-educated mathematician was convinced that a mechanical computer could be built using gears and levers and programmed by punched cards. For a while he was supported by the British government, but he financed most of the development with his personal fortune and that of his sponsor, Ada Augusta, Countess of Lovelace and daughter of Lord Byron. To generate more income he and Lady Lovelace tried to invent a system that would pick horse race winners. Although his computer was never finished, Babbage is regarded as the grandfather of modern computing.

Most numerical analysis textbooks published through the mid 1970s still contained several sections devoted to generating tables and finding errors. Some of the error detection techniques were very clever, involving differences (or differences of differences, etc.) of successive entries, and are worth reading about if you are ever confronted with a suspicious table, or would like some hints to assist you in debugging programs with tabular output.

One final burst of table making occurred in the United States several decades ago. In an attempt to ameliorate the unemployment problems of the Depression, the federal government formed the Works Progress Administration (WPA). The WPA had many programs: construction, sewing, photography, painting, etc. One of the lesser-known was the Mathematical Tables Project, started shortly before World War II.

One of its goals was to hire out-of-work mathematicians but, as with the French project, much of the work was done by people with no special training. In fact, many of the employees could do little more than add. Even negative numbers were a mystery. To overcome this last difficulty, red and black pencils were used, red for negative numbers and black for positive. Around the workroom were giant propaganda-like posters saying, "Black + Black = Black," and more controversially, "Red × Red = Black."

These limitations did not restrict the goals of the project. One of the first tables produced was for e^x. The calculations were based on Taylor series expansions, sometimes to as many as fifty terms, and interpolation. The work was broken down into simple steps on worksheets, something like computer programs. If a particular step required a more advanced operation like multiplication, it was assigned to a higher-level employee. Many of the computations were done by hand since even simple adding machines were in short supply. Much effort was expended to ensure the accuracy of the results, with every calculation being done at least twice, selected entries being computed in two independent ways, and extensive proof-reading. The printed tables were possibly the most accurate ever produced, being almost completely free of errors.

During the war, the Tables Project was involved in special projects for the Department of Defense, being used like a slow programmable calculator. By 1945 the Project had been absorbed by the National Bureau of Standards, and with the invention of the electronic computer, it was in many ways obsolete. However, the tables it produced continued to be of value, and many of them were collected in a book by Milton Abromowitz and Irene Stegun that is still in print and in worldwide use.

4.2 POLYNOMIAL INTERPOLATION

For historical as well as pragmatic reasons, the most important class of interpolating basis functions is the set of algebraic polynomials. Polynomials have the obvious advantage of being easy to evaluate directly. (See Section 7.) They can be easily summed, multiplied, integrated, or differentiated.

Of course, a class of functions can have all of the above properties and still not be satisfactory at approximating functions. Fortunately, we have good reason to believe that *any* continuous function $g(x)$ can be closely approximated on a closed interval by some polynomial $p_n(x)$. This follows from an early result of approximation theory known as the **Weierstrass approximation theorem:**[1] *If g is any continuous function on the finite closed interval $[a, b]$, then for every $\epsilon > 0$ there exists a polynomial $p_n(x)$ of degree $n = n(\epsilon)$ such that*

$$\max_{x \in [a,b]} |g(x) - p_n(x)| < \epsilon.$$

---

[1] Karl Theodor Wilhelm Weierstrass (1815–1897), the most important nineteenth-century German mathematician after Gauss and Riemann, also taught botany, geography, history, German, gymnastics, and calligraphy. He developed the fundamental concepts of the theory of functions. Among his many students were Frobenius, Gegenbauer, Klein, Lie, and Minkowski. He considered geometric proofs in bad taste and rarely used a diagram to clarify a point.

The reader is referred to either of the books by Ralston (1965) or Wendroff (1966) for a detailed proof of this and other results on polynomial interpolation.

Although some proofs of Weierstrass' theorem are constructive, the resulting polynomial is generally of such high degree that it is impractical to use. Furthermore, Weierstrass' theorem tells us nothing about the existence of a satisfactory interpolating polynomial for a given data set. And while it is comforting to know that some polynomial will approximate $g(x)$ to a specified accuracy throughout the interval $[a, b]$, this is no guarantee that such a polynomial will be found by a practical algorithm.

If we select for our basis functions the monomials

$$b_i(x) = x^{i-1}, \qquad i = 1, \ldots, n$$

then the model becomes

$$p_{n-1}(x) = \alpha_1 + \alpha_2 x + \cdots + \alpha_n x^{n-1},$$

with matrix B

$$B = \begin{pmatrix} 1 & x_1 & \cdots & x_1^{n-1} \\ 1 & x_2 & \cdots & x_2^{n-1} \\ \vdots & \vdots & & \vdots \\ 1 & x_n & \cdots & x_n^{n-1} \end{pmatrix}.$$

Example 4.1   Linear Interpolation.

The unique linear interpolant through (x_1, y_1) and (x_2, y_2) is given by

$$p_1(x) = \alpha_1 + \alpha_2 x,$$

where α_1 and α_2 satisfy the two equations

$$\alpha_1 + \alpha_2 x_1 = y_1 \quad \text{and} \quad \alpha_1 + \alpha_2 x_2 = y_2, \quad \text{i.e.,} \quad B = \begin{pmatrix} 1 & x_1 \\ 1 & x_2 \end{pmatrix}.$$

As long as $x_1 \neq x_2$ the matrix B is nonsingular and we can solve for α_1 and α_2. We get

$$p_1(x) = \frac{y_1 x_2 - y_2 x_1}{x_2 - x_1} + \frac{y_2 - y_1}{x_2 - x_1} x.$$

As a specific instance of this, the linear interpolant to $g(x) = \sqrt{x}$ based on its values at $x = 0$ and $x = 0.25$ is $p_1(x) = 2x$. At $x = 1/9$ the interpolation error is $|p_1(1/9) - g(1/9)| = 1/9$.
∎

Example 4.2   Quadratic Interpolation.

Find the quadratic interpolant to $(-1, 2)$, $(1, 1)$, and $(2, 1)$. The interpolant is

$$p_2(x) = \alpha_1 + \alpha_2 x + \alpha_3 x^2.$$

The α's are obtained by requiring p_2 to pass through the three points above, leading to the system of equations

$$
\begin{aligned}
\alpha_1 - \alpha_2 + \alpha_3 &= 2 \\
\alpha_1 + \alpha_2 + \alpha_3 &= 1, \\
\alpha_1 + 2\alpha_2 + 4\alpha_3 &= 1
\end{aligned}
\qquad
B = \begin{pmatrix} 1 & x_1 & x_1^2 \\ 1 & x_2 & x_2^2 \\ 1 & x_3 & x_3^2 \end{pmatrix} = \begin{pmatrix} 1 & -1 & 1 \\ 1 & 1 & 1 \\ 1 & 2 & 4 \end{pmatrix}.
$$

You can verify that the determinant of B,

$$\det(B) = (x_3 - x_2)(x_3 - x_1)(x_2 - x_1) = 6,$$

so B is nonsingular. Solving for the α's gives the interpolant

$$f(x) = \frac{4}{3} - \frac{1}{2}x + \frac{1}{6}x^2. \quad \blacksquare$$

In the two examples above the coefficient matrix B is nonsingular because we have explicitly solved the system. But in general how do we know if it will be possible to do this? For polynomial interpolation, it can be shown that the matrix B satisfies

$$\det(B) = \prod_{1 \le i < j \le n} (x_j - x_i) = (x_n - x_1)(x_n - x_2) \cdots (x_n - x_{n-1}) \cdots (x_3 - x_1)(x_3 - x_2)(x_2 - x_1)$$

which is never zero if the x_i are distinct. Notice that the matrices in Examples 4.1 and 4.2 follow this pattern. Thus as long as no two abscissas are equal the equations for polynomial interpolation always have a nonsingular coefficient matrix, and hence a unique solution. Since a mathematical function must be single valued (it cannot take on two y's for the same x), this condition is perfectly reasonable.

Summarizing

Given n points in the plane with distinct abscissas, there is a unique polynomial of degree at most $n - 1$ passing through these points.

4.3   USING OTHER BASIS FUNCTIONS

In Example 4.1 consider what will happen if we select as basis functions

$$b_1(x) = \frac{x - x_2}{x_1 - x_2}, \qquad b_2(x) = \frac{x - x_1}{x_2 - x_1}.$$

The matrix B then becomes

$$B = \begin{pmatrix} b_1(x_1) & b_2(x_1) \\ b_1(x_2) & b_2(x_2) \end{pmatrix} = \begin{pmatrix} 1 & 0 \\ 0 & 1 \end{pmatrix}$$

i.e., B is the identity, $B = I$ and the equations $B\alpha = y$ are solved by $\alpha_1 = y_1$ and $\alpha_2 = y_2$. We know there is only one linear polynomial interpolating two distinct points, and since both $b_1(x)$ and $b_2(x)$ are linear the interpolant formed from them must be identical to $p_1(x)$ in Example 4.1. Thus we can write $p_1(x)$ either as in Example 4.1 or as

$$p_1(x) = y_1 \frac{x - x_2}{x_1 - x_2} + y_2 \frac{x - x_1}{x_2 - x_1};$$

one is simply an algebraic rearrangement of the other.

Other changes we can make are to interchange b_1 and b_2 so that $b_1(x) = x$, $b_2(x) = 1$, or to replace b_1 and b_2 with their sum and difference, $x + 1$ and $x - 1$. Any replacement of the b's by another independent set which is a linear combination of the first has no effect on the resulting interpolant and is termed **change of representation** or **change of basis**. Changing representations can be useful if it makes the task of generating the interpolant easier or provides special insight. The monomial basis, $b_j(x) = x^{j-1}$ is intuitive but requires solving a system which can often be ill conditioned. In Figure 4.3 we show some of the monomial basis functions, x^k, $k = 0, \ldots, 20$ on $[0, 1]$. On this interval the functions x^{18}, x^{19} and x^{20} are almost identical hence the associated columns of B will be nearly equal. Equal columns correspond to a singular matrix so a high condition number is expected. We might also expect it because the b_{ij} are of very different magnitude, varying from 1 to x_i^{19}, which can be very small. The Lagrange[2] basis, $b_j(x) = l_j(x)$, defined below is a representation change which generates exactly the same polynomial interpolant but for which the interpolation matrix B reduces to the identity. Thus solving the equations becomes trivial.

Assume we have a set of functions $l_1(x), \ldots, l_n(x)$ each of which is a polynomial of degree $n - 1$ and also has the property that

$$l_j(x_i) = \begin{cases} 1 & \text{if } i = j; \\ 0 & \text{otherwise.} \end{cases}$$

That is, l_j takes the value 1 at the point x_j and is zero at all the other x_i. (Notice that b_1 and b_2 displayed at the beginning of this section satisfy these properties.) Any linear combination of the l_j's is still an $(n - 1)$-st degree polynomial. In particular consider

$$p_{n-1}(x) = y_1 l_1(x) + y_2 l_2(x) + \cdots + y_n l_n(x).$$

---

[2] Joseph Louis Lagrange (1736–1813), an Italian from Turin, spent most of his productive life in Paris and Berlin. He was one of the earliest developers of the calculus of variations (a term coined by Euler) and made monumental contributions to mechanics (including the three body problem), theory of numbers, and differential equations. He was sufficiently well regarded that in 1793 he was retained as chairman of the commission to standardize weights and measures (metric system) after Lavoisier, Laplace, Coulomb and Brisson were purged because of the political climate.

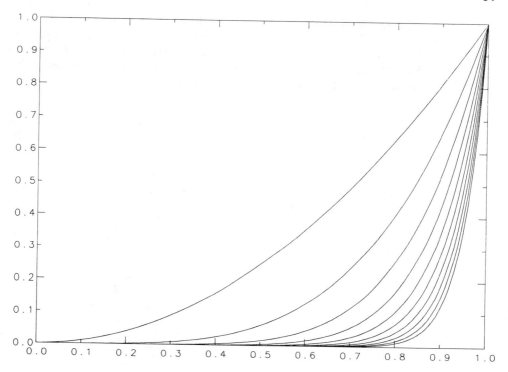

Figure 4.3   Some Monomial Basis Functions on [0,1]

By the properties of l_j it is obvious that

$$p_{n-1}(x_i) = y_1 l_1(x_i) + y_2 l_2(x_i) + \cdots + y_n l_n(x_i) = y_i l_i(x_i) = y_i.$$

Thus p_{n-1} above is an interpolating polynomial, and since we saw that such a polynomial is unique this is equivalent to solving the linear system. The method hinges on finding the l_j's, but these are easy to write:

$$l_j(x) = \frac{(x - x_1)(x - x_2) \cdots (x - x_{j-1})(x - x_{j+1}) \cdots (x - x_n)}{(x_j - x_1)(x_j - x_2) \cdots (x_j - x_{j-1})(x_j - x_{j+1}) \cdots (x_j - x_n)}$$

$$= \prod_{i=1, i \neq j}^{n} (x - x_i) \Big/ \prod_{i=1, i \neq j}^{n} (x_j - x_i)$$

Example 4.3   Lagrange Interpolation to Three Data Points.

Repeat Example 4.2 using the Lagrange representation. We find that

$$p_2(x) = 2 \frac{(x - 1)(x - 2)}{(-1 - 1)(-1 - 2)} + 1 \frac{(x + 1)(x - 2)}{(1 + 1)(1 - 2)} + 1 \frac{(x + 1)(x - 1)}{(2 + 1)(2 - 1)}.$$

The reader should verify that this simplifies to the quadratic of the previous example. ■

Figure 4.4 shows plots of the first five Lagrange basis functions. We must specify the x_i and for illustration these have been taken to be five equally spaced points on [0,1]. To repeat, the Lagrange basis is different for each set of distinct x_i, but the interpolant is identical to that produced by the monomial basis, which is independent of the x_i. For the Lagrange basis the condition number of B is one.

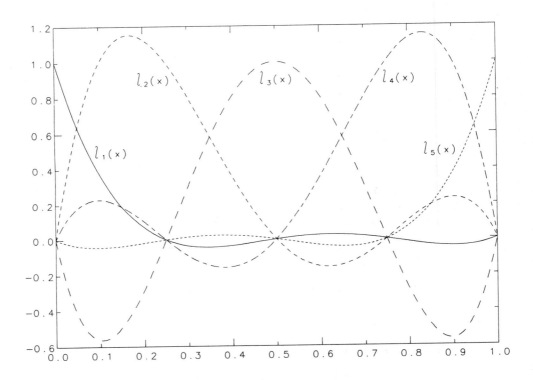

Figure 4.4   Five Lagrange Basis Functions on [0,1]

In general the Lagrange basis is easier to write but more difficult and less efficient to compute with than the monomial basis. But the matrix B associated with monomials is often badly conditioned. Other important changes of representation have different properties. For example, some are designed for the situation in which we get additional data after the interpolant has been formed, and would like to incorporate this in an efficient way.

Usually, the monomial representation of a polynomial interpolant is stored in an array containing its coefficients with respect to the monomial basis. The Lagrange representation of the same interpolant is stored in a pair of arrays, one containing the x_i's which are required to compute $l_j(x)$, and another containing the data values y_j which are the coefficients of $l_j(x)$.

In Example 4.1 if we replace $b_1(x) = 1$ by $b_1(x) = e^x$ we will get an entirely different interpolant, $\alpha_1 + \alpha_2 e^x$. For the two data points of Example 4.1 you can easily check that $\alpha_1 = -0.25/(\exp(0.5) - 1) = -0.3853736, \alpha_2 = -\alpha_1$. This is no longer a straight line, and is not even a polynomial. In this case we say that we are changing the model. Both change of basis and change of model occur frequently, and it is important to be able to distinguish between them.

4.4 HOW GOOD IS POLYNOMIAL INTERPOLATION?

How should we assess the quality of an interpolant? Once we have computed the coefficients α_j one step is to evaluate the interpolant at the data points x_i and verify that the y_i are reproduced to within rounding error. If this fails, either B is ill conditioned or there is a bug in our program. But, in use, an interpolant will probably be evaluated at many other points and it is *not possible to determine its general behavior by knowing only that it reproduces the input data*. The best way, short of analysis, is to evaluate the interpolant at many more points and print or plot the results.

In some situations though, the quality of the interpolant can be analyzed. Let us suppose that the values y_i are exact values of a known function $g(x)$ at the points x_i. Let $p_{n-1}(x)$ be the unique $(n-1)$-st degree polynomial interpolating these n points (x_i, y_i), $i = 1, \ldots, n$. Suppose g has n continuous derivatives for all x. Then it can be proved that for any x

$$g(x) - p_{n-1}(x) = \frac{g^{(n)}(\xi)}{n!}(x - x_1)(x - x_2) \cdots (x - x_n),$$

where ξ is some unknown point between x_1 and x_n. As a practical matter this expression is only useful for simple cases; it sometimes provides an error bound, but mostly it gives us insight and helps to justify the conclusion we make below.

Example 4.4   The Error in Polynomial Interpolation to $\ln x$.

The function $g(x) = \ln x$ is interpolated by a cubic at points $x = 0.4, 0.5, 0.7,$ and 0.8. We wish to bound the error in the interpolant at the point $x = 0.6$.

The Lagrange form of the cubic is

$$p_3(x) = \ln(.4)\frac{(x - .5)(x - .7)(x - .8)}{(.4 - .5)(.4 - .7)(.4 - .8)} + \ln(.5)\frac{(x - .4)(x - .7)(x - .8)}{(.5 - .4)(.5 - .7)(.5 - .8)}$$

$$+ \ln(.7)\frac{(x - .4)(x - .5)(x - .8)}{(.7 - .4)(.7 - .5)(.7 - .8)} + \ln(.8)\frac{(x - .4)(x - .5)(x - .7)}{(.8 - .4)(.8 - .5)(.8 - .7)}.$$

and $p_3(.6) = -0.509975$. The error expression gives

$$\ln(.6) - p_3(.6) = -\frac{6}{\xi^4}\frac{1}{4!}(.6 - .4)(.6 - .5)(.6 - .7)(.6 - .8), \qquad .4 < \xi < .8.$$

So the error is less than

$$\frac{6}{.4^4}\frac{1}{24}0.0004 \approx 0.0039.$$

The actual error is $|p_3(.6) - \ln(.6)| = 0.000851$. For some problems, though, even the error bound provides no information. Example 4.1 is such a case when $y_i = \sqrt{x_i}$. ∎

Now consider what happens if we interpolate at more and more points on a fixed interval from a known function $g(x)$. We hope that the interpolation estimate at other values of x will improve. The error expression is composed of three distinct parts; the factorial and product of point spacing cause the error to decrease with increasing n, but the order of the derivative is increasing. For "most" functions, derivative values increase faster than $n!$. As a result, polynomial interpolants rarely converge to a general continuous function. The mathematically inclined student can consider whether this flies in the face of the Weierstrass Theorem. The practical effect is that a high degree polynomial interpolant can have very bad behavior for x's other than the x_i's and is almost never used above degree 4 or 5.[3]

High degree polynomial interpolation is a bad idea

Example 4.5   Runge's Function.

A detailed analysis of the dangers of polynomial interpolation was first published by C. Runge in 1901. He attempted to interpolate the simple function

$$R(x) = \frac{1}{1 + 25x^2}$$

on the interval $[-1, 1]$ with polynomials and equally spaced points. He discovered that as the degree n of the interpolating polynomial p_n tends toward infinity, $p_n(x)$ diverges in the intervals $0.726\ldots \leq |x| < 1$. This phenomenon is shown graphically in Figure 4.5. Note that in this case, polynomial interpolation worked well in the central portion of the interval. ∎

There are several ways to understand why polynomial interpolation fails for Runge's function. The most direct is to notice that the successive derivatives of $R(x)$, which appear in the expression for the interpolation error, grow rapidly with n. Another way is to consider R as a function of a complex variable. This function has a singularity wherever $1 + 25x^2 = 0$, or at $\pm i/5$. These singularities lurk just off the interpolation interval $[-1, 1]$, close enough to affect the interpolant.

If the data abscissas are not equally spaced, but rather placed nearer to the ends of the interval, then the problem with Runge's function disappears. The resulting polynomial interpolants converge to $R(x)$ for every x in $[-1, 1]$ as n approaches infinity. Unfortunately, this trick does not work in general. A theorem of Faber says that there is no point-placement rule which will work for every continuous function g. For any

---

[3] High degree polynomial interpolation works well if the data points are the values of a function $f(x)$ like $\sin x$ or e^x with the special property that for each fixed z, $|f^{(p)}(z)/p!| \leq M$, $p = 0, 1, \ldots$.

particular function, though, some specific spacing might work. It is also possible to find a general placement rule which will work for all functions with at least two continuous derivatives. For more details see the text by Davis.

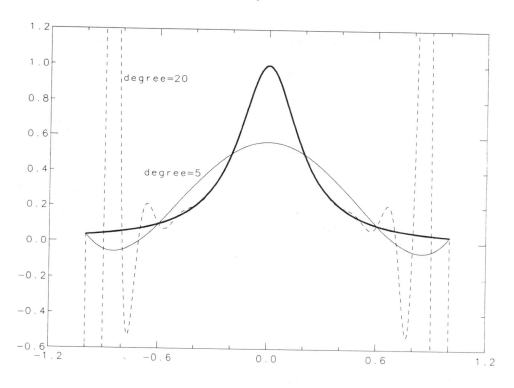

Figure 4.5 Polynomial Interpolation to Runge's Function

4.5 HISTORICAL PERSPECTIVE: RUNGE

Carl David Tolme Runge (1856–1927), was born in Bremen Germany, the fourth and youngest son of a prosperous merchant family. His parents spent most of the 20 years preceding his birth in Havana, Cuba and then retired to Bremen. English was their language of choice and they imbued their son with a British view of the world, particularly an emphasis on sport, self-reliance and fair play. As a young man Runge cut a striking figure, tall, lean with a large and finely sculpted head and with exceptional skill as an ice skater. He spent all of his professional life in Germany—Munich, Berlin, Hanover and finally Göttingen. Max Planck was a warm and close friend through all of Runge's life, but intellectually Runge thought of himself as a disciple of Weierstrass. His early work was under Leopold Kronecker in function theory but he soon after immersed himself in problems of spectroscopy and astrophysics where he spent most of his career. Almost all of Runge's important papers were in these areas but he never ceased to regard himself as a mathematician. His interests gradually focused on precision of data, data reduction, and manipulation.

"Applied Mathematics" as understood and practiced by Runge was different from that of his contemporaries. He was not at all concerned with the rigorous mathematical treatment of models derived from the physical world, and little concerned with the mathematical methods then used in technology. Primarily he wished to treat the theory and practice of numerical computation, with a great deal of emphasis on practice. Some of his methods are still in use today, notably the Runge-Kutta method for solving differential equations. Nevertheless he was not recognized by mathematicians as one of their own, nor by physicists as one of theirs either. As a result Runge did not obtain a worthwhile academic appointment until late in his career. In 1904, after intense lobbying by Planck and Felix Klein, he was appointed to Göttingen as the first (and last) occupant of the first full professorship in Applied Mathematics in Germany. In some sense he was the inventor and at that time the sole practitioner of his discipline. His credibility with professional colleagues increased despite his liberal political views during World War I, and in 1920 Peter Debye recommended him for successor to the chair he was vacating because Runge was "the only person in Göttingen capable of managing the physical institute."

Runge retired in 1925 without ever having a talented student who wished to study his form of applied mathematics, what we would today call numerical analysis. He remained in vigorous health until his death in 1927, leaving two sons, four liberated daughters, and a recollection by his family of grandfather doing handstands at his seventieth birthday party.

4.6 EVALUATION OF POLYNOMIALS

Polynomials are so pervasive in mathematics that one is often faced with the task of evaluating them rapidly from their coefficients. The polynomial

$$p(x) = a_1 x^n + a_2 x^{n-1} + \cdots + a_{n+1}$$

can be evaluated by the Fortran program segment

```
        P = A(N+1)
        DO 10 I = 1,N
          P = P + A(I)*X**(N-I+1)
10      CONTINUE
```

which takes n multiplications, n exponentiations and n additions. A simple technique called **Horner's rule** rewrites $p(x)$ as

$$p(x) = a_{n+1} + x(a_n + x(a_{n-1} + x(\cdots(a_2 + a_1 x)\cdots))).$$

The Fortran for this is

```
P = A(1)
DO 10 I = 1,N
    P = P*X + A(I+1)
10  CONTINUE
```

Horner's rule takes only n multiplications and n additions, and might be familiar to you as "synthetic division." W. G. Horner's name is attached to this method because he presented the rule in a paper (on another subject) in 1819. Actually, the rearrangement was published over 100 years earlier by Isaac Newton. Several generalizations have been published since then, see for example the book by Knuth (1969). Horner's rule should be the method of first choice for evaluation of polynomials at arbitrary points. However, if $p(x)$ is to be evaluated at a sequence of equally spaced points, for example for plotting, other methods are more efficient. See problem P4–10.

4.7 PIECEWISE LINEAR INTERPOLATION

Polynomial interpolation is global, i.e., we use one polynomial function to pass through all the data. Adding data points requires us to increase the polynomial degree and leads to difficulties as we have seen. One alternative which has been popular since the mid 1960's is to use **piecewise polynomials** functions. In this section we introduce these and subsequently generalize the concept to include cubic splines and Hermite cubics which are the most useful of all. A valuable expository and reference text on splines is the book by C. de Boor (1978).[4] In addition to interpolation, which is the topic of this chapter, splines are often used in solving differential equations. For that, a readable introduction is the book by P. Prenter (1975).

In the context of piecewise polynomials the data abscissas are called **knots**, **joints** or **breakpoints**. There are some technical differences, but the three terms are often used interchangeably. A linear piecewise polynomial function $L(x)$, is a function defined for all x with the property that $L(x)$ is a straight line between x_i and x_{i+1}. The definition allows $L(x)$ to be different lines between each pair of adjacent knots. Figure 4.6 illustrates one linear piecewise polynomial. Note that any linear combination of these is still a linear piecewise polynomial.

A linear piecewise polynomial interpolant to the data set (x_i, y_i) is a linear piecewise polynomial with the property

$$L(x_i) = y_i \qquad i = 1, \dots, n.$$

Figure 4.7 shows a linear piecewise polynomial interpolant. It is exactly the "dot to dot" drawing we did as children. Notice that the definition does not say anything about $L(x)$

---

[4] Carl Wilhelm Reinhold de Boor, a brilliant East German immigrant, spent the years 1960–1964 as a research mathematician for General Motors developing a mathematical description of automobile panel shapes. He formulated the concept of B-splines and discovered many algorithms now in worldwide use at CAD installations. He now teaches at the University of Wisconsin in Madison.

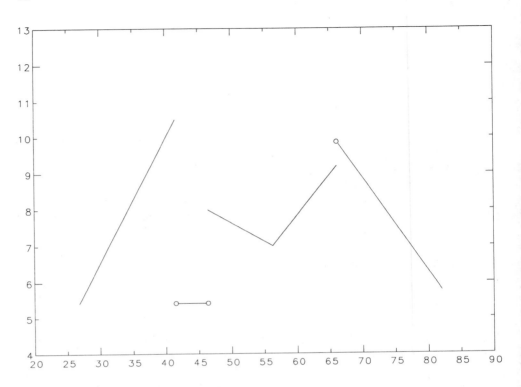

Figure 4.6 A Linear Piecewise Polynomial Function

for $x < x_1$ or $x > x_n$. Thus there are many linear piecewise polynomial interpolants to the same data with different properties on these exterior intervals. Nevertheless on $[x_1, x_n]$ the interpolant is unique. We will restrict our attention to this interval.

Piecewise polynomials may seem unusual, but they are perfectly ordinary functions. For example, it is easy to write a rule for the evaluation of $L(x)$. From Example 4.1 we have

$$L(x) = y_i \frac{x - x_{i+1}}{x_i - x_{i+1}} + y_{i+1} \frac{x - x_i}{x_{i+1} - x_i}, \qquad \text{if } x_i \le x \le x_{i+1}, \qquad i = 1, 2, \ldots, n-1.$$

A linear piecewise polynomial interpolant has the desirable property that if the y_i are values of a known, continuous function, $g(x)$, and if we add more points between x_1 and x_n, the interpolant gets better, i.e., converges to the original function. Furthermore if the data y_i are values of a function $g(x)$ which has a continuous second derivative then it can be proved that

$$|L(x) - g(x)| < \frac{1}{8} h^2 \max |g''(x)| = O(h^2),$$

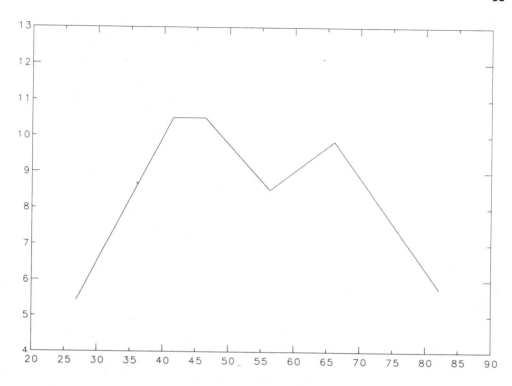

Figure 4.7 A Linear Piecewise Polynomial Interpolant

where h is the maximum spacing between adjacent knots. The importance of this result is that only the second derivative occurs in the error expression, independent of the number of knots. If the knots are equally spaced and we double their number, then the error in the new interpolant should be about $1/4$ the error in the old one. Thus we can make the interpolation error as small as we like by taking sufficiently many knots. In practice, of course, we seldom know the underlying function, and rarely have the luxury of adding more points. Nevertheless, convergence results like this give us confidence in the method, especially when contrasted with the "non-convergence" of polynomial interpolation.

Because $L(x)$ is linear between the knots it can be differentiated there. We get

$$L'(x) = \frac{y_i}{x_i - x_{i+1}} + \frac{y_{i+1}}{x_{i+1} - x_i} = \frac{y_{i+1} - y_i}{x_{i+1} - x_i}, \quad \text{if } x_i < x < x_{i+1}, \quad i = 1, 2, \ldots, n-1.$$

This is the difference quotient approximation to a derivative which is probably familiar to you. It can be applied everywhere except at the knots and provides good estimates of the derivative. It can be proved that

$$|L'(x) - g'(x)| = O(h), \qquad x \neq x_i.$$

Thus the derivative of the interpolant estimates the derivative of the function which generated the data. This is in marked contrast to polynomial interpolation.

Example 4.6   The Error in Piecewise Linear Interpolation to $R(x)$.

How many equally spaced knots are required to generate a linear piecewise polynomial interpolant which has error less than 10^{-5} for Runge's function? We have that

$$g'' = -50(1 + 25x^2)^{-2} \left[\frac{-100x^2}{(1 + 25x^2)} + 1 \right],$$

and by direct computation $|g''| \leq 50$. Thus using the expression for the error above we have that h should be chosen so that

$$\frac{1}{8} 50 h^2 < 10^{-5},$$

or $h < .0013$. As $h = 2/(n - 1)$, about 1540 knots are needed. ∎

4.8   PIECEWISE CUBIC FUNCTIONS

Piecewise linear interpolation solves one problem with polynomial interpolation— convergence, but introduces a separate problem—lack of smoothness; $L(x)$ has corners. To look for an interpolant which is smoother we consider using higher degree piecewise polynomials. We will see that this leads to practical results, as contrasted to polynomials. Piecewise quadratics are considered in problem P4–6, but more useful is the piecewise cubic. A piecewise cubic is a function defined for all x which is a cubic polynomial between adjacent knots. A piecewise cubic interpolant $C(x)$, is a piecewise cubic which interpolates our data. Figure 4.8 shows two piecewise cubic interpolants. In this case we see that even inside $[x_1, x_n]$ the interpolant is not unique. Requiring that the piecewise cubic pass through the data is not enough, but by demanding some smoothness we can obtain a unique result.

Between the knots $C(x)$ is a cubic. It may be differentiated as often as necessary. If there is any lack of differentiability it must occur at the knots. $C(x)$ is already continuous everywhere because it interpolates (see Figure 4.12). A smoother interpolant would have one or more continuous derivative on $[x_1, x_n]$. An **Hermite cubic interpolant**[5] is a piecewise cubic interpolant with a continuous derivative. A **cubic spline** is

---

[5] Charles Hermite (1822–1901), was a French mathematician who worked in many diverse areas including elliptic functions, number theory, quadratic forms, approximation of functions, and differential equations. Although he is less well known today than some of his contemporaries, during his lifetime he exerted a great scientific influence by his correspondence with other prominent mathematicians and was considered an inspiring figure. An honorary member of a great many academies and learned societies with numerous decorations, he spent most of his professional life at the École Polytechnique in Paris. His textbooks in analysis became classics, famous even outside France, and it was his habit to disseminate his knowledge lavishly in correspondence, courses, and short notes. One of the best known facts about Hermite is that he first proved the transcendence of e and provided the basic technique which was ultimately used to prove the transcendence of π. Another indication of the breadth of his interest was that he also studied Sanskrit and ancient Persian.

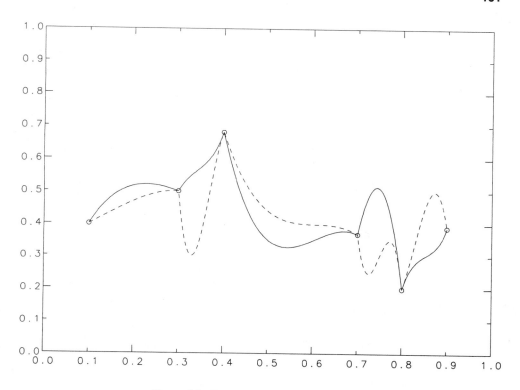

Figure 4.8   Two Piecewise Cubic Interpolants

a piecewise cubic interpolant with two continuous derivatives. Both types are important in applications. In the engineering and science community the term spline was originally synonymous with cubic spline. Today, splines of higher and lower order are known and used. Nevertheless cubic splines are still the most popular and we will restrict our attention to them.

What about requiring three continuous derivatives? Since the third derivative of a cubic is a constant it is easy to show that any piecewise cubic with three continuous derivatives at every knot must be exactly the same cubic in each interval. Since a single cubic can never interpolate more than four points we cannot use this as a general interpolation process.

On each interval $[x_i, x_{i+1}]$ $C(x)$ is a cubic and can be represented by four coefficients. A program using this representation would require an array for the x_i's, and four arrays, a, b, c, and d, for the coefficients of the cubic in each subinterval. We call this the piecewise polynomial representation. For our development we prefer to use a different representation which is more insightful. We will define $2n$ basis functions, $c_i(x)$ and $\hat{c}_i(x)$, $i = 1, \ldots, n$. Each will be a piecewise cubic with a continuous derivative on $[x_1, x_n]$. Thus any linear combination of them has the same properties. Our definition will guarantee that

$$c_i(x_i) = 1, \qquad c_i(x_j) = 0 \qquad j \neq i, \qquad \text{and} \qquad \hat{c}_i(x_j) = 0, \qquad \text{all } i, j.$$

This being the case, the function

$$C(x) = \sum_{i=1}^{n} \big(y_i c_i(x) + d_i \hat{c}_i(x) \big)$$

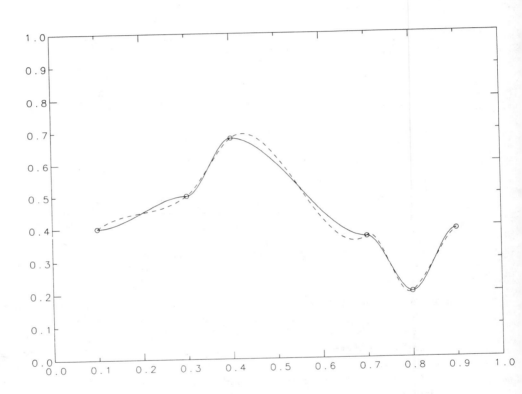

Figure 4.9 Two Hermite Cubic Interpolants with Different Slopes

is an Hermite cubic interpolant for *any* choice of the d_i. We will also arrange that

$$c'_j(x_i) = 0 \qquad \text{all } i, j \qquad \text{and} \qquad \hat{c}'_i(x_i) = 1, \qquad \hat{c}'_i(x_j) = 0 \qquad j \neq i.$$

Then

$$C'(x_k) = \sum_{i=1}^{n} \big(y_i c'_i(x_k) + d_i \hat{c}'_i(x_k) \big) = d_k \hat{c}'_k(x_k) = d_k.$$

Figure 4.10 shows a typical c_i and $\hat{c}_i$. With these things in place it is easy to imagine different Hermite cubic interpolants $C(x)$. They all are piecewise cubics which interpolate the data and have one continuous derivative. At the data values these derivatives are given by d_i. An interpolant in this form is particularly useful if we know the values of the slope of the underlying function at the data points, in addition to the data values. In that case the natural choice for the d_i are exactly these given slopes. Figure 4.9 shows two different $C(x)$, one with all the $d_i = 1$ and the other with $d_i = 0$. These are probably not of practical interest, but illustrate the effect that $\hat{c}_i$ has.

Example 4.7   Hermite Cubic Interpolation to $R(x), d_i = R'(x_i)$.

Determine an Hermite cubic interpolant to Runge's function $R(x)$. Use $R'(x_i)$ for d_i. We omit the details and display the interpolant below. ■

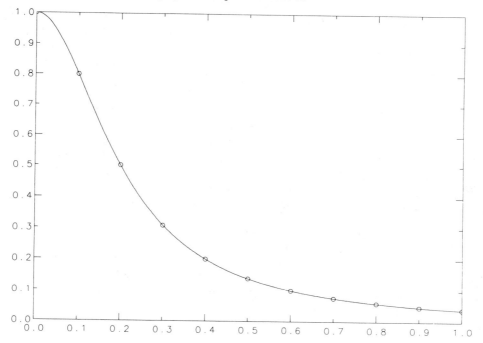

If we do not know the d_i, taking a cue from the the fact that $d_i = C'(x_i)$, we can try to estimate them from the data. For example, there is a unique quadratic through $(x_{i-1}, y_{i-1}), (x_i, y_i)$, and (x_{i+1}, y_{i+1}). Its derivative at x_i is

$$\delta_i \equiv \frac{\Delta_i h_{i-1} + \Delta_{i-1} h_i}{h_{i-1} + h_i}, \qquad \text{where} \qquad \Delta_i = \frac{y_{i+1} - y_i}{h_i}.$$

We can use δ_i to approximate $d_2, \ldots, d_{n-1}$. For d_1 and d_n we can use Δ_1 and Δ_{n-1}, or a more accurate formula such as the derivative (at x_1) of the above quadratic. There are many other ways to estimate derivatives as well.

Example 4.8   Hermite Cubic Interpolation to $R(x), d_i \approx R'(x_i)$.

If we determine an Hermite cubic interpolant to Runge's function $R(x)$, but use the approximations above for the d_i's we will find that the plot of the interpolant cannot be distinguished from the one in Example 4.7. Of course not all data sets give such good results. ∎

This representation requires three arrays, one each for the x_i, y_i, and d_i, and is called the Hermite cubic representation. This is only three-fifths the storage required for the piecewise polynomial representation we mentioned earlier, but since n is rarely more than a few hundred, storage is not a significant consideration. The main advantage of the Hermite cubic representation is the intuitive nature of the coefficients. But when the interpolant is to be evaluated many times (for instance, for plotting) the piecewise polynomial representation may be preferred. In either case, a subroutine which evaluates the interpolant at an input point x must first locate x within one of the subintervals. If the piecewise polynomial representation is being used it is then only necessary to evaluate the cubic whose coefficients are given in that subinterval. If the Hermite cubic representation is used the evaluation routine first decides which c_i and $\hat{c}_i$ will be nonzero in that subinterval (there are normally two of each) and evaluates them at x. Thus evaluating the Hermite cubic representation is slower than the piecewise polynomial representation. Packages for manipulation of piecewise cubics often provide conversion routines from one representation to another.

Routines for piecewise cubic interpolation usually come in pairs. The first computes whichever unknown parameters, d_i's, are required. This routine is normally called only once for each data set. The second routine, the evaluator, can be called as many times as necessary to evaluate $C(x)$ at a point or sometimes at a sequence of points. Often the evaluator will also return values of the first and/or second derivative at the input point(s).

Summarizing

There is no unique Hermit cubic interpolant. Instead, there is an n parameter family of such functions, i.e., piecewise cubics that interpolate n given data values and also have one continuous derivative. The subroutine PCHEZ, described below, can select a member of this family that is "visually pleasing." A discussion and motivation for this idea is presented in Section 12.

4.9   PCHIP, PIECEWISE CUBIC HERMITE INTERPOLATION PACKAGE

PCHIP is a set of subroutines designed by Fritsch and Carlson (1980) to be a flexible tool for manipulating Hermite cubic and spline interpolants. In this section we present two easy-to-use drivers for PCHIP

> PCHEZ: generates a visually pleasing Hermite cubic interpolant or a cubic spline interpolant.

> PCHEV: evaluates the interpolant and derivative of the function generated by PCHEZ.

These routines are designed to be easy-to-use, so they are not exceptionally flexible. But the PCHIP package contains many other routines that can be customized to particular applications. PCHIP also contains routines for the integration of a piecewise cubic. In the example below you will see a call to PCHQA but we defer a discussion of it until Section 9 of Chapter 5. For additional details consult the reference above.

To illustrate we consider the problem of generating an interpolant to Runge's function, similar to the one we displayed in Example 4.7. The abscissa and ordinate values are in arrays X and F, and the intermediate output array D contains the derivatives of the Hermite cubic interpolant or the spline. Once the derivatives have been found, one call to PCHEV will allow evaluation of the interpolant (and its derivative) at an array XE of points, with values returned in the array FE.

```
      REAL    X(21), F(21), D(21), WK(42), FE(101), XE(101), FD(101)
      LOGICAL SPLINE
C
C Arithmetic statement functions for Runge's function and derivative.
C
      R(U)  = 1.0/(1.0+25.0*U*U)
      RP(U) = -50.0*U*R(U)**2
C
C Compute Runge's function at 21 points in [-1,1].
C
      DO 1 I=1,21
         X(I) = -1.0 + (I-1)/10.0
         F(I) = R(X(I))
    1 CONTINUE
      N = 21
      NWK = 42
      SPLINE = .FALSE.
C
C Compute cubic Hermite interpolant because SPLINE is .FALSE.
C
      CALL PCHEZ (N,X,F,D,SPLINE,WK,NWK,IERR)
      IF (IERR .LT. 0) THEN
         WRITE (*,*) 'AN ERROR CALLING PCHEZ, IERR= ',IERR
         STOP
      ENDIF
C
      NE = 101
C
C Evaluate interpolant and derivative at 101 points from -1 to 0.
C
      DO 2 I=1,NE
         XE(I) = -1.0 + (I-1.0)/(NE-1.0)
    2 CONTINUE
      CALL PCHEV (N,X,F,D,NE,XE,FE,FD,IERR)
      IF (IERR .NE. 0) THEN
```

```
          WRITE (*,*) 'AN ERROR CALLING PCHEV, IERR= ',IERR
          STOP
      ENDIF
C
      DO 3 I=1,NE
          ERROR = FE(I) - R(XE(I))
          ERRORD = FD(I) - RP(XE(I))
          WRITE (*,*) XE(I),FE(I),ERROR,ERRORD
    3 CONTINUE
C
C Compute integral over the interval [0,1]
C
      A = 0.0
      B = 1.0
      Q = PCHQA (N,X,F,D,A,B,IERR)
      WRITE (*,*) 'INTEGRAL FROM 0 TO 1 AND IERR ARE ',Q,IERR
C
      STOP
      END
```

*4.10 CUBIC HERMITE INTERPOLATION—DETAILS

Let $h_i = x_{i+1} - x_i$. On each of the subintervals $[x_i, x_{i+1}]$, $i = 1, \ldots, n-1$ we define four cubics

$$c_0^i(x) = \frac{2}{h_i^3}(x - x_{i+1})^2(x - x_i + \frac{h_i}{2}), \qquad \hat{c}_0^i(x) = \frac{1}{h_i^2}(x - x_i)(x - x_{i+1})^2,$$

$$c_1^i(x) = -\frac{2}{h_i^3}(x - x_i)^2(x - x_{i+1} - \frac{h_i}{2}), \qquad \hat{c}_1^i(x) = \frac{1}{h_i^2}(x - x_i)^2(x - x_{i+1}).$$

Now define $c_1(x)$ and $\hat{c}_1(x)$ as

$$c_1(x) = \begin{cases} c_0^1 & \text{on } [x_1, x_2]; \\ 0 & \text{on } [x_2, x_n]; \end{cases} \qquad \text{and} \qquad \hat{c}_1(x) = \begin{cases} \hat{c}_0^1 & \text{on } [x_1, x_2]; \\ 0 & \text{on } [x_2, x_n]. \end{cases}$$

For $i = 2, 3, \ldots, n-1$

$$c_i(x) = \begin{cases} 0 & \text{on } [x_1, x_{i-1}]; \\ c_1^{i-1} & \text{on } [x_{i-1}, x_i]; \\ c_0^i & \text{on } [x_i, x_{i+1}]; \\ 0 & \text{on } [x_{i+1}, x_n]; \end{cases} \qquad \text{and} \qquad \hat{c}_i(x) = \begin{cases} 0 & \text{on } [x_1, x_{i-1}]; \\ \hat{c}_1^{i-1} & \text{on } [x_{i-1}, x_i]; \\ \hat{c}_0^i & \text{on } [x_i, x_{i+1}]; \\ 0 & \text{on } [x_{i+1}, x_n]; \end{cases}$$

and for $i = n$

$$c_n(x) = \begin{cases} 0 & \text{on } [x_1, x_{n-1}]; \\ c_1^{n-1} & \text{on } [x_{n-1}, x_n]; \end{cases} \qquad \text{and} \qquad \hat{c}_n(x) = \begin{cases} 0 & \text{on } [x_1, x_{n-1}]; \\ \hat{c}_1^{n-1} & \text{on } [x_{n-1}, x_n]. \end{cases}$$

Finally, define

$$C(x) = \sum_{i=1}^{n} y_i c_i(x) + d_i \hat{c}_i(x).$$

Figure 4.10 shows a typical c_i and $\hat{c}_i$.

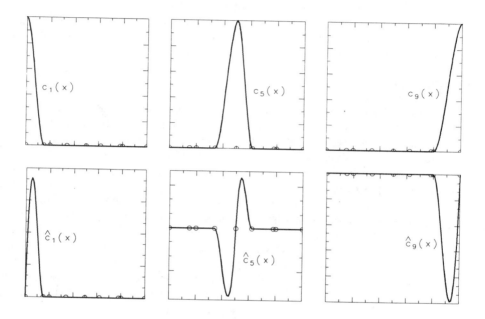

Figure 4.10   Typical Hermite Cubic Basis Functions

Let us try to illustrate the properties of these functions by looking at $\hat{c}_7(x)$. By the definitions above, $\hat{c}_7(x)$ is identically zero when $x \leq x_6$ or $x \geq x_8$. For $x_6 \leq x \leq x_7$, $\hat{c}_7(x) = \hat{c}_1^6(x) = (x - x_6)^2(x - x_7)/h_6^2$. For $x_7 \leq x \leq x_8$, $\hat{c}_7(x) = \hat{c}_0^7(x) = (x - x_7)(x - x_8)^2/h_7^2$. From these expressions we see that $\hat{c}_7(x)$ is well defined for all x and is a piecewise polynomial. Further it is zero at each of the knots. At x_6 it has a double zero from the left, and at x_8 a double zero from the right. Hence its derivative is zero at these points too. At x_7 we can compute the derivative by one of two formulas depending on whether we approach x_7 from the left or right. From the left we have

$$\hat{c}_7'(x_7^-) = \lim_{x \to x_7^-} \hat{c}_7'(x) = \lim_{x \to x_7^-} [\hat{c}_1^6(x)]' = \lim_{x \to x_7^-} \frac{1}{h_6^2}[(x - x_6)^2 + 2(x - x_6)(x - x_7)] = 1.$$

From the right

$$\hat{c}_7'(x_7^+) = \lim_{x \to x_7^+} \hat{c}_7'(x) = \lim_{x \to x_7^+} [\hat{c}_0^7(x)]' = \lim_{x \to x_7^+} \frac{1}{h_7^2}[2(x - x_7)(x - x_8) + (x - x_8)^2] = 1.$$

Since the derivative is the same from either side this shows that $\hat{c}_7'(x_7) = 1$.

With patience it is possible to check that the functions c_i and $\hat{c}_i$ have all the properties listed in Section 8. In particular, they each have one continuous derivative; $C(x)$ does too and is therefore an Hermite cubic interpolant. Actually, $C(x)$ is easy to compute if we know the values of d_i. To evaluate it at any fixed x we note that the functions c_i and $\hat{c}_i$ have at most two intervals as support. So most of the terms in the sum are identically zero; at most four contribute. Thus evaluating C involves first locating x in an appropriate interval, say between x_i and x_{i+1}, evaluating the terms c_i, c_{i+1}, $\hat{c}_i$, $\hat{c}_{i+1}$, multiplying by the requisite y's and d's, and adding.

4.11 CUBIC SPLINES

Another approach to dealing with the d_i's is to force the function $C(x)$ to satisfy additional conditions. This will effectively reduce the number of free parameters. A reasonable requirement is that C be smoother, i.e., that the individual polynomial pieces join together at the knots to give a second continuous derivative. We will see that there are more than enough free parameters to satisfy these requirements, and the process leads to cubic spline interpolation. We know from our derivation of Hermite cubic interpolation that $C(x)$ has a continuous derivative on $[x_1, x_n]$ independent of the d_i. For many applications this is enough, but for others it is not. Furthermore the flexibility afforded by the arbitrariness of the d_i's can also cause difficulties; every problem requires deciding how to set them. Let us see if it is possible to select the d_i so that $C(x)$ has two continuous derivatives. Rather than deriving the formulas for spline interpolation we must first decide if our expectation is realistic. One technique is to count the number of conditions we want $C(x)$ to satisfy and see if there are at least that many free parameters available. This does not prove anything, but the numbers will give us insight. There are $n - 1$ subintervals $x_1 < \cdots < x_n$. On each, $C(x)$ is a cubic defined by four coefficients regardless of which representation we use, or a total of $4(n - 1)$ parameters. In each subinterval $[x_i, x_{i+1}]$ there are two constraints on $C(x)$ because of the interpolation; remember that $C(x_i) = y_i$ and $C(x_{i+1}) = y_{i+1}$. Requiring C' to be continuous adds $n - 2$ constraints (one at each interior knot). The difference between the number of parameters and constraints is $4(n - 1) - 2(n - 1) - (n - 2) = n$. Thus an Hermite cubic interpolant ought to have n free parameters and this is indeed exactly equal to the number of d_i at our disposal. Requiring C'' to be continuous adds $n - 2$ constraints. Thus we conclude that asking for cubic spline interpolation ought to be feasible and in fact, should still leave us with two free parameters.

The preceding counting argument does not tell us how to find a spline interpolant, only that it is plausible that one exists. The mechanics of deriving the interpolant are more complicated and usually involve two steps.

(1) Write the general expression for $C''(x)$. This will involve the unknown d_i's.

(2) Write analytically the requirement that $C''(x)$ be continuous at the $n - 2$ interior knots. This will lead to a system of linear equations for the d_i.

If the system can be solved we will have found the spline. As we anticipated this leads to an underdetermined system, i.e., there are only $n - 2$ equations to find the n d_i's. There are several ways to uniquely specify the spline. The choices you will find most often in subroutine packages involve restricting the behavior of the spline at the endpoints x_1 and x_n. We list some possibilities below. Note that subroutine PCHEZ which is discussed in Section 9 automatically selects item (e) if the input parameter SPLINE is .TRUE.

(a) Specify the derivative of $C(x)$ at x_1 and x_n. That is, set d_1 and d_n to known values or to zero. This is called a **complete cubic spline interpolant**. In many problems, physical considerations enable the endpoint derivatives to be known *a priori*, so this is often a useful option.

(b) Estimate the derivatives from the data and use these values for d_1 and d_n. This has to be done with care. We have known scientists to ruin a spline by using crude guesses for the endpoint derivatives. An acceptable approach is to derive the formula for the cubic polynomial which passes through the first four data points, then differentiate it at x_1 and use that value for d_1. Similarly at x_n.

(c) Force the second derivative of the spline to agree at x_1 with the second derivative of the cubic from (b). Similarly at x_n.

(d) Force $C''(x_1) = 0 = C''(x_n)$. This is called a **natural spline interpolant**. Physically this means that the graph of the spline is a straight line outside the data interval $[x_1, x_n]$.

(e) Force the **not-a-knot** condition: from the construction we know that C'' will be continuous at all the knots (that's what is meant by a spline) but that C''' need not be. We can require that $C'''(x)$ be continuous at x_2. This is equivalent to insisting that the cubic on the first and second subintervals be identical. It is also equivalent to eliminating the knot at x_2 but still requiring that the single cubic between x_1 and x_3 interpolate at the endpoint as well as at x_2. Thus x_2 is an interpolation point but it is not a knot. Similarly at x_{n-1}.

If we use one of these options the system of equations for the d_i's has a matrix of the following form

$$\begin{pmatrix} \times & \times & & & & & \\ \times & \times & \times & & & & \\ & \times & \times & \cdot & & & \\ & & \cdot & \cdot & \cdot & & \\ & & & \cdot & \cdot & \cdot & \\ & & & & \cdot & \times & \times \\ & & & & & \times & \times \end{pmatrix} ;$$

it is symmetric, tridiagonal, nonsingular and can be shown to be well conditioned for any reasonable choice of knots $x_1 < x_2 < \ldots < x_n$. We can solve it using SGEFS of Chapter 3. However, because of the special properties of the matrix, it can be shown that pivoting is unnecessary when Gaussian elimination is used to solve the system. Also, because it is tridiagonal we need not store the full $n \times n$ matrix. Spline programs always take advantage of these facts and do not utilize a general linear system solver. Instead they incorporate a special one.

Symmetric tridiagonal systems which can be solved without pivoting occur frequently in several areas of numerical mathematics. Take a few moments to reflect on the elimination process for such systems. At each stage little work is required: only one element must be eliminated and only one element must be modified. For the usual full systems, elimination requires $O(n^3)$ arithmetic operations; for these special systems this is reduced to $O(n)$. As a result, even the largest systems can be solved quickly.

Some spline programs allow the user to specify other options such as

(a) Forcing $C(x)$ to be periodic, i.e., $C(x_1) = C(x_n)$, $C'(x_1) = C'(x_n)$ and $C''(x_1) = C''(x_n)$.

(b) Forcing the integral of $C(x)$ to take on a specific value, often one,

$$1 = \int_{x_1}^{x_n} C(x)\, dx.$$

Options such as these can destroy the symmetric tridiagonal structure of the matrix for the d_i's and usually are only found in more advanced packages. For most problems there is no best way to choose the extra conditions. On the other hand many data sets produce interpolants which look almost the same regardless of which choice is made.

Once the unknown spline coefficients d_i have been calculated, the spline is completely determined and may be evaluated at any x. These values of d_i are precisely those which give $C(x)$ two continuous derivatives. One point of confusion often occurs when the data y_i are values of a known smooth function $g(x)$, for example Runge's function. It might be expected that the computed d_i would be identically $g'(x_i)$. There is no reason for this to happen. The only time this expectation is fulfilled is in the unusual case when $g(x)$ is itself a spline. In fact in Example 4.7 the Hermite cubic interpolant is definitely not a spline, i.e., it does not have two continuous derivatives.

Example 4.9   Spline Interpolant to $R(x)$.

The spline interpolant to Runge's function (with equally spaced knots and default boundary conditions) looks very much like the one in Example 4.7. For this data almost any reasonable piecewise cubic interpolant can be used. But recall that polynomial interpolation is *not* satisfactory (Figure 4.5). ∎

4.12   PRACTICAL DIFFERENCES BETWEEN SPLINES AND CUBIC HERMITES

Given a specific data set to interpolate, a few general guidelines can be suggested for deciding between these two similar piecewise polynomial interpolations. Of course, if smoothness of the second derivative is required then splines are selected, but usually this is not demanded explicitly. For "nice" data sets there is little qualitative difference between the two interpolants. But it is sometimes possible to notice that the cubic Hermite is less smooth. To obtain a spline we must solve a system of equations for the coefficients. This is fast as mentioned above, but the system must be set up and the time and effort required should be considered. An Hermite cubic interpolant is also defined by coefficients, but these are obtained without solving a linear system. Usually though, the choice between spline or Hermite cubic interpolation does not depend upon the amount of work to compute the d_i. This is because in practical problems the resulting piecewise cubic is evaluated at many points and this is the most significant part of the total work. Of course, the evaluation time is exactly the same for a spline as for an Hermite cubic. If the derivatives of the interpolant or model are known then these may be easily incorporated into the Hermite cubic but not easily into the spline, unless the derivatives occur at the endpoints. If the derivatives are not known then various estimates of them can be computed and used; effort is required here too.

As we have seen, every spline is underdetermined by the requirements of interpolation; extra information must be supplied. All too often programs for spline interpolation do not allow the user to supply this and one may be led to believe that it is not necessary. Rather, the code has built in specific assumptions; a natural spline is the easiest choice, but others are possible too. For example, a program may estimate first derivatives at the two ends from the data and use these for d_1 and d_n. When not required to give any extra conditions, the user should be alert to study the program documentation to discover which defaults are being used.

The theoretical error in either spline or Hermite cubic interpolation can be measured by comparing the interpolant $C(x)$ with a known function which generated the data, much as we did for polynomial or piecewise linear interpolation. If the data y_i are values of a function $g(x)$ which has a continuous fourth derivative, then

$$|C(x) - g(x)| < \text{const} \cdot h^4 \max |g^{(4)}(x)| = O(h^4),$$

where $C(x)$ can be either the cubic spline or the Hermite cubic interpolant based on the same set of knots. We have left unspecified the value of the constant which depends on whether C is a spline or Hermite cubic interpolant, how the two extra conditions are specified in the former case and how the d_i are selected in the latter. The inequality above holds as long as reasonable choices are made for the unspecified values. (Recall that h is the maximum spacing between adjacent knots.) The essential points are that the interpolation gets better as the number of knots increase and their spacing decreases, and that the rate of improvement is proportional to the fourth power of the the knot spacing, much faster than for piecewise linear interpolation. This expression justifies our interest in these piecewise polynomials, but it does *not* imply that $C(x)$ will always look exactly the way we want between our data points. In most problems one set of data values are given, their number can not be increased to study convergence. Only one interpolant is generated and it is either satisfactory or not.

It is often justified to use the derivatives of a spline or Hermite cubic interpolant as approximations to the corresponding derivatives of the underlying model. It is possible to show that if the data come from a function which is sufficiently smooth,

$$|C'(x) - g'(x)| = O(h^3), \quad |C''(x) - g''(x)| = O(h^2), \quad \text{and} \quad |C'''(x) - g'''(x)| = O(h).$$

If $C(x)$ is a spline the first two expressions hold for any x and the last only if x is distinct from a knot. If $C(x)$ is an Hermite cubic interpolant the second and third expressions hold if x is distinct from a knot. Of course, these theoretical results, like the others, also have to be taken with a grain of salt. For example, in Section 9 we saw how to compute a spline interpolant to twenty one equally spaced points from Runge's function $R(x)$ ($h = 0.1$). In that example, the maximum error in the interpolant is about 0.003 and the maximum error in the derivative is about 0.08. We recommend that whenever possible every interpolant be plotted to confirm that it has the physical properties which are desired. The injunction to examine a plot is even more necessary if one is going to differentiate it. (For more detail, see the book by Schultz (1973).)

A tremendous market has been created for spline interpolation algorithms because most users have obtained satisfactory interpolations. These programs have, however, been somewhat oversold. There are many examples of data sets for which spline interpolation (with any set of extra conditions) gives a poor "fit." This occurs mostly in cases where the data undergo rapid changes in a small interval. Figure 4.11 illustrates such a set and shows the natural spline interpolant to it. In this particular case the data were known to come from a physical process which was increasing on the interval, a characteristic lacking in the spline. There is no contradiction between the convergence result given in the preceding paragraph and the practical result shown in this figure.

There are several ways around this problem; we describe briefly two popular methods.

(1) Consider an Hermite cubic interpolant to the data of the previous paragraph. As you now know there exist many possible interpolants, with different d_i's. It might be possible to select from among those one which is always increasing. More generally, papers by Fritsch, Carlson, Brodlie, and others have considered the problem of selecting an Hermite cubic interpolant which is "visually pleasing" in some intuitive sense. An algorithm to do this is comparable in effort to computing a spline. But often the resulting interpolant is much more useful. Figure 4.12 illustrates one of these interpolants to this data. The derivation guarantees that if the data are monotonic the interpolant will be too. If the data are not monotonic (such as the samples from Runge's function) these algorithms allow various options to deal with the region where the data change direction. In that case the computation of the d_i's may not be optimally accurate. Thus this method might be a poor choice for non-monotonic data if the interpolant is to be differentiated.

(2) The original use of the word spline came from the drafting community. There, a spline meant a flexible rod which could be fixed at specific points (called knots) along its length. The draftsman forced the spline to pass through positions on the drawing table by fixing the knots, and then traced the shape of the curve onto

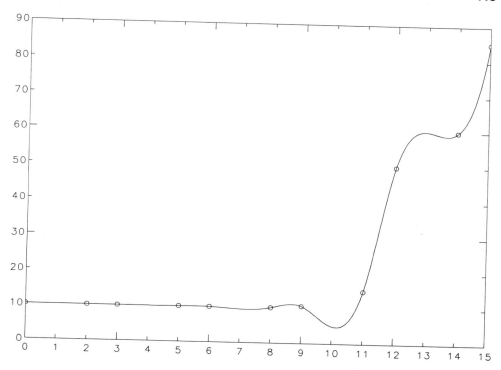

Figure 4.11   A Spline Interpolant to Monotonic Data

the drawing. Our definition of a spline is a mathematical approximation to this physical device. Between 1966 and 1974 Cline and others went back to rethink this approximation with an eye to eliminating some of the problems we have observed. They were guided by the physical notion of an elastic band which goes through rings at the interpolation points, and which can be pulled by its ends to eliminate all unnecessary wiggles. If the pull is strong enough the result is the broken line, or piecewise linear interpolant. Their approach leads to what they have called a "spline under tension." This seems to do a good job interpolating awkward data, and has the further advantage of requiring one intuitive, user-set parameter σ, the "tension." The resulting interpolant is no longer a piecewise polynomial, but involves $\exp(\sigma x)$ and $\exp(-\sigma x)$. This is not a disadvantage, for piecewise polynomial functions are rather arbitrary too. The interested reader can consult the paper by A. Cline (1974).

With either of these techniques it is important for a user to be aware that no interpolation (spline, Hermite cubic, tension spline, etc.) should be treated as a "black box." We repeat the suggestion to examine a plot of the interpolant whenever possible.

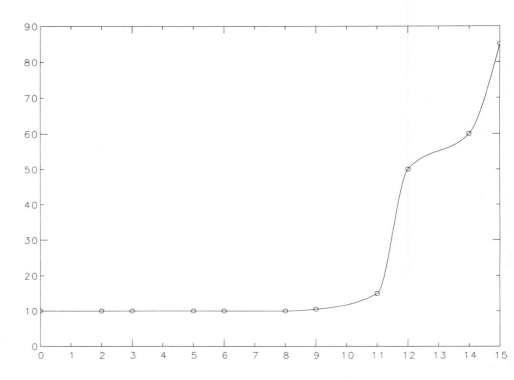

Figure 4.12   Visually Pleasing Interpolant to Monotonic Data

4.13 BÉZIER CURVES

An important problem in computer aided design (CAD) is to find a functional form for a
curve (or surface) that is given graphically. This is often done interactively; a designer
sits at a graphics terminal and tries various parameter values until the resulting function
appears "right." In 1962 Bézier and de Casteljau independently developed a method for
doing this while working on CAD systems for the French car companies, Renault and
Citröen. The Renault software was soon described in several publications by Bézier and
is the reason the underlying theory bears his name. These methods are now established
as the mathematical basis of many CAD systems and have also become a major tool in
computer graphics. The book by Bartels et al (1987) is a good introduction.

   We start by studying a simple set of functions called **Bernstein basis functions**.[6]
Assume that all the data are given in the interval $a \leq x \leq b$. The i-th Bernstein basis

---

   [6] Sergei Natanovich Bernstein [1880–1968], son of a physiology lecturer from Odessa, was educated in
Paris but returned to Russia where he taught in the Ukraine and later in Leningrad and Moscow. His master's
and Ph.D. theses solved the nineteenth and twentieth problems of a famous set by Hilbert. He mainly worked
in partial differential equations, approximation (where he coined the phrase "constructive theory of functions"),
and probability. Bernstein's name is also linked to the Weierstrass approximation theorem because this can be

function is a polynomial of degree n is defined as

$$B_i^n(x) = \binom{n}{i} \frac{(b-x)^{n-i}(x-a)^i}{(b-a)^n}, \qquad i = 0, \ldots, n.$$

Here the notation $\binom{n}{i}$ is the binomial coefficient,

$$\binom{n}{i} \equiv \frac{n!}{i!\,(n-i)!}.$$

For example, for $n = 3$ the four cubic Bernstein basis functions are

$$B_0^3(x) = \frac{(b-x)^3}{(b-a)^3}, \; B_1^3(x) = 3\frac{(b-x)^2(x-a)}{(b-a)^3}, \; B_2^3(x) = 3\frac{(b-x)(x-a)^2}{(b-a)^3}, \; B_3^3(x) = \frac{(x-a)^3}{(b-a)^3}.$$

To actually compute the value of a Bernstein basis function at x we do not use the definition. Rather, we use the iterative algorithm

$$B_0^0(x) = 1, \qquad B_{-1}^n(x) = 0,$$

$$B_i^n(x) = \frac{(b-x)B_i^{n-1}(x) + (x-a)B_{i-1}^{n-1}(x)}{(b-a)}, \qquad 0 \le i \le n$$

$$B_{n+1}^n(x) = 0$$

which is less prone to the growth of roundoff error. We show a subroutine for the evaluation of $B_i^n(x)$ when $a = 0$, $b = 1$ (this is the most important case as we will see below).

```
      SUBROUTINE BP (N,B,X)
C
C         Computes values (at X) of the N+1 Bernstein basis functions
C         of degree N on [0,1]
C
C      Input:   N (integer)        .GE. 0
C               X (real)                Abscissa for evaluation of polynomials
C      Output:  B(0:N) (real array)
C                       B(I)= N!/(I!*(N-I)!) * (1.-X)**(N-I) * X**I
C                       I=0,1,...,N
C
      INTEGER  N,C,R
      REAL B(0:N),X
C
      IF (N .EQ. 0) THEN
         B(0) = 1.0
```

proved by a clever use of the classical probability theorem, called the Weak Law of Large Numbers. The key step in the short proof is to use the binomial distribution, which turns out to be a Bernstein basis function in disguise.

```
        ELSE IF (N .GT. 0) THEN
           DO 2 C=1,N
              IF (C .EQ. 1) THEN
                  B(1) = X
              ELSE
                  B(C) = X*B(C-1)
              END IF
C
              DO 1 R=C-1,1,-1
                  B(R) = X*B(R-1) + (1.0-X)*B(R)
       1      CONTINUE
C
              IF (C .EQ. 1) THEN
                  B(0) = 1.0-X
              ELSE
                  B(0) = (1.0-X)*B(0)
              END IF
       2   CONTINUE
           END IF
           RETURN
           END
```

Since each B_i^n is a polynomial of degree n, so is the linear combination

$$P_n(x) = \sum_{i=0}^{n} p_i B_i^n(x),$$

and we can try to find the p_i so that P_n interpolates at $n+1$ points, $P_n(x_i) = y_i$. (In this section we assume that there are $n+1$ rather than n data points, and number them for convenience from 0 to n.) This can be done by solving a set of $n+1$ equations for the p_i as described in Section 1. Using these $B_i^n(x)$ instead of the monomial or Lagrange basis, we have a change of basis in the sense of Section 3.

What is the relation between the coefficients p_i and the y_i? Using the monomial representation, the coefficients of x^i have no simple relation to the data. To say this another way, it is difficult to guess the appearance of a high degree polynomial, such as $3 - x + 2x^2 + x^3 + 4x^4 - 1.3x^5$, by looking at its coefficients. In the Lagrange representation the coefficients are *exactly* the ordinates of the data points but even this does not help us guess its appearance since we know that a polynomial can fluctuate wildly between the interpolated values. For $P_n(x)$ the first and last coefficients (p_0 and p_n) are also equal to the data values; because $B_i^n(a) = 0$ if $i > 0$ and $B_0^n(a) = 1$, p_0 must be equal to y_0. Similarly, $p_n = y_n$. The fascinating thing about this representation is that the remaining coefficients p_i are *almost* as simply related to the data y_i and can be easily used to guess the shape of the curve. The relationship is so strong between p_i and the curve's shape that it has been exploited to make these curves major design tools. By that we mean that if p_i is obtained in any way, *even by guessing*, we can predict intuitively what $P_n(x)$ will look like before we evaluate it. For polynomial interpolation knowing that the interpolant passes through the data is not a good predictor of the appearance of this function. Now we can generate the interpolant by guessing at the p_i, and then looking at $P_n(x)$. If it is not satisfactory we make minor adjustments in the p_i and repeat. The advantage of this approach is that it can be applied even when interpolation is not the final goal. In many cases the CAD designer does not have any data to interpolate, but only a general idea of what shape curve is desired.

How should we guess at the p_i which will generate an acceptable shape? The basic idea is to specify the p_i graphically. (Refer to Figure 4.13.) To this end we associate each p_i with a point in the plane, (t_i, p_i), called a **control point**, and we define $t_i = x_0 + i(x_n - x_0)/n = a + i(b - a)/n$, so the t_i are equally spaced on the interval of interest. Notice $t_0 = x_0$ and $t_n = x_n$ but the remaining t_i are not usually equal to x_i. For each trial set of control points we can form the Bézier **polygon**; starting with (t_0, p_0) connect successive control points with straight line segments and finally connect (t_n, p_n) back to (t_0, p_0). It can be shown that this polygon provides a rough idea of the shape of $P_n(x)$. It is also known that $P_n(x)$ is variation diminishing. That means that if the p_i are monotonic, so is $P_n(x)$; if they are convex (concave) so is $P_n(x)$. In addition, $P_n(x)$ lies entirely within the convex hull formed from the control points. (A set S is convex if a line from any two points in S remains in S. The convex hull is the smallest convex set containing all the control points.) Often the convex hull is the same as the region enclosed by the Bézier polygon. Thus moving a control point up or down has a direct and intuitive effect on the function P_n.

Usually the number and position of control points is not known in advance but only decided by the minimum number required to get a satisfactory curve. This is not an interpolation because we cannot tell in advance exactly what points $P_n(x)$ will pass through, except the endpoints. But in practice we can make $P_n(x)$ pass through any point we like by adjusting the control points. If the designer actually wants the curve to pass through a specific point he locates its position graphically and some minor extensions to the theory will force the curve to interpolate there.

Figure 4.13 illustrates a Bézier polygon formed from some control points and the resulting $P_n(x)$.

One difficulty with this approach is that it is not possible to move the control points to the right or left. Once n is chosen, t_i is fixed at $t_i = a + i(b - a)/n$. This is counterintuitive to the designer, who sees the natural effect of up/down movements and wants the analogous ability to move the control points in any direction. To allow this we require a generalization. Let $\mathbf{p}_i$ be a point in the plane, and define the vector **Bézier curve**

$$\mathbf{P}_n(x) = \sum_{i=0}^{n} \mathbf{p}_i B_i^n(x).$$

As x varies from a to b, $\mathbf{P}_n(x)$ follows a trajectory in the plane beginning with $\mathbf{p}_0$ and ending at $\mathbf{p}_n$. In this form $\mathbf{p}_i$ can be an arbitrary point. There is no need to assume anything about the t_i, or even that $\mathbf{p}_i$ lies to the left of $\mathbf{p}_{i+1}$. Thus $\mathbf{P}_n(x)$ can describe a much more complicated shape. For example, if $\mathbf{p}_0 = \mathbf{p}_n$, then $\mathbf{P}_n(x)$ will be a closed

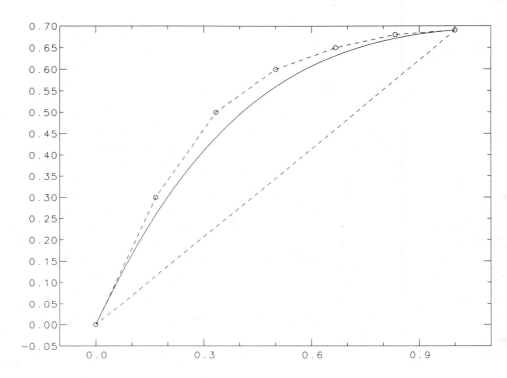

Figure 4.13   Control Points, A Bézier Polygon, and $P_n(x)$.

curve. For a vector Bézier curve the numbers a and b are arbitrary and are usually taken to be 0 and 1. In other words x is simply a parameter that varies along the length of the curve. In the literature, the adjective "vector" is usually omitted. Figure 4.14 illustrates such a Bézier curve.

Example 4.10 $\mathbf{P}_n$ versus P_n.

The three quadratic Bernstein basis functions on the interval $[0, 1]$ are $(1 - x)^2$, $2x(1 - x)$ and x^2. Consider the three values $p_0 = 1, p_1 = 0, p_2 = 1$. In the case of $P_n(x)$ these are associated with the control points, $(0, 1)$, $(1/2, 0)$, $(1, 1)$ and give

$$P_2(x) = (1 - x)^2 + 0 \cdot 2x(1 - x) + x^2 = 1 - 2x + 2x^2.$$

Note that when $x = 3/4$, $P_2(x) = 20/32$. On the other hand, for $\mathbf{P}_n(x)$ there are no constraints on the control points, and for illustration we take $\mathbf{p}_0 = (0, 1)$, $\mathbf{p}_1 = (3/4, 0)$, and $\mathbf{p}_2 = (1, 1)$. Then the Bézier curve is

$$\mathbf{P}_2(x) = \left(0 \cdot (1 - x)^2 + 3/4 \cdot 2x(1 - x) + 1 \cdot x^2, \ (1 - x)^2 + x^2\right) = \left(1 - x/2 - x^2/2, \ 1 - 2x + 2x^2\right).$$

When $x = 3/4$, $\mathbf{P}_2(x) = (27/32, 20/32)$, which does not have abscissa value 3/4. In this form x is only a parameter; $\mathbf{P}_2(x)$ traces a curve between the first and last control point.

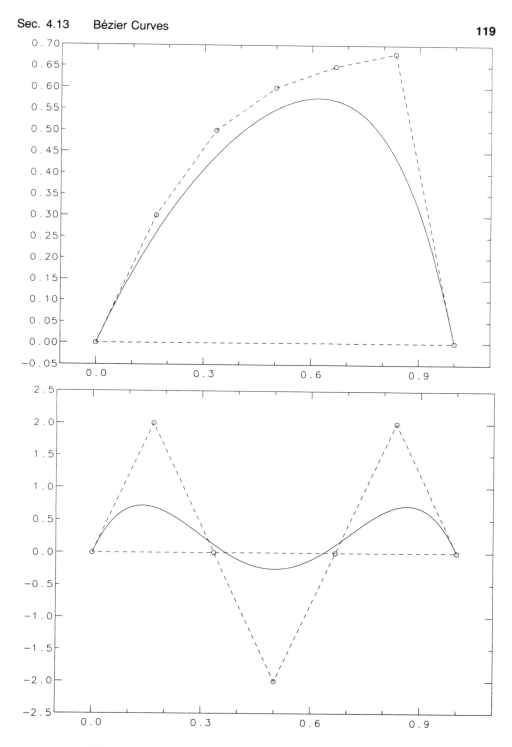

Figure 4.13 (*cont.*)   Control Points, A Bézier Polygon, and $P_n(x)$.

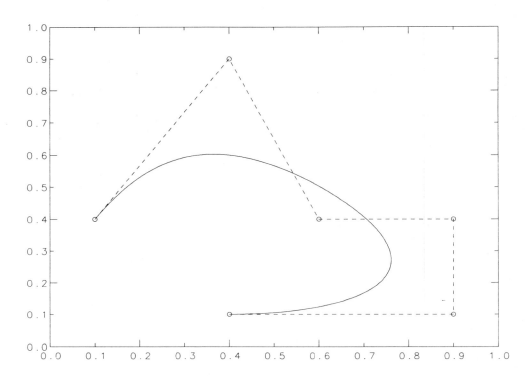

Figure 4.14 A Bézier Curve

In fact, when $\mathbf{P}_2$ has abscissa value 3/4, its ordinate value is about 0.536. The two curves, P_2 and $\mathbf{P}_2$ are plotted in Figure 4.15. On each curve we indicate the point on the curve corresponding to $x = 3/4$. Finally, we plot the convex hulls formed from each of the two sets of control points. As a check of your understanding, compute $\mathbf{P}_2(x)$ with the control points $(0, 1)$, $(1/2, 0)$, $(1, 1)$ and verify that the curve traced out by $\mathbf{P}_2(x)$ is identical to $P_2(x)$. ■

*4.14 B-SPLINES

From Section 11 we know that a spline interpolant can be written

$$C(x) = \sum y_i c_i(x) + d_i \hat{c}_i(x),$$

where the d_i are obtained by solving a system of linear equations. But note that neither the $c_i(x)$ nor the $\hat{c}_i(x)$ are themselves splines because they are not twice differentiable.

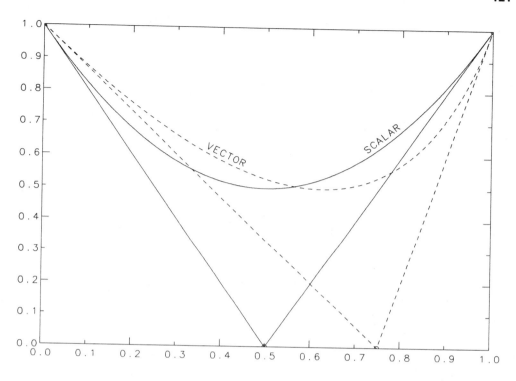

Figure 4.15 $P_n(x)$ versus $\mathbf{P}_n(x)$.

For many applications it would be useful to be able to write the interpolant as a linear combination of functions which are also splines

$$C(x) = \sum a_i N_i^3(x).$$

The most heavily used and flexible way to do this is when the $N_i^3(x)$ are **B-splines**. The superscript 3 refers to "cubic" B-splines; there are other types and this is a common notation. The graph of a B-spline resembles the graph of one of the $c_i(x)$ in Figure 4.10 in that it is zero on most of the interval, and that each B-spline looks like a shifted version of any other if the knots are uniformly spaced. Figure 4.16 shows typical B-splines. Notice that a B-spline is always non-negative and is nonzero over at most four intervals. Thus for any fixed x the above sum for $C(x)$ contains at most four nonzero terms. This leads to banded matrices in applications. For example, the spline interpolant $C(x)$ can be found in terms of B-splines, with the coefficients a_i found by solving a banded system similar to the one mentioned in Section 11.

A cubic B-spline is a spline. Thus it is a piecewise polynomial with two continuous derivatives. For some applications it is essential to master the details of the construction

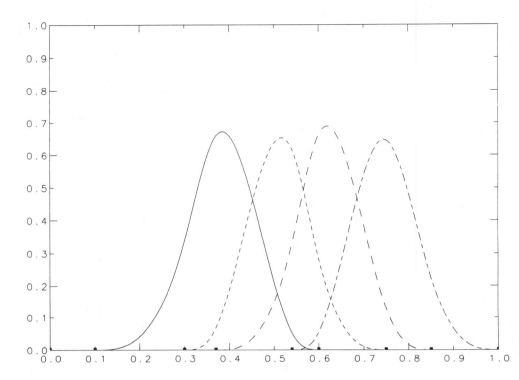

Figure 4.16   Typical Cubic B-splines With Distinct Knots

of B-splines. But in this text we shall discuss them in a more superficial way, and refer the interested reader to either of the references above.

Subroutines which manipulate B-splines usually require as input an array of knots. Setting these is one of the most confusing aspects for most users. The confusion stems from the fact that the definition of a B-spline allows several knots to coincide. While this may not seem natural for a problem in interpolation it allows the theory to deal with functions with fewer than two continuous derivatives. If $C(x)$ is a linear combination of B-splines with two equal knots, say at the point η, then C is a function with two continuous derivatives everywhere except at η where it has only one. If three knots are at η then $C(x)$ is only continuous there, and if there are four knots at η then $C(x)$ has a jump. Using this generalization allows B-spline interpolation programs to deal with the important problem of forming an interpolant composed of two smooth sections joined at a point which is less smooth, for example an automobile body panel which has a crease. Figure 4.17 shows some B-splines in cases where some of the knots are coincident.

The rules for using knots with B-splines are not difficult. To find a cubic spline interpolant through n data points by the use of B-splines you will need $n + 6$ knots. There is some flexibility in selecting the knot positions, but a common approach is to

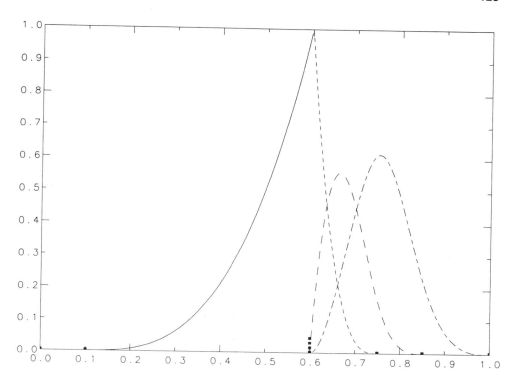

Figure 4.17   Typical Cubic B-splines with Coincident Knots

place four knots at each endpoint, x_1 and x_n, and one knot at every interior data point. If your interpolant is to have fewer than two derivatives at a point, add one or more extra knots there. There will be four fewer B-splines than knots, (thus at least $n + 2$ B-splines) and the output of the interpolation routine will be an array a_i of the same number of B-spline coefficients. The interpolant will pass through the data, have the correct number of derivatives and also satisfy two extra conditions. If there are no coincident interior knots the spline obtained in this way will have two continuous derivatives and be identical to the one derived from the Hermite cubic representation if the same extra conditions are used. The coefficients of these B-splines are found by solving a system of linear equations similar to the one in Section 11.

One of the most interesting things about B-splines is that there is a remarkable relationship between B-splines and Bézier curves, and the B-spline coefficients have a physical significance that is analogous to the coefficients of a Bézier curve. Specifically, in analogy with Section 13, Bézier-B-splines are defined as

$$\mathbf{B}(x) = \sum_{i=0}^{n+1} \mathbf{p}_i N_i^3(x),$$

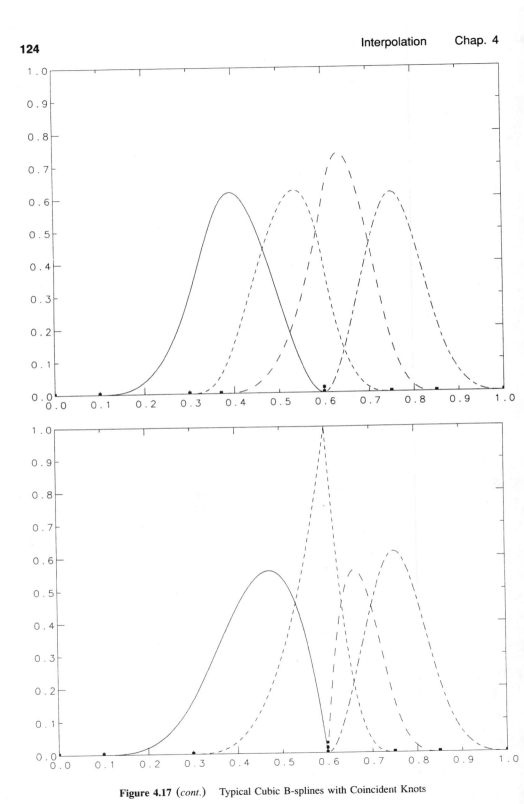

Figure 4.17 (*cont.*)   Typical Cubic B-splines with Coincident Knots

where $\mathbf{p}_i$ are arbitrary points in the plane and the B-spline knots are equally spaced in the interval $[0, 1]$. Now plot the control points $\mathbf{p}_i$, $i = 0, \ldots, n + 1$ and form the convex hull, first of $\mathbf{p}_0$, $\mathbf{p}_1$, $\mathbf{p}_2$, $\mathbf{p}_3$, then of $\mathbf{p}_1$, $\mathbf{p}_2$, $\mathbf{p}_3$, $\mathbf{p}_4$, and so on, in each case using four points, adding one new point on the right and deleting one on the left. The curve of $\mathbf{B}(x)$ can be shown to lie in the union of these polygons. Figure 4.18 shows some control points, the region described by these polygons, and the B-spline curve $\mathbf{B}(x)$. Thus altering the B-spline coefficients has an effect exactly similar to that of altering the control points in a Bézier curve. (See the book by Barnhill and Riesenfeld (1979).)

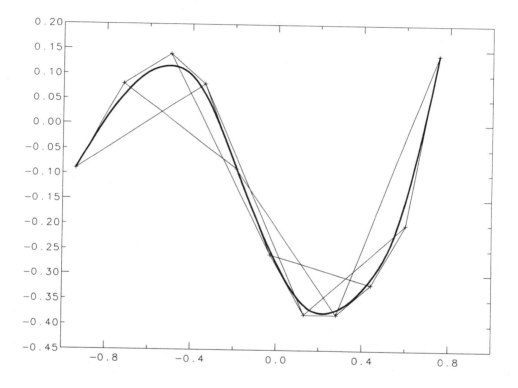

Figure 4.18   Bézier B-spline

4.15   PROBLEMS

P4–1.–Generate $n = 11$ data points by taking

$$\left.\begin{array}{l} x_i = \frac{i-1}{10} \\[2mm] y_i = \mathrm{erf}(x_i) \end{array}\right\} \quad i = 1, \ldots, 11.$$

Obtain the values of erf(x) from published tables or a subroutine on your system.

(a) Try to interpolate the data using a tenth degree polynomial. Set up the matrix as described in Section 2 and use SGEFS to estimate its condition. Even if the condition number is large and the coefficients are innacurate, it is still possible that values of the interpolating polynomial will not be badly contaminated with roundoff. Why? Find the interpolating polynomial and use Horner's method to evaluate it at points between the data points. Repeat using the Lagrange representation instead. Compare the values in both cases with those of erf(x). What is the maximum error?

(b) Use PCHEZ to interpolate using both a cubic Hermite and a cubic spline interpolant. Use PCHEV to evaluate these interpolants at the same points as in (a) and compare your results.

P4–2.–The following figures from the US Census Bureau give the population of the United States

Year	Population
1900	75,994,575
1910	91,972,266
1920	105,710,620
1930	122,775,046
1940	131,669,275
1950	150,697,361
1960	179,323,175
1970	203,235,298

(a) There is a unique seventh degree polynomial interpolating the data. We consider four representations of this polynomial

$$\sum_{j=0}^{7} a_j t^j, \qquad \sum_{j=0}^{7} b_j (t - 1900)^j$$
$$\sum_{j=0}^{7} c_j (t - 1935)^j, \qquad \sum_{j=0}^{7} d_j \left((t - 1935)/35)\right)^j.$$

In exact arithmetic it doesn't matter which is used, but some of these are computationally more satisfactory than others. For each of these polynomials set up the 8 by 8 matrix to find the coefficients and estimate the condition of the matrix using SGEFS. If the matrix is not too ill conditioned, use SGEFS to find the coefficients. How well do the computed polynomials reproduce the original data? Explain why plots of the last three polynomials may look similar even if the coefficients have little or no significance (IND=-10).

(b) Interpolate the data by the best-conditioned polynomial interpolant found in (a) and also by a cubic Hermite using PCHEZ. Plot the polynomial and the piecewise polynomial at one year intervals between 1900 and 1980. They should agree fairly well up to 1970 but their character between 1970 and 1980 is quite different. The 1980 census figure is 226,547,082. Which interpolant gives the most accurate prediction?

P4–3.–Here are the data that were used in Figure 4.11.

x	y
0	10.0
2	10.0
3	10.0
5	10.0
6	10.0
8	10.0
9	10.5
11	15.0
12	50.0
14	60.0
15	85.0

Interpolate them with a cubic spline using `PCHEZ`. Is the interpolant monotonic? Repeat with `SPLINE = .FALSE.`

P4–4.–Parametric interpolation: In computerized typography we are faced with the problem of finding an interpolant to points which lie on a path in the plane, for example a printed capital S. Such a shape cannot be represented as a function of x because it is not single valued. One approach is to number the points $\mathbf{P}_1, \ldots, \mathbf{P}_n$ as we traverse the curve. Let d_i be the (straight-line) distance between $\mathbf{P}_i$ and $\mathbf{P}_{i+1}$, $i = 1, \ldots, n-1$ and let $t_i = \sum_{j=1}^{i-1} d_j$, $i = 1, \ldots, n$. That is, $t_1 = 0, t_2 = d_1, t_3 = d_1 + d_2$, etc. If $\mathbf{P}_i = (x_i, y_i)$, consider the two sets of data $\{(t_i, x_i)\}$ and $\{(t_i, y_i)\}$, $i = 1, \ldots, n$. We can interpolate each of these independently to generate the functions $f(t)$ and $g(t)$ respectively. Then $\mathbf{P}(t) \equiv (f(t), g(t))$, $0 \leq t \leq t_n$ is a point in the plane and as t increases from 0 to t_n, $\mathbf{P}(t)$ interpolates the data and, with luck, reproduces the shape of the letter too. We can think of t as a parameter that varies as we move along the curve.

Using drafting paper draw an S and measure eight points $\mathbf{P}_i$ from it. Interpolate the data by `PCHEZ`

(a) with `SPLINE = .FALSE.`

(b) `SPLINE = .TRUE.`

Plot the results and see how faithfully they follow the letter. If the shape were a closed curve, for example the letter O, better results would be obtained by requiring the interpolant to be periodic.

P4–5.–Suppose you were given n data points, $(x_i, f(x_i))$, $i = 1, \ldots, n$, and desired to find points where $f'(x) = 0$.

(a) Why would it not be a good idea to interpolate by $p(x)$, a polynomial of degree $n-1$, and then solve $p'(x) = 0$?

(b) Describe an efficient algorithm for solving this problem using cubic interpolatory splines.

(c) Write a program using your method and `PCHEZ`. Try your program on the data in Problem P4–3.

P4–6.–This problem shows why piecewise quadratics are not much used. A piecewise quadratic function $Q(x)$ is a function defined for all x which is a quadratic between adjacent knots. If the knots are $x_1 < x_2 < \cdots < x_n$ then between x_i and x_{i+1}, $Q(x)$ is defined by three parameters and can be written as

$$Q(x) = a_i + b_i(x - x_i) + c_i(x - x_i)(x - x_{i+1}) \qquad x_i \leq x \leq x_{i+1}, \quad i = 1, \ldots, n-1.$$

Requiring $Q(x)$ to interpolate at each knot, e.g., $Q(x_i) = y_i$ determines a_i and b_i.

(a) Show that $Q(x)$ will interpolate if it can be written in each subinterval as

$$Q(x) = y_i + \frac{y_{i+1} - y_i}{x_{i+1} - x_i}(x - x_i) + c_i(x - x_i)(x - x_{i+1}),$$

and conclude that $Q(x)$ is continuous for all x.

(b) Show that the interpolating Q will also have one continuous derivative if

$$c_i = \frac{1}{x_i - x_{i+1}} \left[c_{i-1}(x_i - x_{i-1}) + \frac{y_i - y_{i-1}}{x_i - x_{i-1}} - \frac{y_{i+1} - y_i}{x_{i+1} - x_i} \right]$$

$$i = 2, \ldots, n-1.$$

If we specify c_1 this uniquely determines all the other c_i and hence $Q(x)$.

(c) Consider using such a technique to interpolate two data sets differing only in the first data value, y_1 and $\tilde{y}_1$. Call the interpolants $Q(x)$ and $\widetilde{Q}(x)$. If we use the same value for c_1 in both and if the x_i are equally spaced, show that

$$\tilde{c}_1 = c_1, \qquad \tilde{c}_i = c_i + (-1)^i \left(\frac{y_1 - \tilde{y}_1}{h^2} \right), \qquad i = 2, \ldots, n-1,$$

and thus that

$$\widetilde{Q}(x) = Q(x) + (-1)^i \left(\frac{y_1 - \tilde{y}_1}{h^2} \right) (x - x_i)(x - x_{i+1}), \qquad x_i \leq x \leq x_{i+1},$$

$$i = 1, 2, \ldots, n-1.$$

(d) Conclude that a small change in the first data value propagates its effect, undiminished, through the entire domain of the interpolant. Although this can be avoided, see de Boor (1978), p. 75, piecewise quadratics are used far less frequently than piecewise cubics.

P4–7.–Inverse Interpolation: In the usual setting for interpolation you are given values x_i, y_i and try to find a function $y = f(x)$. However, if the y_i were monotonic you could just as readily find a function $x = g(y)$. Once such a function were determined it would then be easy to estimate values of x corresponding to "special" values of y, for example the x corresponding to $y = 0$. This technique is called **inverse interpolation**.

For use in statistical computations it is desired to compute a table of the percentiles of the error function $\text{erf}(x)$, that is, numbers x_i so that

$$\frac{i}{10} = \text{erf}(x_i), \qquad 0 < i < 10.$$

Obtain sixteen values of $\text{erf}(x)$ for $0 \leq x \leq 1.5$ from published tables or from a subroutine. By looking at the numbers estimate the nine percentiles. Use PCHEZ to obtain a cubic spline interpolant in the form $x = g(y)$. Graph this function. Then use PCHEV to evaluate $g(y)$ at $y = 0, .1, .2, \ldots, .9$. Compare these results to your earlier estimates.

P4–8.–Bézier curves.

(a) Using the iterative algorithm given in Section 13, compute and plot the four cubic Bernstein basis functions on $[0, 1]$.

(b) Write a program to plot a Bézier curve $\mathbf{P}_n(x)$ from given points in the plane. If you have access to a graphics terminal try to write your program so you can enter the control points graphically. Use your program to approximate the shape of capital S of P4–4.

P4–9.–The techniques in this and the next problem would be of special interest to anyone who wanted to draw circles quickly, for example in a heavily used graphics package, or in a hardware implementation. The task is to generate two arrays $X(I)$, $Y(I)$, $I=1,\ldots,N$, of points on the first quadrant of the unit circle. A program segment which does this is:

```
        N = 90
        R = PI/180.0
        DO 1 I=0,90
            RAD = R*I
            X(I) = COS(RAD)
            Y(I) = SIN(RAD)
1       CONTINUE
```

but this requires the evaluation of both a sine and a cosine for each point on the circle. As an alternative consider the four Bézier control points, $\mathbf{p}_0 = (1, 0)$, $\mathbf{p}_1 = (1, 0.552)$, $\mathbf{p}_2 = (0.552, 1)$, $\mathbf{p}_3 = (0, 1)$.

(a) Show that the associated vector Bézier curve is

$$\mathbf{P}(t) = (1 - 1.344t^2 + 0.344t^3, 1.656t - 0.312t^2 - 0.344t^3).$$

(b) As t varies from 0 to 1, $\mathbf{P}(t)$ traces a curve in the plane which approximates a quarter circle. Using Horner's rule evaluate $\mathbf{P}(t)$ at 100 points in $[0, 1]$ and compare the values with those on an exact circle. What is the maximum error? Since the parametric value t is not proportional to angle some thought is required to decide how to measure error. Is this method faster than the program given above? Why?

P4–10.–The purpose of this problem is to show that if a polynomial is to be calculated at an sequence of equally spaced points there are methods which are much faster than Horner's rule. From a table of N equally spaced values (x_i, f_i) we can compute $N - 1$ first differences $\Delta f_i = f_{i+1} - f_i$. From these we can compute $N - 2$ second differences $\Delta^2 f_i = \Delta f_{i+1} - \Delta f_i$, etc.

These are often arranged in a table such as this one

TABLE 4.1   TYPICAL DIFFERENCE TABLE

x	$f(x)$	$\Delta f \cdot 10^3$	$\Delta^2 f \cdot 10^6$	$\Delta^3 f \cdot 10^7$	$\Delta^4 f \cdot 10^{11}$
20	0.229314955248				
		.701747247			
22	0.230016702495		.602297		
		.702349544		−.1944	
24	0.230719052039		.600353		.4
		.702949897		−.1940	
26	0.231422001936		.598413		.3
		.703548310		−.1937	
28	0.232125550246		.596476		
		.704144786			
30	0.232829695032				

(a) Write a program that will verify the last four columns of the table above.

(b) Difference tables can be used to look for errors in tables. In that case the data available correspond to the columns labelled x and $f(x)$ above. For this problem $f(x)$ is a polynomial and we are instead given the upper diagonal of the table, and we are asked to fill in the remaining entries. We can get the elements on the next diagonal with only additions,

$$f_1 = f_0 + \Delta f_0, \quad \Delta f_1 = \Delta f_0 + \Delta^2 f_0, \ldots .$$

If $f(t) = a + bt + ct^2 + dt^3$ and $f_i = f(x_0 + i \cdot \delta)$ show that if $x_0 = 0$

$$\Delta f_0 = b\delta + c\delta^2 + d\delta^3$$
$$\Delta^2 f_0 = 2c\delta^2 + 6d\delta^3$$
$$\Delta^3 f_0 = 6d\delta^3$$
$$\Delta^i f_0 = 0, \qquad i > 3.$$

(In general if f_i are equally spaced values of a polynomial of degree p then the differences above the p-th are zero.) Use these formulas to compute the top diagonal for the cubics in the preceding problem. Take $\delta = 0.01$.

(c) Once this diagonal is computed, show that $\mathbf{P}(t)$ from problem P4–9 can be evaluated at equally spaced points with only additions, in a loop of the form

$$f \leftarrow f + \Delta f,$$
$$\Delta f \leftarrow \Delta f + \Delta^2 f,$$
$$\Delta^2 f \leftarrow \Delta^2 f + \Delta^3 f.$$

Compare the work to Horner's rule.

P4–11.–The following scheme combines interpolation and the flavor of Bézier approximation. Let us consider a set of $n+5$ points in the plane, and denote them $\mathbf{p}_i$, $i = -2, \ldots, n+2$, $\mathbf{p}_i = (x_i, y_i)$. Define new points $\mathbf{p}_{i+1/2}$ by

$$\mathbf{p}_{i+1/2} = (1/2 + w)(\mathbf{p}_i + \mathbf{p}_{i+1}) - w(\mathbf{p}_{i-1} + \mathbf{p}_{i+2}), \qquad i = -1, 0, \ldots, n, \qquad w > 0.$$

Add these points to the original set, throw out the first and last points, $\mathbf{p}_{-2}$ and $\mathbf{p}_{n+2}$, renumber the points from -2 to $2n+2$ and repeat. It has been shown that if $0 < w < 0.25$, the set of points approaches a smooth curve. Since the original data points are always retained (except for the first and last points), the curve interpolates the data.

(a) Implement this algorithm on the twenty one points from Runge's function. Generate enough extra points so you can produce a smooth plot of the resulting curve.

(b) Repeat (a) with the data from problem P4–3. The number w is something like a "tension;" $w = 0$ generates the straight line segments between the points, $w > 0$ "loosens" the curve although it still interpolates the original points. A minor difficulty with this algorithm is that it does not generate new points near the ends of the data. You can get around this for Runge's function by adding points to the left and right of your original data. Another possibility is to replicate the first and last data point so that they each appear twice among the $\mathbf{p}_i$.

(c) Repeat (a) with your data from problem P4–4. (The paper by Dyn (1987) gives much more mathematical detail on this algorithm.)

P4–12.–This problem illustrates two dimensional interpolation on a grid. Suppose we are given a set of data values $g_{ij} \approx g(x_i, y_j)$, $1 \leq i, j \leq N$, and wish to find a bivariate function $f(x, y)$ such that $f(x_i, y_j) = g_{ij}$. A simple technique is given here. Find an interpolant for each row, that is for each fixed i find a function $f_i(y)$ satisfying $f_i(y_j) = g_{ij}, j = 1, \ldots, N$. To evaluate the bivariate interpolant at a point (r, s) first evaluate each of the one dimensional interpolants at s, e.g., compute $f_i(s), i = 1, \ldots, N$. Then find the interpolant through $(x_i, f_i(s)), i = 1, \ldots, N$, and evaluate that function at the point r. Use this technique on $g_{ij} = \exp(x_i^2 \cdot y_j)$, $x_i = 1/10$, $y_j = j/10$, $i, j = 0, \ldots, 10$. Evaluate the interpolant at the midpoints of all the grid squares and print out the interpolation error.

P4–13.–The Chebyshev Polynomials. Let us consider polynomial interpolation on a fixed interval, say $-1 \leq x \leq 1$, to a given function, for example Runge's function $R(x)$, with a fixed number of points n. The interpolation error (see Section 4) is composed of a part that involves $R^{(n)}$, the n-th derivative of the underlying function, and a part that involves the product $(x - x_1)(x - x_2) \cdots (x - x_n)$. The latter is independent of R but depends upon the points. Suppose that *a priori* we want to bound the error that will occur when we evaluate the interpolant at an arbitrary point on $[-1, 1]$. This can be no larger than

$$\max_{-1 \leq \xi \leq 1} |R^{(n)}(\xi)| \cdot \max_{-1 \leq x \leq 1} |(x - x_1) \cdots (x - x_n)|/n!.$$

The only term that we can control is the point spacing, so let us consider moving the points x_i to reduce this product.

(a) Let $n = 21$ and take the points x_i to be equally spaced on $[-1, 1]$. Plot the polynomial $t_n(x) = (x - x_1)(x - x_2) \cdots (x - x_n)$ and find its maximum absolute value. Experiment with different values for the x_i and see how much you can reduce the maximum value of $|t_n(x)|$.

(b) Consider the following two sequences of polynomials $T_0(x) = 1$, $T_1(x) = x$, $T_{n+1}(x) = 2xT_n(x) - T_{n-1}(x)$, $n > 0$, and $\widetilde{T}_n(x) = T_n(x)/(2^{n-1})$. The T_n are called the **Chebyshev** polynomials. Compute and plot $\widetilde{T}_0, \ldots, \widetilde{T}_6(x)$.

(c) By using trigonometric identities show that $\cos(n+1)\theta = 2\cos\theta\cos n\theta - \cos(n-1)\theta$. By formally setting $x = \cos\theta$ and $\cos n\theta = T_n(x)$, verify that this is the same as the recursion in (b). This suggests that another definition for $T_n(x)$ is $T_n(x) = \cos n(\arccos x)$. Using *this* expression compute and plot $\widetilde{T}_0, \ldots, \widetilde{T}_6(x)$ and verify that you get the same values as in (b).

(d) Show that $T_n(x)$ has simple zeros at the n points $x_k = \cos[(2k-1)\pi/(2n)]$, $k = 1, \ldots, n$. Using these points plot $t_n(x)$. Why must its maximum absolute value be 2^{1-n}? It is known that no other selection of points yields a t_n with smaller maximum absolute value.

(e) Use the zeros of $T_{21}(x)$ as interpolating points for Runge's function and plot the resulting interpolant. Is there an improvement over using equally spaced points?

P4–14.–B-splines: Let $t_0, \ldots, t_{14}$ be fifteen knots on $[0, 1]$, four at each end and the remaining seven at $i/8$, $i = 1, \ldots, 7$. Define $N_i^0(x)$ as

$$N_i^0(x) = \begin{cases} 1, & t_i \leq x < t_{i+1}; \\ 0, & \text{elsewhere} \end{cases} \qquad i = 0, 1, \ldots, 13.$$

(If $t_i = t_{i+1}$ we take $N_i^0(t_i) = 1$.) Define $N_i^1(x)$ as

$$N_i^1(x) = (x - t_i)\frac{N_i^0(x)}{t_{i+1} - t_i} + (t_{i+2} - x)\frac{N_{i+1}^0(x)}{t_{i+2} - t_{i+1}}, \qquad i = 0, 1, \ldots, 12.$$

Define $N_i^2(x)$ as

$$N_i^2(x) = (x - t_i)\frac{N_i^1(x)}{t_{i+2} - t_i} + (t_{i+3} - x)\frac{N_{i+1}^1(x)}{t_{i+3} - t_{i+1}}, \qquad i = 0, 1, \ldots, 11.$$

Finally define $N_i^3(x)$ as

$$N_i^3(x) = (x - t_i)\frac{N_i^2(x)}{t_{i+3} - t_i} + (t_{i+4} - x)\frac{N_{i+1}^2(x)}{t_{i+4} - t_{i+1}}, \qquad i = 0, 1, \ldots, 10.$$

(a) Use the definitions above to compute and plot the eleven cubic B-splines. Remember to use the convention that whenever $0/0$ occurs in the definitions above, the fraction is set to zero.

(b) Pick control points $\mathbf{p}_i$ in the plane so that

$$\sum_{i=0}^{10} \mathbf{p}_i N_i^3(x)$$

approximates e^x. Plot the control polygon.

4.16 PROLOGUES: PCHEZ, **AND** PCHEV

```
        SUBROUTINE PCHEZ(N,X,F,D,SPLINE,WK,LWK,IERR)
C***BEGIN PROLOGUE  PCHEZ
C***DATE WRITTEN    870821    (YYMMDD)
C***REVISION DATE   870908    (YYMMDD)
C***CATEGORY NO.   E1B
C***KEYWORDS  CUBIC HERMITE MONOTONE INTERPOLATION, SPLINE
C             INTERPOLATION, EASY TO USE PIECEWISE CUBIC INTERPOLATION
C***AUTHOR   KAHANER, D.K., (NBS)
C            SCIENTIFIC COMPUTING DIVISION
C            NATIONAL BUREAU OF STANDARDS
C            GAITHERSBURG, MARYLAND 20899
C            (301) 975-3808
C***PURPOSE  Easy to use spline or cubic Hermite interpolation.
C***DESCRIPTION
C
C
C         PCHEZ:  Piecewise Cubic Interpolation, Easy to Use.
C
C      From the book "Numerical Methods and Software"
C          by  D. Kahaner, C. Moler, S. Nash
C                 Prentice Hall 1988
C
C      Sets derivatives for spline (two continuous derivatives) or
C      Hermite cubic (one continuous derivative) interpolation.
C      Spline interpolation is smoother, but may not "look" right if the
C      data contains both "steep" and "flat" sections.  Hermite cubics
C      can produce a "visually pleasing" and monotone interpolant to
C      monotone data. This is an easy to use driver for the routines
C      by F. N. Fritsch in reference (4) below. Various boundary
C      conditions are set to default values by PCHEZ. Many other choices
C      are available in the subroutines PCHIC, PCHIM and PCHSP.
C
C      Use PCHEV to evaluate the resulting function and its derivative.
C
C - - - - - - - - - - - - - - - - - - - - - - - - - - - - - - - - - -
C
C  Calling sequence:   CALL  PCHEZ (N, X, F, D, SPLINE, WK, LWK, IERR)
C
C      INTEGER  N, IERR,  LWK
C      REAL  X(N), F(N), D(N), WK(LWK)
C      LOGICAL SPLINE
C
C   Parameters:
C
C     N - (input) number of data points.  (Error return if N.LT.2 .)
```

```
C                    If N=2, simply does linear interpolation.
C
C         X - (input) real array of independent variable values.  The
C             elements of X must be strictly increasing:
C                    X(I-1) .LT. X(I),   I = 2(1)N.
C             (Error return if not.)
C
C         F - (input) real array of dependent variable values to be inter-
C             polated.  F(I) is value corresponding to X(I).
C
C         D - (output) real array of derivative values at the data points.
C
C         SPLINE - (input) logical variable to specify if the interpolant
C             is to be a spline with two continuous derivaties
C             (set SPLINE=.TRUE.) or a Hermite cubic interpolant with one
C             continuous derivative (set SPLINE=.FALSE.).
C           Note: If SPLINE=.TRUE. the interpolating spline satisfies the
C             default "not-a-knot" boundary condition, with a continuous
C             third derivative at X(2) and X(N-1). See reference (3).
C                  If SPLINE=.FALSE. the interpolating Hermite cubic will be
C             monotone if the input data is monotone. Boundary conditions are
C             computed from the derivative of a local quadratic unless this
C             alters monotonicity.
C
C         WK - (scratch) real work array, which must be declared by the calling
C             program to be at least 2*N if SPLINE is .TRUE. and not used
C             otherwise.
C
C         LWK - (input) length of work array WK. (Error return if
C             LWK.LT.2*N and SPLINE is .TRUE., not checked otherwise.)
C
C         IERR - (output) error flag.
C             Normal return:
C                 IERR = 0  (no errors).
C             Warning error:
C                 IERR.GT.0  (can only occur when SPLINE=.FALSE.) means that
C                     IERR switches in the direction of monotonicity were detected
C                     When SPLINE=.FALSE.,  PCHEZ guarantees that if the input
C                     data is monotone, the interpolant will be too. This
C                     warning is to alert you to the fact that the input data
C                     were not monotone.
C             "Recoverable" errors:
C                 IERR = -1  if N.LT.2 .
C                 IERR = -3  if the X-array is not strictly increasing.
C                 IERR = -7  if LWK is less than 2*N and SPLINE is .TRUE.
C                 (The D-array has not been changed in any of these cases.)
C                 NOTE:  The above errors are checked in the order listed,
C                     and following arguments have **NOT** been validated.
```

```
C
C - - - - - - - - - - - - - - - - - - - - - - - - - - - - - - - - - - - -
C***REFERENCES  1. F.N.FRITSCH AND R.E.CARLSON, 'MONOTONE PIECEWISE
C                  CUBIC INTERPOLATION,' SIAM J.NUMER.ANAL. 17, 2 (APRIL
C                  1980), 238-246.
C               2. F.N.FRITSCH AND J.BUTLAND, 'A METHOD FOR CONSTRUCTING
C                  LOCAL MONOTONE PIECEWISE CUBIC INTERPOLANTS,' LLNL
C                  PREPRINT UCRL-87559 (APRIL 1982).
C               3. CARL DE BOOR, A PRACTICAL GUIDE TO SPLINES, SPRINGER-
C                  VERLAG (NEW YORK, 1978).  (ESP. CHAPTER IV, PP.49-62.)
C               4. F.N.FRITSCH, 'PIECEWISE CUBIC HERMITE INTERPOLATION
C                  PACKAGE, FINAL SPECIFICATIONS', LAWRENCE LIVERMORE
C                  NATIONAL LABORATORY, COMPUTER DOCUMENTATION UCID-30194,
C                  AUGUST 1982.
C***ROUTINES CALLED  PCHIM,PCHSP
C***END PROLOGUE  PCHEZ
C
      SUBROUTINE PCHEV(N,X,F,D,NVAL,XVAL,FVAL,DVAL,IERR)
C***BEGIN PROLOGUE  PCHEV
C***DATE WRITTEN   870828   (YYMMDD)
C***REVISION DATE  870828   (YYMMDD)
C***CATEGORY NO.   E3,H1
C***KEYWORDS  CUBIC HERMITE OR SPLINE DIFFERENTIATION,CUBIC HERMITE
C             EVALUATION,EASY TO USE SPLINE OR CUBIC HERMITE EVALUATOR
C***AUTHOR  KAHANER, D.K., (NBS)
C             SCIENTIFIC COMPUTING DIVISION
C             NATIONAL BUREAU OF STANDARDS
C             ROOM A161, TECHNOLOGY BUILDING
C             GAITHERSBURG, MARYLAND 20899
C             (301) 975-3808
C***PURPOSE  Evaluates the function and first derivative of a piecewise
C             cubic Hermite or spline function at an array of points XVAL,
C             easy to use.
C***DESCRIPTION
C
C
C         PCHEV:  Piecewise Cubic Hermite or Spline Derivative Evaluator,
C                 Easy to Use.
C
C     From the book "Numerical Methods and Software"
C          by  D. Kahaner, C. Moler, S. Nash
C               Prentice Hall 1988
C
C
C     Evaluates the function and first derivative of the cubic Hermite
C     or spline function defined by  N, X, F, D, at the array of points XVAL.
C
C     This is an easy to use driver for the routines by F.N. Fritsch
C     described in  reference (2) below. Those also have other capabilities.
C
```

```
C - - - - - - - - - - - - - - - - - - - - - - - - - - - - - - -
C
C  Calling sequence: CALL  PCHEV (N, X, F, D, NVAL, XVAL, FVAL, DVAL, IERR)
C
C      INTEGER  N, NVAL, IERR
C      REAL  X(N), F(N), D(N), XVAL(NVAL), FVAL(NVAL), DVAL(NVAL)
C
C    Parameters:
C
C      N - (input) number of data points.  (Error return if N.LT.2 .)
C
C      X - (input) real array of independent variable values.  The
C            elements of X must be strictly increasing:
C               X(I-1) .LT. X(I),  I = 2(1)N. (Error return if not.)
C
C      F - (input) real array of function values.  F(I) is
C            the value corresponding to X(I).
C
C      D - (input) real array of derivative values.  D(I) is
C            the value corresponding to X(I).
C
C   NVAL - (input) number of points at which the functions are to be
C            evaluated. ( Error return if NVAL.LT.1 )
C
C   XVAL - (input) real array of points at which the functions are to
C            be evaluated.
C
C            NOTES:
C              1. The evaluation will be most efficient if the elements
C                 of XVAL are increasing relative to X;
C                 that is,   XVAL(J) .GE. X(I)
C                 implies    XVAL(K) .GE. X(I),  all K.GE.J .
C              2. If any of the XVAL are outside the interval [X(1),X(N)],
C                 values are extrapolated from the nearest extreme cubic,
C                 and a warning error is returned.
C
C   FVAL - (output) real array of values of the cubic Hermite function
C            defined by  N, X, F, D  at the points  XVAL.
C
C   DVAL - (output) real array of values of the first derivative of
C            the same function at the points  XVAL.
C
C   IERR - (output) error flag.
C            Normal return:
C               IERR = 0  (no errors).
C            Warning error:
C               IERR.GT.0  means that extrapolation was performed at
C                  IERR points.
C
```

```
C              "Recoverable" errors:
C                 IERR = -1  if N.LT.2 .
C                 IERR = -3  if the X-array is not strictly increasing.
C                 IERR = -4  if NVAL.LT.1 .
C              (Output arrays have not been changed in any of these cases.)
C                 NOTE:  The above errors are checked in the order listed,
C                     and following arguments have **NOT** been validated.
C                 IERR = -5  if an error has occurred in the lower-level
C                     routine CHFDV.  NB: this should never happen.
C                     Notify the author **IMMEDIATELY** if it does.
C
C - - - - - - - - - - - - - - - - - - - - - - - - - - - - - - - - - - -
C***REFERENCES  1. F.N.FRITSCH AND R.E.CARLSON, 'MONOTONE PIECEWISE
C                  CUBIC INTERPOLATION,' SIAM J.NUMER.ANAL. 17, 2 (APRIL
C                  1980), 238-246.
C               2. F.N.FRITSCH, 'PIECEWISE CUBIC HERMITE INTERPOLATION
C                  PACKAGE, FINAL SPECIFICATIONS', LAWRENCE LIVERMORE
C                  NATIONAL LABORATORY, COMPUTER DOCUMENTATION UCID-30194,
C                  AUGUST 1982.
C***ROUTINES CALLED  PCHFD
C***END PROLOGUE  PCHEV
```

5

Numerical Quadrature

5.1 INTRODUCTION

In this chapter we will be interested in solving the following,

$$\text{Evaluate}\quad I = \int_a^b f(x)\, dx.$$

This is one of the two fundamental problems in calculus. It is closely related to the problem of solving a differential equation (see Chapter 8), and the techniques used are based on interpolation (Chapter 4).

As an example of where such a problem might arise, consider the analysis of measurement errors in scientific experiments. Suppose that a surveyor is measuring mountain terrain as part of a highway construction project. The equipment is accurate (say) to the nearest foot. What is the probability that a particular measurement overestimates the true value by less than 2 feet? If the measurement errors have a standard normal distribution, i.e., they follow a "bell" curve, then the desired probability is given by

$$\frac{1}{\sqrt{2\pi}} \int_0^2 e^{-x^2/2} dx.$$

This problem is closely related to the computation of the error function, erf, which we have seen in preceding chapters. There is no closed form expression for the value of this integral, and it must be estimated using numerical methods.

The evaluation of integrals is called **quadrature**, a word that comes from the Latin *quadratura*, meaning the act of making square or squaring. An ancient geometry problem, quadrature of the circle, was to construct a square with area equal to that of a given circular area. The solution had to be obtained in a finite number of steps by drawing pictures with a straight edge and compass. About two thousand years later it was realized that this could not be done and the name was attached to any difficult or time-consuming activity, "as hard to find thy cure As circles puzling Quadrature" (1652). Subsequently, quadrature became synonymous with finding areas and volumes. In this and subsequent chapters we will use this term. The word **integration** is reserved for the solution of differential equations or the evaluation of an indefinite integral. The two problems share some common ideas. However, an integral is a single number representing an area or volume and has no directionality, whereas the solution of a differential equation is a function, and often represents an evolutionary process, such as marching forward from an initial point.

The techniques used to solve quadrature problems are often based on interpolation. Instead of trying to evaluate $\int_a^b f(x)dx$ directly, suppose that we first evaluate $f(x)$ at selected points $x_i \in [a, b]$. Let $p(x)$ be the interpolating polynomial for the points $(x_i, f(x_i))$. Then if $p(x) \approx f(x)$, it should also be true that $\int_a^b p(x)dx \approx \int_a^b f(x)dx$. Since it is easy to integrate polynomials and interpolation techniques are available (see Chapter 4), this is a computationally feasible and efficient approach.

Methods differ in the way they select the points x_i, in the number of points selected, and in the way the points are used to construct interpolating polynomials. For example, some methods split the interval $[a, b]$ into pieces, and work on each piece separately. The simplest rules use equally-spaced points, but much more effective techniques can be obtained through more elaborate schemes. In certain problems, the value of the function will only be known at certain prespecified points; in this case no choice of x_i will be possible (see Section 8).

A good piece of quadrature software should not only evaluate the integral I, but should also estimate the error in the result. The error estimates are not only of value to the user of the software, but they also can assist in the evaluation of the integral. The error estimate can indicate that the function $f(x)$ is changing rapidly in one portion of the interval $[a, b]$ and thus guide the software into working harder to evaluate that portion of the integral. This topic is discussed in Section 6.

The above remarks apply mainly to the case of a problem on a finite interval, where the function $f(x)$ is continuous. If the function is not continuous, and especially if it has a singularity (such as $f(x) = \ln x$ at $x = 0$), then it will not be possible to accurately approximate $f(x)$ by a polynomial. Also, if the interval is infinite, then the integral of any non-zero polynomial over that interval will be infinite. In these cases special strategies must be employed to solve quadrature problems. These are discussed in Section 9.

So far we have only mentioned quadrature problems in one dimension. Integrals in two, three, or higher dimensions are also common, but are more difficult to evaluate. In two or three dimensions, the above techniques can be adapted, but the resulting methods are more complex. They are also much harder to derive. These approaches are discussed in Section 10. For high-dimensional problems it is usually necessary to use Monte Carlo

methods, based on random selection of the points x_i. Such methods do not converge
rapidly, but their cost does not depend heavily on the dimension of the problem (see
Section 11).

In elementary courses in calculus, the evaluation of integrals is one of the most
important topics. The emphasis there is on techniques for obtaining closed-form expressions for indefinite integrals. The computer analogue of this is **symbolic computation**,
that is, to have the computer produce a formula in terms of the endpoints a, b and the
subexpressions of f. There has been remarkable progress in this area. The Macsyma
system, J. Moses (1971), is an excellent tool for those problems for which explicit expressions exist for the integrand and for which closed form solutions, if they exist, will
be especially insightful. For example, Macsyma will discover that

$$\int_0^b \frac{dt}{t^4 + 1} = \frac{\sqrt{2}}{4} \left(\frac{1}{2} \ln \left[\frac{b^2 + \sqrt{2b+1}}{b^2 - \sqrt{2b+1}} \right] + \arctan \left[\frac{\sqrt{2b}}{1 - b^2} \right] \right).$$

However, for many engineering problems efficiency is essential, and we may find that
evaluation of the right hand side for a particular b is more expensive than approximating
the integral by a well chosen method. This chapter only deals with approximate numerical
methods.

5.2   ONE DIMENSIONAL QUADRATURE RULES AND FORMULAS

We derive here the simplest quadrature rules, based on interpolation at a few points.
First we summarize the terminology we will use. An n-point quadrature **formula** is

$$I = \int_a^b f(x)\, dx = \sum_{i=1}^n w_i f(x_i) + R_n.$$

The w_i and x_i are called quadrature **weights** and **nodes**, and R_n the **remainder** or **error**.
The weights and nodes depend upon a, b and n but not f. The sum above is thought of
as an approximation to I and is often called a quadrature **rule**. To estimate an integral
we evaluate *only* the rule, since the remainder term usually involves expressions which
are not available to us, for example derivative values of the integrand $f(x)$. An important
property of our formulas is that they are linear. By that we mean that the rule estimate
of the integral of $f + g$ can be obtained by adding the rule estimates of the integrals of
f and g.

A quadrature rule is a generalized Riemann sum. Recall that a Riemann sum is
obtained by dividing $[a, b]$ into n subintervals of width Δ_i, evaluating $f(x)$ once in each
subinterval and forming $\sum \Delta_i f(x_i)$. The quadrature rule above looks like a Riemann
sum if we identify w_i with Δ_i. Looks can be deceiving however, as we have not said
that w_i need to be positive or even that x_i are required to be in the interval $[a, b]$. For
the most part these distinctions are only of interest to specialists and all the rules we will
discuss are Riemann sums. On the other hand a few practical and famous rules are not;
for example, some w_i can be negative (see problem P5–6).

Over the years, thousands of quadrature rules have been studied. In Section 2.1 we discuss the three simplest: midpoint, trapezoid and Simpson's rules. You may have seen them before in an elementary calculus book; they are easy to derive but are rarely used directly. Mostly they are generalized or used in concert with some more sophisticated strategy, such as compounding or adaption, which are discussed in Sections 4 and 6. But these rules are far from the best. In Section 2.1 we also discuss the more effective Gauss rules, but delay the most effective Gauss-Kronrod rules to Section 5. As both of these are more difficult to derive and their derivations are not of much practical use, we will omit the details and suggest that interested readers consult the references.

With so many rules coming and going, it is difficult to know which to select for any particular problem. The main focus of this text is that practical scientists should use the general purpose software included here for the majority of their tasks. Thus it is unlikely that you will program your own application of any of these rules when a subroutine like Q1DA is appropriate (see Section 7). But to understand why this routine works as well as it does, you must appreciate that its underlying rules are fundamentally of high quality. Of course, in those special cases where you need to program your own algorithm it will be necessary to have a basic grasp of which rules are the best to use. Finally, as general purpose software is scarce for infinite regions, several rules are presented in Section 9 for such cases.

5.2.1 Elementary Formulas: Midpoint, Trapezoid, Simpson and Gauss

The formulas presented here are elementary. They are not useful directly, but many practical methods are based upon them. They can also be interpreted geometrically, and you should refer to Figure 5.1 while you read this section.

The midpoint and trapezoid formulas are derived as follows. The area under $f(x)$ on the interval $[a, b]$ may be approximated by the area under the line segment joining $(a, f(a))$ with $(b, f(b))$. This line has equation

$$l(x) = f(a)\frac{x - b}{a - b} + f(b)\frac{x - a}{b - a}$$

and the area of the trapezoid formed by the four points $(a, 0)$, $(b, 0)$, $(b, f(b))$, $(a, f(a))$ is

$$\frac{b - a}{2}f(a) + \frac{b - a}{2}f(b)$$

This is a quadrature rule obtained by simple geometric reasoning. We now derive it again, by a method which also gives an expression for the error.

Expand $f(x)$ in a Taylor series about the midpoint $m = (a + b)/2$

$$f(x) = f(m) + f^{(1)}(m)(x - m) + f^{(2)}(m)(x - m)^2/2$$
$$+ f^{(3)}(m)(x - m)^3/6 + f^{(4)}(m)(x - m)^4/24 + \cdots.$$

Integrating this from a to b gives

$$I = f(m)(b-a) + f^{(2)}(m)\frac{(b-a)^3}{24} + f^{(4)}(m)\frac{(b-a)^5}{1920} + \cdots.$$

This is called the **midpoint**, or one-point rectangle formula. Now in the Taylor expansion set $x = a$, then $x = b$ and add the two series. The odd-order derivative terms drop out. Rearranging to solve for $f(m)$ gives

$$f(m) = \frac{1}{2}[f(a) + f(b)] - f^{(2)}(m)(b-a)^2/8 - f^{(4)}(m)(b-a)^4/384 + \cdots.$$

Substituting this into the midpoint formula and combining like terms gives

$$I = \frac{(b-a)}{2}[f(a) + f(b)] - f^{(2)}(m)\frac{(b-a)^3}{12} - f^{(4)}(m)\frac{(b-a)^5}{480} + \cdots.$$

This is the **two-point trapezoid formula**, which has nodes $x_1 = a$, $x_2 = b$ and weights $w_1 = w_2 = (b-a)/2$, and is the same as the rule derived geometrically above.

The **three-point Simpson's rule** can also be derived by geometrical reasoning akin to that for the trapezoid rule. From Section 3 of Chapter 4 the quadratic $Q(x)$ through the three points $(a, f(a))$, $(m, f(m))$, and $(b, f(b))$ is

$$Q(x) = f(a)\frac{(x-m)(x-b)}{(a-m)(a-b)} + f(m)\frac{(x-a)(x-b)}{(m-a)(m-b)} + f(b)\frac{(x-a)(x-m)}{(b-a)(b-m)}.$$

Integrating Q from a to b gives

$$\int_a^b Q(x)\,dx = f(a)w_1 + f(m)w_2 + f(b)w_3,$$

$$\text{where} \quad w_1 = \int_a^b \frac{(x-m)(x-b)}{(a-m)(a-b)}\,dx, \quad \text{etc.}$$

You should check that

$$w_1 = \frac{b-a}{6}, w_2 = \frac{4(b-a)}{6}, w_3 = \frac{b-a}{6},$$

which is Simpson's rule. Also note that once we have integrated Q, the formula for $Q(x)$ is no longer needed. Simpson's rule was first given by Cavalieri in 1639, and later by James Gregory (1668) and Thomas Simpson (1743). It is also known as the parabolic rule.

Figure 5.1 illustrates the midpoint, trapezoid, and three-point Simpson rules.

Just as for the trapezoid rule, we can also get Simpson's formula, which includes the error term, by a more algebraic approach. Add two times the midpoint formula to the trapezoid formula and divide the result by three to get

$$I = \frac{b-a}{6}[f(a) + 4f(m) + f(b)] - f^{(4)}(m)\frac{(b-a)^5}{2880} + \cdots.$$

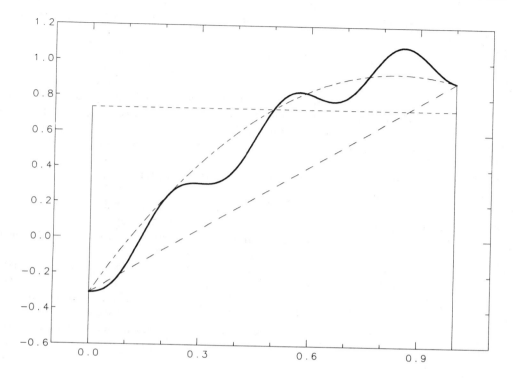

Figure 5.1   Rectangle, Trapezoid and Simpson's Rule

Example 5.1   Estimating erf(1) by Various Rules.

We will estimate erf(1) with the midpoint, trapezoid and Simpson rules

$$I = \text{erf}(1) = \frac{2}{\sqrt{\pi}} \int_0^1 \exp(-t^2)\, dt = .84270079294971484 \ldots .$$

The one point midpoint, two point trapezoid and three point Simpson's rule estimates are

$$I \approx \frac{2}{\sqrt{\pi}} e^{-\frac{1}{4}} = .8788 \ldots$$

$$I \approx \frac{2}{\sqrt{\pi}} \left(\frac{1}{2}\right) [1 + e^{-1}] = .7717433 \ldots$$

$$I \approx \frac{2}{\sqrt{\pi}} \left(\frac{1}{6}\right) [1 + 4e^{-\frac{1}{4}} + e^{-1}] = .84310283 \ldots$$

respectively. ■

We note that if we apply the trapezoid rule to a linear polynomial $f(x) = \alpha + \beta x$, then

$$I = \frac{(b-a)}{2}[f(a) + f(b)].$$

In this case the remainder is zero. This should be expected because of the way we introduced this rule at the beginning of the section. Similarly the error in the three-point Simpson's rule is zero if f is any cubic polynomial.

An n point quadrature rule is said to be of **polynomial degree** d if its remainder R_n is zero for every polynomial of degree d, but not for some polynomial of degree $d+1$. The trapezoid and midpoint rule are of polynomial degree one. Simpson's rule is of polynomial degree three.

A quadrature rule is seldom used to integrate a polynomial, but we expect that rules of higher polynomial degree are better, at least if our integrand is polynomial-like. Thus high-degree rules are preferred, but in practice the details of strategy and implementation are often more important than the specific rule that is used.

Our first approach to deriving Simpson's rule illustrates a general technique which is useful for deriving many other quadrature rules: the function $f(x)$ is replaced by an interpolant. The interpolant is integrated analytically, resulting in a quadrature rule. This can then be used without explicitly having an expression for the interpolant.

For example, if we integrate the Lagrange representation of the polynomial interpolant $p_{n-1}(x)$ through the points $(x_i, f(x_i))$, $i = 1, \ldots, n$ we get (see Section 3 of Chapter 4)

$$\int_a^b p_{n-1}(x)\, dx = \sum_{i=1}^n f(x_i) w_i, \qquad \text{where} \qquad w_i = \int_a^b l_i(x)\, dx.$$

Rules of this type are called **interpolatory**. A famous example of an interpolatory rule occurs when we take the x_i to be equally spaced on the interval of integration; the quadrature rule obtained in this way is called a **Newton-Cotes** rule. The trapezoid and Simpson's rule are low degree Newton-Cotes rules. Unfortunately, high-degree rules of this type have negative weights; see Problem P5–6. This is usually considered an undesirable property because the integration of a positive function can lead to catastrophic cancellation, as discussed in Section 4 of Chapter 2. This ought not to come as too much of a surprise; we saw in Chapter 4 that high-degree polynomial interpolants can look bad and the integral of such an interpolant is not likely to produce a good approximation to the integral we desire. On the other hand interpolatory quadrature can be useful when the interpolant is not a polynomial of high degree. We will use this technique in Section 8 to derive a quadrature formula based on an Hermite Cubic interpolant for use in data integration.

We now sketch a derivation of the two-point Gaussian rule. Any two point rule has the general form

$$I \approx w_1 f(x_1) + w_2 f(x_2).$$

Among all such rules can we find one with the highest polynomial degree? The trapezoid rule is one candidate, but we can do much better. We guess that if w_1, w_2, x_1 and x_2 are at our disposal we can find some selection of these numbers allowing us to integrate an arbitrary cubic, which has four coefficients. If this is successful then that rule will exactly integrate each of the particular cubics, $f(x) = 1 = (x - m)^0$, $f(x) = (x - m)$, $f(x) = (x - m)^2$ and $f(x) = (x - m)^3$, where $m = (a + b)/2$. Conversely, because the rule is linear, if it is exact for these four integrands, it is exact for any cubic, i.e., is of degree 3. Thus we ask that the approximate equality '$\approx$' be replaced by equality whenever $f(x)$ is one of these functions. This gives

$$f(x) = (x - m)^0 \quad \Rightarrow \quad (b - a) = w_1 + w_2$$

$$f(x) = (x - m)^1 \quad \Rightarrow \quad 0 = w_1(x_1 - m) + w_2(x_2 - m)$$

$$f(x) = (x - m)^2 \quad \Rightarrow \quad \frac{(b - a)^3}{12} = w_1(x_1 - m)^2 + w_2(x_2 - m)^2$$

$$f(x) = (x - m)^3 \quad \Rightarrow \quad 0 = w_1(x_1 - m)^3 + w_2(x_2 - m)^3$$

This is a system of four equations for the four numbers w_1, w_2, x_1, x_2. The equations are nonlinear and cannot be solved by the methods of Chapter 3. A general set of nonlinear equations might not have any solution, or there might be many solutions. If there is a solution, there is no reason to think that the numbers we get for x_i need be real, because note that these quantities appear as squared terms. Even if there is a real solution the x_i might lie outside the interval $[a, b]$ or the w_i might be negative. Fortunately, for this particular system none of these terrible things occur. The unique solution can be shown to be

$$x_1 = \frac{a + b}{2} + \frac{b - a}{2\sqrt{3}}, \qquad x_2 = \frac{a + b}{2} - \frac{b - a}{2\sqrt{3}}, \qquad w_1 = w_2 = \frac{b - a}{2},$$

which is the **two-point Gaussian quadrature rule**

$$I \approx G_2 = \frac{b - a}{2} \left[f\left(\frac{a + b}{2} - \frac{b - a}{2\sqrt{3}} \right) + f\left(\frac{a + b}{2} + \frac{b - a}{2\sqrt{3}} \right) \right].$$

This rule is of polynomial degree three.

Example 5.2   Gauss Rule Approximation to erf(1).

This gives

$$\text{erf}(1) = 0.8427007929\ldots \approx G_2 = \frac{1}{\sqrt{\pi}} \left[\exp\left(-\frac{1}{4} \left[1 - \frac{1}{\sqrt{3}} \right]^2 \right) + \exp\left(-\frac{1}{4} \left[1 + \frac{1}{\sqrt{3}} \right]^2 \right) \right]$$

$$= .84244189\ldots.$$

Notice that this is more accurate than the three point Simpson rule. ∎

As mentioned in Section 2 a two-point formula is not often useful. But the idea behind the two-point Gaussian formula can be extended to any number of nodes. That is, we try to find a formula

$$\int_a^b f(x)\,dx = \sum_{i=1}^{n} w_i f(x_i) + R_n,$$

where neither the x_i nor the w_i are prescribed in advance, but determined so that the rule is exact for polynomials of as high degree as possible. A unique Gaussian quadrature formula exists for every n. It is of polynomial degree $2n - 1$ and there is no other n-point formula of higher degree. In this sense it is the best n-point formula. R_n can be shown to be

$$R_n = \frac{(b-a)^{2n+1}(n!)^4}{(2n+1)[(2n)!]^3} f^{(2n)}(\xi)$$

which applies as long as $f(x)$ has $2n$ continuous derivatives on $[a, b]$. The number ξ is unknown, except that $a < \xi < b$. Gaussian rules are among the most useful in practice. We list some of their properties below.

1. The weights and nodes are irrational numbers, with a few exceptions. We have seen that when $n = 2$ the weights are $w_1 = w_2 = (b - a)/2$. In addition when n is odd there is a node at the midpoint, $(a + b)/2$. Gaussian rules are usually provided in a subroutine which has them precomputed for a selection of n's. The book by Abramowitz and Stegun (1965) gives them in tabular form. Table 5.1 lists three Gaussian rules for you to examine. In a hand calculation Gaussian rules are difficult to use, but on a computer there is no reason to require simple numbers for the weights or nodes.

2. Gaussian rules are "open." That is, no node occurs at either endpoint a or b. This property is shared by the midpoint rule, which is the one-point Gauss rule. This is convenient because integrand difficulties are usually at endpoints. For example

$$I_1 = \int_0^1 \exp\left(-\frac{1}{x^2}\right)\,dx, \qquad I_2 = \int_0^1 \frac{\sin x}{x}\,dx$$

involve perfectly well-behaved integrands, but some special programming would be required to use any quadrature rule which evaluates $f(0)$. Strictly speaking, the integrands above are singular because they are not defined at $x = 0$. But in both cases $\lim_{x \to 0} f(x)$ exists, respectively zero and one, and serves as a reasonable definition for the integrand at the left endpoint. Such a singularity is called "removable."

3. The sets of nodes of two Gaussian rules are almost disjoint. That is, the nodes for an n-point rule are distinct from those for an m-point rule. The exception is that whenever n is odd there is a node at the center of the interval.

4. The derivation of Gaussian rules suggests that if $f(x)$ is like a polynomial, a Gaussian rule will do a good job integrating it. In fact, it can be shown that among all rules using n function evaluations, the n-point Gaussian rule is likely to produce the most accurate estimate, at least if the integrand is smooth. It is also true that Gaussian rules work well even for integrands with non-removable singularities in the function or one of its derivatives. The function $f(x) = x^2 \ln x$ has an unbounded second derivative on the interval $[0, 1]$. The expression for the error in G_2 given above provides no information because it involves $f^{(4)}$ which does not exist at $x = 0$. But that should not be taken to mean that the error is large. An easy calculation shows that $G_2 = -.1085\ldots$, with error $.0023\ldots$. The two-point trapezoid and three-point Simpson rules have errors $.1111\ldots$ and $.0044\ldots$.

5. Gaussian rules are interpolatory. That is, if we take the n nodes x_i from an n-point Gaussian formula, and the n associated function values $f(x_i)$, and form the unique $n - 1$ degree polynomial which interpolates these data, the number obtained by integrating this interpolant on $[a, b]$ is identical to the value obtained by applying the Gaussian formula to f. This is a case where interpolatory quadrature works well and is related to the fact that the Gaussian nodes bunch up near the endpoints and are non-uniformly spaced on $[a, b]$. It is possible to prove that the weights in Gaussian quadrature are always positive, and increasing the number of points almost always improves the accuracy of the estimate.

6. There is an intimate relationship between Gauss quadrature and **orthogonal polynomials**, a topic of great interest to physicists. The n Gauss nodes are the zeros of the n-th Legendre polynomial. In fact, what we have termed Gaussian quadrature is often referred to as Gauss Legendre quadrature because of this association.

TABLE 5.1 SOME GAUSSIAN QUADRATURE
RULES FOR $\int_{-1}^{1} f(x)\, dx$

Nodes	Weights
$n = 2$	
± 0.577350269189625	1.000000000000000
$n = 4$	
± 0.861136311594052	0.347854845137454
± 0.339981043584856	0.652145154862546
$n = 8$	
± 0.960289856497536	0.101228536290376
± 0.796666477413627	0.222381034453374
± 0.525532409916329	0.313706645877887
± 0.183434642495650	0.362683783378362

Gaussian quadrature is much more general than our presentation might lead you to believe. Two particularly useful generalizations are given in Sections 4 and 9.4.

5.3 CHANGE OF INTERVAL

A computer subroutine which applies a quadrature rule usually allows the user to input
the interval endpoints a, b. But tabulated rules such as those in Table 5.1 will only be
given for a fixed interval, often either $[0, 1]$ or $[-1, 1]$. Assume we have a tabulated
formula

$$\int_\alpha^\beta f(x)\, dx = \sum_{i=1}^n w_i f(x_i) + R_n.$$

(In Table 5.1, $\alpha = -1, \beta = 1$.) If the integral to be evaluated is over the interval $[\alpha, \beta]$
then the tabular values can be used exactly as they appear. On the other hand, suppose
what is required is an estimate of

$$I = \int_a^b g(t)\, dt$$

on some other interval. We can still utilize the tabulated values after making a change
of variable from $[\alpha, \beta]$ in x to $[a, b]$ in t. As long as α, β, a, and b are finite, set
$t = [(b - a)x + a\beta - b\alpha]/(\beta - \alpha)$. Then

$$I = \left(\frac{b - a}{\beta - \alpha}\right) \int_\alpha^\beta g\left(\frac{(b - a)x + a\beta - b\alpha}{\beta - \alpha}\right)\, dx$$

$$= \left(\frac{b - a}{\beta - \alpha}\right) \sum_{i=1}^n w_i g\left(\frac{(b - a)x_i + a\beta - b\alpha}{\beta - \alpha}\right) + \mathcal{R}_n.$$

The change of variable from t to x transforms the interval $[a, b]$ into $[\alpha, \beta]$. There
are many different transformations that could have been used, although this one is exceptionally simple because it is linear. If $P(t)$ is a polynomial of degree d in t then
$P([(b - a)x + a\beta - b\alpha]/(\beta - \alpha))$ is a polynomial of the same degree in x. Thus this
change of variable does not affect the polynomial degree of the formula. If R_n is zero
when $f(x)$ is a polynomial of degree d, then $\mathcal{R}_n$ is also zero when $g(x)$ is a polynomial
of degree d. In later sections we will see some examples of nonlinear transformations.

Example 5.3 Changing Intervals.

Compute the four point approximation to

$$I = \int_1^2 x^x\, dx$$

using Table 5.1. With $\alpha = -1, \beta = 1, a = 1, b = 2$, and $g(t) = t^t$, we have

$$I \approx \frac{1}{2} \sum_{i=1}^4 w_i g\left(\frac{x_i + 3}{2}\right) = \frac{1}{2} \sum_{i=1}^4 w_i \left(\frac{x_i + 3}{2}\right)^{(x_i + 3)/2} = 2.05044\ldots,$$

with w_i, x_i from Table 5.1 ($n = 4$). ∎

5.4   COMPOUND QUADRATURE RULES AND ERROR ESTIMATES

We stated in Section 2 that one, two, or three point rules are seldom used directly. They are usually applied by combining them, as we now illustrate. Let $[a, b]$ be divided into s subintervals, called **panels**, by the points $a = t_0 < t_1 < t_2 < \cdots < t_{s-1} < t_s = b$. Since

$$I = \int_a^b f(x) \, dx = \sum_{i=0}^{s-1} \int_{t_i}^{t_{i+1}} f(x) \, dx,$$

we may estimate I by applying a quadrature rule to each panel. If the same rule is used on each, and if the points t_i are uniformly spaced on $[a, b]$ the result is called a **compound** quadrature rule. *Any* rule can be compounded, but the two most popular are derived from the trapezoid and Simpson formulas.

We now develop the compound trapezoid formula. Let $h = (b - a)/s$ be the panel width. Then $t_i = a + ih$. Applying the two point trapezoid formula to the intervals $[a + ih, a +(i + 1)h]$ and adding gives a rule with $s + 1$ points,

$$\int_a^b f(x) \, dx = h \left[\frac{f(a)}{2} + f(a+h) + f(a+2h) + \cdots + f(b-h) + \frac{f(b)}{2} \right]$$

$$- \frac{h^3}{12} \left[f^{(2)}(m_1) + f^{(2)}(m_2) + \cdots + f^{(2)}(m_s) \right] + \cdots$$

or

$$I = T_{s+1} + R_{s+1}$$

where $m_i = (t_{i-1} + t_i)/2$. In practice only the rule T_{s+1} is computed. At best the remainder can be estimated, but rarely by computing the derivatives which appear. If $f^{(2)}$ is smooth enough we can approximate the sum of second derivatives by $s \cdot f^{(2)}(\xi)$ and then

$$R_{s+1} \approx -\frac{h^3}{12}[sf^{(2)}(\xi)] + \cdots \approx -\frac{(b-a)}{12} h^2 f^{(2)}(\xi).$$

We will shortly present one method to estimate this without computing $f^{(2)}$.

The compound Simpson formula follows in a similar way. We set $h = (b-a)/2s$, $t_i = a + 2ih$, and obtain

$$I = \frac{h}{3}[f(a) + 4f(a+h) + 2f(a+2h) + 4f(a+3h) + \cdots + 4f(b-h) + f(b)]$$

$$-(b-a)\frac{h^4}{180} f^{(4)}(\xi) + \cdots$$

This is a rule with $2s + 1$ points.

In the literature the term "trapezoid rule" can refer to either the two-point or the compound rule. The context makes clear which is to be considered. A similar remark applies to Simpson's rule. Finally, notice that the polynomial degree of a compound rule is the same as that of the basic rule.

There are several ways to estimate errors in compound rules. A quadrature rule is not useful unless there is some way to estimate or bound the remainder R_n. For simple integrals like the one defining $\text{erf}(x)$ the first few derivatives of the integrand are easy to obtain. Thus the error in the compound trapezoid rule when it is used to estimate $\text{erf}(1)$ is

$$|R_{s+1}| = \frac{1}{12s^2}\frac{2}{\sqrt{\pi}}|e^{-\xi^2}(4\xi^2 - 2)| < \frac{1}{12s^2}\frac{2}{\sqrt{\pi}}2 = \frac{1}{3s^2\sqrt{\pi}}.$$

If we use a 100-point rule the integral estimate will be in error by no more than 10^{-4}. Of course, the actual error might be much less than 10^{-4} (see Example 5.4).

For complicated problems it can be inconvenient or difficult to compute high-order derivatives. The example above might give the reader a false impression about the usefulness of remainders containing derivatives. In any realistic problem, the remainder must be estimated from numerical data. A typical error estimation algorithm is described below, and another is given in Section 5. Error estimates derived analytically are too complicated or too pessimistic. In many applications the integrand is not even given by a formula, but its values are obtained from a lengthy computation which is impossible to analyze.

A common procedure is to estimate a remainder by combining two or more quadrature rules. In the simplest case we apply the two rules and take the absolute value of the difference as an estimate of the error in the less accurate one. For example, we could apply the compound trapezoid rule and the compound Simpson's rule and use the difference to estimate the error in the former. There are many other possibilities. The discussion below shows how to use two trapezoid rules with different numbers of panels. In the next section we consider two more rules, a Gauss-Kronrod pair. The two important ideas are

(a) Error estimation procedures have a large heuristic component. Selecting and combining two rules is as much an art as a science.

(b) It is significantly more expensive to obtain both an integral estimate and an error estimate than it is to obtain only the integral estimate.

Compound rules are useful because they offer a simple method for obtaining rules with any number of nodes. The advantage of these rules stems from the effect on the remainder of increasing s. Examining R_{s+1} for the compound trapezoid rule we see that increasing the number of panels changes $h = (b-a)/s$ and also changes the point ξ at which $f^{(2)}$ is computed but does not change the order of the derivative. There is a similarity between this discussion and the one in Section 5 of Chapter 2 where we introduced the concept of extrapolation. The same ideas can be used here. Thus if we apply the compound trapezoid rule twice, first with s and then $2s$ panels, the remainders are

$$|R_{s+1}| = \frac{b-a}{12}h^2|f^{(2)}(\xi_1)|, \qquad |R_{2s+1}| = \frac{b-a}{12}\left(\frac{h}{2}\right)^2|f^{(2)}(\xi_2)|.$$

Assuming the values of $f^{(2)}$ are equal at ξ_1, ξ_2,

$$|R_{2s+1}| = \frac{|R_{s+1}|}{4}.$$

Doubling the number of panels results in an approximate fourfold reduction in the error. Because of the h^2 term this formula is called "second order." The compound Simpson's rule is "fourth order." Doubling the number of panels reduces the error by a factor of about sixteen. We saw a similar reduction in the interpolation error when we studied piecewise polynomial interpolation in Sections 8 and 11 of Chapter 4. Indeed, there is an intimate relationship between piecewise polynomial interpolation and quadrature rules. We carry this further in Section 9.

Example 5.4   Compound Trapezoid Rule Estimate of erf(1).

Table 5.2 illustrates the compound trapezoid rule for a few values of s when applied to the integral for erf(1). It also gives the factor by which the error is reduced as s is doubled. This number should be about 0.25. ∎

TABLE 5.2   COMPOUND TRAPEZOID RULE ESTIMATES OF erf(1)

No. of intervals s	Trap. Rule	Error, R_{s+1}	R_{s+1}/R_{2s+1}
2	.82526295	1.74×10^{-2}	—
4	.83836777	4.33×10^{-3}	.2485
8	.84161922	1.08×10^{-3}	.2496
16	.84243051	2.70×10^{-4}	.2499
32	.84263323	6.76×10^{-5}	.2500
64	.84268390	1.69×10^{-5}	.2500
128	.84269657	4.22×10^{-6}	.2500

One practical application of these ideas is to error estimation. If we compute two estimates using the compound trapezoid rule with s and $2s$ panels, we expect that $|R_{2s+1}| \approx |R_{s+1}|/4$. Denoting by T_{s+1}, T_{2s+1} the computed compound trapezoid rule estimates,

$$I = T_{s+1} + R_{s+1} = T_{2s+1} + R_{2s+1}$$

we have

$$|T_{s+1} - T_{2s+1}| = |R_{s+1} - R_{2s+1}| \approx \frac{3}{4}|R_{s+1}|.$$

Thus 4/3 of the difference between two successive estimates provides an estimate of the error in T_{s+1}. We would expect T_{2s+1} to have about 1/4-th this error. A practical, but conservative procedure would then use T_{2s+1} as an integral estimate but $\frac{4}{3}|T_{s+1} - T_{2s+1}|$ as an error estimate.

There are several issues associated with developing a reliable computer program using this technique.

(1) The trapezoid and Simpson formula are "closed," requiring $f(a)$ and $f(b)$. They cannot be applied to integrals which have non-removable singularities, such as

$$\int_0^1 \ln x \cos x^2 \, dx \qquad \text{or} \qquad \int_0^1 \frac{e^{\sin x}}{\sqrt{x}} \, dx.$$

Both integrals exist, that is, the area under each curve is finite, even though the integrands become infinite at the left endpoint. Many practical problems involve integrable singularities such as these. This difficulty could be avoided by using the compound midpoint rule (see Problem P5–2b), or a compound Gauss rule.

(2) Look back at the derivation of the trapezoid or Simpson's rule in Section 2.1. The argument hinged on the existence of a Taylor expansion. If f has a singularity on $[a, b]$, such as $f = \sqrt{x}$ on $[0, 1]$, a Taylor expansion will not be valid on the whole interval and the derivation of the error, i.e., the trapezoid formula is unjustified. A similar remark applies to the compound rules, but for these there is the further assumption that it is reasonable to approximate the sum of a large number of derivative terms by s times a derivative at one point. In fact, the error estimation procedure above requires that $f^{(2)}$ does not change much for different values of ξ_i. If f has large oscillations or other rapid changes this assumption is invalid. For either of these two cases ($f^{(2)}$ does not exist on all of $[a, b]$ or it varies dramatically) we can still apply the compound trapezoid rule, but cannot make the same inferences about the error. In general, a Riemann sum quadrature rule such as the trapezoid rule leads to a Riemann sum compound rule. Such rules can be made arbitrarily accurate as we increase the number of nodes. But the rate of convergence depends on the smoothness of f. Specifically, doubling the number of panels may *not* decrease the error by a factor of four.

(3) If we already have a compound trapezoid estimate requiring n nodes ($n-1$ panels) and we want a more accurate estimate there are three natural choices.

    (a) Increase the number of points by one, T_{n+1} (n panels).
    (b) Double the number of points, T_{2n} ($2n - 1$ panels).
    (c) Double the number of panels, T_{2n-1} ($2n - 2$ panels).

Of these, T_{n+1} is only slightly more accurate than T_n, while T_{2n} and T_{2n-1} are about four times as accurate (for smooth f). However, a quick sketch will convince you that given T_n, computing T_{n+1} requires $n - 1$ *new* evaluations of f, T_{2n} requires $2n - 1$, but T_{2n-1} only requires $n - 1$. In fact, if h is the spacing between the nodes of T_n,

$$T_{2n-1} = \frac{1}{2}T_n + \frac{h}{2}\left[f(a + \frac{h}{2}) + f(a + \frac{3h}{2}) + \cdots + f(b - \frac{3h}{2}) + f(b - \frac{h}{2}) \right].$$

Thus software which implements the compound trapezoid rule almost always uses (c) along with the formula above.

During the 1960's subroutines based on compound rules were developed which dealt with difficulties (1) and (2). One of the most prolific lines of research used extrapolation techniques applied to the error expansion of the compound trapezoid rule and may be found under the heading "Romberg" quadrature in several of the references given in Chapter 1. But since 1970, the adaptive quadrature algorithms described in Section 6 have become the most popular methods for general subroutine use. The interest in extrapolation has not faded though, and more advanced techniques have been described which can be used in concert with adaptive algorithms. See for example the paper by E. De Doncker and I. Robinson (1984).

5.5 GAUSS-KRONROD QUADRATURE RULES

In Section 4 we saw that for the compound trapezoid rule, doubling the number of panels gives us two estimates, T_n and T_{2n-1}. For smooth functions T_{2n-1} is about four times as accurate as T_n, with a total amount of work equal to $2n - 1$ evaluations. In Section 2.1 we pointed out that Gauss formulas are "best" for a given number of nodes, and we wonder if it would be possible to generate a pair of them to produce somehow both an integral and an error estimate. If G_n denotes the n point Gauss rule, we expect that G_{2n-1} will be far more accurate and we could use $|G_n - G_{2n-1}|$ as an error estimate. But we remarked earlier that none of the Gauss rules share nodes except at the midpoint. Thus the total work for G_n, and G_{2n-1} is $3n - 1$ evaluations (or $3n - 2$ if n is odd). A similar situation occurs using G_n and G_{2n}. If we use G_n and G_{n+1} the total work is $2n + 1$, comparable to the compound trapezoid rule, but G_{n+1} is only slightly more accurate than G_n.

In 1965 the Russian computer scientist Kronrod considered the problem of estimating errors in Gaussian quadrature rules in this spirit. Kronrod's idea was to start with G_n and find a new formula which used all of these same evaluation points as well as some others. If

$$G_n = \sum_{i=1}^{n} w_i f(x_i),$$

Kronrod considered

$$K_{2n+1} = \sum_{i=1}^{n} a_i f(x_i) + \sum_{j=1}^{n+1} b_j f(y_j).$$

Notice that both G_n and K_{2n+1} share n nodes. The latter has $n + 1$ additional ones, and all different weights, a_i, b_j. The Gaussian rule G_n is of polynomial degree $2n-1$. Kronrod found the $3n + 2$ a's, b's and y's so that K_{2n+1} was of degree $3n + 1$. Table 5.3 lists K_{15} to 15 digits. The two rules (G_n, K_{2n+1}) are called a **Gauss-Kronrod pair**. The cost to compute the pair is $2n + 1$ function values. This is the same as for G_{2n+1} which

TABLE 5.3　THE (7,15) GAUSS-KRONROD PAIR
ON $[-1, 1]$.

Seven-Point Gaussian Nodes	Weights
$\pm$0.94910 79123 42758	0.12948 49661 68870
$\pm$0.74153 11855 99394	0.27970 53914 89277
$\pm$0.40584 51513 77397	0.38183 00505 05119
$\pm$0.00000 00000 00000	0.41795 91836 73469

Fifteen-Point Kronrod Nodes	Weights
$\pm$0.99145 53711 20813	0.02293 53220 10529
$\pm$0.94910 79123 42759	0.06309 20926 29979
$\pm$0.86486 44233 59769	0.10479 00103 22250
$\pm$0.74153 11855 99394	0.14065 32597 15525
$\pm$0.58608 72354 67691	0.16900 47266 39267
$\pm$0.40584 51513 77397	0.19035 05780 64785
$\pm$0.20778 49550 07898	0.20443 29400 75298
0.00000 00000 00000	0.20948 21410 84728

is of degree $4n + 1$ but provides no error estimate. It is also the same cost as the two rules G_n and G_{n+1} which do. But K_{2n+1} should be much more accurate than G_{n+1}. We could take as an error estimate for G_n the difference $|G_n - K_{2n+1}|$ and return K_{2n+1} as an integral estimate. This is a perfectly reasonable approach. However, experience has shown that this difference severely overestimates the error in the far more accurate K_{2n+1}. Experimentation has suggested that

$$\text{Integral Estimate} = K_{2n+1}$$
$$\text{Error Estimate} = \left(200|G_n - K_{2n+1}|\right)^{1.5}$$

is realistic, but still conservative. (Actually, most computer implementations scale the error estimate slightly to reflect the magnitude of f, but for $f \approx 1$ the error estimate given here is correct.) At first glance the error estimate might appear larger than $|G_n - K_{2n+1}|$, but the latter is usually much less than 1.0 and the 1.5 power makes it smaller yet.

If we apply a double precision implementation of this rule to estimate erf(1), after transforming the interval from $[-1, 1]$ to $[0, 1]$, we find that

$$G_7 = 0.842700792948824892$$
$$K_{15} = 0.842700792949714861$$
$$\left(200|G_7 - K_{15}|\right)^{1.5} = 2 \cdot 10^{-15}.$$

The actual error in K_{15} is about $2 \cdot 10^{-17}$. By comparison, the compound trapezoid rule is far less accurate. In Table 5.2 of Example 5.4 the least error is $4 \cdot 10^{-6}$ and this requires 128 evaluations.

A Gauss-Kronrod pair, along with the above error estimate, is currently one of the most effective methods for calculating general integrals. The pair (G_7, K_{15}) is standard. When combined with an adaptive quadrature algorithm we obtain reliable and efficient general purpose subroutines.

The positive remarks above notwithstanding, readers should be aware that a totally numeric estimate such as this one can still be wrong. There exist relatively well behaved functions for which both these estimates are arbitrarily bad, that is, of a wrong order of magnitude and of a wrong sign. Changing to another rule or another error estimate or both may reduce the likelihood of a poor result but cannot eliminate the possibility.

5.6   AUTOMATIC AND ADAPTIVE QUADRATURE ALGORITHMS

An **automatic quadrature algorithm** takes as input f, $[a, b]$ and an accuracy request ϵ and produces a result Q and an error estimate E which we hope will satisfy

$$|Q - I| < E < \epsilon.$$

No explicit assumptions are made about f except that an external procedure is provided to calculate $f(x)$ for any given x. Most scientists would prefer that an algorithm attempt to satisfy a *relative error request*

$$|Q - I| < E \cdot |I| < \epsilon \cdot |I|$$

and many programs, such as Q1DA in this chapter, try to do this. But some special logic is required in case $I \approx 0$, and in this text we discuss only the first, or absolute error request above.

An automatic quadrature algorithm can be developed from the compound trapezoid rule. Compute T_{s+1} with $s = 1, 2, 4, 8, 16, \ldots$ until $\frac{4}{3}|T_{2s+1} - T_{s+1}| < \epsilon$. For this automatic quadrature algorithm the evaluation points of f are uniformly spaced on $[a, b]$. Different integrands only cause the algorithm to terminate earlier or later.

Automatic quadrature algorithms are always expensive to use compared to the best that can be achieved by studying the particular problem and taking advantage of its features or the intended use of the calculation. Several hundred percent extra work is common, and an order of magnitude or more can also occur. But they are convenient and for many problems the total computer time is still small enough to justify the expense. In particular this often occurs at the initial stages of a computational project. As the project progresses, an assessment needs to be made of the benefit in human and computer time of additional analysis.

An **adaptive automatic quadrature algorithm** is a special kind of automatic quadrature algorithm which seeks to tailor the evaluation points to each individual integrand. The two important components of this are a "local quadrature module" (LQM) and an overall strategy.

An LQM is a procedure with inputs f and an interval $[\alpha, \beta]$, and outputs $Q_{[\alpha, \beta]}$ and $E_{[\alpha, \beta]}$ which we refer to as the local quadrature and local error estimates. The examples below illustrate two distinct ways to provide local quadrature and local error estimates.

Example 5.5   A Simple Local Quadrature and Local Error Estimate.

$$Q_{[\alpha,\beta]} = T_{15}$$

$$E_{[\alpha,\beta]} = \frac{4}{3}|T_{15} - T_8|$$

The local error estimate is based upon the derivation in Section 4. ∎

Example 5.6   A Realistic Local Quadrature and Local Error Estimate.

$$Q_{[\alpha,\beta]} = K_{15}$$

$$E_{[\alpha,\beta]} = \left(200|G_7 - K_{15}|\right)^{1.5}$$

This local error estimate is based upon the formula given in Section 5. ∎

Local quadrature modules need to be reliable. Usually they have some technique to analyze roundoff or other pathological behavior. Nevertheless they are thought of as "building blocks" rather than as complete algorithms.

One of the best overall strategies, called **globally adaptive**, operates as follows. The initial interval $[a, b]$ is sent to the LQM. If $E_{[a,b]} < \epsilon$ we quit, returning $Q = Q_{[a,b]}$, $E = E_{[a,b]}$. Otherwise subdivide $[a, b]$ in half sending each to the LQM. If $E = E_{[a, \frac{a+b}{2}]} + E_{[\frac{a+b}{2}, b]} < \epsilon$ quit returning E and $Q = Q_{[a, \frac{a+b}{2}]} + Q_{[\frac{a+b}{2}, b]}$. Otherwise subdivide the half with the larger local error estimate. At a general step in the algorithm the original interval $[a, b]$ has been replaced by a set of subintervals of different sizes. If the sum of all the local error estimates exceeds ϵ, the subinterval with largest local error estimate is subdivided. Eventually the sum is less than ϵ, in which case we quit, returning the sum of the local quadrature estimates as the result Q, and the sum of the local error estimates as E.

This strategy follows from the premise that on a small enough subinterval the integrand will be well behaved and the local quadrature module will give an accurate local quadrature estimate and small local error estimate. When $f(x)$ is badly behaved at x_0 that point will become trapped in smaller and smaller subintervals. On such subintervals the local quadrature estimate will eventually be small too, proportional to the size of the subinterval.

In a practical algorithm some provision is required for preventing intervals from getting too small, either by limiting the number of halvings or by more sophisticated strategies. It is also necessary to consider the effect of roundoff error in the evaluation of the integrand, which can occur even on a large subinterval. Attempts to detect this are largely ad hoc and differ from one routine to another. A good general reference for these algorithms, which also contains programs is the book by R. Piessens (1983).

The value of an integral will not be altered by changing the integrand at a finite number of points. Nevertheless the value returned by an automatic quadrature algorithm is completely determined by the integrand values at some finite set. No program of this type can rigorously guarantee that either its integral or error estimate have *any* accuracy. It is easy to find a problem which will cause any particular routine to either give up or produce incorrect results. This can happen if significant integrand features are missed by the sampled values $f(x_i)$. For example, suppose we wish to estimate

$$I = \frac{1}{\sqrt{2\pi}} \int_{-200,000}^{200,000} t^2 \exp(-t^2/2)\, dt \approx 1.0.$$

This will result in an underflow error in the EXP routine (-10^{10} is too small an argument for any exponential subroutine). If we get around this by setting underflows to zero our routine will ultimately almost certainly return with an integral estimate of 0. On a total interval of length $400,000$, the integrand is essentially zero on all but one hundredth of one percent. Unless serendipity comes to our aid the automatic quadrature algorithm will not sample there. There is no substitute for an alert user! (See also Problem P5–4).

5.7 SUBROUTINES Q1DA AND QK15

The automatic quadrature subroutine Q1DA is designed for general problems. The user must provide a main program that calls Q1DA and also provide a Fortran function named F defining the integrand. Q1DA is part of a set consisting of Q1DA, Q1DAX, and GL15T. The latter routine requests the evaluation of F at appropriate points on the integration interval. Q1DAX is a globally adaptive quadrature algorithm calling the local quadrature module GL15T. Q1DA is a "driver" for Q1DAX. No computation is done by Q1DA; its function is to set certain variables and arrays to standard, default values and then call Q1DAX. The use of drivers is common in mathematical software. Q1DA returns a quadrature estimate Q, an error estimate E, an error flag signifying success or failure, and a count of the number of integrand evaluations, KF. The count measures the amount of work that was required.

Often physical insight suggests a "natural frequency" for the problem which can be directly related to expected work. If we wish to integrate $e^x \sin^2 100x$ on $[0, \pi]$, there are 50 cycles of the sine. Even the most optimistic user would expect a few evaluations in each cycle. Thus if KF is returned as 100 or less, something is probably wrong either in the problem set-up or the use of the program.

Q1DA follows the general outline given in Section 6 with two enhancements.

(1) Randomization. The algorithm described in Section 6 has the interval $[a, b]$ halved at the first subdivision. Q1DA selects the first subdivision point randomly, near but not at $(b + a)/2$. All subsequent subdivisions are bisections. This causes all the integrand evaluation points to be slightly different each time the program is called. If the integrand has sharp spikes, calling Q1DA two or three times and comparing the results is a good way to reduce the possibility of missing some important feature.

(2) Singularity weakening at endpoints. Practical integrals often have singularities. Most of these occur at endpoints. One of the best ways to deal with these is to make a transformation of the variable of integration. If we make the change of variable $x = p(t)$, $dx = p'(t)\, dt$ then we have

$$I = \int_a^b f(x)\, dx = \int_\alpha^\beta f(p(t))p'(t)\, dt = \int_\alpha^\beta g(t)\, dt$$

where $p(\alpha) = a$, $p(\beta) = b$. Although the integral value, I, is the same for the integral of f as it is for g, the latter can be easier to evaluate numerically. Q1DA uses the following for $p(t)$,

$$p(t) = b - (b - a)u^2(2u + 3), \qquad \text{with } u = \frac{t - b}{b - a}$$

Note that $p(a) = a$, $p(b) = b$, and $p'(a) = p'(b) = 0$. If $f(x)$ has a singularity at a or b, $g(t)$ may not. For example, if $f(x) = \ln x$, $[a, b] = [0, 1]$

$$\int_0^1 \ln x\, dx = \int_0^1 \ln[t^2(3 - 2t)]6t(1 - t)\, dt$$

Even though it is more complicated, the second integral is easier to evaluate numerically. Q1DA makes this change of variable automatically. If f is not singular at an endpoint changing variables does not have much effect. Also, since $|p'(t)| \leq p'((a + b)/2) = 3/2$, the new integrand is slightly amplified near the midpoint at the same time that it is damped at the endpoints.

Q1DAX has a few extra features which Q1DA does not take advantage of. These include the ability to call the subroutine again and ask for more accuracy without repeating the preceding calculation, and to initialize the computation with more than one interval (useful if the integrand has a jump within the integration interval–normally Q1DAX would waste a great deal of time trying to locate this special point).

A global adaptive quadrature algorithm has one drawback—the number of intervals grows. Storage must be provided for these intervals and a difficult problem can use up all the available space. When that happens there are likely to be subintervals which already have small local error estimates and are not going to contribute much to further progress. In fact it is possible these will never be subdivided further. Recall the interval to be subdivided is that one with largest local error estimate. Q1DAX throws out intervals with small local error estimate when it runs out of space. This is often just enough to allow a computation nearing completion to finish. Q1DA allocates space for 50 intervals, enough for many practical problems. The user can specify this value explicitly in Q1DAX.

Q1DAX also provides as output the largest and smallest values of the integrand which it encountered. This can be useful if you would like to generate an integrand plot.

The following program illustrates the use of Q1DA.

```
C Typical problem setup for Q1DA
C
         A = 0.0
         B = 1.0
C Set interval endpoints to [0,1]
         EPS = 0.001
C Set accuracy request for three digits
         CALL Q1DA (A,B,EPS,R,E,KF,IFLAG)
         WRITE(*,*) A,B,EPS,R,E,KF,IFLAG
         END
C
C
         FUNCTION F(X)
C Define integrand F
         F = SIN(2.*X)-SQRT(X)
         RETURN
         END
C
C For this sample program the output is
C
C   0.0 1.0 .001   .041406750   .69077E-07 30 0
```

For many quadrature problems it is unnecessary or undesirable to use a sophisticated strategy. What is often needed is a routine which is simple and hence quick. QK15 is an local quadrature module which is an implementation of the (G_7, K_{15}) Gauss Kronrod pair described in Section 5. QK15 is from Quadpack, a unified collection of subroutines for one dimensional quadrature, that is described in the book by Piessens (1983). Quadpack contains subroutines for automatic quadrature as well as implementations of six different Gauss Kronrod pairs. QK15 returns a local quadrature estimate RESULT, and local error estimate ABSERR, for an arbitrary function f and finite interval (a, b). The transformations of Section 3 are performed internally, relieving the user of doing them. One advantage of such a local quadrature module is that it will always use a predictable amount of work, fifteen integrand evaluations. QK15 also uses these values to provide estimates of two other related integrals,

$$\int_a^b |f|\, dx, \qquad \text{and} \approx \int_a^b \left| f - \frac{\text{RESULT}}{b-a} \right| dx.$$

QK15 is like Q1DA in that it requires that the user write a Fortran function which will evaluate the integrand for an arbitrary x. Q1DA does not allow the user to select the function name, it must be F. But QK15 is more flexible. (Q1DAX has similar flexibility.) You can select any name for this function, but your choice must appear in *at least three* places, (1) in an EXTERNAL statement in the routine which calls QK15, (2) as the name of the function that you write, and (3) in the argument list of each call to QK15. In

Fortran, if one of the arguments to a subroutine or function is the name of a routine, this must be handled differently than other types of parameters, and the EXTERNAL statement warns Fortran to do this. Here is an example of the use of QK15 which illustrates the EXTERNAL statement.

```
C       Compute erf(1), i.e. integral of 2/sqrt(pi) * exp(-x*x) from 0 to 1.0
C
        EXTERNAL FUNC
        REAL A,B,RESULT,ABSERR,RESABS,RESASC,PI
C
        A = 0.0
        B = 1.0
        PI = 4.0*ATAN(1.0)
        CALL QK15(FUNC,A,B,RESULT,ABSERR,RESABS,RESASC)
        WRITE(*,*) ' QK15 ESTIMATE OF ERF(1) '
        WRITE(*,*) ' 2.0/SQRT(PI)*RESULT,        ABSERR'
        WRITE(*,*) 2.0/SQRT(PI)*RESULT,ABSERR
        STOP
        END
C
        REAL FUNCTION FUNC(X)
          REAL X
          FUNC = EXP(-X*X)
          RETURN
        END
C
C       Output (from IBM PC/AT) is
C
C   QK15 ESTIMATE OF ERF(1)
C   2.0/SQRT(PI)*RESULT,        ABSERR
C       0.842701            0.50229e-4
C
```

Notice that each evaluation of the integrand requires a call to the function FUNC. A more efficient way is to make FUNC a subroutine and pass to it an array of abscissas. This replaces fifteen function calls by one subroutine call. Then, in FUNC the evaluation at each abscissa is independent, and on some computers the Fortran compiler may be able to organize the calculation to compute several of them in parallel. Unfortunately, none of the available general purpose quadrature software has been written to take advantage of this, and we will continue to use the traditional mechanism here.

5.8 DATA INTEGRATION

Suppose we are given a sequence of (x, y) pairs

$$(x_i, y_i), \qquad x_1 < x_2 < \cdots < x_n$$

which are thought of as accurate values of some (unknown) function $f(x)$,

$$f(x_i) = y_i.$$

The problem is to compute

$$I = \int_{x_1}^{x_n} f(x) \, dx.$$

As the data are accurate, a reasonable approach is to interpolate to the data in the sense of Chapter 4. That is, to construct a known function $g(x)$ such that $g(x_i) = y_i$ and then compute

$$I \approx I' = \int_{x_1}^{x_n} g(x) \, dx.$$

The most heavily used interpolants are polynomials and piecewise polynomials, such as piecewise linear polynomials, Hermite cubics or splines. Piecewise polynomials lead to simple expressions for the quadrature rule. These are worked out below. However, if the data contain substantial amounts of experimental or computational error, then interpolation is inappropriate and the techniques described in Chapter 6 are recommended instead.

Let $g(x)$ be the function that connects each pair of data points with a line segment. This is the **piecewise linear interpolant** defined in Section 7 of Chapter 4 and denoted $L(x)$ there.

$$I \approx I' = \int_{x_1}^{x_n} g(x) \, dx = \sum_{i=1}^{n-1} \int_{x_i}^{x_{i+1}} g(x) \, dx.$$

Since g is linear between x_i and x_{i+1} it can be integrated exactly, giving

$$I \approx I' = \frac{1}{2} \sum_{i=1}^{n-1} (x_{i+1} - x_i)(y_i + y_{i+1})$$

$$= \frac{1}{2}[(x_2 - x_1)y_1 + (x_3 - x_1)y_2 + (x_4 - x_2)y_3 + \cdots$$
$$+ (x_n - x_{n-2})y_{n-1} + (x_{n-1} - x_n)y_n].$$

Note that if the x_i are equally spaced this is the compound trapezoid rule, otherwise it is a more general trapezoid rule. We can interpret this rule as giving the *exact* integral of the function $g(x)$ which is the piecewise linear interpolant through the data.

*5.8.1 Data Integration: Hermite Cubic Quadrature

If we use as an interpolant the Hermite cubic $c(x)$ defined in Section 10 of Chapter 4,

$$g(x) \equiv c(x) = \sum_{i=1}^{n} y_i c_i(x) + d_i \hat{c}_i(x),$$

then

$$I \approx I' = \int_{x_1}^{x_n} c(x)\,dx = \sum_{i=1}^{n} y_i \int_{x_1}^{x_n} c_i(x) + d_i \int_{x_1}^{x_n} \hat{c}_i(x).$$

Because formulas for the c_i and $\hat{c}_i$ are known explicitly it is possible to compute these integrals in closed form. The result can be shown to be

$$I' = \sum_{i=1}^{n} \alpha_i y_i + \beta_i d_i,$$

where

$$\alpha_i = \begin{cases} (x_2 - x_1)/2, & \text{if } i = 1; \\ (x_{i+1} - x_{i-1})/2, & \text{if } i = 2, \ldots, n-1; \\ (x_n - x_{n-1})/2, & \text{if } i = n, \end{cases}$$

and

$$\beta_i = \begin{cases} (x_2 - x_1)^2/12, & \text{if } i = 1; \\ (x_{i+1} - x_{i-1})(x_{i+1} - 2x_i + x_{i-1}), & \text{if } i = 2, \ldots, n-1; \\ -(x_n - x_{n-1})^2/12, & \text{if } i = n. \end{cases}$$

In practice we do not know the d_i's. But if these are determined by either spline interpolation as in Chapter 4 or by another method, then this rule provides accurate estimates.

We can interpret the formula for Hermite cubic quadrature as follows. Note that the first sum above, $\sum \alpha_i y_i$, is exactly the generalized trapezoid rule. That represents the area under the straight line segments connecting the y_i's. Hermite cubic interpolation is smoother. Between the line connecting y_i with y_{i+1} and the cubic connecting them is a small cap (or cup). The second sum $\sum \beta_i d_i$ is the area of these regions.

The only important special case occurs when the data are equally spaced. All the β_i's will be zero except for β_1, β_n, and $\sum \alpha_i y_i$ becomes the compound trapezoid rule. Thus we have the important result that for equally spaced data

$$I' = h\left[f(x_1)/2 + f(x_2) + \cdots + f(x_{n-1}) + f(x_n)/2\right] + \frac{h^2}{12}(d_1 - d_n),$$

Compound trapezoid rule + two end corrections = Hermite cubic quadrature.

Note that *any* set of interior derivatives d_i produces identical integral estimates although different interpolants.

If the physical model suggests that the underlying function $f(x)$ is periodic with $f'(x_1) = f'(x_n)$, then the end corrections disappear leaving exactly the compound trapezoid rule. The derivation in Section 4 gives no hint that periodicity of $f(x)$ is a distinguished case. It can be shown that the compound trapezoid rule is often exceptionally accurate in the special situation when the integrand is a smooth periodic function which is given at equally spaced points.

Example 5.7   Compound Trapezoid Rule for a Periodic Function.

The function $f(x) = 1/(1 + .5\sin 10\pi x)$ is periodic, of period one fifth. This is not a data integration problem *per se*, i.e., we can evaluate $f(x)$ everywhere. But its periodicity suggests that equally spaced evaluations and the trapezoid rule would be a good bet. The compound trapezoid estimates for the integral on [0, 1] with 2, 4, 8 and 16 points are 1.0, 1.16666650, 1.15476180 and 1.15470050. The last estimate is in error by 4 in the last place. The Gauss quadrature estimates of this integral with the same numbers of points are far less accurate. ■

5.8.2 Subroutine PCHQA

For many problems the formulas in Sections 8–8.1 are sufficiently simple that it is possible to program them directly. But because the integration depends upon manipulation of Hermite cubics, the PCHIP package which was described in Section 9 of Chapter 4 provides subroutines for this purpose too. In this text we have included an easy-to-use driver PCHQA. This subroutine requires that the user provide input arrays containing the abscissas and data values as well as the derivatives d_i at these points. It also must be given an interval $[a, b]$, contained within the data, $x_1 \leq a \leq b \leq x_n$ on which to perform the integration. Thus you can use PCHQA even if your integration interval does not begin or end exactly at a data point. Usually, you will call PCHEZ first to compute the unknown devivatives (see Section 9 of Chapter 4) but PCHQA can also be called directly. In the common situation when the data are equally spaced and $[a, b] = [x_1, x_n]$ we know from Section 8.1 that $d_2, \ldots, d_{n-1}$ are not involved in the integral estimate and they can be safely set to zero. One disadvantage of PCHQA is that the data are always treated as if they were unequally spaced, and you must input all the abscissas. An example of its use can be found in the sample program in Section 9 of Chapter 4.

5.9   INFINITE AND SEMI-INFINITE INTERVALS

The Quadpack subroutine QAGI can evaluate many infinite integrals. It is one of only a few pieces of quality software for infinite intervals, and none are as reliable or general as finite interval routines. The best choice of technique depends heavily on the characteristics of the particular problem under consideration. The basic methods described here are in common use, with Truncation (Section 9.1) and Weight Functions (Section 9.4) the most popular. Unfortunately it is not possible to cite one technique or rule which works so well that it can be used exclusively.

5.9.1 Truncation

If either endpoint a, b or both can be infinite, a common approach is to truncate the limits and use a standard method on the finite part. Sometimes it is possible to analyze the tails of the integrand and show that their contribution is negligible, but more often several different intervals are selected and it is verified computationally that the finite integral estimate is not changing. This can give wrong answers in a general situation, but often works if some physical insight can be brought to the problem.

Example 5.8   An Infinite Integral by Truncation.

Compute

$$I = \int_0^\infty \exp(-x) \cos^2(x^2) \, dx = 0.70260\ldots.$$

We have (denoting this integrand by $f(x)$)

$$I - \int_0^A f(x) \, dx = \int_A^\infty f(x) \, dx < \int_A^\infty \exp(-x) \, dx = \exp(-A).$$

Thus the error in neglecting the tail is no more than $\exp(-A)$. Figure 5.2 shows this integrand, and illustrates the effect of truncating the interval. ■

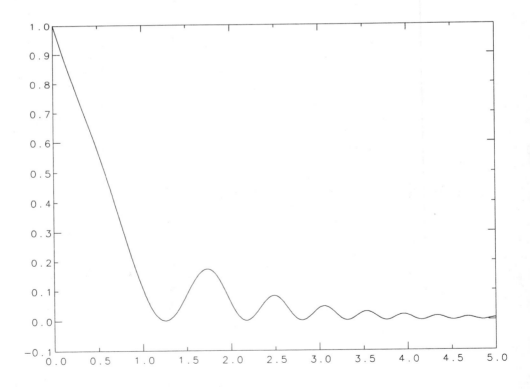

Figure 5.2   Integrand of Example 5.8

5.9.2 Transformation

By setting $x = p(t)$ we can transform the interval. Typical $p(t)$'s for the interval $[0, \infty)$ are $x = -\ln t$ or $x = t/(1-t)$. For the former we get

$$\int_0^\infty f(x)\,dx = -\int_1^0 f(-\ln t)\,\frac{dt}{t} = \int_0^1 f(-\ln t)\,\frac{dt}{t}.$$

For Example 5.8 this is

$$I = \int_0^1 \cos^2(\ln^2 t)\,dt.$$

This integrand oscillates infinitely often between the values 0 and 1 on any interval containing the origin. If instead we had used the transformation $x = p(t) = -2\ln t$, the integrand would be $2t\cos^2(4\ln^2 t)$. This also oscillates, but it is modulated by the line $y = t$ and hence goes to zero at the origin. Ignoring a small interval $[0, \epsilon]$ leaves us with a more tractable problem. On the other hand, Q1DA can do either of these transformed integrals about equally well. For example, asking for five digits in the first transformed integral on $[0, 1]$, Q1DA estimates $I = 0.7026013 \pm 6 \cdot 10^{-6}$ with 870 integrand evaluations. Using the second transformation, the estimate is $I = 0.7026032 \pm 7 \cdot 10^{-6}$ with 930 function evaluations.

Transformations must be applied with great care or the result will not be an easier integral, only a finite one. On the other hand, judicious transformations can make almost magical improvements. This is still an active research area. The paper by K. Murota (1982) is a useful reference on these techniques.

*5.9.3 Trapezoid Rule for an Infinite Interval

If we select a spacing h, the trapezoid rule for semi-infinite and doubly-infinite intervals is

$$\int_0^\infty f(x)\,dx \approx h\sum_{k=1}^\infty f(kh),$$

$$\int_{-\infty}^\infty f(x)\,dx \approx h\sum_{k=-\infty}^\infty f(kh).$$

In practice we must not only pick h but also a finite limit N for the sum. One strategy is to compute a sequence of estimates with decreasing h and increasing N, in such a way that Nh^2 is approximately constant. These rules are remarkably efficient for many problems, and in particular for those integrands which decay rapidly to zero at $-\infty$ and $+\infty$. For example, if we consider the problem $I = 1/\sqrt{\pi}\int_{-\infty}^\infty \exp(-x^2)\,dx$, picking $N = 10$ and $h = 1/\sqrt{N}$, the trapezoid rule estimate is accurate to $2 \cdot 10^{-6}$. On the other hand for the integrand $1/(1 + 10x^2)$, $N = 100$ produces an estimate which is only accurate to 0.01, and $N = 1000$ is accurate to $6 \cdot 10^{-3}$.

The success of the method depends upon the endpoint behavior of the integrand $f(x)$. Consequently, several researchers have proposed using a change of variable $x = p(t)$ which leaves the interval infinite but forces the new integrand $p'(t)f(p(t))$ to go to zero more rapidly at the endpoints. This is similar in philosophy to the transformation used in Q1DA. Recall there, $p'(a) = p'(b) = 0$. For example, the change of variable

$$x = p(t) = \exp(t) - \exp(-t)$$

transforms the interval $(-\infty, \infty)$ into itself, but may cause more rapid decay of the integrand at $\pm\infty$. The integral becomes

$$\int_{-\infty}^{\infty} f(x)\, dx = \int_{-\infty}^{\infty} (e^t + e^{-t})f(e^t - e^{-t})\, dt.$$

If we apply this transformation to the integrand $1/(1 + 10x^2)$ and then use the trapezoid rule, 100 points gives an estimate which is accurate to $8 \cdot 10^{-5}$, and $N = 1000$ gives 15 digits. Sometimes this transformation can be applied more than once.

5.9.4 Weight Functions and Gauss Laguerre Quadrature

Certain functions commonly turn up as part of the integrand in infinite intervals. Two of the most frequently occuring are $\exp(-x)$ and $\exp(-x^2)$. Denote by $q(x)$ either of these two. The **weight function** technique is to find formulas of the form

$$I = \int_a^b q(x)f(x)\, dx = \sum_{i=1}^{n} w_i f(x_i) + R_n.$$

The limits a, b may be finite or infinite. Thus the weights and nodes are thought of as incorporating analytic information about $q(x)$. This generalizes Section 2.1 where we implicitly took $q(x) = 1$.

As an example of the weight function approach we look for a rule of the form

$$\int_0^{\infty} \exp(-x)f(x)\, dx \approx w_1 f(x_1) + w_2 f(x_2).$$

As in Section 2.1, we think of the two nodes and two weights as unknowns and seek to find them so that the approximation is an equality for $f(x) = 1$, x, x^2 and x^3. This leads, as in that section, to a system of four nonlinear equations

$$i! = \int_0^{\infty} \exp(-x)x^i\, dx = w_1(x_1)^i + w_2(x_2)^i \qquad i = 0, 1, 2, 3.$$

These equations have a unique solution. Many of the same conclusions apply as in Section 2.1. For the two-point rule we find

$$\int_0^{\infty} \exp(-x)f(x)\, dx \approx \frac{2+\sqrt{2}}{4}f(2-\sqrt{2}) + \frac{2-\sqrt{2}}{4}f(2+\sqrt{2}).$$

This is the two-point **Gauss Laguerre** quadrature rule because the nodes are the zeros of the quadratic Laguerre polynomial[1]. Three of these quadrature rules are listed in Table 5.4, but most scientific program libraries provide a subroutine that will generate the rules for arbitrary n.

Similar rules for arbitrary n and general weight function $q(x)$ have been derived, and there is a complete theory, at least if $q(x)$ is always positive. These rules work well if f can be approximated by a polynomial, and sometimes in more general situations. For other examples of these formulas see the book by P. Davis and P. Rabinowitz (1984).

Example 5.9   Infinite Integral by Gauss Laguerre.

Compute approximately

$$I = 1.046\ldots = \int_0^\infty \exp(-x)x^{1.2}\, dx.$$

The two point Gauss Laguerre rule gives

$$I \approx \frac{2+\sqrt{2}}{4}(2-\sqrt{2})^{1.2} + \frac{2-\sqrt{2}}{4}(2+\sqrt{2})^{1.2} = 1.089\ldots. \quad \blacksquare$$

*5.9.5 The tanh rule

The infinite trapezoid rule works so well that a realistic approach to evaluating an integral on a *finite* interval is to convert it to an infinite interval in such a way that the transformed integrand dies out at the infinite endpoints! One specific transformation which has been studied is

$$x = p(t) = \tanh(\frac{t}{2}) = \frac{e^{t/2} - e^{-t/2}}{e^{t/2} + e^{-t/2}},$$

which leads to

$$I = \int_{-1}^1 f(x)\, dx = \int_{-\infty}^\infty f(p(t))p'(t)\, dt = 2\int_{-\infty}^\infty \frac{1}{(e^{t/2} + e^{-t/2})^2} f\left(\frac{e^{t/2} - e^{-t/2}}{e^{t/2} + e^{-t/2}}\right)\, dt.$$

This is then approximated by

$$I \approx h \sum_{k=-N}^N p'(kh)f(p(kh)) = 2h \sum_{k=-N}^N \frac{1}{(e^{kh/2} + e^{-kh/2})^2} f\left(\frac{e^{kh/2} - e^{-kh/2}}{e^{kh/2} + e^{-kh/2}}\right),$$

---

[1] Edmond Nicols Laguerre (1834–1886), was a French mathematician, who spent all of his professional career at the École Polytechnique in Paris. Most of his work was in what we now call analytic geometry although he made important advances in analysis. In addition to the quadrature formula above there is a class of differential equations associated with his name. He was said to be a quiet, gentle man who was passionately devoted to his research, his teaching, and the education of his two daughters.

TABLE 5.4 SOME GAUSS LAGUERRE QUADRATURE RULES FOR $\int_0^\infty \exp(-x)f(x)\,dx$

Nodes	Weights
$n = 2$	
0.585786437626905	0.853553390593274
$0.341421356237310 \times 10^1$	0.146446609406726
$n = 4$	
0.322547689619392	0.603154104341634
$0.174576110115835 \times 10^1$	0.357418692437800
$0.453662029692113 \times 10^1$	$0.388879085150054 \times 10^{-1}$
$0.939507091230113 \times 10^1$	$0.539294705561327 \times 10^{-3}$
$n = 8$	
0.170279632305101	0.369188589341638
0.903701776799380	0.418786780814343
$0.225108662986613 \times 10^1$	0.175794986637172
$0.426670017028766 \times 10^1$	$0.333434922612157 \times 10^{-1}$
$0.704590540239347 \times 10^1$	$0.279453623522567 \times 10^{-2}$
$0.107585160101810 \times 10^2$	$0.907650877335821 \times 10^{-4}$
$0.157406786412780 \times 10^2$	$0.848574671627253 \times 10^{-6}$
$0.228631317368893 \times 10^2$	$0.104800117487151 \times 10^{-8}$

Table 5.5 tanh QUADRATURE NODES AND WEIGHTS FOR N=5

Nodes	Weights
0.000000	.893459
±0.713098	.439127
±0.945434	.094844
±0.990649	.016631
±0.998428	.002807
±0.999737	.000471

with a suggested step

$$h = \pi\sqrt{\frac{2}{N}} - \frac{1}{N}.$$

This is a strange transformation, as the resulting rule on $[-1, 1]$ is not exact for any polynomials—not even constants—for any positive h and finite N. Nevertheless it is remarkably accurate. To give a feeling for the numbers we consider $N = 5, h = \pi\sqrt{0.4}$. Table 5.5 lists $p(kh)$ and $2hp'(kh)$ which are the quadrature nodes and weights. Note that the rule has $2N + 1$ nodes. These are symmetric with respect to the origin, and the origin is itself a node.

We see from the table that most of the nodes cluster near ± 1. This can lead to computational pitfalls: integrands which contain factors such as $1 - x$ may be impossible to evaluate directly, on some machines, because of underflow; and many integrands will need special evaluation at nodes near ± 1 to avoid excessive loss of significance.

Example 5.10   Finite Integral by tanh Rule.

Compute by the tanh rule

$$\int_{-1}^{1} \exp(-x^2)\ln(1 - x)\, dx.$$

We get the following results for various values of N and h.

N	h	Rule
5	1.787	$-.306715$
10	1.304	$-.315748$
20	0.943	$-.316688$
30	0.778	$-.316713$
40	0.677	$-.316714$

The estimate for $N = 40$ (81 integrand evaluations) is correct to all the digits printed. ∎

*5.10 DOUBLE INTEGRALS

If D is a domain in the x-y plane, we often want to compute

$$I = \iint_D f(x, y)\, dA.$$

This is a much more difficult problem than the evaluation of a one-dimensional integral. It is still an active research topic and many gaps remain in our knowledge.

In analogy to the definition in Section 2, an n-point quadrature formula is

$$I = \sum_{i=1}^{n} w_i f(x_i, y_i) + R_n.$$

The formula is of degree d if $R_n = 0$ whenever f is any bivariate polynomial of degree d, but is nonzero for some polynomial of degree $d + 1$. A bivariate polynomial of degree d is a linear combination of terms $x^p y^q$ with $p + q \leq d$.

If D is a rectangle, triangle, circle or other standard domain, families of quadrature formulas exist as we will illustrate in Section 10.1. More general domains sometimes can be dealt with by one of the following techniques.

(1) Embedding the domain in a larger rectangle and altering the definition of $f(x, y)$ so it is zero outside of D. Then the integral on the rectangle is equal to the integral on D. Unfortunately, the new integrand now has a jump discontinuity when thought of as a function on the rectangle.

(2) Approximation of D by the union of triangles. If the boundary of D is nonlinear we may need a large number of small triangles. Since there will be some error on each triangle, the trick is to leave out enough of D so that the omission error is the same magnitude as the quadrature error on the union of triangles.

(3) Transforming D into a more standard region. This is less general, but in the next section we show how an integral on a triangle can be transformed into one on a square.

*5.10.1 Product Rules for Rectangles and Triangles

If

$$\int_\alpha^\beta f(x)\, dx = \sum_{i=1}^n w_i f(x_i) + R_1$$

and

$$\int_a^b g(y)\, dy = \sum_{j=1}^m p_j g(y_j) + R_2,$$

then

$$\int_a^b \int_\alpha^\beta f(x, y)\, dx\, dy = \int_a^b \left[\sum w_i f(x_i, y) + R_1 \right] dy$$

$$= \sum w_i \int_a^b f(x_i, y)\, dy + \int_a^b R_1\, dy$$

$$= \sum_{i=1}^n w_i \left[\sum_{j=1}^m p_j f(x_i, y_j) + R_2 \right] + \int_a^b R_1\, dy$$

$$= \sum_{i=1}^n \sum_{j=1}^m w_i p_j f(x_i, y_j) + \sum_{i=1}^n w_i R_2 + \int_a^b R_1\, dy$$

$$= \sum_{i=1}^n \sum_{j=1}^m w_i p_j f(x_i, y_j) + \mathcal{R}.$$

This is called a **product formula** because it uses nm integrand evaluation points. Figure 5.3 shows a $2 \times 3 = 6$ point product rule on a rectangle formed from a three-point Gauss quadrature and a two-point Gauss quadrature. The quadrature weight associated with a node in the plane is the product of the two one-dimensional weights.

Figure 5.3   Product Rule Composed of Two-Point and Three-Point Gauss Rules

If the one-dimensional formulas are of polynomial degree d_1 and d_2 then the product formula will integrate polynomials $x^p y^q$ with $p \leq d_1$, $q \leq d_2$. Thus a product formula is exact for all bivariate polynomials of degree $\min(d_1, d_2)$, although it will also integrate some polynomials of degree $d_1 + d_2$. Product rules are easy to derive and hence are popular. Their disadvantage is the requirement of a larger number of nodes than necessary. For example, a product rule using two three-point Gaussian rules has nine nodes and is of bivariate degree five. A non-product rule of degree five exists with only seven nodes.

A surprising number of regions can be transformed into a rectangle. Sometimes other difficulties are introduced but the technique is still valuable. To illustrate, suppose we want to compute the integral on a triangle

$$I = \iint_\Delta f(x, y) \, dA \qquad \Delta = \{ (x, y) : 0 \leq x \leq 1, \, 0 \leq y \leq x \}.$$

Let us make a change of variable

$$u = x, \quad v = y/x.$$

Then the original integral becomes

$$I = \int_0^1 \int_0^x f(x, y) \, dy \, dx = \int_0^1 \int_0^1 f(u, uv) \, u \, dv \, du.$$

We can now apply a product rule for the integral on a square. This is an algebraic trick, but as with all transformations its usefulness depends upon the problem. For example, Figure 5.4 illustrates the nodes on the triangle when the three-point Gaussian quadrature rule is used in each dimension. Such rules place a large fraction of the nodes in the upper portion of the triangle leaving the lower portion sparsely sampled. For this reason triangle product rules with large numbers of points are seldom used.

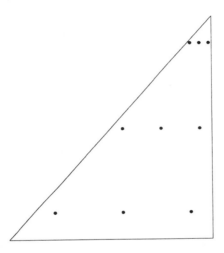

Figure 5.4 Product Rule on Triangle

5.10.2 Using One-Dimensional Programs for Double Integrals

Until recently, there were no high quality subroutines for double integrals, but for some time we have had good one-dimensional routines. Thus it was natural to ask if the one dimensional routines could be used to evaluate a double integral. In this section we describe a technique which is still in widespread use.

Assume we have two subroutines for one-dimensional quadrature
SUBROUTINE Q1 (G,A1,B1,EPS1,RES1,ERR1,KF1,IFLAG1)
SUBROUTINE Q2 (F,A2,B2,EPS2,RES2,ERR2,KF2,IFLAG2)
In problem P5–11a you will use Q1DA and a copy which you rename Q1, but the method works with any two reliable routines. Our task is to figure out what functions to take for the one dimensional integrands F and G and how to set the input error requests EPS1 and EPS2. If I can be written in the form

$$I = \int_a^b \left[\int_\alpha^\beta f(x,y)\, dy \right] dx,$$

we can think of the term in brackets as a function $g(x)$,

$$I = \int_a^b g(x)\, dx, \qquad g(x) = \int_\alpha^\beta f(x,y)\, dy,$$

and can use the two subroutines together to evaluate the double integral. Given x, we can use Q2 to evaluate the integral defining $g(x)$. We will get

$$\text{RES2} = g(x) - E_2,$$

with, we hope,

$$|E_2| < \text{ERR2} < \text{EPS2}.$$

The numbers RES2, ERR2, and E_2 depend on x. Now use Q1 to evaluate the integral of g. Of course we are not really integrating $g(x)$ but rather $\text{RES2} = g(x) - E_2$. The output of Q1 is

$$\text{RES1} = \int_a^b [g(x) - E_2(x)] \; dx - E_1,$$

with

$$|E_1| < \text{ERR1} < \text{EPS1}.$$

Thus

$$I = \int_a^b g(x) \; dx = \text{RES1} + \int_a^b E_2(x) \; dx + E_1.$$

We want the magnitude of the last two terms to be less than an absolute accuracy ϵ. Since

$$\left| \int_a^b E_2(x) \; dx + E_1 \right| < (b-a)\text{EPS2} + \text{EPS1},$$

any combination of EPS1 and EPS2 making the right-hand side less than ϵ would seem to be acceptable. Actually we have not taken into consideration the amount of work being performed by Q1 and Q2. It is intuitively obvious, and can be proven, that if the inner integral is computed inaccurately then the outer subroutine Q1 will think that it is integrating a rough function. This will result in extra work and occasionally a complete failure of Q1. A practical rule of thumb is that the inner integral be computed about a factor of ten more accurately than the outer. We use

$$\text{EPS1} = 0.9\epsilon, \qquad \text{EPS2} = \frac{\epsilon}{10(b-a)}.$$

Then an error estimate is given by

$$\text{ERR1} + (b-a) \max_x \text{ERR2}.$$

Example 5.11 Double Integral by a Pair of One-Dimensional Programs.

The following program estimates a two dimensional integral on a rectangle.

$$I = \int_0^1 \int_0^2 \exp(-x^2 y^2) \, dx \, dy.$$

```
C Double integral by two one dimensional subroutines
C

      EXTERNAL G
      COMMON    ERMAX, XX
C

      ERMAX = 0.0
      EPS   = 1.E-4
      EPS1  = 9.0/10.0 * EPS
      A1    = 0.0
      B1    = 1.0
      CALL Q1 (G,A1,B1,EPS1,RES1,ERR1,KF1,IFLAG1)
C

      ERR   = ERR1+(B1-A1)*ERMAX
      WRITE (*,*) RES1, ERR, KF1, IFLAG1

      STOP
      END
C
C

      FUNCTION G(X)
        EXTERNAL F
        COMMON    ERMAX, XX
C

        XX    = X
        EPS2  = 0.1 * 1.E-4
        A2    = 0.0
        B2    = 2.0
        CALL Q2 (F,A2,B2,EPS2,RES2,ERR2,KF2,IFLAG2)
C
C      Test IFLAG2
         IF (IFLAG2.NE.0)
     *         WRITE (*,*) 'ERROR IN Q2, IFLAG, KF2 = ', IFLAG2, KF2
C
```

```
          ERMAX = MAX (ERMAX,ERR2)
          RETURN
          G   = RES2
      END
C
C

      FUNCTION F(Y)
         COMMON  ERMAX, X
C

         F   = EXP (-X*X*Y*Y)
         RETURN
      END    ■
```

Notice both the COMMON statement and the assignment statement XX=X within G. These are needed because F, the integrand of Q2, needs to know the value of X that is the input to G. Use of COMMON is the closest that Fortran gets to allowing 'global variables.' The variables X in FUNCTION F(Y) and XX in the main program and in FUNCTION G(X) refer to the same memory location. The COMMON is required here since F only has one argument and we have to get the value of X to it. In Fortran, a variable cannot be in an argument list and in COMMON, so we declare another variable XX that is in COMMON and set its value to X.

*5.10.3 Non-Product Rules and Automatic Two-Dimensional Programs

A non-product rule is any quadrature rule which is not the result of applying a one-dimensional rule to successive dimensions. Non-product rules are usually more efficient for the same accuracy than product rules, hence there is strong motivation to discover them. A natural way to look for them is to say: if we use a rule with m nodes how can we select the nodes and weights so that the rule integrates bivariate polynomials of as high degree as possible? In one dimension this is the argument we used for Gaussian rules. In two and higher dimensions the situation is much more complicated. With m nodes there may be several rules with different nodes or weights which integrate the same degree polynomial, and no way to integrate higher degree polynomials without more nodes. For another region, the results might be completely different for the same m. For most regions we simply do not know what the best rules are. A reference to these techniques is the book by A. H. Stroud, (1972). Because triangles are important in applications there are detailed references specific to them. An excellent manual containing advice on selection and use is the report by J. N. Lyness (1983).

We have seen that Gauss-Kronrod pairs are useful for one dimensional quadrature. Some work has also been done to extend these ideas to two dimensions. At this time there is one success to report, Laurie (1980), began with the 7-node degree-5 rule on a triangle and added 12 nodes (for a total of 19) to obtain a rule of degree 8. It is known that to integrate all polynomials of degree less than or equal to 8 on a triangle requires at least 16 nodes, so Laurie's rules require 3 more nodes but also allows the error to be estimated. Figure 5.5 shows the placement of the 7-node degree-5 rule and the additional 12 nodes which give Laurie's degree 8 rule.

The algorithm and program segment in Section 10.2 have been used for many years because of the availability of one-dimensional automatic programs. Because we now have some non-product rules which generate error estimates, a few truly two-dimensional programs have made their appearance. The earliest of these used a rectangle as a basic region but newer ones utilize triangles instead. The overall strategy is global adaptation. One of the most attractive features of the global algorithm is that it makes perfectly good sense in two or more dimensions. Instead of intervals we think of triangles, and bisection is replaced by some simple scheme for subdividing a triangle into two (or four) subtriangles. The storage requirements for these programs are substantially greater than for one-dimensional programs, and are also greater than for the program in Section 10.2. But compared to using two one-dimensional subroutines there is usually an overall gain in efficiency in terms of the number of integrand evaluations. Furthermore these programs can handle much more difficult integrals. For a typical and well done program, see the paper by E. de Doncker and I. Robinson (1984).

5.11 MONTE CARLO METHODS

During World War II the term "Monte Carlo" was introduced by von Neumann and Ulam at Los Alamos in the context of simulation of neutron diffusion in fissionable material. The idea was known to statisticians as early as 1910 (e.g. W. S. Gosset who wrote under the pseudonym "Student"). The name, however, was coined by Nick Metropolis. He states that Enrico Fermi used the idea, by hand, to predict the results of experiments to astonished colleagues at the University of Rome in the early 1930's. The method was also used to approximate some complicated multidimensional integrals. Since then, Monte Carlo has been generalized in over 3000 articles and books. We present here a simple example, based upon the earliest ideas for the approximate evaluation of a one dimensional integral. For some other illustrations see Section 9 of Chapter 10. Two good references are the book by R. Rubinstein (1981), and the paper by S. Haber (1970).

We consider sample-mean Monte Carlo, which is easy to describe and implement if we have a source of uniformly distributed random numbers. The generation of pseudo random numbers is presented in Chapter 10. We select N random points $0 \leq U_i \leq 1$ and scale each to $[a, b]$, $u_i = a + U_i(b - a)$. Then we compute

$$\theta_N = (b - a)\frac{1}{N} \sum_{i=1}^{N} f(u_i).$$

This is $(b - a)$ times the average of f. Intuition suggests that this will approximate the integral I and elementary statistical analysis confirms that. To use Monte Carlo to estimate an integral on a two dimensional region we compute the average of the integrand evaluated at N points randomly chosen in the region and then multiply by the area.

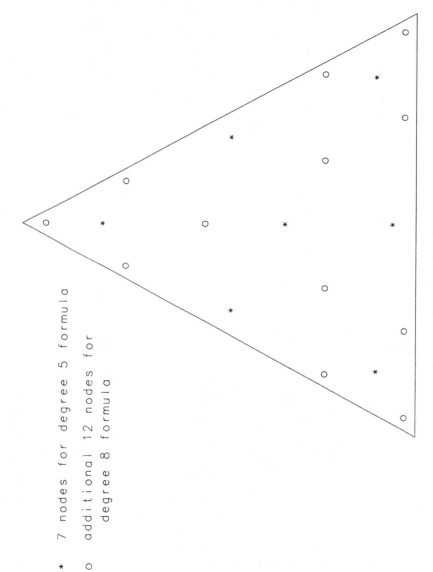

* 7 nodes for degree 5 formula

o additional 12 nodes for
  degree 8 formula

Figure 5.5 Non-Product Rule for a Triangle

How accurate are these estimates? Since the u_i are random there is nothing in principle to prevent them from all falling into any particular little subinterval of $[a, b]$ in a single experiment or sampling. Thus all we can say with certainty is that θ_N will lie somewhere between the maximum and minimum values of f—a rather useless fact. However, it is not likely that all the u_i will fall in one small subinterval, particularly if N is large. Thus there is still the possibility of obtaining probabilistic statements about the accuracy of θ_N—for example, statements that with a certain high probability, $|\theta_N - I|$ will be smaller than some given number. In fact, it follows from simple considerations in probability theory that θ_N, which is a random variable, has its mean value equal to the integral I, while its standard deviation is given by

$$\sigma(\theta_N) = |b - a| N^{-1/2} \sigma(f),$$

where $\sigma(f)$ is a constant, which can be estimated, depending on f but not N. Further, if we suppose, as is commonly done, that θ_N is close to being normally distributed for practical values of N, we may say that there are about nineteen chances out of twenty that

$$|\theta_N - I| \leq 2|b - a| \sigma(f) N^{-1/2} = O(N^{-1/2}).$$

In this sense we say that the error of θ_N goes to zero like $N^{-1/2}$ as N increases.

Roughly speaking, to get an extra decimal place of accuracy it is necessary to increase N by a factor of 100. This does not seem impressive, especially in the light of some of the examples in this chapter. What is not obvious is that "sample-mean" Monte Carlo can be applied in more than one dimension to just about any function and region as long as the volume of the region is known and it is possible to conveniently choose points randomly inside it. The basic result about $N^{-1/2}$ reduction of the error applies without regard to the dimension of the region, and with few assumptions about the smoothness of f. All that is required is that all integrals which appear actually exist—f need not even be continuous.

In one dimension, of course, more rapid convergence than $N^{-1/2}$ can usually be obtained. If the integrand is smooth this can be done easily; and if it is not, its singularities—which occur at isolated points—may be avoided by various devices. So Monte Carlo methods are almost never used for one dimension. In two or more dimensions, singularities occur not only at points but also along curves or surfaces of complicated shape and in these cases can rarely be removed. Thus for discontinuous functions of several variables the Monte Carlo method is one of the few good methods.

Example 5.12 Monte Carlo Evaluation of a One-Dimensional Integral.

Using sample-mean Monte Carlo, we estimate the integral of Example 5.8, Section 9.1 after applying each of the transformations $x = -\ln t$ and $x = -2 \ln t$.

N	Estimate with $x = -\ln t$	Estimate with $x = -2\ln t$
100	0.713998	0.710285
1000	0.710525	0.734007
10000	0.712493	0.705209
100000	0.700767	0.698311
1000000	0.703014	0.702628

Notice that the Monte Carlo method does not prefer either transformation; after $1,000,000$ points both estimates are good to about 3–4 digits, which agrees with $N^{-1/2}$ convergence. Recall that $I = 0.70260\ldots$. The reader may be shocked to consider that one million points are required for this problem. Indeed this is excessive for a one-dimensional integral. On the other hand if the integral were six-dimensional we would expect comparable errors. But in six dimensions, one million points only amounts to an average of ten per dimension. Many high-dimensional integrals routinely require millions of evaluations. ∎

Two generalizations of Monte Carlo methods are worth noting here. (1) Even the earliest practitioners realized that it should be possible to reduce the amount of work by limiting the selection of points to regions where the integrand was changing most rapidly. This can be done automatically, and in a sense makes the algorithms adaptive. Unfortunately, it is much more difficult to write programs to do this than to implement the basic method and simplicity has always been an attractive feature of Monte Carlo; hence this enhancement is not always used. (2) Since 1970 there has been active work in using "quasi-random" or number-theoretic methods for multiple integrals. The idea is that by picking special points which are not random but have certain other good theoretical properties it is possible to get estimates which converge like N^{-1} rather than Monte Carlo's $N^{-1/2}$. The integrand must satisfy certain assumptions which are more restrictive than before, but in practice this does not seem to be an impediment. Readers who would like to explore these ideas further can consult either of the two references given earlier.

5.12 HISTORICAL PERSPECTIVE: ULAM (1909–1984) AND VON NEUMANN (1903–1957)

John von Neumann and Stanislaw Marcin Ulam, American immigrants from Hungary and Poland, had profound effects on our activities during World War II. Working together at the top secret Manhattan Project on the mesas of Los Alamos, New Mexico, they devised the Monte Carlo method for finding approximate answers to problems in hydrodynamics which were too complex for exact mathematical treatment or too time consuming for traditional computation. The original problem was, "how far would neutrons travel through different shielding materials?" Most of the basic information was known: a neutron's average distance of travel in a given substance before colliding with an atomic nucleus, its chances of being repelled or absorbed by the nucleus, etc. But it wasn't possible to derive a precise equation for predicting the result of a long sequence of such events, even though the individual probability of each was easily calculable. Ulam and von Neumann suggested that the answer could be obtained by obtaining a large sample of neutrons and tracing the passage of each through the shielding material. They proposed to do this, not by actual experiment, but by the construction of artificial life histories. A neutron's journey would be broken into discrete steps such as collision, absorption, collision, etc. Which one would occur on each step could be calculated by use of a table of random numbers. The life history would be composed of a number of steps, and the entire sample would comprise millions of such histories. The calculation had to be done on a computer, and the development of these machines was proceeding in parallel with the war effort. Today the Monte Carlo method is still an important tool in studying nuclear reactors and many other phenomena as well.

Individually Ulam and von Neumann, pure mathematicians with traditional European training, made many other contributions to applied problems. Ulam for instance is credited with proving the inadequacy of the plan for the first hydrogen bomb and suggesting the idea that eventually worked successfully. The "inventor" of the digital computer is still hotly debated, but there is less debate as to who was the first to realize the full potential of the computer. This was von Neumann's great achievement. In fact, Herman Goldstine, a colleague, states that "von Neumann's great status in the world of the physical and social sciences was sufficient so that when he told people to compute digitally ... they believed him [and] that this in large measure accounted for the early acceptance of the digital computer. He caused it all to happen at a rate which was much accelerated over what it would have been." His proposals to use computers in weather forecasting, oil reservoir study and hydrodynamics, years before it was practical to do so, had a major impact on the funding for institutions engaged in research on these subjects.

"Johnny" has been called the most important and wide ranging mathematician of the twentieth century. He certainly was the quickest. His ability to absorb and digest an enormous amount of extremely diverse information was exceptional. In a profession where quick minds are somewhat commonplace his amazing rapidity was proverbial. He has been compared in ability to Gauss and there is hardly a single important part of mathematics of the 1930's with which he had not at least a passing acquaintance, and the same is probably true of theoretical physics. One reliable story describes von Neumann wandering into a colleague's office while the latter was discussing with a graduate student a computation which the student had just performed over the span of several months. Von Neumann listened to the conversation, thought briefly about the problem and without realizing it had already been computed explained what the results had to be. The student stepped in to agree with most of the remarks but to point out that some of the figures were slightly in error. A moment's thought forced von Neumann, who normally did not mind showing off brilliance or special ingenuity, to adjust his figures too, but he left the office astonished that this new "kid" had duplicated his efforts.

Von Neumann, the son of a well-to-do banker, was privately tutored in mathematics when his teachers realized that the conventional Gymnasium program was a waste of his time. His early career involved theoretic studies in logic, formal systems, and foundations of mathematics, prophetically useful for his later interest in computers. In 1928 he founded a new branch of science: the theory of games. The conceptual center of the new discipline was the proof, which he first advanced, that for all games of strategy in a certain class, there existed at least one optimum line of play that would in the long run guarantee the greatest possible minimization of loss. As presented and codified by von Newmann, the theory of games found rapid acceptance and use in economics, warfare, and many branches of the social sciences. He enthusiastically encouraged the development of game playing programs—in particular chess. An early one written at Los Alamos had to be simplified to run on existing equipment. It used a smaller board and had no bishops. Appropriately, it was called "anticlerical." During the period 1925–1940 he seemed to be advancing at a breathless pace on all fronts of logic and analysis at once, not to speak of mathematical physics. Many of his papers in operator theory have yet to be improved upon and his work in spectral theory is still the basis of nonrelativistic quantum theory. His work was sufficiently impressive that in 1933 he was invited to join the Institute of Advanced Study at Princeton University, of which he was the youngest permanent member at that time.

His health began to fail in 1955. This period was one of great trial for von Neumann, as his famous mental facilities waned with his illness. He died of cancer two years later.

5.13 PROBLEMS

P5–1.–The integral defining the error function

$$\text{erf}(x) = \frac{2}{\sqrt{\pi}} \int_0^x \exp(-t^2) \, dt$$

is fairly easy to evaluate numerically.

(a) Write a program which uses Q1DA to print a table of erf(x) for x = 0.0, 0.1, 0.2, ..., 1.9, 2.0. Take EPS = 10^{-5}. As a rule you should never ask for more accurate answers than the values of the integrand will permit. The latter is accurate to a small factor (5 or 10) times ϵ_{mach}. Compare your table with published values or with values available from a reliable subroutine on your computer. The integral is to be computed for different values of the upper limit. Each call to Q1DA is independent. A more efficient way is to treat the problem as a differential equation; see Problem P8–1 of Chapter 8.

(b) Repeat the computations of (a) with subroutine QK15. Compare the accuracy of the results and of the error estimates.

P5–2.–Compute approximations to π from the integral

$$\pi = \int_0^1 \frac{4}{1 + x^2} \, dx.$$

(a) Use the compound trapezoid and compound Simpson rules with 2, 4, 8, 16, 32, 64 and 128 panels. Tabulate the error in each case. How does the error decrease as the number of panels is doubled? (The program that you write for the compound trapezoid rule should make use of remark (3) in Section 4.) Depending on the value of ϵ_{mach} on your computer, you may find that there is no further improvement by doubling the number of panels. The error might even get worse. How could this be possible?

(b) Using the method of Section 4 derive the compound midpoint formula. Compare this with the compound trapezoid formula with respect to (i) openness, (ii) remainder, (iii) work (iv) existence of remark analogous to (3) in Section 4. Repeat the computations of (a) with the compound midpoint rule.

(c) Using the same points as in (a), estimate the integral either by Hermite cubic quadrature or by using PCHQA. How does the error decrease as the number of panels is doubled? Note: As the points are equally spaced the only derivatives that are needed are d_1 and d_n. Either compute them directly by differentiating the integrand or approximate them as explained in Chapter 2.

(d) Use the 2, 4 and 8 point Gauss quadrature formulas to estimate the integral. Divide the interval of integration into two equal panels and use the 4 point Gauss rule on each. Compare the results of this compound rule (using 8 points) with the 8 point Gauss estimate.

(e) Use QK15 to estimate the integral and error. How reliable is your error estimate?

(f) Use Q1DA with various values of EPS. This is an easy integral for the program. As long as EPS is well above ϵ_{mach}, the number of integrand evaluations should only depend weakly on EPS.

P5–3.–Repeat Problem P5–2 for the integral

$$-\frac{4}{9} = \int_0^1 \sqrt{x}\, \ln x\, dx.$$

P5–4.–If Q1DA were used to evaluate a function which is identically zero on $[0, 1]$ the result would be zero. Where will the integrand be evaluated? Try to determine this by using Q1DA on the function

```
REAL FUNCTION F (X)
REAL X
PRINT *, X
F = 0.0
RETURN
END
```

Call the numbers printed $x_1, x_2, \ldots, x_k$. What would Q1DA estimate for the integral

$$\int_0^1 (x - x_1)^2 (x - x_2)^2 \cdots (x - x_k)^2\, dx?$$

Explain your answer. How does Q1DA try to mitigate this problem?

P5–5.–Consider the following quadrature problems.

(a) Use Q1DA with EPS $= 10^{-4}$ on the integral

$$\int_0^1 f(x)\, dx, \qquad f(x) = \begin{cases} \exp(x), & \text{if } 0 < x \le \tau; \\ \sin x, & \text{if } \tau < x < 1. \end{cases}$$

with $\tau = 0.5$ and $\tau = 0.3$. Explain the differences (if any) in the amount of work required. If Q1DA did not randomize but made the first subdivision at the midpoint, would either of these problems be easier?

(b) In (a) with $\tau = 0.3$ determine the smallest subinterval used by Q1DA. If all of $[0, 1]$ was subdivided into intervals of this length what would be the total number of evaluations of $f(x)$?

P5–6.–The 11-point Newton-Cotes quadrature rule on $[0, 1]$ is

$$\int_0^1 f(x)\, dx \approx \sum_{i=0}^{10} w_i f(i/10)$$

with the w_i determined by requiring that the sum be exact for $f(x) = 1, x, x^2, \ldots, x^{10}$.

(a) Use SGEFS to find the weights w_i. Is the resulting rule a Riemann sum?

(b) Apply the rule in (a) to the integrals in P5–2, P5–3 and P5–5. This will show you that a negatively weighted quadrature rule can be useful too.

(c) If you have access to a book of tables, such as Abramowitz and Stegun, compare the values of the weights that you computed to the exact values. What is the largest error in the weights? Repeat (b) with the exact weights. How different are the estimates from those in (b)?

P5–7.–Using the data from problem P4–3, compute an estimate of the integral using

(a) the generalized trapezoid rule,

(b) PCHQA. You will need values for the derivatives d_i. Obtain them from the output of PCHEZ.

P5–8.–Select twenty non-equally-spaced points on $[0, 1]$ of the form $(x_i, \text{erf}(x_i))$.

(a) Use PCHEZ to compute the spline interpolant and then use PCHQA to compute its integral from x_1 to x_n.

(b) If you did not have PCHQA one possible way to integrate the spline produced in (a) would be to use Q1DA with each call to the Fortran function F(X) resulting in a call to PCHEV. Write such a program and compare your results with those in (a). If you have access to timing routines for your computer, use these to time the two approaches. What conclusions can you draw?

P5–9.–Which of the following problems would require many function evaluations or result in a failure for Q1DA with EPS $= 10^{-4}$? Try to answer first without running the program, then perform the calculation and compare your results. Call QK15 in the same program. How reliable are the error estimates which it returns?

(a) $\int_0^1 \exp(x^2)\, dx$

(b) $\int_0^1 x^x\, dx$

(c) $\int_0^1 \sin 50x\, dx$

(d) $\int_0^{10} \ln x\, dx$

(e) $\int_0^1 x^{-1/2} \exp(x^2) \, dx$

(f) $\int_1^5 (x-1)^{0.2} \cdot (x^2+1)^{-1} \, dx$

P5-10.–Estimate the value of

$$\int_0^\infty \exp(-x)\cos^2 x \, dx$$

(a) Truncate the integral and use Q1DA on the finite part.

(b) Try the transformation $x = -\ln t$ on this integral, and use Q1DA on the new integral. Repeat with the transformation $x = t/(1-t)$ and compare your results.

(c) Use the 2, 4 and 8 point Gauss Laguerre quadrature rules from Table 5.4 to estimate this integral. Compare your results to (a) and (b).

(d) Use the semi-infinite trapezoid rule. Select several values of N and h.

(e) Make the transformation $x = \exp(t)$ and use the infinite trapezoid rule. Select several values of N and h.

P5-11.–Consider the following:

(a) Compute

$$\iint_D \exp(x^2 y^2) \, dA \qquad D = \{ (x,y) : 0 \le x \le 1, \, 0 \le y \le 1 \}$$

using Q1DA and a renamed copy, Q1. Hints: (1) The sequence of calls will be MAIN, Q1, ..., G, Q1DA, Q1DAX, GL15T, F. Thus Q1 cannot call Q1DAX or GL15T. (2) In addition to altering the names be sure that Q1 does not call F. (3) Test your program first on a function whose integral you know, for example $f(x,y) = 1$.

(b) Using the program in (a) compute the integral of the same function for the domain D that is a quarter circle,

$$D = \left\{ (x,y) : 0 \le x \le 1, \, 0 \le y \le \sqrt{1-x^2} \right\}.$$

(c) Using Monte Carlo integration compute the integrals of $f(x,y) = 1$, and $f(x,y) = \exp(x^2 y^2)$ on a quarter circle. Compare the number of evaluations and the amount of programming effort with (a). Hint: Use UNI from Chapter 10. In particular, see problem P10-5.

P5-12.–Verify the values given in Example 5.10. Then use Q1DA on this problem. Which technique requires more work?

5.14 PROLOGUES: QK15, Q1DA AND PCHQA

```
        SUBROUTINE QK15(F,A,B,RESULT,ABSERR,RESABS,RESASC)
C***BEGIN PROLOGUE   QK15
C***DATE WRITTEN     800101    (YYMMDD)
C***REVISION DATE    870530    (YYMMDD)
```

```
C***CATEGORY NO.   H2A1A2
C***KEYWORDS  15-POINT GAUSS-KRONROD RULES
C***AUTHOR  PIESSENS, ROBERT, AND DE DONCKER, ELISE,
C               APPLIED MATH. AND PROGR. DIV. - K. U. LEUVEN
C***PURPOSE  To compute I = Integral of F over (A,B), with error estimate
C               and    J = integral of ABS(F) over (A,B)
C***DESCRIPTION
C
C          From the book "Numerical Methods and Software"
C                 by  D. Kahaner, C. Moler, S. Nash
C                      Prentice Hall 1988
C
C          Real version
C
C          PARAMETERS ON ENTRY
C             F       - Real
C                       Function subprogram defining the integrand
C                       FUNCTION F(X). The actual name for F needs to be
C                       Declared E X T E R N A L in the calling program.
C
C             A       - Real: Lower limit of integration
C
C             B       - Real: Upper limit of integration
C
C          PARAMETERS ON RETURN
C             RESULT - Real: Approximation to the integral I
C                      Result is computed by applying the 15-POINT
C                      KRONROD RULE (RESK) obtained by optimal addition
C                      of abscissae to the 7-POINT GAUSS RULE(RESG).
C
C             ABSERR - Real: Estimate of the modulus of the absolute error,
C                      which should not exceed ABS(I-RESULT)
C
C             RESABS - Real: Approximation to the integral J
C
C             RESASC - Real: Approximation to the integral of ABS(F-I/(B-A))
C                      over (A,B)
C***REFERENCES    PIESSENS R. ET. AL., "QUADPACK: A SUBROUTINE PACKAGE FOR
C                    AUTOMATIC INTEGRATION" SPRINGER, BERLIN 1983.
C***ROUTINES CALLED  R1MACH
C***END PROLOGUE  QK15
```

link mach con
external

```
      SUBROUTINE Q1DA(A,B,EPS,R,E,KF,IFLAG)
C***BEGIN PROLOGUE  Q1DA
C***DATE WRITTEN   821018   (YYMMDD)
C***REVISION DATE  870525   (YYMMDD)
C***CATEGORY NO.   H2A1A1
C***KEYWORDS   ADAPTIVE  QUADRATURE, AUTOMATIC  QUADRATURE
```

```
C***AUTHOR  KAHANER, DAVID K., SCIENTIFIC COMPUTING DIVISION, NBS.
C***PURPOSE  Approximates one dimensional integrals of user defined
C              functions, easy to use.
C
C***DESCRIPTION
C       Q1DA IS A SUBROUTINE FOR THE AUTOMATIC EVALUATION
C            OF THE DEFINITE INTEGRAL OF A USER DEFINED FUNCTION
C            OF ONE VARIABLE.
C
C       From the book "Numerical Methods and Software"
C            by  D. Kahaner, C. Moler, S. Nash
C                Prentice Hall 1988
C
C       A R G U M E N T S    I N    T H E    C A L L    S E Q U E N C E
C
C       A
C       B       (INPUT) THE ENDPOINTS OF THE INTEGRATION INTERVAL
C       EPS     (INPUT) THE ACCURACY TO WHICH YOU WANT THE INTEGRAL
C               COMPUTED.  IF YOU WANT 2 DIGITS OF ACCURACY SET
C               EPS=.01, FOR 3 DIGITS SET EPS=.001, ETC.
C               EPS MUST BE POSITIVE.
C       R       (OUTPUT) Q1DA'S BEST ESTIMATE OF YOUR INTEGRAL
C       E       (OUTPUT) AN ESTIMATE OF ABS(INTEGRAL-R)
C       KF      (OUTPUT) THE COST OF THE INTEGRATION, MEASURED IN
C               NUMBER OF EVALUATIONS OF YOUR INTEGRAND.
C               KF WILL ALWAYS BE AT LEAST 30.
C     IFLAG (OUTPUT) TERMINATION FLAG...POSSIBLE VALUES ARE
C          0    NORMAL COMPLETION, E SATISFIES
C                   E<EPS  AND  E<EPS*ABS(R)
C          1    NORMAL COMPLETION, E SATISFIES
C                   E<EPS, BUT E>EPS*ABS(R)
C          2    NORMAL COMPLETION, E SATISFIES
C                   E<EPS*ABS(R), BUT E>EPS
C          3    NORMAL COMPLETION BUT EPS WAS TOO SMALL TO
C                   SATISFY ABSOLUTE OR RELATIVE ERROR REQUEST.
C
C          4    ABORTED CALCULATION BECAUSE OF SERIOUS ROUNDING
C                   ERROR.  PROBABLY E AND R ARE CONSISTENT.
C          5    ABORTED CALCULATION BECAUSE OF INSUFFICIENT STORAGE.
C                   R AND E ARE CONSISTENT.
C          6    ABORTED CALCULATION BECAUSE OF SERIOUS DIFFICULTIES
C                   MEETING YOUR ERROR REQUEST.
C          7    ABORTED CALCULATION BECAUSE EPS WAS SET <= 0.0
C
C          NOTE...IF IFLAG=3, 4, 5 OR 6 CONSIDER USING Q1DAX INSTEAD.
C
C     W H E R E    I S    Y O U R    I N T E G R A N D ?
C
```

```
C          YOU MUST WRITE A FORTRAN FUNCTION, CALLED F, TO EVALUATE
C          THE INTEGRAND.  USUALLY THIS LOOKS LIKE...
C                  FUNCTION F(X)
C                    F=(EVALUATE THE INTEGRAND AT THE POINT X)
C                    RETURN
C                  END
C
C
C     T Y P I C A L    P R O B L E M    S E T U P
C
C          A=0.0
C          B=1.0          (SET INTERVAL ENDPOINTS TO [0,1])
C          EPS=0.001          (SET ACCURACY REQUEST FOR 3 DIGITS)
C          CALL Q1DA(A,B,EPS,R,E,KF,IFLAG)
C          END
C          FUNCTION F(X)
C            F=SIN(2.*X)-SQRT(X)          (FOR EXAMPLE)
C            RETURN
C          END
C     FOR THIS SAMPLE PROBLEM, THE OUTPUT IS
C   0.0    1.0     .001     .041406750    .69077E-07    30     0
C
C   R E M A R K   I.
C
C          A SMALL AMOUNT OF RANDOMIZATION IS BUILT INTO THIS PROGRAM.
C          CALLING Q1DA A FEW TIMES IN SUCCESSION WILL GIVE DIFFERENT
C          BUT HOPEFULLY CONSISTENT RESULTS.
C
C   R E M A R K    II.
C
C          THIS ROUTINE IS DESIGNED FOR INTEGRATION OVER A FINITE
C          INTERVAL.  THUS THE INPUT ARGUMENTS A AND B MUST BE
C          VALID REAL NUMBERS ON YOUR COMPUTER.  IF YOU WANT TO DO
C          AN INTEGRAL OVER AN INFINITE INTERVAL SET A OR B OR BOTH
C          LARGE ENOUGH SO THAT THE INTERVAL [A,B] CONTAINS MOST OF
C          THE INTEGRAND.  CARE IS NECESSARY, HOWEVER.  FOR EXAMPLE,
C          TO INTEGRATE EXP(-X*X) ON THE ENTIRE REAL LINE ONE COULD
C          TAKE A=-20., B=20. OR SIMILAR VALUES TO GET GOOD RESULTS.
C          IF YOU TOOK A=-1.E10 AND B=+1.E10 TWO BAD THINGS WOULD
C          OCCUR. FIRST, YOU WILL CERTAINLY GET AN ERROR MESSAGE FROM
C          THE EXP ROUTINE, AS ITS ARGUMENT IS TOO SMALL.  OTHER
C          THINGS COULD HAPPEN TOO, FOR EXAMPLE AN UNDERFLOW.
C          SECOND, EVEN IF THE ARITHMETIC WORKED PROPERLY Q1DA WILL
C          SURELY GIVE AN INCORRECT ANSWER, BECAUSE ITS FIRST TRY
C          AT SAMPLING THE INTEGRAND IS BASED ON YOUR SCALING AND
C          IT IS VERY UNLIKELY TO SELECT EVALUATION POINTS IN THE
C          INFINITESMALLY SMALL INTERVAL [-20,20] WHERE ALL THE
C          INTEGRAND IS CONCENTRATED, WHEN A, B ARE SO LARGE.
```

(handwritten annotations): $8 * Atan(1.0) = \ldots 2\pi$ →write statement

```
C
C    M O R E   F L E X I B I L I T Y
C
C            Q1DA IS AN EASY TO USE DRIVER FOR ANOTHER PROGRAM, Q1DAX.
C            Q1DAX PROVIDES SEVERAL OPTIONS WHICH ARE NOT AVAILABLE
C               WITH Q1DA.
C
C***REFERENCES  (NONE)
C***ROUTINES CALLED  Q1DAX
C***END PROLOGUE  Q1DA


      REAL FUNCTION PCHQA(N,X,F,D,A,B,IERR)
C***BEGIN PROLOGUE  PCHQA
C***DATE WRITTEN  870829   (YYMMDD)
C***REVISION DATE  870829   (YYMMDD)
C***CATEGORY NO.  E3,H2A2
C***KEYWORDS  EASY TO USE CUBIC HERMITE OR SPLINE INTEGRATION
C            NUMERICAL INTEGRATION, QUADRATURE
C***AUTHOR  KAHANER, D.K., (NBS)
C            SCIENTIFIC COMPUTING DIVISION
C            NATIONAL BUREAU OF STANDARDS
C            ROOM A161, TECHNOLOGY BUILDING
C            GAITHERSBURG, MARYLAND 20899
C            (301) 975-3808
C***PURPOSE  Evaluates the definite integral of a piecewise cubic Hermite
C            or spline function over an arbitrary interval, easy to use.
C***DESCRIPTION
C
C            PCHQA:  Piecewise Cubic Hermite or Spline Integrator,
C                    Arbitrary limits, Easy to Use.
C
C            From the book "Numerical Methods and Software"
C                    by  D. Kahaner, C. Moler, S. Nash
C                        Prentice Hall 1988
C
C    Evaluates the definite integral of the cubic Hermite or spline
C    function defined by  N, X, F, D  over the interval [A, B].  This
C    is an easy to use driver for the routine PCHIA by F.N. Fritsch
C    described in reference (2) below. That routine also has other
C    capabilities.
C- - - - - - - - - - - - - - - - - - - - - - - - - - - - - - - - - -
C
C  Calling sequence:
C
C            VALUE = PCHQA (N, X, F, D, A, B, IERR)
C
C    INTEGER  N, IERR
C    REAL  X(N), F(N), D(N), A, B
```

```
C
C    Parameters:
C
C       VALUE - (output) VALUE of the requested integral.
C
C       N - (input) number of data points.  (Error return if N.LT.2 .)
C
C       X - (input) real array of independent variable values.  The
C              elements of X must be strictly increasing:
C                  X(I-1) .LT. X(I),   I = 2(1)N.
C              (Error return if not.)
C
C       F - (input) real array of function values.  F(I) is
C              the value corresponding to X(I).
C
C       D - (input) real array of derivative values.  D(I) is
C              the value corresponding to X(I).
C
C       A,B - (input) the limits of integration.
C              NOTE:  There is no requirement that [A,B] be contained in
C                     [X(1),X(N)].  However, the resulting integral value
C                     will be highly suspect, if not.
C
C       IERR - (output) error flag.
C              Normal return:
C                 IERR = 0  (no errors).
C              Warning errors:
C                 IERR = 1  if  A  is outside the interval [X(1),X(N)].
C                 IERR = 2  if  B  is outside the interval [X(1),X(N)].
C                 IERR = 3  if both of the above are true.  (Note that this
C                           means that either [A,B] contains data interval
C                           or the intervals do not intersect at all.)
C              "Recoverable" errors:
C                 IERR = -1  if N.LT.2 .
C                 IERR = -3  if the X-array is not strictly increasing.
C                 (Value has not been computed in any of these cases.)
C                 NOTE:  The above errors are checked in the order listed,
C                        and following arguments have **NOT** been validated.
C
C***REFERENCES  1. F.N.FRITSCH AND R.E.CARLSON, 'MONOTONE PIECEWISE
C                  CUBIC INTERPOLATION,' SIAM J.NUMER.ANAL. 17, 2 (APRIL
C                  1980), 238-246.
C               2. F.N.FRITSCH, 'PIECEWISE CUBIC HERMITE INTERPOLATION
C                  PACKAGE, FINAL SPECIFICATIONS', LAWRENCE LIVERMORE
C                  NATIONAL LABORATORY, COMPUTER DOCUMENTATION UCID-30194,
C                  AUGUST 1982.
C***ROUTINES CALLED  PCHIA
C***END PROLOGUE  PCHQA
```

6

Linear Least-Squares Data Fitting

6.1 INTRODUCTION

Consider the following experiment: Water is being pumped through a container to which an amount of dye has been added. Every few seconds the concentration of dye is measured in the water leaving the container. It is expected that the concentration of dye will decrease linearly over time. The results are graphed in Figure 6.1.

Notice that the data points do not lie on a straight line. This is not so unexpected. The measuring equipment may not be perfectly accurate, it may not be possible to interpret the measurements exactly, and the mixing may not behave exactly as predicted. To determine the rate at which the concentration decreases, the experimenter would have to approximate the data by a straight line, a line that "best" approximated the data in some sense. One such approximation is drawn.

Such experiences are common. One reason for wanting to know the mixing rate is to be able to predict how other experiments will behave. In other circumstances, we might want to model the inflation rate in the economy, the spread of a disease, or the population of a country. When the data is "noisy," that is, full of random errors, then approximating the data by a simple function allows us to study the trends in the data; this is called **smoothing**.

There are two qualitatively different reasons for wanting to find an approximating line to the data.

(1) The mixing rate, α is needed, for example to determine if the dye injection equipment is being overloaded.

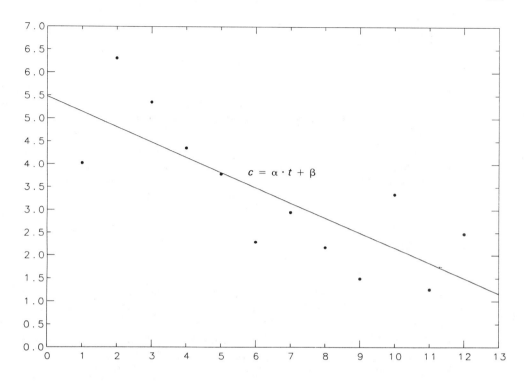

Figure 6.1 Mixing Experiment

(2) The approximating line is needed to predict the concentration of dye at times for which there are no measurements. In this case we are not particularly interested in the values of α and β, only in the values of the fitting function.

In other times, this work was performed by fortune tellers and magicians who would consult crystal balls and Tarot cards for wisdom. The ancients would ask questions of the Oracle at Delphi, as described in Shakespeare's play *The Winter's Tale*. Some of the romance has been taken away from data fitting, but the improved scientific foundations offer some compensation.

Before we go further, let us make the data fitting problem more explicit. Suppose that we are given data (t_i, b_i), $i = 1, \ldots, m$, representing some underlying function $b(t)$ for which $b_i = b(t_i)$. In the experiment above, the t_i's represent the times where measurements were taken and the b_i's represent the dye concentration. We assume that some model for the data is given. In words, we have the relationship

$$\text{observation} = \text{model} + \text{error}.$$

More specifically, we assume that the model has the form

$$b_i \approx x_1 \phi_1(t_i) + x_2 \phi_2(t_i) + \cdots + x_n \phi_n(t_i),$$

where the functions $\phi_j(t)$ are the given **model functions**. In our example where the model is a straight line, we assume that

$$b_i \approx x_1 + x_2 t_i$$

so that $\phi_1(t) = 1$ and $\phi_2(t) = t$. The coefficients x_j are called the **parameters** of the model, and it is these numbers that are to be determined. If the model were perfect and there were no measurement errors, then we could replace $\approx$ with $=$; this will rarely happen.

The model above is called a **linear model** because it is a linear combination of the model functions $\phi_i(t)$. This does not mean that the model functions are linear. Even if our data were to be approximated by a quadratic polynomial so that $b(t) \approx x_1 + x_2 t + x_3 t^2$, then the model would still be linear even though one of the model functions $\phi_3(t) = t^2$ is a nonlinear function of t. The crucial notion is that the model is a linear function of the parameters x. "Nonlinear" models are also useful, and they are discussed in Chapter 9. A common example is an exponential model such as

$$b(t) \approx x_1 e^{x_2 t}.$$

Here the model is a nonlinear function of the parameter x_2.

Most of the models discussed so far have polynomial model functions, where the data are approximated by a polynomial in the variable t. Any functions can be used as model functions. In the case of cubic-Hermite interpolation, the model functions would be the piecewise cubics $c_i(t)$ and $\bar{c}_i(t)$ described in Section 10 of Chapter 4. Models can also involve more than one independent variable. This can occur, for example, in the analysis of census data where the independent variables might include a person's age, income, and family size. In this case, the model functions would be $\phi_1(t_i) = \text{age}(i)$, $\phi_2(t_i) = \text{income}(i)$, and $\phi_3(t_i) = \text{family-size}(i)$.

We can express our data-fitting problem more concisely using matrix/vector notation. Define the $m \times n$ matrix A by

$$A_{ij} = \phi_j(t_i),$$

and let b and x be the vectors of observations and parameters, respectively. Then the above conditions can be written as

$$b \approx Ax, \quad \text{or} \quad b - Ax \approx 0;$$

$b - Ax$ is called the vector of **residuals**.

Example 6.1   Forming the Data Matrix.

Suppose that the model is a quadratic polynomial $b(t) \approx x_1 + x_2 t + x_3 t^2$ and that data are available for four values of t: $t_1 = 1$, $t_2 = 2$, $t_3 = 3$, $t_4 = 4$. The modelling functions are $\phi_1(t) = 1$, $\phi_2(t) = t$, and $\phi_3(t) = t^2$. Thus the coefficient matrix is

$$A = \begin{pmatrix} 1 & t_1 & t_1^2 \\ 1 & t_2 & t_2^2 \\ 1 & t_3 & t_3^2 \\ 1 & t_4 & t_4^2 \end{pmatrix} = \begin{pmatrix} 1 & 1 & 1 \\ 1 & 2 & 4 \\ 1 & 3 & 9 \\ 1 & 4 & 16 \end{pmatrix}. \quad \blacksquare$$

The parameters will be chosen by making the residuals $b - Ax$ as small as possible. A common approach, and the one taken here, is to solve

$$\min_x \sum_{j=1}^{m} [(b - Ax)_j]^2.$$

For the model in Example 6.1 this would take the form

$$\min_{x_1, x_2, x_3} \sum_{j=1}^{4} [b_j - (x_1 + x_2 t_j + x_3 t_j^2)]^2.$$

Because we are *minimizing* the sum of *squares*, this is called **least squares data fitting**. Using vector norms (see Section 2.1 of Chapter 3), this problem is equivalent to solving

$$\min_x \| b - Ax \|_2^2 \equiv (b - Ax)^T (b - Ax)$$

where the Euclidean 2-norm is used.

If there are an equal number of data points and model functions, the matrix A is square. If A is also nonsingular, the solution to the least squares problem is the interpolant, i.e., the residual is zero. Thus we see that when posed as a matrix problem, least-squares data fitting includes as a special case the problem of solving linear equations $Ax = b$, if A is square and nonsingular. The methods and software described in this chapter can be used to solve linear equations, but they use about twice as many arithmetic operations as Gaussian elimination. However, the least-squares techniques will produce somewhat more accurate solutions in rounded arithmetic. In addition they remove the need for pivoting, making the algorithms better suited for many parallel and vector computers. On a parallel computer with many processors the pivoting steps can require more time than the arithmetic and hence slow down the algorithm.

The least-squares problem can be interpreted graphically as minimizing the vertical distance from the data points to the model. Underlying this idea is the assumption that all the errors in the approximation correspond to errors in the observations b_i. If there are also errors in the independent variables t_i then it may be more appropriate to minimize the Euclidean distance from the data to the model. This is illustrated in Figure 6.2. This can be especially useful if the graph of the model is steep, such as near a singularity in a nonlinear model. This is illustrated in the figure. In this case the steepness of the model will make the least-squares error large even though the data point is close to the graph, and the fit will be visually satisfactory. Minimizing the Euclidean distance from the model is referred to as **total least squares** or **orthogonal distance regression**. It is beyond the scope of this book, but details can be found in Golub and Van Loan (1983).

One final point concerns notation. In this chapter we discuss the least-squares problem in the form

$$\min_x \| Ax - b \|_2$$

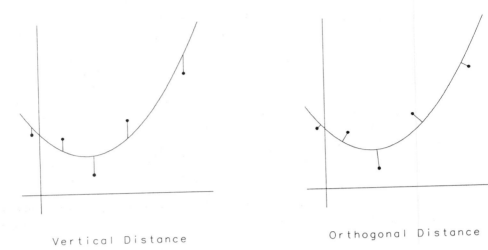

Vertical Distance   Orthogonal Distance

Figure 6.2   Vertical-distance and Orthogonal-distance Models

whereas statisticians use the notation

$$\min_{\beta} \| X\beta - y \|_2 .$$

The two problems are exactly the same, only the names have been changed: $A \to X$, $x \to \beta$, $b \to y$. Even though this is a simple cosmetic change, it can cause confusion when trying to learn about data fitting. In this book, the main concern is the computation of the parameters in the model and not the statistical properties of the solution, so we use the notation $\| Ax - b \|_2$ common among numerical analysts. This notation also helps to emphasize the relation between data fitting and solving linear equations $Ax = b$.

6.1.1 Weighted Least-Squares Data Fitting

In many applications, the data points will not be equally important. This is often because some data points are known to be more accurate than others. For example, in an astronomical model of the Milky Way galaxy, it would be easier to get accurate measurements from our own sun than from stars many light years away.

This information can be incorporated in the least-squares problem by weighting the data points. Instead of solving

$$\min_{x} \sum_{j=1}^{m} [(b - Ax)_j]^2$$

for the parameters x, we would instead solve

$$\min_{x} \sum_{j=1}^{m} [w_j(b - Ax)_j]^2$$

where w_j is a weight reflecting the importance of the data point. The more important the data point, the larger the weight. If the error in the j-th data point is approximately e_j, then choose $w_j = 1/e_j$. Thus the smaller the error, the larger the weight. Sometimes the data are all accurate to the same number of digits, i.e., have the same relative error. In that case a good choice of w_j is $w_j = 1/b_j$, as long as b_j is nonzero. Weighting can improve the statistical properties of the solution (see Section 2 below).

Any software for solving an unweighted least squares problem can be used to solve a weighted problem by scaling the observation vector b and the coefficient matrix A. We multiply the j-th row of A as well as the j-th observation b_j by w_j, and then solve

$$\min_{x} \|b_w - A_w x\|_2^2 .$$

The coefficients in the new problem are defined by

$$(b_w)_j = w_j b_j, \quad (A_w)_{jk} = w_j A_{jk}.$$

Of course, the parameters obtained from the model will change as the weights change.

Example 6.2   Weighted Least-Squares Data Fitting

Consider the model from Example 6.1, with observations $b_1 = 2$, $b_2 = 7$, $b_3 = 9$, $b_4 = 6$. Suppose that the errors in the data are estimated to be $e_1 = 1/4$, $e_2 = 1$, $e_3 = 1/2$, $e_4 = 1/3$. If we use the weights $w_j = 1/e_j$ then the weighted least-squares problem

$$\min_{x} \|b_w - A_w x\|_2^2$$

has coefficients

$$A_w = \begin{pmatrix} 4 & 4 & 4 \\ 1 & 2 & 4 \\ 2 & 6 & 18 \\ 3 & 12 & 48 \end{pmatrix} \quad \text{and} \quad b_w = \begin{pmatrix} 8 \\ 7 \\ 18 \\ 18 \end{pmatrix}.$$

The weights are $w_1 = 4$, $w_2 = 1$, $w_3 = 2$, $w_4 = 3$. ∎

*6.1.2 Data Fitting with Other Norms

Least-squares data fitting is based on compromise: it is not possible for the model to pass through all the data points, and so the 2-norm is used as a mediator to choose the parameters. Before least-squares techniques were developed, different rules were used to determine the fit. Many of these rules correspond to using a different norm as a mediator.

Some common choices of norms are

$$\min_x \sum_{j=1}^{m} |(b - Ax)_j|, \qquad \text{(the 1-norm)}$$

and

$$\min_x \max_j |(b - Ax)_j|. \qquad \text{(the } \infty\text{-norm)}$$

The optimal parameters for these norms will be different than those for the 2-norm. Computing with these norms is also more difficult (the solutions can be obtained using linear programming). They do have useful properties, however. For example, the 1-norm is less sensitive to the presence of "outliers" (spurious data points). As a result, they continue to be used for data fitting in certain applications. The different norms are illustrated in a simple case in Example 6.3.

Example 6.3   Estimation with Different Norms.

Suppose that we are given observations b_i and we want to approximate them by a constant x, i.e., $b_i = x + \text{error}_i$. The estimators in the three cases are

1-norm: median of the b_i's

2-norm: average of the b_i's

∞-norm: $\frac{1}{2}(\min_i b_i + \max_i b_i)$

Consider the data set $\{ b_i \} = \{ 1, 2, 5, 9, 12 \}$. The three estimators are

1-norm: 5 (the middle number in the set)

2-norm: $5.8 = (1 + 2 + 5 + 9 + 12)/5$

∞-norm: $6.5 = \frac{1}{2}(1 + 12)$

Suppose that a mistake had been made collecting the data, so that the fifth number was thought to be 112 and not 12. Then the estimators are

1-norm: 5 (the middle number in the set is still the same)

2-norm: $25.8 = (1 + 2 + 5 + 9 + 112)/5$

∞-norm: $56.5 = \frac{1}{2}(1 + 112)$

Notice that the 1-norm estimate is unaffected by the error. ∎

The least-squares (2-norm) approach is by far the most commonly used technique for data fitting. It is the simplest computationally. It also has important statistical interpretations (in many cases it produces the maximum-likelihood estimate of the parameters). This is described in more detail in the book by Draper and Smith (1981). For these reasons it is the approach discussed here.

6.2   EXPLORING DATA

Analyzing data consists of two stages: determining the model and computing the parameters. In this book we will be mainly concerned with the second stage, and assume that the model is available. This is often reasonable. For the mixing experiment above, the behavior of the dye concentration could be predicted theoretically and only the mixing rate would need to be determined experimentally.

However in many instances, particularly in the social sciences, the data are used to help determine the model. For example, a sociologist might collect data from the residents of a city, such as income, education level, age, sex, weight, race, and occupation. If the sociologist were trying to study the incomes of the residents, the following model might be used

$$\text{income}_i \approx x_1 \text{education}_i + x_2 \text{age}_i + x_3 \text{sex}_i + x_4 \text{weight}_i + x_5 \text{race}_i + x_6 \text{occupation}_i.$$

All the data have been used to set up the model. However, it is not clear that a person's weight will have anything to do with their income, and the presence of this term in the model could affect the values of the parameters x_j, and hence distort the conclusions drawn from the data. The sociologist would like to include in the model only those terms that are significant, *based on the data*.

When exploring data, the questions that arise include: (1) Does the model adequately explain the data? In other words, should additional terms be included in the model? (2) Are any of the model terms redundant, and hence can be ignored? (3) How accurate are the parameters? (4) How accurate are predictions made from the model?

Answers to these questions are based on statistical tests, some of which are discussed below. However, they rely on an important assumption about the data: that the errors in the model are randomly distributed. More specifically, the model errors should be independent and normally distributed. If this assumption is not satisfied, the conclusions drawn from the data may be unreliable.

Exploring data can be a challenging and time-consuming project, and may sometimes require the assistance of an experienced statistician. However, many problems can be detected by examining a plot of the residuals. Such a picture is given in Figure 6.3 for the mixing example above.

Essentially, there should be no patterns visible in the residual plot. The residuals should have scattered values, and there should not be clumps of residuals. The magnitude of the residuals should not change within the plot. Loosely speaking, they should look "random." Some examples of suspicious residual plots are given in Figure 6.4.

Further information can be obtained by scaling the residuals by their standard deviation s, a measure of the spread in their values. The **standard deviation** of the residuals is defined by the formula

$$s = \frac{\|b - Ax^*\|_2}{\sqrt{m - n}},$$

where x^* is the solution of the least-squares problem, and the scaled residuals are then

$$(b - Ax^*)/s.$$

Approximately 95% of the scaled residuals should lie in the interval $[-2, 2]$. If a value is outside this range, that particular data point should be examined more carefully since

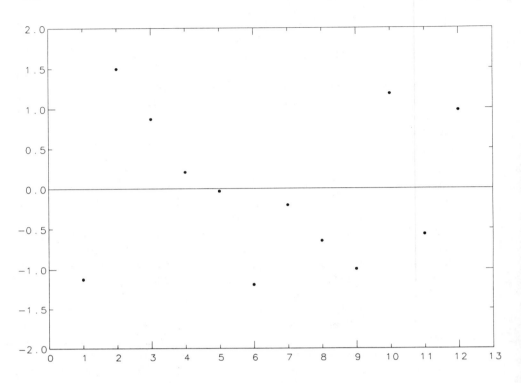

Figure 6.3 Mixing Experiment: Residual Plot

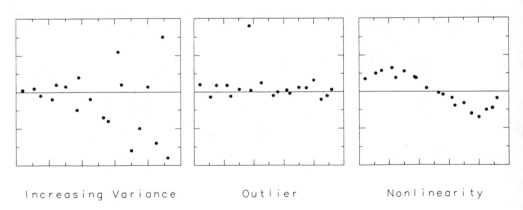

Figure 6.4 Suspicious Residual Plots

it may indicate an error in the data or an inadequacy in the model. Points with large residuals can exert a strong influence on the values of the parameters and so should be treated with care. Be aware, though, that influential points need not have large residuals. For the mixing example, $s = .993$ and so the residual plot is virtually the same as the scaled residual plot.

If the residuals are not satisfactory, there may be problems with the data or the model or both. There are many ways to correct these problems. For a thorough discussion, see the books by Chatterjee and Price (1977), and by Belsley, Kuh, and Welsch (1981).

Statistical software packages will often provide plots of the residuals, and will perform auxiliary calculations to help determine if the least-squares model is appropriate for the data. Some of these calculations are discussed below. Many of these calculations depend on the assumptions about the residuals mentioned above, and the results may not be accurate in other circumstances.

1. Standard deviation of b—a measure of the variability in the observations b. The larger the value of the standard deviation, the greater the spread in the values of b. If the observations were normally distributed, then about 95% of the observations would lie within two standard deviations of the average value.

2. Standard errors of x—estimates, the same as the standard deviation, of the errors in the parameters x. The larger the value, the larger the error in the parameter is likely to be.

3. Confidence intervals—probabilistic bounds on the errors in the parameters. Confidence intervals are determined from the standard errors in the parameters together with a desired probability that the bounds are correct. Under the assumptions mentioned above, the true value of the parameter will lie within the interval with, say, 90% or 95% probability.

4. R^2—indicates how much of the variation in the data is explained by the model. R^2 is in the interval $[0, 1]$. If R^2 is near 1, then the model appears to be adequate. A smaller value of R^2 may indicate that there are terms missing in the model. This calculation is only meaningful if the model contains a constant term.

5. Normal probability plot of residuals—a graph of the residuals against the normal probability distribution. This helps to indicate if the residuals have the appropriate distribution for least-squares data fitting. Ideally this graph would be the straight line $y = x$. Actual data will not produce this straight line, but should not be too far from it.

6. Durbin-Watson statistic—can indicate if there is degeneracy in the data (see Section 7). This number is in the range $[0, 4]$. If $d = 2$ then there is no evidence of degeneracy. The further d is from 2, the firmer the evidence of degeneracy.

7. Variance-covariance matrix Y—measures the variability and interdependence of the parameters x. The diagonal entry Y_{ii} is the **variance** of the i-th parameter x_i; the variance is defined as the square of the standard deviation. The off-diagonal entry Y_{ij} is the **covariance** of parameters x_i and x_j. If the two parameters are completely independent, their covariance will be zero. If the two are related (for example, if one increases whenever the other decreases) it will be non-zero. Ide-

ally the parameters will be independent, and the off-diagonal entries will be small. Computing the variance-covariance matrix is moderately expensive, since it involves the inverse of $(A^T A)$ where A is the coefficient matrix in the data-fitting problem.

These calculations are applied to the mixing experiment in Example 6.4 below.

Example 6.4   Statistical Analysis of Mixing Experiment.

The data for the mixing experiment are given in the table below.

TABLE 6.1   MIXING EXPERIMENT

t	b
1.0	4.0225
2.0	6.3095
3.0	5.3522
4.0	4.3553
5.0	3.7861
6.0	2.2947
7.0	2.9492
8.0	2.1732
9.0	1.4921
10.0	3.3424
11.0	1.2596
12.0	2.4732

We use the linear model described in Section 1, with model functions $\phi_1(t) = 1$ and $\phi_2(t) = t$. The coefficient matrix A has entries $A_{ij} = \phi_j(t_i)$ and so

$$
A = \begin{pmatrix}
1 & 1.0 \\
1 & 2.0 \\
1 & 3.0 \\
1 & 4.0 \\
1 & 5.0 \\
1 & 6.0 \\
1 & 7.0 \\
1 & 8.0 \\
1 & 9.0 \\
1 & 10.0 \\
1 & 11.0 \\
1 & 12.0
\end{pmatrix}.
$$

With this model, we obtain the following values for the statistical parameters discussed above.

$$\text{standard deviation of } b = \quad 1.527,$$

$$\text{standard deviation of } b - Ax^* = \quad 0.993,$$

$$\text{parameter } x_1 = \quad 5.478,$$

$$\text{parameter } x_2 = -0.332,$$

$$\text{standard error of } x_1 = \quad 0.611,$$

$$\text{standard error of } x_2 = \quad 0.083,$$

$$95\% \text{ confidence interval for } x_1 = [\quad 4.116, \quad 6.840],$$

$$95\% \text{ confidence interval for } x_2 = [-0.517, -0.147],$$

$$R^2 = \quad 0.616,$$

$$\text{Durbin-Watson statistic} = \quad 2.090,$$

$$\text{variance-covariance matrix} = \begin{pmatrix} 0.3735 & -0.0448 \\ -0.0448 & 0.0069 \end{pmatrix}.$$

These results are quite satisfactory. The covariance of the two parameters (-0.0448) is small as desired. The Durbin-Watson statistic is almost equal to the ideal value 2. The standard errors in the parameters are not too large relative to their estimated values. The R^2 statistic indicates that the straight-line model explains about two-thirds of the variability in the data. We can conclude that the linear least-squares model is appropriate in this case. ∎

6.3 THE NORMAL EQUATIONS

There are many different algorithms for computing a set of coefficients which give a minimum sum of squares. One possibility is to use calculus. Writing the problem in terms of vectors

$$r^2 = \|b - Ax\|_2^2 = (b - Ax)^T(b - Ax)$$
$$= b^Tb - b^TAx - x^TA^Tb + x^TA^TAx$$
$$= b^Tb - 2b^TAx + x^TA^TAx.$$

Our goal is to minimize this function of x. For this to have a minimum at the solution the first derivatives with respect to x will be zero. The derivatives of this function are

$$-2A^Tb + 2A^TAx,$$

and so the solution must satisfy the system of linear equations

$$(A^TA)x = (A^Tb).$$

These equations are called the **normal equations**.[1]

Example 6.5 Use of the Normal Equations.

As an example, consider fitting a straight line to the data $(1, 1)$, $(2, 1.5)$, $(3, .75)$, and $(4, 1.25)$. The model is $b_i \approx x_1 + x_2 t_i$ with model functions $\phi_1(t) = 1$ and $\phi_2(t) = t$. Thus

$$A = \begin{pmatrix} 1 & 1 \\ 1 & 2 \\ 1 & 3 \\ 1 & 4 \end{pmatrix}, \quad b = \begin{pmatrix} 1.00 \\ 1.50 \\ 0.75 \\ 1.25 \end{pmatrix}.$$

Forming the normal equations gives

$$A^T A x = \begin{pmatrix} 4 & 10 \\ 10 & 30 \end{pmatrix} \begin{pmatrix} x_1 \\ x_2 \end{pmatrix} = A^T b = \begin{pmatrix} 4.50 \\ 11.25 \end{pmatrix}.$$

The solution is $x = (1.125, 0)^T$ so that the "best" line through the data has formula $b(t) = 1.125$. ∎

The normal equations are a set of linear equations, and so they can be solved using Gaussian elimination as long as the coefficient matrix $A^T A$ is invertible. It is possible to show that the solutions to the normal equations are exactly the solutions to the least-squares problem. In principle, the normal equations could be solved using SGEFS from Chapter 3. But the coefficient matrix $A^T A$ is symmetric, and the time and storage required by SGEFS can be cut in half. Moreover, the matrix can be shown to be positive definite, and so no pivoting is needed. Consequently, by using a variant of Gaussian elimination intended for positive-definite, symmetric matrices, the normal equations can be solved with less than half the effort required by SGEFS.

However, there is a major disadvantage to using the normal equations. It turns out that the matrix $A^T A$ often has a high condition number, so that no matter how the normal equations are actually solved, errors in the data and roundoff errors introduced during the solution are excessively magnified in the computed coefficients. As a rule of thumb, the condition number of $A^T A$ is the *square* of the condition number of A.

As an illustration, suppose that $\epsilon_{\text{mach}} = 10^{-6}$ and that $\text{cond}(A) = 1000$, not a particularly large value. Then the error in the solution from Gaussian elimination applied to $Ax = b$ would be about $1000 \times 10^{-6} = 10^{-3}$, so that about three digits would be correct. On the other hand, if we try to solve $A^T A x = A^T b$ the error would be about $1000^2 \times 10^{-6} = 1$, so that the solution might have no accurate digits.

In the extreme situation where the basis functions $\phi_i(t)$ are linearly dependent, it can be shown that $A^T A$ is singular, and the condition number can be regarded as infinite. Methods which avoid the high condition number inherent in the normal equations are also

---

[1] If this is unfamiliar, it can be derived as follows, without matrix/vector notation. The least-squares function is $\sum_{i=1}^{m} [b_i - \sum_{j=1}^{n} A_{ij} x_j]^2$. The first derivative of this function with respect to x_k is $2 \sum_{i=1}^{m} (-A_{ik})[b_i - \sum_{j=1}^{n} A_{ij} x_j] = -2 \sum_{i=1}^{m} (A^T)_{ki} b_i + 2 \sum_{i=1}^{m} (A^T)_{ki} [\sum_{j=1}^{n} A_{ij} x_j] = -2 \sum_{i=1}^{m} (A^T)_{ki} b_i + 2 \sum_{i=1}^{m} (A^T)_{ki} (Ax)_i$. These two terms are the k-th components of $-2A^T b$ and $2A^T Ax$, respectively.

methods which can better detect linear dependence among the basis functions. Gaussian elimination and its variants are not well suited to detecting such linear dependence. The condition estimation imbedded in SGEFS is some help, but it can only warn of trouble—it cannot suggest a cure.

The normal equations are not the recommended approach for general least squares problems. The most reliable methods are based on matrix factorizations using "orthogonal" matrices. These are somewhat more expensive, but in return they produce more accurate solutions. In addition, they enable the solution of singular systems of equations and underdetermined systems of equations and have applications to the solution of constrained optimization problems. These techniques are discussed in the next section.

Nevertheless, there are situations where the normal equations offer an advantage. It is tempting to use the normal equations when there are many more data points than parameters; for example, if there 10, 000 data points and only 2 parameters, then $A^T A$ is only 2×2 while A is $10,000 \times 2$. Even though the routine in this chapter requires that A be stored, the approach based on orthogonal factorizations can be made as storage efficient as the normal equations.

A decision about which method to use may be difficult to make in advance. The correct choice may depend on such factors as the condition of the matrix, the size of the residuals at the solution, and the form of the data. For use in a general purpose software library, software based on orthogonal factorizations is considered to be the best choice.

6.4 ORTHOGONAL FACTORIZATIONS

Our goal is to reduce general least-squares problems to simpler ones, where the solution can be easily computed. Since linear systems of equations are special cases of linear least-squares problems, it is not surprising that "simpler" has much the same meaning for both problems.

Example 6.6   Triangular Least-Squares Problem.

Consider the following problem with 3 parameters and 5 observations:

$$\text{minimize } \|Ax - b\|_2^2$$

where

$$A = \begin{pmatrix} 3 & 4 & 1 \\ 0 & 2 & 2 \\ 0 & 0 & 7 \\ 0 & 0 & 0 \\ 0 & 0 & 0 \end{pmatrix}, \quad b = \begin{pmatrix} 14 \\ 10 \\ 21 \\ 6 \\ 2 \end{pmatrix}.$$

Because the bottom part of the matrix A is zero, this problem splits into

$$\|Ax - b\|_2^2 = \left\| \begin{pmatrix} 3 & 4 & 1 \\ 0 & 2 & 2 \\ 0 & 0 & 7 \end{pmatrix} \begin{pmatrix} x_1 \\ x_2 \\ x_3 \end{pmatrix} - \begin{pmatrix} 14 \\ 10 \\ 21 \end{pmatrix} \right\|_2^2 + \left\| \begin{pmatrix} 6 \\ 2 \end{pmatrix} \right\|_2^2 = \|Rx - b_{(1)}\|_2^2 + \|-b_{(2)}\|_2^2,$$

with

$$R = \begin{pmatrix} 3 & 4 & 1 \\ 0 & 2 & 2 \\ 0 & 0 & 7 \end{pmatrix}, \quad b_{(1)} = \begin{pmatrix} 14 \\ 10 \\ 21 \end{pmatrix}, \quad b_{(2)} = \begin{pmatrix} 6 \\ 2 \end{pmatrix}.$$

The second term $\|b_{(2)}\|_2$ is independent of x and so cannot be reduced. The first term can be made zero by by back-substitution on the 3×3 system $Rx = b_{(1)}$, giving

$$x_3 = 21/7 = 3,$$

$$x_2 = (10 - 2x_3)/2 = 4/2 = 2,$$

$$x_1 = (14 - 4x_2 - 1x_3)/3 = 3/3 = 1.$$

Thus $x^* = (1, 2, 3)^T$ is the solution of the least-squares problem. ∎

If the least-squares problem has a "triangular" coefficient matrix, i.e., a matrix with zeroes below the main diagonal as in the example above, then it can be solved via back-substitution. The algorithm is the same as for linear equations. For an $m \times n$ system, the first n components of the residual will be zero and the remaining $m - n$ components cannot be controlled by the parameters.

Gaussian elimination can be applied to a rectangular matrix A to reduce it to upper triangular form. However this will not help in solving the least-squares problem since the solution of the reduced problem will not be the same as the solution of the original problem. This is illustrated in Example 6.7.

Example 6.7   Gaussian Elimination for Rectangular Systems.

Consider the 3×2 problem

$$\text{minimize} \left\| \begin{pmatrix} 1 & 0 \\ 2 & 1 \\ 3 & 0 \end{pmatrix} \begin{pmatrix} x_1 \\ x_2 \end{pmatrix} - \begin{pmatrix} 1 \\ 3 \\ 5 \end{pmatrix} \right\|_2^2 .$$

The solution of this problem (obtained via the techniques described below) is $x^* = (1.6, -.2)^T$ and $\|Ax^* - b\|_2 \approx .63$. We can apply Gaussian elimination to this system: subtract 2 times the first row from the second, and 3 times the first row from the third. The reduced system is

$$\text{minimize} \left\| \begin{pmatrix} 1 & 0 \\ 0 & 1 \\ 0 & 0 \end{pmatrix} \begin{pmatrix} x_1 \\ x_2 \end{pmatrix} - \begin{pmatrix} 1 \\ 1 \\ 2 \end{pmatrix} \right\|_2^2 .$$

with solution $\bar{x} = (1, 1)^T$ and $\|A\bar{x} - b\|_2 = 2$. This is much larger than the optimal value of .63, and the parameters are far from the correct values. The difficulty arises because elimination techniques do not preserve the value of the 2-norm; they do not produce equivalent least-squares problems. ∎

Another type of transformation, called an **orthogonal** transformation, can be used here. An orthogonal transformation is an $n \times m$ matrix P satisfying $P^T P = I$, where

I is the $n \times n$ identity matrix. A simple example of an orthogonal transformation is the identity matrix since $I^T I = I \times I = I$. The matrix P need not be square; for example $P = \frac{1}{2}(1, 1, 1, 1)^T$ is orthogonal since $P^T P = \frac{1}{4}(1^2 + 1^2 + 1^2 + 1^2) = 1$. Another orthogonal transformation is

$$P = \begin{pmatrix} 3/5 & -4/5 \\ -4/5 & 3/5 \end{pmatrix}$$

which is also a symmetric matrix. P is an example of a special kind of orthogonal transformation called a **Householder transformation**. These are described in the following section; they are the main tools that we will use to solve least squares problems.

Orthogonal transformations are of value because they preserve the 2-norm:

$$\|Px\|_2 = \sqrt{(Px)^T (Px)} = \sqrt{x^T P^T P x} = \sqrt{x^T I x} = \sqrt{x^T x} = \|x\|_2 .$$

Since

$$\min_x \|Ax - b\|_2 = \|P(Ax - b)\|_2 = \|(PA)x - (Pb)\|_2 ,$$

our approach to solving a least squares problem will be to find a sequence of orthogonal transformations P that reduces A to triangular form, $A \rightarrow R$, where R is an upper triangular matrix. Each member of the sequence will be a Householder transformation. The reduction of A to upper-triangular form can be interpreted as a matrix factorization: we will factor $A = QR$ as the product of an orthogonal and an upper-triangular matrix. This is often referred to as a QR **factorization**.

The most important property of this approach is accuracy—the solutions obtained via QR factorization will usually be much more accurate than those from the normal equations. Surprisingly the QR factorization is almost as efficient as the normal equations—if A is an $m \times n$ matrix the QR method requires $n^2(m - n/3)$ floating-point operations; to form and solve the normal equations takes $\frac{1}{2}n^2(m + n/3)$ operations. Hence the improved accuracy of the QR method does not carry a high price.

*6.4.1 Householder Transformations

A Householder transformation P is an $m \times m$ matrix of the following form:

$$P = I - 2\frac{vv^T}{v^T v}$$

where v is a non-zero $m \times 1$ column vector. See Example 6.8.

Example 6.8   Householder Transformation.

If we construct a Householder transformation using $v = (1, 2)^T$ then

$$
\begin{aligned}
P &= \begin{pmatrix} 1 & 0 \\ 0 & 1 \end{pmatrix} - 2 \times \frac{1}{1 \cdot 1 + 2 \cdot 2} \begin{pmatrix} 1 \\ 2 \end{pmatrix} (1 \quad 2) \\
&= \begin{pmatrix} 1 & 0 \\ 0 & 1 \end{pmatrix} - \frac{2}{5} \begin{pmatrix} 1 & 2 \\ 2 & 4 \end{pmatrix} \\
&= \begin{pmatrix} 3/5 & -4/5 \\ -4/5 & 3/5 \end{pmatrix}. \quad \blacksquare
\end{aligned}
$$

It is easily checked that any matrix of this type is symmetric. In addition, any P of this form is automatically orthogonal:

$$
\begin{aligned}
P^T P &= P \cdot P \\
&= \left(I - 2\frac{vv^T}{v^T v} \right) \left(I - 2\frac{vv^T}{v^T v} \right) \\
&= I - 4\frac{vv^T}{v^T v} + 4\frac{vv^T}{v^T v} \cdot \frac{vv^T}{v^T v} \\
&= I - 4\frac{vv^T}{v^T v} + 4\frac{v(v^T v)v^T}{(v^T v)^2} \\
&= I - 4\frac{vv^T}{v^T v} + 4\frac{vv^T}{v^T v} \\
&= I.
\end{aligned}
$$

For the 2×2 matrix in Example 6.8 a simple calculation shows that indeed $P \times P = I$.

Householder transformations are important because they make it easy to reduce a matrix to upper-triangular form. If $a = (a_1, \ldots, a_m)^T$ is any vector then it is possible to find a vector v so that the Householder transformation that is constructed from v zeroes out all but one component of a:

$$
P \begin{pmatrix} a_1 \\ a_2 \\ \vdots \\ a_m \end{pmatrix} = \begin{pmatrix} \alpha \\ 0 \\ \vdots \\ 0 \end{pmatrix}.
$$

Since orthogonal transformations preserve norms, we must have that $\alpha = \pm \|a\|_2$. We will describe how to choose v below.

To show how we reduce A to triangular form, suppose that A is a 5×3 matrix

$$
A = \begin{pmatrix} \times & \times & \times \\ \times & \times & \times \\ \times & \times & \times \\ \times & \times & \times \\ \times & \times & \times \end{pmatrix},
$$

where $\times$ represents some arbitrary number. If we let a be the first column of A and choose P as above, then

$$PA = \begin{pmatrix} \alpha & \times & \times \\ 0 & \times & \times \\ 0 & \times & \times \\ 0 & \times & \times \\ 0 & \times & \times \end{pmatrix}.$$

The elements marked $\times$ have been modified by P.

At the second step choose a as the bottom four elements of the second column of PA and construct a corresponding Householder transformation $\bar{P}$. Then

$$P_1 PA = \begin{pmatrix} 1 & 0 \\ 0 & \bar{P} \end{pmatrix} \begin{pmatrix} \alpha & \times & \times \\ 0 & \times & \times \\ 0 & \times & \times \\ 0 & \times & \times \\ 0 & \times & \times \end{pmatrix} = \begin{pmatrix} \alpha & \times & \times \\ 0 & \bar{\alpha} & \times \\ 0 & 0 & \times \\ 0 & 0 & \times \\ 0 & 0 & \times \end{pmatrix}.$$

To complete the reduction, at the third step choose a as the bottom three elements of the last column of the transformed matrix and construct a Householder transformation $\tilde{P}$:

$$P_2 P_1 PA = \begin{pmatrix} 1 & 0 & 0 \\ 0 & 1 & 0 \\ 0 & 0 & \tilde{P} \end{pmatrix} \begin{pmatrix} \alpha & \times & \times \\ 0 & \bar{\alpha} & \times \\ 0 & 0 & \times \\ 0 & 0 & \times \\ 0 & 0 & \times \end{pmatrix} = \begin{pmatrix} \alpha & \times & \times \\ 0 & \bar{\alpha} & \times \\ 0 & 0 & \tilde{\alpha} \\ 0 & 0 & 0 \\ 0 & 0 & 0 \end{pmatrix}.$$

Since $\bar{P}$ and $\tilde{P}$ are orthogonal, so are P_1 and P_2 as well as $P_2 P_1 P$. Thus A has been reduced to upper-triangular form by the premultiplication by an orthogonal matrix.

Before a numerical example can be given, it is necessary to know how to construct the Householder transformations. Let a be the vector we wish to transform. We desire that

$$Pa = \alpha(1, 0, \ldots, 0)^T.$$

So

$$Pa = \left(I - 2\frac{vv^T}{v^T v} \right) a$$

$$= a - \left(2\frac{v^T a}{v^T v} \right) v$$

$$= \alpha e_1$$

where $e_1 = (1, 0, \ldots, 0)^T$. Rearranging gives

$$\left(2\frac{v^T a}{v^T v} \right) v = a - \alpha e_1.$$

In other words, v is a multiple of $a - \alpha e_1$. If v is multiplied by a non-zero constant, then its Householder transformation does not change. (Why?) This means that we can choose $v = a - \alpha e_1$. Since $\alpha = \pm \|a\|_2$, $v = a \mp \|a\|_2 e_1$.

Example 6.9   Determining the Householder Transformation.

As an example, let $a = (3, 0, 4)^T$. Then $\|a\|_2 = (3^2 + 0^2 + 4^2)^{\frac{1}{2}} = 5$ and

$$v = \begin{pmatrix} 3 \\ 0 \\ 4 \end{pmatrix} \mp 5 \begin{pmatrix} 1 \\ 0 \\ 0 \end{pmatrix}.$$

There are two choices for the vector v, both of which lead to valid Householder transformations; in this case $v = (-2, 0, 4)^T$ or $v = (8, 0, 4)^T$. Since subtraction can sometimes lead to rounding-error problems, it is safer to choose the sign of $\|a\|_2$ to avoid subtraction. In this case we would prefer to choose $v = (8, 0, 4)^T$. Forming the Householder transformation gives

$$P = I - 2\frac{vv^T}{v^Tv} = \begin{pmatrix} -0.6 & 0 & -0.8 \\ 0 & 1 & 0 \\ -0.8 & 0 & 0.6 \end{pmatrix},$$

and

$$Pa = \begin{pmatrix} -0.6 & 0 & -0.8 \\ 0 & 1 & 0 \\ -0.8 & 0 & 0.6 \end{pmatrix} \begin{pmatrix} 3 \\ 0 \\ 4 \end{pmatrix} = \begin{pmatrix} -5 \\ 0 \\ 0 \end{pmatrix}$$

as desired. ■

We can now use these techniques to solve a least-squares problem. This is described in Example 6.10.

Example 6.10   Solving a Least-Squares Problem with Householder Transformations.

Consider

$$\min_x \|Ax - b\|_2$$

with

$$A = \begin{pmatrix} 3 & -1 \\ 0 & 0 \\ 4 & 7 \end{pmatrix}, \quad b = \begin{pmatrix} 0 \\ 18 \\ 25 \end{pmatrix}.$$

First we reduce A to upper triangular form. The first step involves the first column of A, $a = (3, 0, 4)^T$, the vector used in the previous example. Applying the Householder transformation to A gives

$$PA = \begin{pmatrix} -5 & -5 \\ 0 & 0 \\ 0 & 5 \end{pmatrix}.$$

The second step takes a as the bottom of the second column of PA so that $a = (0,5)^T$. Then $\|a\|_2 = 5$ and

$$v = \begin{pmatrix} 0 \\ 5 \end{pmatrix} \mp 5 \begin{pmatrix} 1 \\ 0 \end{pmatrix}.$$

It doesn't matter how the sign is chosen in this case; we take $v = (5,5)^T$ making the Householder transformation

$$\bar{P} = \begin{pmatrix} 0 & -1 \\ -1 & 0 \end{pmatrix}.$$

Applying this transformation to PA gives the upper-triangular matrix

$$\begin{pmatrix} 1 & 0 \\ 0 & \bar{P} \end{pmatrix} PA = \begin{pmatrix} 1 & 0 & 0 \\ 0 & 0 & -1 \\ 0 & -1 & 0 \end{pmatrix} \begin{pmatrix} -5 & -5 \\ 0 & 0 \\ 0 & 5 \end{pmatrix} = \begin{pmatrix} -5 & -5 \\ 0 & -5 \\ 0 & 0 \end{pmatrix}.$$

To solve the least-squares problem, the Householder transformations must also be applied to the right-hand side b:

$$\begin{pmatrix} 1 & 0 \\ 0 & \bar{P} \end{pmatrix} Pb = \begin{pmatrix} 1 & 0 \\ 0 & \bar{P} \end{pmatrix} \begin{pmatrix} -20 \\ 18 \\ 15 \end{pmatrix} = \begin{pmatrix} -20 \\ -15 \\ -18 \end{pmatrix}.$$

Then the solution is obtained by solving the triangular system

$$\begin{pmatrix} -5 & -5 \\ 0 & -5 \end{pmatrix} \begin{pmatrix} x_1 \\ x_2 \end{pmatrix} = \begin{pmatrix} -20 \\ -15 \end{pmatrix}$$

via back-substitution, giving $x^* = (1,3)^T$. The residual at the solution is $Ax^* - b = (0, -18, 0)^T$. ∎

The reduction of A to upper-triangular form can be interpreted as a matrix factorization. At each stage of the reduction, the matrix is multiplied by an orthogonal matrix built from a Householder matrix. If the final upper-triangular matrix is called R and A is an $m \times n$ matrix then

$$R = P_n P_{n-1} \cdots P_2 P_1 A.$$

Each of the matrices P_i satisfies $P_i \cdot P_i = I$ so that

$$A = P_1 P_2 \cdots P_{n-1} P_n R.$$

Let $Q = P_1 \cdots P_n$. Since the product of orthogonal matrices is orthogonal, Q is orthogonal, and we have factored $A = QR$ as the product of an orthogonal and an upper-triangular matrix. In the example above,

$$A = \begin{pmatrix} 3 & -1 \\ 0 & 0 \\ 4 & 7 \end{pmatrix} = \begin{pmatrix} -0.6 & 0.8 & 0 \\ 0 & 0 & -1 \\ -0.8 & -0.6 & 0 \end{pmatrix} \begin{pmatrix} -5 & -5 \\ 0 & -5 \\ 0 & 0 \end{pmatrix} = QR.$$

Although this factorization has been derived here via Householder transformations, a related result can be obtained using the Gram-Schmidt orthogonalization procedure applied to the columns of A, an approach discussed in many texts on linear algebra.

6.5 **SUBROUTINE** SQRLS

Subroutine SQRLS is a driver for routines from Linpack, a collection of subroutines that is described in more detail in Chapter 3. SQRLS forms the QR factorization of an $m \times n$ matrix A and then solves a least-squares problem using this factorization. If several problems must be solved, all of which have the same coefficient matrix but different right-hand sides, then A need only be factored once, and SQRLS can be told to skip this step for the subsequent right-hand sides.

Many of the arguments will be familiar. SQRLS requires the problem data A, M, N, B, and returns the array of parameters X and the array of residuals RSD. Some of the other arguments require a bit of explanation.

LDA is the leading dimension of the matrix A as declared in the main program, and is discussed in Section 3 of Chapter 3. SQRLS requires several extra storage arrays JPVT, QRAUX, and WORK; the latter is a work array. The output arrays JPVT and QRAUX are used to store pivoting and other information used in the QR factorization; when problems with multiple right-hand sides are being solved, they must not be modified or used for other purposes since they are needed to determine the orthogonal transformations used in the factorization.

The input argument TOL, and the output value KR are related. They are used to determine if a least-squares problem is degenerate. This decision is based on the size of the rounding errors which are assumed to be proportional to TOL. To make this an effective estimate, the elements in the matrix A should be roughly the same size. One way to ensure this is to divide each row of A by its largest element (note that the corresponding component of B must also be scaled by this value). If this is done then it is sensible to set TOL = ϵ_{mach}. In many least-squares problems the data are inaccurate since they come from measurement equipment of limited accuracy. In this case the measurement errors will dominate the rounding errors and it may be sensible to set TOL to be the relative measurement error of the elements in the matrix A. It may also be worthwhile to consider orthogonal distance models (see Section 1).

If the input value of TOL leads SQRLS to conclude that the problem is degenerate then some of the solution coefficients will be arbitrary (the problem does not have a unique solution). SQRLS will set the arbitrary coefficients to zero and then the remaining ones will be uniquely determined. The number of uniquely-determined parameters is equal to the rank of the matrix (the number of independent columns), which is estimated by SQRLS and returned in the argument KR; N − KR parameters are arbitrary and are set to zero. You should *always* examine KR after a call to SQRLS. If it is less than N the model is degenerate. Perhaps the data need to be scaled or the choice of model functions rethought. It is a source of concern and should not be ignored!

The techniques used in SQRLS are similar to those described in the previous sections. However, some modifications are made so that degenerate and underdetermined problems can be solved. We have been tacitly assuming that the number of data points is greater than the number of parameters (i.e., $m \geq n$). This is generally true for data fitting problems, but there are other applications where $m < n$ and hence the solution is not uniquely determined; see Section 9. We have also been assuming that the model is nondegenerate (that A has full rank, that the upper-triangular matrix from the QR factorization is invertible). The least-squares approach is still sensible for underdetermined and degenerate problems, and the QR factorization can be adjusted to handle them. The routine SQRLS can still be used in these cases. For degenerate problems pivoting is required, much as partial pivoting was necessary in Gaussian elimination. If the coefficient matrix A changes slightly then it is possible to update the QR factorization much more efficiently than it is to recompute it from scratch. There are other routines in Linpack for updating the QR factorization.

The variance-covariance matrix $(A^T A)^{-1}$ can also be computed using Linpack subroutines. If you are working in single precision, the correct routine to use is SPODI. This is described in more detail in the Linpack manual (see the Bibliography).

A simple least-squares problem is solved below. It is based on the quadratic model discussed in Example 6.1. The output from the program is

```
COEFFICIENT MATRIX
     1.000000      1.000000      1.000000
     1.000000      2.000000      4.000000
     1.000000      3.000000      9.000000
     1.000000      4.000000     16.000000
     1.000000      5.000000     25.000000
RIGHT-HAND SIDE
     1.000000      2.300000      4.600000      3.100000      1.200000
RANK OF MATRIX =            3
PARAMETERS
    -3.020001      4.491428     -0.728571
RESIDUALS
     0.257143     -0.748572      0.702858     -0.188571     -0.022857
```

The main program is

```
      PARAMETER (MM = 5, NN = 3)
      REAL       A(MM,NN), B(MM), X(NN), QRAUX(NN), WORK(NN), TOL
      INTEGER    JPVT(NN)
      DATA       B / 1.0, 2.3, 4.6, 3.1, 1.2/
C
C SET UP LEAST-SQUARES PROBLEM
C     QUADRATIC MODEL, EQUALLY-SPACED POINTS
C
      M = 5
      N = 3
      DO 10 I = 1,M
         A(I,1) = 1.0
         DO 10 J = 2,N
            A(I,J) = A(I,J-1)*I
10    CONTINUE
      TOL = 1.E-6
      WRITE (*,*)   ' COEFFICIENT MATRIX'
```

```
      WRITE (*,800)  ((A(I,J),  J = 1,N),  I = 1,M)
      WRITE (*,*)    ' RIGHT-HAND SIDE'
      WRITE (*,810)  (B(I),  I = 1,M)
C
C SOLVE LEAST-SQUARES PROBLEM
C
      ITASK = 1
      CALL SQRLS (A, MM, M, N, TOL, KR, B, X, B, JPVT, QRAUX, WORK,
     *            ITASK, IND)
C
C PRINT RESULTS
C
      IF (IND .NE. 0) WRITE (*,*) ' ERROR CODE =', IND
      WRITE (*,*)    ' RANK OF MATRIX =', KR
      WRITE (*,*)    ' PARAMETERS'
      WRITE (*,800)  (X(J),  J = 1,N)
      WRITE (*,*)    ' RESIDUALS'
      WRITE (*,810)  (B(I),  I = 1,M)
C
      STOP
800   FORMAT (3F12.6)
810   FORMAT (5F12.6)
      END
```

The documentation for SQRLS can be found at the end of the chapter.

6.6 HISTORICAL PERSPECTIVE: GAUSS

Carl Friedrich Gauss is considered to be one of the greatest mathematicians of all time, comparable in stature to Archimedes and Newton. Through his publications and note-books, he transformed nineteenth century mathematics with investigations into number theory, geometry, and analysis. His fame extended to other scientists with his discoveries in astronomy, and to the general public with his work in electromagnetism and geodesy.

Gauss was born in 1777 into a poor family. His parents had little education—only his father could read and write. Gauss was sent to an undistinguished school, but he astonished the teacher in his first arithmetic class at age 8 by solving instantly a busy-work problem of summing the first 100 integers. His abilities eventually brought him to the attention of the Duke of Brunswick, who adopted Gauss as his protegé, financing his education and continuing to support him until the Duke's death in 1806 during the Napoleonic wars.

As a teenager, Gauss made important advances in geometry (solving a 2000-year-old problem due to Euclid), algebra (providing a rigorous proof of the fundamental theorem of algebra), and number theory (work published in his *Disquisitiones Arithmeticae* or *Arithmetical Investigations*). In each case, he not only solved a particular problem, but also gave complete discussions of the topics and new methods of solution which had a far-reaching influence on mathematics.

In 1801, Gauss became interested in applications of mathematics. Twenty years earlier the seventh planet, Uranus, had been discovered by Sir William Herschel. The German philosopher Hegel considered seven to be a philosophically satisfying number, and sarcastically criticized astronomers who in his view were wasting their time searching for an eighth planet. Almost as soon as Hegel's thoughts were published, on January 1, 1801 Giuseppe Piazzi discovered the asteroid Ceres.

Ceres was only visible for forty days before it was lost to view behind the sun. Astronomers wanted to know where in the sky it would reappear. At that time, astronomers could calculate orbits for comets and the larger planets, but their techniques were inadequate for predicting the orbit of Ceres. However Gauss, using three observations, extensive analysis, and the method of least squares, was able to determine the orbit with such accuracy that Ceres was easily found when it reappeared in late 1801. This work gave Gauss extraordinary fame among scientists throughout Europe. (There were also mild side effects in his personal life: Gauss named his eldest son after Piazzi, and his other children from his first marriage after the astronomers Olbers and Harding.)

Gauss was one of the first to use least-squares techniques, but long before their invention, scientists were trying to model data. Many of the applications were to astronomy, as here in the study of Ceres. Several estimates of its position were made but, because the telescopes and other tools were not very accurate, the measurements would not agree and some approximate value would have to be accepted as the "true" position.

Another common application would be in map-making, where surveyors would measure the countryside to determine the distances between landmarks. The measurements would not be exact, either because of limitations of the equipment or because of mistakes made in taking the measurements. This would lead to inconsistencies in the data which would have to be resolved before the map could be drawn.

Many approaches were suggested, going back at least to sixteenth-century Europe and probably much earlier. They were generally based on properties of the residuals, the vector $Ax - b$ in our notation. One of the first ideas was to choose the parameters so that the sum of the residuals was zero. This technique was in use for over two hundred years, with various embellishments to ensure that the solution was computable and unambiguous. One idea was to use the 1-norm of the residuals to break ties. In 1783 Laplace suggested using the ∞-norm to determine the solution. It is somewhat surprising that these approaches were the first to be considered. Scientists at that time were very much concerned with practical calculations, but these norms are difficult to work with computationally. However, they are intuitively appealing as techniques to reconcile measurement errors, and this may have made them popular and acceptable.

The concept of a least-squares estimator was first discovered in 1795 by the eighteen-year-old Gauss. He needed a practical tool for analyzing survey data, and later used the technique on a regular basis for astronomical calculations. Gauss thought it such a simple idea that he did not publish it; he assumed that it must have been discovered many years earlier. In fact the first published description of least squares appeared in 1805, in an appendix of Legendre's book *New Methods for Determining the Orbits of Comets*. Gauss later wrote several times on the properties of least-squares estimators, justifying their use via newly-developed ideas in probability theory.

When Gauss published a description of his astronomical calculations, Legendre accused him of stealing his ideas on least squares. Legendre was wrong, but Gauss refused to provide evidence that he had been using the technique for a decade. This was not the only time that Gauss anticipated the work of others. Gauss set very high standards for his own work. He would not publish unless the results were complete and elegant, the ideas had far-reaching applicability, and the proofs were rigorous. His sons reported that Gauss discouraged them from going into science because he did not want any second-rate work associated with his name.

After his death, Gauss' notebooks and unpublished manuscripts revealed many results that, had they been circulated, would have saved other scientists decades of work. During his life, mathematicians would often send Gauss their papers for comment, and frequently his response would be a polite compliment along with the remark that he had made the same discoveries himself years earlier. This did not endear him to students and other mathematicians.

After the Duke's death, and hence the end of Gauss' financial support, Gauss became professor of astronomy at the University of Göttingen, a position he held for the rest of his life. He took the job in part because he would have light teaching duties. He was repelled by the idea of teaching mathematics largely because it meant drilling ill-prepared and unmotivated students in the most elementary manipulations.

His later work included studies in probability, differential geometry (work that later led to Einstein's theories of relativity), electromagnetism, non-Euclidean geometry, and geodesy. As an outgrowth of his work in electricity and magnetism, in 1833 he was the first to transmit sentences via electric telegraph, five years before Morse.

Gauss married twice, once in 1805 and again in 1810. The death of his first wife plunged him into a loneliness from which he never recovered. He dominated his daughters and quarreled with his younger sons, who emigrated to the United States. He had a robust, long life, but a troubled one. He died on February 23, 1855.

*6.7 DEGENERATE LEAST-SQUARES PROBLEMS

A degenerate least-squares problem is one where the solution is not unique. This is primarily a problem with the data; it usually means that there are not enough data points to distinguish all of the model functions. However, it may sometimes indicate a deficiency in the model. A simple example is given below.

Example 6.11 A Degenerate Least-Squares Problem.

Consider the model

$$b(t) \approx x_1 \cdot (1) + x_2 \cdot (t) + x_3 \cdot (2t + 1).$$

The third model function $2t + 1$ is a linear combination of the others so that there will be infinitely many ways to approximate a given data set. To see this, notice that

$$\alpha \cdot [1 \cdot (1) + 2 \cdot (t) - 1 \cdot (2t + 1)] = 0$$

for any α, and so any set of parameters $x = (x_1, x_2, x_3)$ can be replaced by $x + \alpha(1, 2, -1)$ without changing the fit. (That is, the value of $b(t)$ will be unaffected by the particular value of α chosen.) ■

In general, there will be degeneracy if the modelling functions are linearly dependent, i.e., if some non-zero linear combination of the modelling functions equals zero, as happens for this example. More precisely, the problem will be degenerate if the columns of the coefficient matrix A are linearly dependent. When the modelling functions are linearly dependent, the residuals will still be unique even if the parameters are not uniquely determined. Thus any solution to the least-squares problem will produce the same fit to the data. This happens in the simple example above. If the main purpose of the least-squares approximation is to smooth a data set, then degeneracy may not be a serious difficulty. When it is important to obtain parameter estimates, however, degeneracy can be a major problem.

Since degenerate problems have many solutions instead of just one solution, it might seem that the problem should be easier to solve. In fact, degeneracy makes life more difficult for us. In exact arithmetic there is a sharp distinction between degenerate and non-degenerate problems, but in rounded arithmetic it may be hard to decide if a problem is degenerate or not. If we make a mistake, the parameters will be vastly inaccurate. Also, the closer a problem comes to being degenerate, the worse the effects of the mistake will be. If our least squares problem is

$$\min \left\| \begin{pmatrix} 1 & 1 \\ 0 & 10^{-k} \end{pmatrix} \begin{pmatrix} x_1 \\ x_2 \end{pmatrix} - \begin{pmatrix} 1 \\ 1 \end{pmatrix} \right\|_2 ,$$

then the solution is $x = (1 - 10^k, 10^k)^T$. Suppose that these coefficients were obtained from the QR factorization of another matrix. Then the entry 10^{-k} might only represent rounding errors from the factorization and so, to within the precision of the computer, this might be a degenerate problem. If we decided this were the case, then we would solve instead

$$\min \left\| \begin{pmatrix} 1 & 1 \\ 0 & 0 \end{pmatrix} \begin{pmatrix} x_1 \\ x_2 \end{pmatrix} - \begin{pmatrix} 1 \\ 1 \end{pmatrix} \right\|_2 .$$

This problem has many solutions. Typically we would choose $x_2 = 0$ to eliminate the ambiguity, giving $x = (1, 0)^T$. The two solutions are vastly different. As k gets larger, and hence as 10^{-k} gets closer to zero, the more different they become.

In a sense the difficulty here is not computational. Even if the least-squares problem is degenerate, there is no problem forming the QR factorization of the coefficient matrix A. The only difference is that the upper triangular matrix R will no longer be invertible, since it will have some zero entries on the diagonal. However, we would like to use the diagonal entries of R to determine if a least-squares problem is degenerate.

Even in exact arithmetic the QR factorization of a matrix may fail to reveal near-degeneracy. The $n \times n$ matrix

$$A = \begin{pmatrix} 1 & -1 & -1 & \cdots & -1 \\ & 1 & -1 & \cdots & -1 \\ & & 1 & \cdots & -1 \\ & & & \ddots & \vdots \\ & & & & 1 \end{pmatrix}$$

is already upper-triangular so it would not be changed by the QR factorization. It appears innocuous. The diagonal entries are all equal to one, the determinant is equal to one, and the matrix is clearly nonsingular. However it is very close to being singular. If we set $A_{n,1} = 2^{-(n-2)}$ the matrix becomes singular.

To improve our ability to detect degeneracy we introduce pivoting into the factorization. Row pivoting was used in Gaussian elimination when solving linear equations. Here *column* pivoting is used. At the first stage the "largest" column is determined by computing the 2-norms of all the columns of the matrix. This column is moved to the front of the matrix by interchanging (pivoting) it with the first column. Then the first step of the orthogonal factorization is performed, zeroing all but the first element of the new first column. At the beginning of each succeeding iteration, the "largest" remaining column (in the 2-norm) is moved to the front of the matrix and then the orthogonal transformation is applied as usual. At the later iterations only the bottom parts of the columns are considered. Thus at the second step, we only look at elements 2 through m, and at the third step elements 3 through m, etc. The resulting matrix R will have diagonal entries that decrease monotonically in absolute value, and will be a more accurate indicator of degeneracy.

Example 6.12   The QR Factorization with Pivoting.

When pivoting is applied to

$$A = \begin{pmatrix} 1 & -1 & -1 & -1 & -1 \\ & 1 & -1 & -1 & -1 \\ & & 1 & -1 & -1 \\ & & & 1 & -1 \\ & & & & 1 \end{pmatrix}$$

the last column is the largest, and will be interchanged with the first column at step 1. Ultimately, the QR factorization produces

$$R = \begin{pmatrix} 2.2361 & 0.8944 & 0.4472 & 0 & -0.4472 \\ 0 & 1.7889 & 0.3354 & 0 & -0.3354 \\ 0 & 0 & -1.6394 & 0 & 0.4194 \\ 0 & 0 & 0 & -1.4142 & 0.7071 \\ 0 & 0 & 0 & 0 & 0.1078 \end{pmatrix}.$$

The final diagonal entry is now small ($.1078 \approx .125 = 2^{-(n-2)}$). ∎

The factorization with column pivoting can be represented as

$$A = QRP$$

where P is a permutation matrix that records the column interchanges made. The matrix R will have the form

$$R = \begin{pmatrix} R_{(1)} & R_{(2)} \\ 0 & 0 \end{pmatrix}$$

where $R_{(1)}$ is upper triangular and $R_{(2)}$ is just some non-zero rectangular matrix. The number of non-zero rows in R is the *rank* of the matrix A, the number of linearly independent columns.

This factorization can be used to obtain a reduced least-squares problem

$$\|Ax - b\|_2 = \|RPx - Q^T b\|_2 .$$

To simplify the formulas, we define new parameters by $y = Px$ and new data by $c = Q^T b$. We also split y and c into pieces as in the matrix R, i.e., $y = (y_{(1)}, y_{(2)})^T$ and $c = (c_{(1)}, c_{(2)})^T$. Then we obtain

$$\|Ax - b\|_2^2 = \left\| \begin{pmatrix} R_{(1)} & R_{(2)} \\ 0 & 0 \end{pmatrix} \begin{pmatrix} y_{(1)} \\ y_{(2)} \end{pmatrix} - \begin{pmatrix} c_{(1)} \\ c_{(2)} \end{pmatrix} \right\|_2^2$$
$$= \|R_{(1)}y_{(1)} + R_{(2)}y_{(2)} - c_{(1)}\|_2^2 + \|-c_{(2)}\|_2^2 .$$

The second term $-c_{(2)}$ cannot be affected by the parameters y. However the first term can be made zero in infinitely many ways. The components of $y_{(2)}$ can be chosen arbitrarily; then $y_{(1)}$ is determined by solving

$$R_{(1)}y_{(1)} = c_{(1)} - R_{(2)}y_{(2)}.$$

This is an upper-triangular system and is solved using back-substitution. Typically we set $y_{(2)} = 0$. In summary:

> Factor $A = QRP$
> Form $c = Q^T b$
> Solve $R_{(1)}y_{(1)} = c_{(1)}$ and set $y = (y_{(1)}, 0)^T$
> Set $x = P^T y$, i.e., re-order the variables.

The algorithm is much the same as in the non-degenerate case. It is applied in Example 6.13.

Example 6.13   Solving a Degenerate Least-Squares Problem.

To illustrate the techniques consider the problem

$$A = \begin{pmatrix} 4 & 2 & 3 \\ 0 & 1 & 5 \\ 0 & 0 & 0 \\ 0 & 0 & 0 \end{pmatrix} \quad b = \begin{pmatrix} 8 \\ 2 \\ 5 \\ 7 \end{pmatrix}.$$

The matrix A is already in appropriate form and so we will not apply the QR factorization. The matrices $R_{(1)}$ and $R_{(2)}$ are

$$R_{(1)} = \begin{pmatrix} 4 & 2 \\ 0 & 1 \end{pmatrix} \quad R_{(2)} = \begin{pmatrix} 3 \\ 5 \end{pmatrix}.$$

To obtain the parameters we solve

$$R_{(1)}y_{(1)} = \begin{pmatrix} 4 & 2 \\ 0 & 1 \end{pmatrix} y_{(1)} = c_{(1)} = \begin{pmatrix} 8 \\ 2 \end{pmatrix}$$

giving $y_{(1)} = (1, 2)^T$ and hence $x = (1, 2, 0)^T$. If we were to choose another value for $y_{(2)}$, say $y_{(2)} = 3$, then we would obtain the parameters from

$$R_{(1)}y_{(1)} = \begin{pmatrix} 4 & 2 \\ 0 & 1 \end{pmatrix} y_{(1)} = c_{(1)} - R_{(2)}y_{(2)} = \begin{pmatrix} 8 \\ 2 \end{pmatrix} - \begin{pmatrix} 3 \\ 5 \end{pmatrix}(3) = \begin{pmatrix} -1 \\ -13 \end{pmatrix}$$

giving $y_{(1)} = (25/4, -13)^T$ and $x = (25/4, -13, 3)^T$. In both cases the residual vector is $(0, 0, -5, -7)^T$. ■

*6.8   THE SINGULAR-VALUE DECOMPOSITION

Although the QR factorization with column pivoting is almost always an effective tool for detecting degeneracy, it is not foolproof. Certain rare nearly-degenerate problems slip through its grip. The most reliable tool available is the singular-value decomposition (SVD), another orthogonal matrix factorization. Its reliability carries a price—it requires 5–10 times as many arithmetic operations as the QR factorization. Also, it is not possible to update the SVD efficiently when the data change. We will only give a brief discussion here of the SVD and its more interesting uses. For more complete information see Golub and Van Loan (1983). The routine SSVDC from Linpack can be used to compute the SVD.

If A is an $m \times n$ matrix with $m \geq n$ then one form of the SVD is

$$A = U\Sigma V^T$$

where U and V^T are orthogonal and Σ is diagonal. That is, $U^TU = I_m$, $VV^T = I_n$, U is $m \times m$, V is $n \times n$, and

$$
\Sigma = \begin{pmatrix}
\sigma_1 & & & & & \\
& \sigma_2 & & & & \\
& & \ddots & & & \\
& & & \sigma_{n-1} & & \\
& & & & \sigma_n & \\
& & \mathbf{0} & & &
\end{pmatrix}
$$

is an $m \times n$ diagonal matrix (the same dimensions as A). In addition $\sigma_1 \geq \sigma_2 \geq \cdots \geq \sigma_n \geq 0$. The σ_i's are called the **singular values** of A. The smallest singular value σ_n is the distance in the 2-norm from A to the nearest degenerate matrix. The number of non-zero singular values is equal to the rank of the matrix. Thus if A is singular then at least $\sigma_n = 0$. In practice, singular values are rarely exactly zero, but if A is "nearly singular" some of the singular values will be small. The ratio σ_1/σ_n can be regarded as a condition number of the matrix A. It is not the same condition number we discussed in Chapter 3, but it has many of the same properties and is usually about the same order of magnitude numerically. An example of the singular-value decomposition is given below.

Example 6.14   The Singular-Value Decomposition.

If

$$
A = \begin{pmatrix}
1 & 2 & 3 \\
4 & 5 & 6 \\
7 & 8 & 9 \\
10 & 11 & 12
\end{pmatrix}
$$

then its SVD is

$$
A = U\Sigma V^T = \begin{pmatrix}
.1409 & .8247 & -.4202 & -.3513 \\
.3439 & .4263 & .2985 & .7816 \\
.5470 & .0278 & .6638 & -.5093 \\
.7501 & -.3706 & -.5420 & .0790
\end{pmatrix}
\begin{pmatrix}
\mathbf{25.4624} & 0 & 0 \\
0 & \mathbf{1.2907} & 0 \\
0 & 0 & \mathbf{0} \\
0 & 0 & 0
\end{pmatrix}
$$
$$
\begin{pmatrix}
.5045 & .5745 & .6445 \\
-.7608 & -.0571 & .6465 \\
.4082 & -.8165 & .4082
\end{pmatrix}.
$$

For emphasis, the singular values have been printed in bold face. Notice that $\sigma_3 = 0$, indicating that the matrix is not of full rank. ∎

The singular-value decomposition is about one hundred years old. It was discovered independently by Beltrami in 1873 and Jordan in 1874 for the case of square matrices. The technique was extended to rectangular matrices by Eckart and Young in the 1930's. However, its use as a computational tool is much more recent, dating back only to the 1960's. This is not so hard to understand when it is realized that the computation of the SVD requires a variety of sophisticated numerical techniques. The development of the SVD as a practical tool is primarily due to Gene Golub who, in a series of papers, demonstrated its usefulness and feasibility in a wide variety of applications.

*6.8.1 The SVD for Solving Linear Least-Squares Problems

The SVD has many uses. First of all, it can be used to solve least-squares problems. We have

$$\|Ax - b\|_2 = \|U\Sigma V^T x - b\|_2$$
$$= \|U^T(U\Sigma V^T x - b)\|_2 \qquad \text{(Since } U \text{ is orthogonal)}$$
$$= \|\Sigma V^T x - U^T b\|_2 .$$

Denoting $U^T b$ by d and $V^T x$ by z, we have

$$\|Ax - b\|_2^2 = \|\Sigma z - d\|_2^2 = \left\| \begin{pmatrix} \sigma_1 z_1 - d_1 \\ \sigma_2 z_2 - d_2 \\ \vdots \\ \sigma_n z_n - d_n \\ -d_{n+1} \\ -d_{n+2} \\ \vdots \\ -d_m \end{pmatrix} \right\|_2^2 = (\sigma_1 z_1 - d_1)^2 + \cdots + (\sigma_n z_n - d_n)^2 \\ + d_{n+1}^2 + \cdots + d_m^2.$$

As long as none of the singular values are zero we can uniquely select the z_i's (or equivalently the x_i's) to reduce this to its minimum value

$$\|Ax - b\|_2^2 = d_{n+1}^2 + \cdots + d_m^2.$$

In this case the least squares problem has a unique solution However, if $\sigma_n = 0$ then *any* choice of z_n is allowed and all choices give exactly the same residual sum of squares

$$\|Ax - b\|_2^2 = d_n^2 + d_{n+1}^2 + \cdots + d_m^2.$$

In that case the least squares problem is *not* uniquely solvable. Whenever a $\sigma_i = 0$, the usual convention is to set the corresponding $z_i = 0$ too. It turns out that the singular values are nonzero if and only if the basis functions ϕ_j are linearly independent at the data points. So if some of the basis functions are nearly dependent one or more singular value will be close to zero. Since a zero singular value implies degeneracy and non-uniqueness, you should not be surprised to learn that small singular values are a symptom of ill conditioning. This shows itself in large changes in the computed solution when either the data or the arithmetic changes a little.

    The proper use of the SVD involves a tolerance reflecting the accuracy of the original data and the floating point arithmetic being used. Any σ_j greater than the tolerance is acceptable and the corresponding z_j is computed from $z_j = d_j / \sigma_j$. Any σ_j less than the tolerance are regarded as negligible, and the corresponding z_j is set to zero.

   The tolerance in the SVD plays a role that is similar to TOL in SQRLS. Increasing the tolerance leads to larger residuals but gives results that are less likely to change if we alter the data. Decreasing the tolerance leads to smaller residuals and gives results that are more sensitive to changing the data. Neglecting σ_j's less than the tolerance has the effect of decreasing the condition number. Since the condition number is an error magnification factor, this results in a more reliable determination of the least squares parameters. The cost of this increased reliability is a possible increase in the size of the residuals.

Example 6.15   Data Fitting with the SVD.

   Suppose we want to solve a least-squares problem with the 4×3 matrix from Example 6.12 and with right-hand side $b = (1, 0, 0, 0)^T$. By observation we see that

$$U^T b = \begin{pmatrix} 0.1409 \\ 0.8247 \\ -0.4202 \\ -0.3513 \end{pmatrix},$$

so if none of the singular values were zero we could reduce the residual sum of squares to $d_m^2 = d_4^2 = 0.3513^2$. But we know that $\sigma_3 = 0$ from the preceding example, so the best we can do is $d_3^2 + d_4^2 = 0.4202^2 + 0.3513^2$. This will occur if $z_1 = d_1/\sigma_1$, $z_2 = d_2/\sigma_2$, $z_3 = 0$, and then $V^T x = z$, or $x = Vz$. The calculations give $z_1 = 0.005533$, $z_2 = 0.6390$, $z_3 = 0$, and

$$x_{SVD} = \begin{pmatrix} -0.4833 \\ -0.0333 \\ 0.4167 \end{pmatrix}. \quad \blacksquare$$

   A more realistic example occurs using the U.S. Census data from problem P4–2. There are $m = 8$ data points. The values of t_i are $1900, \ldots, 1970$, and the corresponding values of y_i in units of a million people, are about 75.99, 91.97,..., 203.21. We will try to fit these by a quadratic,

$$b(t) \approx c_1 + c_2 t + c_3 t^2.$$

Using the single-precision Linpack routine SSVDC on an IBM personal computer with a tolerance of 0.0 (i.e., include all the singular values) we find that the least-squares problem has coefficients

$$c_1 = -0.372 \times 10^5, \qquad c_2 = 0.368 \times 10^2, \qquad c_3 = -0.905 \times 10^{-2},$$

and the 8×3 matrix has singular values

$$\sigma_1 = 0.106 \times 10^8, \qquad \sigma_2 = 0.648 \times 10^2, \qquad \sigma_3 = 0.346 \times 10^{-3}.$$

Using double precision gives about the same singular values, but the coefficients are found to be about

$$c_1 = 0.373 \times 10^5, \qquad c_2 = -.402 \times 10^2, \qquad c_3 = 0.108 \times 10^{-1}.$$

The signs of the two sets of coefficients do not even agree. In fact, when the model is used to predict the population in 1980, the coefficients obtained with double precision predict 227.78 million, whereas the coefficients obtained with single precision predict 145.21 million. The single precision coefficients are clearly useless, but how about the others? How reliable are they? For this problem the condition number is $\sigma_1/\sigma_3 = 0.306 \times 10^{11}$, which is a signal that there is some difficulty. For t between 1900 and 1970, the three basis functions are nearly linearly dependent.

To improve the situation we recognize that $\sigma_3 \approx 0$, and set it to zero. Then the coefficients we obtain in double precision are

$$c_1 = -0.167 \times 10^{-2}, \qquad c_2 = -0.162 \times 10^1, \qquad c_3 = -0.871 \times 10^{-3},$$

and in single precision

$$c_1 = -0.166 \times 10^{-2}, \qquad c_2 = -0.162 \times 10^1, \qquad c_3 = -0.869 \times 10^{-3}.$$

Now they are in much better agreement. Moreover, they are much smaller, which means that there will be less cancellation in evaluation of the quadratic. The predicted populations in 1980 are 212.91 million and 214.96 million. The effect of single precision is still noticed, but the results are no longer disastrous. (Another approach to modelling the census data can be found in Problem P4–2.)

When the least-squares problem is solved using the SVD, the solution x_{SVD} has a valuable property: it will be the solution of minimal length. That is, if $\bar{x}$ is any set of parameters that minimizes $\|Ax - b\|_2$, then $\|x_{SVD}\|_2 \leq \|\bar{x}\|_2$.

*6.8.2 The SVD for Data Compression

Suppose that a satellite in space is taking photographs of Jupiter to be sent back to earth. The satellite digitizes the picture by subdividing it into tiny squares called **pixels** or picture elements. Each pixel is represented by a single number that records the average light intensity of the photograph in that square. If each photograph were divided into 500×500 pixels the satellite would have to send 250,000 numbers to earth for each picture. This would take a great deal of time and would limit the number of photographs that could be transmitted. It may be possible to approximate this matrix with a "simpler" matrix which requires less storage. A reasonable meaning of a simple matrix is one of relatively low rank, that is, with only a few independent columns. In addition to saving storage, such matrices may provide important insight into particular problems.

Consider this 500×500 array of numbers as a matrix A. We will compute its singular-value decomposition and use it to obtain an approximation to the picture.

To derive the approximation, we let u_i and v_i be the i-th columns of U and V, respectively. Then it is not difficult to check that the singular-value decomposition can be written as

$$A = U\Sigma V^T = \sum_{i=1}^n \sigma_i u_i v_i^T.$$

The matrix $u_i v_i^T$ is the **outer product** of a column of U with the corresponding column of V. Each can be stored using only $m + n$ locations rather than mn locations. For Example 6.12 we obtain

$$
A = \begin{pmatrix} 1 & 2 & 3 \\ 4 & 5 & 6 \\ 7 & 8 & 9 \\ 10 & 11 & 12 \end{pmatrix} = U\Sigma V^T
$$

$$
= 25.4624 \begin{pmatrix} .1409 \\ .3439 \\ .5470 \\ .7501 \end{pmatrix} (\,.5045 \quad .5745 \quad .6445\,)
$$

$$
+ \; 1.2907 \begin{pmatrix} .8247 \\ .4263 \\ .0278 \\ -.3706 \end{pmatrix} (-.7608 \quad -.0571 \quad .6465\,)
$$

$$
+ \; 0 \quad \begin{pmatrix} -.4202 \\ .2985 \\ .6638 \\ -.5420 \end{pmatrix} (\;\; .4082 \quad -.8165 \quad .4082\,).
$$

To compress the data, the smaller singular values are set to zero. If only 10 singular values were used, then the approximation would be

$$
A = \sum_{i=1}^{n} \sigma_i u_i v_i^T \approx \sum_{i=1}^{10} \sigma_i u_i v_i^T.
$$

There is no need to form the approximation matrix explicitly.

We can apply this idea to the satellite photograph. To obtain an approximation to the picture we set all the small singular values to zero; to maintain an acceptable picture we might only have to keep the 10–20 largest singular values. The approximate picture only depends on the first 10–20 columns of U and V and the corresponding singular values; the rest of the coefficients will be multiplied by zero and can be ignored. Thus instead of 250,000 numbers, the approximate picture would depend on only 10,000–20,000 numbers. Many more pictures could be sent to earth. Of course, the usefulness of this procedure depends on the particular matrix and the distribution of its singular values.

This idea is discussed in more detail in an article by Andrews and Patterson (1975). It is illustrated in Figure 6.5.

*6.9 THE NULL-SPACE PROBLEM

Most of our attention has been directed at over-determined systems of equations, that is, least-squares problems where the number of data points is greater than the number

Original Data (Fingerprint)   Using 1 singular value   Using 2 singular values

Using 4 singular values   Using 10 singular values   Using 20 singular values

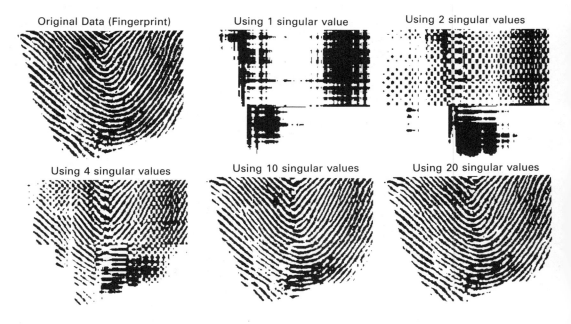

Figure 6.5   Approximating Pictures with the SVD

of parameters. If the problem is underdetermined and there are multiple solutions to the system of equations, then it may be important to determine not just one solution but also to characterize all solutions to the system.

An important application is in constrained optimization. Suppose we had to solve

$$\text{minimize} \quad F(x_1, x_2),$$
$$\text{subject to} \quad x_1 + x_2 = 1.$$

The constraint allows us to simplify this by setting $x_2 = 1 - x_1$ giving

$$\text{minimize} \quad F(x_1, 1 - x_1).$$

Not only is this problem unconstrained, we have also decreased the number of variables by one.

In general, suppose we wish to solve the n-dimensional problem

$$\text{minimize} \quad F(x),$$
$$\text{subject to} \quad Ax = b.$$

We would like to reduce this to an unconstrained problem with fewer variables by using the constraints to determine some variables in terms of the others. There will usually be fewer constraints than there are variables, since otherwise the solution would be completely determined by the constraints and there would be no need for an optimization problem. Here we want to determine all points x that satisfy the constraints.

Using routine SQRLS we can find some solution $\tilde{x}$ of $Ax = b$. (If the equations have no solution, the minimization problem does not make sense.) If $\bar{x}$ is any other solution to $Ax = b$, then

$$A(\bar{x} - \tilde{x}) = A\bar{x} - A\tilde{x} = b - b = 0.$$

Thus any solution of $Ax = b$ can be written as $\bar{x} = \tilde{x} + d$ where d satisfies $Ad = 0$. We say that d lies in the **null space** of A. The task is to determine all vectors that lie in the null-space of A.

For the constraint $x_1 + x_2 = 1$ the coefficient matrix is $A = (1, 1)$. A particular solution of the constraint is $\tilde{x} = (0, 1)^T$. The general solution can be written as $\bar{x} = \tilde{x} + d$ where $d = (\alpha, -\alpha)^T$ and α is arbitrary. Note that $Ad = 0$. Our original optimization problem then becomes

$$\min_{\alpha} F(\tilde{x} + \alpha d) = F(\alpha, 1 - \alpha),$$

an unconstrained problem in one variable α.

Suppose that we have the QR factorization of A^T

$$A^T = QR = (Q_{(1)} \quad Q_{(2)}) \begin{pmatrix} R_{(1)} \\ 0 \end{pmatrix},$$

where $R_{(1)}$ is a square, upper-triangular matrix. Rearranging this equation produces

$$AQ = (AQ_{(1)} \quad AQ_{(2)}) = R^T = (R_{(1)}^T \quad 0)$$

and so $AQ_{(2)} = 0$. Thus the columns of $Q_{(2)}$ are in the null space of A. In fact it can be shown that any vector in the null space of A is a linear combination of the columns of $Q_{(2)}$.

Example 6.16   Determination of the Null Space.

For the constraint matrix

$$A = \begin{pmatrix} 1 & 2 & 3 & 4 \\ 5 & 6 & 7 & 8 \end{pmatrix}$$

the QR factorization of A^T is

$$Q = \begin{pmatrix} -0.1826 & -0.8165 & -0.4001 & -0.3741 \\ -0.3651 & -0.4082 & 0.2546 & 0.7970 \\ -0.5477 & 0 & 0.6910 & -0.4717 \\ -0.7303 & 0.4082 & -0.5455 & 0.0488 \end{pmatrix} \quad R = \begin{pmatrix} -5.4772 & -12.7802 \\ 0 & -3.2660 \\ 0 & 0 \\ 0 & 0 \end{pmatrix}$$

and so $Q_{(2)}$ is given by the final two columns of Q:

$$Q_{(2)} = \begin{pmatrix} -0.4001 & -0.3741 \\ 0.2546 & 0.7970 \\ 0.6910 & -0.4717 \\ -0.5455 & 0.0488 \end{pmatrix}.$$

The calculations were performed using Linpack in 16-digit arithmetic even though only four digits are given here. To check that the null-space has been correctly determined we form

$$A \cdot Q_{(2)} = \begin{pmatrix} 0.5551 \times 10^{-16} & 0.1388 \times 10^{-16} \\ 0 & 0.2776 \times 10^{-16} \end{pmatrix},$$

all of whose entries are at the level of round-off. ∎

We now apply this idea to the constrained optimization problem. If we have some solution $\bar{x}$ to the constraint equations $Ax = b$, then the general solution can be written as $\tilde{x} = \bar{x} + Q_{(2)}p$ where p is an arbitrary vector. Thus the constrained optimization problem can be reduced to

$$\min_{p} F(\bar{x} + Q_{(2)}p).$$

If the original problem had n variables and m linearly-independent constraints, then the new problem would have only $n - m$ variables and would be unconstrained.

Although we have only discussed problems where the constraints are a set of linear equations, the ideas presented here can also be applied to inequality constraints of the form $Ax \leq b$ and hence also to linear programming problems, for example.

6.10 PROBLEMS

P6–1.–Generate 11 data points by taking $t_i = (i - 1)/10$ and $b_i = \text{erf}(t_i)$, $i = 1, \ldots, 11$. Compute $\text{erf}(t)$ using the techniques from Problem P1–1.

(a) Fit the data in a least-squares sense with polynomials of degrees from 1 to 10. Compare the fitted polynomial with $\text{erf}(t)$ for a number of values of t between the data points, and see how the maximum error depends on n, the number of coefficients in the polynomial.

(b) Since $\text{erf}(t)$ is an odd function of t, that is, $\text{erf}(t) = -\text{erf}(-t)$, it is reasonable to fit the same data by a linear combination of odd powers of t,

$$\text{erf}(t) \approx c_1 t + c_2 t^3 + \cdots + c_n t^{2n-1}.$$

Again, see how the error between data points depends on n. Since t varies over $[0, 1]$ in this problem, it is not necessary to consider using other basis polynomials.

(c) Polynomials are not particularly good approximants for $\text{erf}(t)$ because they are unbounded for large t, whereas $\text{erf}(t)$ approaches 1 for large t. So, using the same data points, fit a model of the form

$$\text{erf}(t) \approx c_1 + e^{-t^2}(c_2 + c_3 z + c_4 z^2 + c_5 z^3)$$

where $z = 1/(1 + t)$. How does the error between data points compare with the polynomial models?

P6–2.–This problem uses least-squares techniques to approximate the census data from Problem P4–2.

(a) Fit the census data by polynomials of various degrees. Use the fits to predict the 1980 population. How is the predicted population affected by your choice of basis polynomials?

By your choice of tolerance for negligible singular values? By the precision of arithmetic if you have a choice?

(b) Try to fit the census data by a quadratic

$$b(t) \approx c_1 + c_2 t + c_3 t^2$$

using the normal equations approach. What is the condition number of the resulting matrix? What is the predicted 1980 population?

P6–3.–Consider the following data, obtained from a physical experiment at intervals of one second, with the first observation being taken at time $t = 1.0$:

$t : 1 - 9$	$t : 10 - 18$	$t : 19 - 25$
5.0291	7.5677	14.5701
6.5099	7.2920	17.0440
5.3666	10.0357	17.0398
4.1272	11.0708	15.9069
4.2948	13.4045	15.4850
6.1261	12.8415	15.5112
12.5140	11.9666	17.6572
10.0502	11.0765	
9.1614	11.7774	

We will try to fit the data by various models as a way of learning more about the data.

(a) Fit the data by a straight line using SQRLS. Scale the residuals as described in Section 2, and plot the residuals (either using graphics software or by hand). One of the data points has a much larger residual than the others. We suspect that it does not fit with the rest of the data, that is, it is an **outlier**.

(b) Discard the outlier, and fit the data again by a straight line. Again, scale the residuals and plot them. What pattern do you notice in the residual plot? Do the residuals appear random?

(c) To get rid of the trends in the residuals, fit the data with a new model

$$y(t) \approx x_1 \cdot 1 + x_2 \cdot t + x_3 \cdot \sin t.$$

Plot the scaled residuals. Do they appear random now?

P6–4.–An outlier can sometimes have a dramatic effect on the parameters in a model, and this is why it is important to check for such points, and determine if they are correct. To illustrate this point, we will consider the following artificial data set:

$$t_i = \begin{cases} i, & i = 1, 2, \ldots, 10, \\ 20, & i = 11, \end{cases} \qquad y_i = \begin{cases} 0, & i = 1, 2, \ldots, 10, \\ M, & i = 11. \end{cases}$$

We will experiment with various values of M. Fit the data to a straight line using SQRLS, for $M = 0, 5, 10, 15, 20$. How rapidly do the parameters change as a function of M? Using a scaled residual plot, is it possible to identify point 11 as a potential outlier?

P6–5.–Consider the following data set:

t	y	Error bound
0.00	20.00	20.00
0.25	51.58	24.13
0.50	68.73	26.50
0.75	75.46	27.13
1.00	74.36	26.00
1.25	67.09	23.13
1.50	54.73	18.50
1.75	37.98	12.13
2.00	17.28	4.00

Notice that the error bound varies with each data point. Fit the data by a quadratic polynomial using the weighting technique discussed in Section 1.

P6–6.–The following data represent mortality rates (death rates per hundred thousand) for individuals age 20–45 in turn of the century England (read down each column)

ages 20–26	ages 27–33	ages 34–40	ages 41–45
431	499	746	956
409	526	760	1014
429	563	778	1076
422	587	828	1134
530	595	846	1024
505	647	836	
459	669	916	

(a) Plot the data. Use SQRLS to fit a straight line and plot it along with the data. Do you think that the data are well represented by a straight line?

(b) The plot suggests that the data could be represented by different straight lines over the age intervals [20, 28], [28, 39], and [39, 45]. Use SQRLS again to fit the data to these three lines, and plot them on the same graph as in (a) Since we have not made any assumptions about the relationships between these lines you can determine the fits by treating the data on each subrange completely independently.

(c) The fit you get in (b) will not be continuous at 28 or at 39. One way to force continuity is to pick model functions that have this property. Since three straight lines need six coefficients, and continuity at 28 and 39 imposes two conditions we expect that we will need $6 - 2 = 4$ model functions. The four functions we suggest you use are plotted in Figure P6.6, and are labelled $l_i(x)$, $i = 1, \ldots, 4$. Each of these is defined and continuous on $20 \le x \le 45$, and so any linear combination is too. Using these model functions set up and solve the least

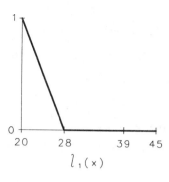

$l_1(x)$

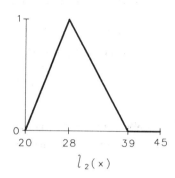

$l_2(x)$

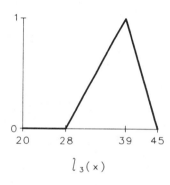

$l_3(x)$

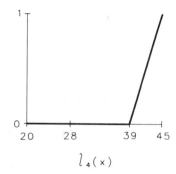

$l_4(x)$

Figure P6.6

square problem with SQRLS. Plot the fit on the same graph as in (a) and (b). Which of these three produces the best fit?

P6–7.–Consider the following data set

t	y
0.00	20.00
0.25	51.58
0.50	68.73
0.75	75.46
1.00	74.36
1.25	67.09
1.50	54.73
1.75	37.98
2.00	17.28

Suppose we wished to fit the data to the following model

$$y(t) \approx x_1 \cdot 1 + x_2 \sin t + x_3 \cos t + x_4 \sin 2t + x_5 \cos 2t.$$

In addition, suppose that the parameters must satisfy the following two linear constraints

$$x_1 + x_2 + x_3 + x_4 + x_5 = 1,$$
$$3x_1 + 2x_2 + x_3 = 4.$$

We will use the techniques of Section 9 to solve this problem.

(a) Determine a particular solution $\bar{x}$ to the constraints by solving the linear system

$$x_1 + \ x_2 = 1,$$
$$3x_1 + 2x_2 = 4,$$

using SGEFS, and setting $\bar{x} = (x_1, x_2, 0, 0, 0)^T$.

(b) Let

$$B = \begin{pmatrix} 1 & 1 & 1 & 1 & 1 \\ 3 & 2 & 1 & 0 & 0 \end{pmatrix}$$

be the coefficient matrix of the constraints. Verify that $B^T = QR$, where

$$Q = \begin{pmatrix} -0.4472141 & -0.6902676 & -0.3827515 & -0.2975130 & -0.2975130 \\ -0.4472141 & -0.3067856 & 0.1717229 & 0.5815458 & 0.5815458 \\ -0.4472141 & 0.0766956 & 0.8048115 & -0.2705498 & -0.2705498 \\ -0.4472141 & 0.4601789 & -0.2968903 & 0.4932585 & -0.5067396 \\ -0.4472141 & 0.4601789 & -0.2968903 & -0.5067396 & 0.4932585 \end{pmatrix}$$

$$R = \begin{pmatrix} -2.2360649 & -2.6832657 \\ 0 & -2.6076736 \\ 0 & 0 \\ 0 & 0 \\ 0 & 0 \end{pmatrix},$$

and where Q is an orthogonal matrix. (This factorization can be computed using the Linpack routines called by SQRLS.)

(c) The general solution of the constraints can be written in the form

$$x = \bar{x} + Q_{(2)}d$$

using the notation of Section 9. Using this formula, write the original least-squares problem *without any constraints* in terms of the new parameters d.

(d) Solve this least-squares problem for d using SQRLS, and then determine the parameters x that solve the original constrained least-squares problem.

P6-8.–Many cyclic phenomena in nature can be modelled by a sinusoid and its higher harmonics:

$$y(t) = \sum_{i=1}^{n} A_i \sin\left[\frac{2\pi}{(T/i)}(t + \phi_i)\right].$$

The numbers A_i, T, and ϕ_i are called the **amplitude, period,** and **phase** respectively of the i-th harmonic. Given data, fitting to this model in the least squares sense is a nonlinear problem which cannot be solved by the methods of this chapter, although techniques we will study in Chapter 9 can be applied. However, in many situations the period T is known or can be guessed, leaving only the amplitude and phase as unknowns. This still appears to be nonlinear because ϕ_i is inside the sine.

(a) By using a formula for the sine of a sum show that

$$A_i \sin \left[\frac{2\pi}{(T/i)}(t + \phi_i) \right] = a_i \cos \left(\frac{2\pi t}{T/i} \right) + b_i \sin \left(\frac{2\pi t}{T/i} \right),$$

with $a_i \equiv A_i \sin(\theta_i), b_i \equiv A_i \cos(\theta_i), \theta_i \equiv 2\pi \phi_i/(T/i)$. Thus if a_i and b_i can be found then

$$A_i = \sqrt{a_i^2 + b_i^2}, \qquad \phi_i = \frac{T/i}{2\pi} \tan^{-1}(a_i/b_i).$$

Since the a_i and b_i appear linearly, SQRLS can be used to find them.

(b) There is interest in studying the affect of the burning of fossil fuels on weather. One interesting set of measurements is the concentration of carbon dioxide in the atmosphere. The data file CO2.DAT on the software disk gives the CO_2 concentration (in parts per million) as measured at the Mauna Loa Observatory (Hawaii) each month from 1958 to 1974. Plot the data. You should observe that there is a cyclical component of about twelve months ($T = 12$). There is also an upward trend to the data which cannot be captured with only sines and cosines. Why is this trend reasonable? By examining records of total world fossil fuel production from the mid 1860s, it has been determined that a suitable model for the data is

$$y(t) = B + d\exp(\alpha t) + \sum_{i=1}^{n} A_i \sin \left[\frac{2\pi}{(12/i)}(t + \phi_i) \right],$$

where t is in months,(i.e., $t = 1$ for the first point, $t = 2$ for the second, etc.) $\alpha \approx 0.0037/\text{month}$, and the other parameters are unknown. Experiment by first fixing n and then use the technique of (a) and SQRLS to find the other parameters. Plot the fit and the residuals as you vary $n \leq 4$. Are the residuals "random" or do you see some regularity?

6.11 PROLOGUE: SQRLS

```
        SUBROUTINE SQRLS (A,LDA,M,N,TOL,KR,B,X,RSD,JPVT,QRAUX,WORK,
     *                    ITASK,IND)
C***BEGIN PROLOGUE   SQRLS
C***DATE WRITTEN    870911    (YYMMDD)
C***REVISION DATE   871016    (YYMMDD)
C***CATEGORY NO.   D9
C***KEYWORDS   LEAST SQUARES,OVERDETERMINED,LINEAR EQUATIONS
```

```
C***AUTHOR  STEPHEN NASH (GEORGE MASON UNIVERSITY)
C***PURPOSE
C***PURPOSE  SQRLS solves an overdetermined, underdetermined or singular
C            system of linear equations in least square sense.  The
C            solution is obtained using a QR factorization of the
C            M  by  N  coefficient matrix  A.
C***DESCRIPTION
C    From the book "Numerical Methods and Software"
C       by  D. Kahaner, C. Moler, S. Nash
C          Prentice Hall 1988
C
C    SQRLS IS USED TO SOLVE IN A LEAST SQUARES SENSE
C    OVERDETERMINED, UNDERDETERMINED AND SINGULAR LINEAR SYSTEMS .
C    THE SYSTEM IS  A*X  APPROXIMATES  B  WHERE  A  IS  M  BY  N.
C    B  IS A GIVEN  M-VECTOR, AND  X  IS THE  N-VECTOR TO BE COMPUTED.
C    A SOLUTION  X  IS FOUND WHICH MINIMIMZES THE SUM OF SQUARES (2-NORM)
C    OF THE RESIDUAL,  A*X - B .
C    THE NUMERICAL RANK OF A IS DETERMINED USING THE TOLERANCE TOL.
C
C    SQRLS USES THE LINPACK SUBROUTINE SQRDC TO COMPUTE THE QR
C    FACTORIZATION, WITH COLUMN PIVOTING, OF AN  M  BY  N  MATRIX  A .
C    FOR MORE INFORMATION, SEE CHAPTER 9 OF THE REFERENCE BELOW.
C
C
C       ON ENTRY
C
C          A      REAL (LDA,N) .
C                 THE MATRIX WHOSE DECOMPOSITION IS TO BE COMPUTED.
C                 IN A LEAST SQUARES DATA FITTING PROBLEM, A(I,J) IS THE
C                 VALUE OF THE J-TH BASIS (MODEL) FUNCTION AT THE I-TH
C                 DATA POINT.
C
C          LDA    INTEGER.
C                 THE LEADING DIMENSION OF A .
C
C          M      INTEGER.
C                 THE NUMBER OF ROWS OF A .
C
C          N      INTEGER.
C                 THE NUMBER OF COLUMNS OF  A .
C
C          TOL    REAL.
C                 A RELATIVE TOLERANCE USED TO DETERMINE THE NUMERICAL
C                 RANK.  THE PROBLEM SHOULD BE SCALED SO THAT ALL THE
C                 ELEMENTS OF  A   HAVE ROUGHLY THE SAME ABSOLUTE ACCURACY
C                 EPS.  THEN A REASONABLE VALUE FOR  TOL  IS ROUGHLY  EPS
C                 DIVIDED BY THE MAGNITUDE OF THE LARGEST ELEMENT.
C
```

```
C        JPVT  INTEGER(N)
C        QRAUX REAL(N)
C        WORK  REAL(N)
C                    THREE AUXILIARY ARRAYS USED TO FACTOR THE MATRIX A.
C                    (NOT REQUIRED IF ITASK .GT. 1)
C
C        B     REAL(M)
C              THE RIGHT HAND SIDE OF THE LINEAR SYSTEM.
C              IN A LEAST SQUARES DATA FITTING PROBLEM B(I) CONTAINS THE
C              VALUE OF I-TH OBSERVATION.
C
C        ITASK INTEGER.
C              IF ITASK=1, THEN SQRLS FACTORS THE MATRIX A AND
C                          SOLVES THE LEAST SQUARES PROBLEM.
C              IF ITASK=2, THEN SQRLS ASSUMES THAT THE MATRIX A
C                          WAS FACTORED WITH AN EARLIER CALL TO
C                          SQRLS, AND ONLY SOLVES THE LEAST SQUARES
C                          PROBLEM.
C
C    ON RETURN
C
C        X     REAL(N) .
C              A LEAST SQUARES SOLUTION TO THE LINEAR SYSTEM.
C
C        RSD   REAL(M) .
C              THE RESIDUAL, B - A*X .  RSD MAY OVERWRITE  B .
C
C        IND   INTEGER
C              ERROR CODE:  IND = 0:  NO ERROR
C                           IND = -1: N .GT. LDA   (FATAL ERROR)
C                           IND = -2: N .LT. 1     (FATAL ERROR)
C                           IND = -3: ITASK .LT. 1 (FATAL ERROR)
C
C        A     CONTAINS THE OUTPUT FROM SQRDC.
C              THE TRIANGULAR MATRIX  R  OF THE QR FACTORIZATION IS
C              CONTAINED IN THE UPPER TRIANGLE AND INFORMATION NEEDED
C              TO RECOVER THE ORTHOGONAL MATRIX  Q  IS STORED BELOW
C              THE DIAGONAL IN  A  AND IN THE VECTOR QRAUX .
C
C        KR    INTEGER.
C              THE NUMERICAL RANK.
C
C        JPVT  THE PIVOT INFORMATION FROM SQRDC.
C
C    COLUMNS JPVT(1),...,JPVT(KR) OF THE ORIGINAL MATRIX ARE LINEARLY
C    INDEPENDENT TO WITHIN THE TOLERANCE TOL AND THE REMAINING COLUMNS
C    ARE LINEARLY DEPENDENT.  ABS(A(1,1))/ABS(A(KR,KR))  IS AN ESTIMATE
C    OF THE CONDITION NUMBER OF THE MATRIX OF INDEPENDENT COLUMNS,
```

```
C      AND OF  R .  THIS ESTIMATE WILL BE .LE. 1/TOL .
C
C      USAGE....
C        SQRLS CAN BE EFFICIENTLY USED TO SOLVE SEVERAL LEAST SQUARES
C      PROBLEMS WITH THE SAME MATRIX A.  THE FIRST SYSTEM IS SOLVED
C      WITH ITASK = 1.  THE SUBSEQUENT SYSTEMS ARE SOLVED WITH
C      ITASK = 2, TO AVOID THE RECOMPUTATION OF THE MATRIX FACTORS.
C      THE PARAMETERS  KR, JPVT, AND QRAUX MUST NOT BE MODIFIED
C      BETWEEN CALLS TO SQRLS.
C
C***REFERENCES  DONGARRA, ET AL, LINPACK USERS GUIDE, SIAM, 1979
C***ROUTINES CALLED  SQRANK,SQRLSS,XERROR
C***END PROLOGUE  SQRLS
```

7

Solution of Nonlinear Equations

7.1 INTRODUCTION

In Chapter 3 we learned how to solve linear equations. In many applications, linear equations are effective at approximating the behavior of a "real-world" problem. But there are limits to what linear models can do. In biological models, for example, animal populations often satisfy equations of the form

$$\text{population}(t) = 2e^{.1t},$$

where t represents time and the numbers 2 and .1 are constants that indicate the initial size of the population and its rate of growth, respectively. In this equation, the relationship between time and population is nonlinear. If we wished to know when a population of moose, say, became equal to 1000, we would have to solve the nonlinear equation

$$1000 - 2e^{.1t} = 0$$

for t. For this simple example the solution could be written explicitly in terms of logarithms; but if the model became more complex

$$\text{population}(t) = 2e^{.1t} - .05t^2 - .001t^{1.3}$$

this would not be possible.

In general we will be interested in solving

$$f(x) = 0.$$

For the above example, $f(x) = 1000 - 2e^{.1x}$. Any nonlinear equation can be put in this form by a simple rearrangement of terms. Much of our interest will be in equations of one variable, but we will also consider systems of n equations in n unknowns

$$f_1(x_1, \ldots, x_n) = 0,$$
$$f_2(x_1, \ldots, x_n) = 0,$$
$$\vdots$$
$$f_n(x_1, \ldots, x_n) = 0.$$

If we let $x = (x_1, \ldots, x_n)^T$ and $f(x) = (f_1(x), \ldots, f_n(x))^T$, then we can represent this system of equations in the notation $f(x) = 0$ as well.

Systems of nonlinear equations arose in the derivation of Gaussian quadrature rules (see Section 2.1 of Chapter 5). They would also arise, for example, in more complex population models involving more than one animal. If moose and elk in a region were competing for the same food supply then we might have

$$\text{moose}(t) = 2e^{.1t} - .4\text{elk}(t),$$
$$\text{elk}(t) = 5e^{.2t} - .6\text{moose}(t).$$

If we then asked when the population of moose would become equal to 1000, and what the corresponding elk population would be, we would solve

$$1000 - 2e^{.1t} + .4\text{elk}(t) = 0,$$
$$\text{elk}(t) - 5e^{.2t} + .6(1000) = 0.$$

This is a system of two nonlinear equations in two unknowns.

There are some fundamental differences between linear and nonlinear equations. First of all, any nonsingular system of linear equations has a unique solution. This is not true for nonlinear equations. There may be no real solutions as in

$$f(x) = \sin x + 2 = 0,$$

or multiple solutions as in

$$f(x) = \sin x = 0.$$

It is even possible to derive more contrived examples where every point in an interval is a solution.

Second, a system of linear equations can be solved using a fixed, finite number of arithmetic operations, about $n^3/3$ for Gaussian elimination applied to a system of n equations. Only certain special nonlinear equations can be solved. If $f(x)$ is a polynomial in one variable of degree at most four, then there are formulas for the zeroes (some of which may be complex numbers). The quadratic formula is well known, and Cardano's formula for cubic equations is discussed in the exercises at the end of this chapter. However, as Galois showed around 1830, if the ordinary operations of arithmetic and root extraction are used, there are no formulas for the zeroes of general polynomials of degree five or higher. Since any finite sequence of computer calculations corresponds to a formula of some sort, it will not be possible to solve general nonlinear equations, even one-variable equations. Quite a dilemma!

We get around this difficulty by broadening our definition of "solve." For simplicity, consider a single nonlinear equation in one unknown. Let x^* satisfy $f(x^*) = 0$. We will say that $\bar{x}$ "solves" $f(x) = 0$ if either

$$|f(\bar{x})| \approx 0 \quad \text{or} \quad |\bar{x} - x^*| \approx 0.$$

Thus either $\bar{x}$ nearly satisfies the nonlinear equation or $\bar{x}$ is approximately equal to a solution of the equation. These are not always equivalent definitions, and sometimes one will be preferable to the other. The meaning of $\approx$ has been left vague; it will usually depend on the accuracy of the computer arithmetic. Since computer arithmetic is subject to rounding errors, these definitions should not appear unnatural. In rounded arithmetic there may be no floating-point number x such that $f(x) = 0$, and hence there might be no solution to the equation if we insisted on the strict definition of "solve."

The second definition here may seem strange since it depends on the knowledge of a point x^* satisfying $f(x^*) = 0$. If we know of such a point, why are we trying to solve the equation? And if we don't know such a point, how can we test this condition? Well, if the function $f(x)$ is continuous we may not need to know x^*. Suppose that we have two points $a < b$ such that $f(a) \cdot f(b) < 0$. Then since $f(x)$ is continuous there must be a point x^* in the interval $[a, b]$ satisfying $f(x^*) = 0$. Thus any point x in this interval satisfies $|x - x^*| \leq |b - a|$. If a is close to b then the above condition will be satisfied.

In many important applications the two definitions of "solve" are equivalent. Suppose for example that $f'(x)$ is continuous and bounded, i.e. $|f'(x)| \leq L$. If we expand in a Taylor series

$$|f(\bar{x})| = |f(x^*) + f'(\eta)(\bar{x} - x^*)| = |f'(\eta)| \cdot |\bar{x} - x^*| \leq L|\bar{x} - x^*|,$$

where η is some point between $\bar{x}$ and x^*. Thus $|f(\bar{x})| \approx 0$ if $|\bar{x} - x^*| \approx 0$. If we also assume that $|f'(x)| \geq c > 0$ then we can conclude in addition that

$$|\bar{x} - x^*| \leq \frac{1}{c}|f(\bar{x})|,$$

so that if $|f(\bar{x})| \approx 0$ then $|\bar{x} - x^*| \approx 0$ too.

Our methods for solving nonlinear equations will be iterative, that is, the methods will construct a sequence of approximate solutions $x_1, x_2, \ldots, x_n, x_{n+1}, \ldots$ to the equation $f(x) = 0$. It is not usually possible to guarantee in advance that the method will converge to an approximate solution of the nonlinear equation. However, if we know points a and b satisfying $f(a) \cdot f(b) < 0$ then not only will it be possible to guarantee convergence, but it will also be possible to predict the maximum number of iterations required to estimate a solution to any desired accuracy. This is not an unrealistic assumption. Frequently information from the real-world application will provide the desired upper bounds (in the case of the moose population example above, we might know the initial population of the herd). If not, a few trial evaluations of the function might determine the bounds.

For systems of equations the situation is more difficult. As before, there may be no solutions or multiple solutions. There is in general no algorithm to find a solution in finite time, and we must accept approximate solutions as a substitute. The methods for solving the problem will be iterative. However, it is much harder to guarantee convergence to a solution of a system of equations than it is for a single equation, or to guarantee the existence of a solution.

In judging the efficiency of algorithms for solving nonlinear equations, it is usually assumed that the function $f(x)$ is expensive to evaluate, so expensive that the cost of evaluating $f(x)$ dominates all other arithmetic costs of the algorithm. For example, if evaluating $f(x)$ involves solving a differential equation or integrating a function then this will certainly be true for any of the algorithms discussed in this chapter. Thus, although every effort is made to make the algorithms as efficient as possible, extra calculations will sometimes be performed if they offer the hope of eliminating an extra function evaluation. For this reason, if the nonlinear function is easy to evaluate, the algorithms of this chapter may run more slowly than less sophisticated methods. However, if the nonlinear function is expensive to evaluate, or if the nonlinear equation is difficult to solve (for example, if it is badly conditioned; see Chapter 3), then the algorithms here are likely to out-perform their simpler relatives.

One of the most famous nonlinear equation problems is the finding of the zeroes of a polynomial. This problem is famous, but its practical significance does not measure up to its grand reputation. There are few situations where the zeroes of a polynomial are required. The techniques of this chapter can be applied to the special case of polynomials. If more efficient algorithms are required they can be found in the ACM collection of algorithms (see the Bibliography). The methods are designed to converge rapidly, and find all zeroes (both real and complex) of polynomials with either real or complex coefficients. For more information see the paper by Jenkins and Traub (1970).

The eigenvalues of a matrix A can be defined as the solutions to the polynomial equation $\det(A - xI) = 0$. This polynomial should not be formed explicitly, since its roots can be sensitive to small errors in the coefficients (see Section 4 of Chapter 3). More appropriate methods can be found in Eispack (see the manual by Smith et al. (1976)). On the other hand, it is possible to compute the roots of a polynomial by finding the eigenvalues of an appropriate matrix. This is not efficient in terms of arithmetic or storage costs, but it works.

One common way that systems of nonlinear equations arise is in function minimization, but this approach is not recommended. If we are trying to solve

$$\min F(x_1, \ldots, x_n)$$

and $F(x)$ is continuously differentiable, then at the solution x^* the vector of first derivatives will be zero:

$$\left.\frac{\partial F(x)}{\partial x_1}\right|_{x=x^*} = 0,$$

$$\left.\frac{\partial F(x)}{\partial x_2}\right|_{x=x^*} = 0,$$

$$\vdots$$

$$\left.\frac{\partial F(x)}{\partial x_n}\right|_{x=x^*} = 0.$$

So if we can find an x^* that solves this system of nonlinear equations we will have found a possible solution to the minimization problem. It might also be a maximum or a point of inflection. However, it is not usually a good idea to solve the minimization problem using the techniques in this chapter. The related techniques discussed in the optimization chapter (Chapter 9) are more appropriate. Using optimization methods, convergence to an approximate solution can be guaranteed under mild assumptions. Also, the storage and arithmetic costs of the optimization algorithm will be lower.

Nonlinear equations also arise in data-fitting problems with nonlinear models (see Chapter 6). There we have

$$f_1(x_1, \ldots, x_n) = 0,$$
$$f_2(x_1, \ldots, x_n) = 0,$$

$$\vdots$$

$$f_m(x_1, \ldots, x_n) = 0,$$

where $m > n$. Since there are more equations than unknowns, there is usually no solution to the system. Just as in the case of linear least squares, a compromise is necessary, and an approximate solution is determined by solving

$$\min \sum_{i=1}^{m} [f_i(x_1, \ldots, x_n)]^2.$$

Because we are minimizing the sum of squares of a set of nonlinear functions, this is called a **nonlinear least-squares problem**. In this case also, the techniques discussed in Chapter 9 are more appropriate for computing the solution.

In this chapter we will usually only be interested in finding a single real solution of one or more nonlinear equations. Determining all solutions of a nonlinear equation is still a subject of research, unless it is a polynomial equation.

7.2 METHODS FOR COMPUTING REAL ROOTS

There are many algorithms for solving nonlinear equations, but almost all of them follow the same principle. Instead of solving the nonlinear equation directly, we solve a sequence of simpler problems, and hope that the sequence of simpler solutions tends toward the solution of the original problem. For us, the simpler problems will usually be linear equations, and they correspond to approximating the nonlinear function by a straight line. The methods will differ in the quality of the linear approximation, the rate of convergence to the solution of the original problem, the reliability of the approach, and the assumptions made about the nonlinear function. For one-variable problems it would also be possible to approximate the nonlinear equation by a sequence of quadratic equations, but this approach will not be examined in detail here.

7.2.1 The Bisection Method

One of the simplest and most reliable methods for solving $f(x) = 0$ is the **bisection** method. To use bisection, two points a and b must be given such that $a < b$ and $f(a) \cdot f(b) < 0$; thus $f(x)$ changes sign on the interval $[a, b]$. It is usual but not essential to assume that $f(x)$ is continuous on the interval $[a, b]$ so that $f(x^*) = 0$ for some $x^* \in [a, b]$. The bisection method systematically reduces the interval $[a, b]$, splitting it in half at every iteration, until it is smaller than some tolerance (tol$_1$), and hence the value of the solution x^* is determined to within that tolerance. If at some step $|f(x)|$ is small (less than tol$_2$), the iteration is terminated. The algorithm is:

1. If $b - a \leq$ tol$_1$ then stop.
2. Set $m = \frac{1}{2}(a + b)$, the midpoint of the interval $[a, b]$. Compute $f(m)$. If $|f(m)| \leq$ tol$_2$ then stop.
3. If $f(a) \cdot f(m) < 0$ then set $b \leftarrow m$, else set $a \leftarrow m$. Go to step 1.

Notice how both interpretations of "solve" are incorporated into the algorithm. If we accidentally find a zero, then the algorithm will terminate in step 2. Otherwise we will continue until the interval of uncertainty $[a, b]$ is as small as desired.

Example 7.1   The Bisection Method.

Here are a few iterations of the bisection algorithm for the function $f(x) = x^3$ on the interval $[a, b] = [-1, 2]$:

$$\begin{aligned}
\text{Set up:} \quad & a = -1, \ b = 2, \ f(a) = -1, \ f(b) = 8. \\
\text{Iteration 1:} \quad & m = (-1 + 2)/2 = 1/2, \ f(m) = 1/8, \\
& b \leftarrow 1/2, \quad [a, b] = [-1, 1/2]. \\
\text{Iteration 2:} \quad & m = (-1 + 1/2)/2 = -1/4, \ f(m) = -1/64, \\
& a \leftarrow -1/4, \quad [a, b] = [-1/4, 1/2]. \\
\text{Iteration 3:} \quad & m = (-1/4 + 1/2)/2 = 1/8, \ f(m) = 1/512, \\
& b \leftarrow 1/8, \quad [a, b] = [-1/4, 1/8]. \\
\text{etc.} \quad & \blacksquare
\end{aligned}$$

At each iteration the interval is divided in half. If we take the midpoint m to be the estimate of the solution x^*, then the error $|m - x^*|$ is also reduced by up to half at each iteration. In other words, one binary digit of the solution is obtained at each iteration. For a 32-bit floating-point number with 24-bit mantissa, if the initial interval were [1, 2], full accuracy would be obtained in 24 iterations. At each iteration the function $f(x)$ is evaluated once and so only 24 function evaluations would be required. (If the function were evaluated at every floating-point number in the interval then $2^{24} = 16777216$ evaluations would be required, about $700,000$ times as much work as for bisection.)

If we let e_i denote the error at iteration i, $e_i = m - x^*$, then for the bisection method

$$\frac{|e_{i+1}|}{|e_i|} \approx \tfrac{1}{2}.$$

In general a method is said to converge with rate r if

$$\lim_{i \to \infty} \frac{|e_{i+1}|}{|e_i|^r} = C$$

where C is some finite non-zero constant. Using the notation of Section 5 of Chapter 2, we write $|e_{i+1}| = O(|e_i|^r)$. For the bisection method $r = 1$ and $C = \tfrac{1}{2}$. If $r = 1$ the rate of convergence is called **linear**; if $r > 1$ the rate is **superlinear**; if $r = 2$ the rate is **quadratic**.

Often the rate of convergence r matters more than the constant C, with larger values of r being more desirable. However the constant may also be important. If the rate of convergence were linear with $C \geq 1$ then there would be no guarantee that $e_i \to 0$, i.e., the method might not converge. (Why?) Also a linearly-convergent method with $C \approx 0$ might initially converge faster than a quadratically-convergent method with a large constant. The following table illustrates these ideas in the ideal situation where $e_{i+1} = Ce_i^r$ for various values of C and r. In all cases $e_0 = 1$. The zero entries in the table represent underflow.

Notice that although larger values of r ultimately give rapid convergence, linear rates of convergence are respectable if the constant C is small. However if the constant C is near one then linear convergence will be unacceptably slow.

The above description of the bisection method is traditional. Bisection can also be interpreted as approximating $f(x)$ by a particular linear function at each iteration and finding a zero of the linear approximation. The approximation is the line passing through the two points $(a, \text{sign } f(a))$ and $(b, \text{sign } f(b))$. If $f(a) < 0$ and $f(b) > 0$ then this line has equation

$$y = -1 + 2(x - a)/(b - a)$$

and $y = 0$ if $x = m = \tfrac{1}{2}(a + b)$. This is a crude approximation to $f(x)$. It completely ignores the values of the function $f(x)$ and only looks at its sign. In Example 7.1 at iteration 1, $a = -1$ and $b = 2$, so the linear approximation is $y = -1 + 2(x + 1)/3$. Setting $y = 0$ and solving gives $x = \tfrac{1}{2}$ as before. See Figure 7.1.

i	$r = 1$ $C = .1$	$r = 1$ $C = .5$	$r = 1$ $C = .95$	$r = 1.3$ $C = .95$	$r = 1.6$ $C = .95$	$r = 2$ $C = .95$
1	1.0(−01)	5.0(−01)	9.5(−01)	9.5(−01)	9.5(−01)	9.5(−01)
2	1.0(−02)	2.5(−01)	9.0(−01)	8.9(−01)	8.8(−01)	8.6(−01)
3	1.0(−03)	1.2(−01)	8.6(−01)	8.1(−01)	7.7(−01)	7.0(−01)
4	1.0(−04)	6.2(−02)	8.1(−01)	7.3(−01)	6.2(−01)	4.6(−01)
5	1.0(−05)	3.1(−02)	7.7(−01)	6.3(−01)	4.4(−01)	2.0(−01)
6	1.0(−06)	1.5(−02)	7.4(−01)	5.2(−01)	2.6(−01)	3.9(−02)
7	1.0(−07)	7.8(−03)	7.0(−01)	4.1(−01)	1.1(−01)	1.5(−03)
8	1.0(−08)	3.9(−03)	6.6(−01)	2.9(−01)	2.8(−02)	2.1(−06)
9	1.0(−09)	2.0(−03)	6.3(−01)	1.9(−01)	3.1(−03)	4.1(−12)
10	1.0(−10)	9.8(−04)	6.0(−01)	1.1(−01)	9.0(−05)	1.6(−23)
11	1.0(−11)	4.9(−04)	5.7(−01)	5.5(−01)	3.2(−07)	2.5(−46)
12	1.0(−12)	2.4(−04)	5.4(−01)	2.2(−01)	3.9(−11)	6.0(−92)
13	1.0(−13)	1.2(−04)	5.1(−01)	6.7(−01)	2.1(−17)	0.0(−00)
14	1.0(−14)	6.1(−05)	4.9(−01)	1.4(−01)	1.9(−27)	0.0(−00)
15	1.0(−15)	3.0(−05)	4.6(−01)	1.9(−04)	1.7(−43)	0.0(−00)
16	1.0(−16)	1.5(−05)	4.4(−01)	1.4(−05)	3.5(−69)	0.0(−00)
17	1.0(−17)	7.6(−06)	4.2(−01)	4.5(−07)	0.0(−00)	0.0(−00)
18	1.0(−18)	3.8(−06)	4.0(−01)	5.3(−09)	0.0(−00)	0.0(−00)
19	1.0(−19)	1.9(−06)	3.8(−01)	1.7(−11)	0.0(−00)	0.0(−00)
20	1.0(−20)	9.5(−07)	3.6(−01)	9.2(−15)	0.0(−00)	0.0(−00)

It is possible to construct better approximations by using the function values. The methods we will discuss now have higher rates of convergence (larger values of r above) but may converge slowly if a good initial guess of the solution is not available, and may not converge at all unless safeguards are incorporated into the algorithm. However, in many cases they will perform better than bisection, converging rapidly to the solution from a wide range of starting guesses. These methods generally assume that $f(x)$ has one or more continuous derivatives, even in the cases where the derivatives do not enter into the calculations explicitly.

7.2.2 Newton's Method

One of the best general methods for solving $f(x) = 0$ is Newton's method. Given x_i, some estimate of the solution x^*, Newton's method approximates $f(x)$ by its tangent line at the point x_i. The zero of the tangent line is taken as the new estimate of x^*. This is illustrated in Figure 7.2. Newton's method often works as in the picture, with the estimates converging rapidly to the solution.

To derive the formulas for Newton's method, we expand $f(x)$ in a Taylor series about the point x_i:

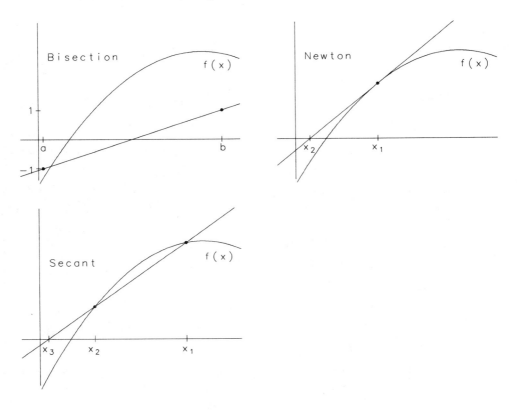

Figure 7.1   Linear Approximations to a Nonlinear Function

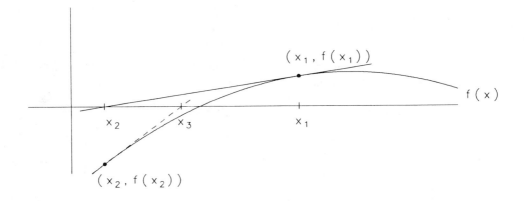

Figure 7.2   Newton's Method

$$f(x) = f(x_i) + f'(x_i)(x - x_i) + \cdots.$$

The tangent line is given by the first two terms in the series

$$y = f(x_i) + f'(x_i)(x - x_i)$$

and setting $y = 0$ gives

$$x_{i+1} = x_i - \frac{f(x_i)}{f'(x_i)}.$$

The iteration formula was given in substantially this form by Raphson in 1690, but the method is commonly called Newton's because he had suggested a similar process a few years earlier.

Example 7.2   Newton's Method.

If we apply Newton's method to the problem $f(x) = x^2 - 2 = 0$ with $x_0 = 1$ then (since $f'(x) = 2x$)

Iteration 0: $x_1 = x_0 - (x_0^2 - 2)/(2x_0) = 1 - (1 - 2)/2 = 3/2$

Iteration 1: $x_2 = 3/2 - (9/4 - 2)/3 = 17/12$

Iteration 2: $x_3 = 17/12 - (289/144 - 2)/(17/6) = 577/408 \approx 1.414216$

The solution of the nonlinear equation is $x^* = \sqrt{2} \approx 1.414214$. If the Newton iteration formula for $f(x) = x^2 - 2$ is rearranged slightly it becomes

$$x_{i+1} = \frac{1}{2}\left(x_i + \frac{2}{x_i}\right)$$

which is the "pencil and paper" formula for computing square roots that you may have been taught long ago. In rounded arithmetic it is less satisfactory than the iteration formula derived above (see Section 3). ■

Newton's method is a desirable method because it converges quickly, with a quadratic rate of convergence. This is not difficult to show. We expand $f(x)$ in a Taylor series, again about the point x_i

$$f(x) = f(x_i) + f'(x_i)(x - x_i) + \tfrac{1}{2}f''(\eta)(x - x_i)^2,$$

where η is some unknown point between x and x_i. If we let $x = x^*$ the solution, then $f(x) = f(x^*) = 0$:

$$0 = f(x^*) = f(x_i) + f'(x_i)(x^* - x_i) + \tfrac{1}{2}f''(\eta)(x^* - x_i)^2.$$

We now divide this equation by $f'(x_i)$ and rearrange

$$x^* - \left(x_i - \frac{f(x_i)}{f'(x_i)}\right) = \frac{f''(\eta)}{2f'(x_i)}(x^* - x_i)^2.$$

By the formulas for the Newton iteration, the left-hand side is just $x^* - x_{i+1}$. If we define the errors to be $e_i = x^* - x_i$ and $e_{i+1} = x^* - x_{i+1}$, then

$$e_{i+1} = C_i e_i^2,$$

where $C_i = f''(\eta)/2f'(x_i)$. If the iteration converges then

$$\lim_{i \to \infty} \frac{|e_{i+1}|}{|e_i|^2} = C$$

with $C = |f''(x^*)/2f'(x^*)|$. This is just the definition of quadratic convergence.

If we apply Newton's method to $f(x) = x^2 - 2 = 0$ then we obtain the following results

x_i	e_i	e_{i+1}/e_i^2	C
1.500000000000	8.6(−02)	0.5000	0.3536
1.416666666667	2.5(−03)	0.3333	0.3536
1.414215682745	2.1(−06)	0.3529	0.3536
1.414213452374	1.6(−12)	0.3536	0.3536

Since the nonlinear function is so simple, we obtain complete agreement with the theoretical result.

Newton's method does not always work this well. It may not converge. For example if $f'(x_i) = 0$ the method is not defined. If $f'(x_i) \approx 0$ then there can be difficulties, since the new estimate x_{i+1} may be a much worse approximation to the solution than x_i is. This is illustrated in Figure 7.3. However, it is possible to slightly modify Newton's method so that failures of this type cannot occur. This is discussed in more detail in Chapter 9.

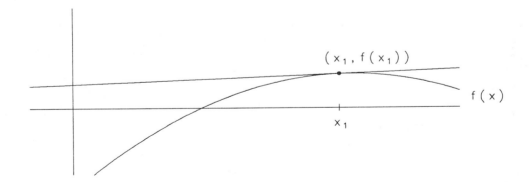

Figure 7.3   Failure of Newton's Method

Another disadvantage of Newton's method is that it requires the computation of $f'(x)$. This may be time-consuming, difficult, or impossible, especially if evaluating $f(x)$ requires (say) the evaluation of an integral, the solution of a differential equation, or if the value of the function is based on the performance of a piece of physical equipment. One way around this is to approximate $f'(x)$ using finite differences as described in Chapter 2. The resulting method will behave almost exactly like Newton's method with exact derivative values, but has the disadvantage of requiring extra function evaluations. Another idea is to use a different method that does not require derivative values. One such method is discussed below.

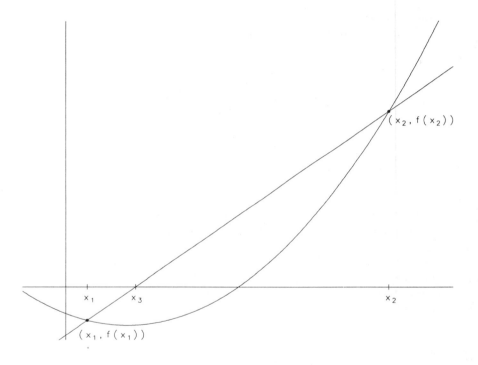

Figure 7.4   Secant Method

7.2.3 The Secant Method

The **secant method** constructs a linear approximation to $f(x)$ using *two* previous iterates x_i and x_{i-1}. The secant line passing through the points $(x_i, f(x_i))$ and $(x_{i-1}, f(x_{i-1}))$ is determined, and the zero of this line is taken to be the new estimate of the solution. This is illustrated in Figure 7.4.

To derive the formula for the secant method, consider the equation of the line through $(x_i, f(x_i))$ and $(x_{i-1}, f(x_{i-1}))$:

$$y = f(x_i) + (x - x_i)\frac{f(x_i) - f(x_{i-1})}{x_i - x_{i-1}}.$$

Setting $y = 0$, letting $x = x_{i+1}$, and rearranging we obtain

$$x_{i+1} = x_i - f(x_i)\frac{x_i - x_{i-1}}{f(x_i) - f(x_{i-1})}.$$

This is the formula for the secant method. Notice that it corresponds to the formula for Newton's method, but where we have used the approximation

$$f'(x_i) \approx \frac{f(x_i) - f(x_{i-1})}{x_i - x_{i-1}}.$$

By using this approximation we remove the requirement for derivative calculations, but we also slow the convergence of the algorithm.

If we apply the secant method to the problem $f(x) = x^2 - 2 = 0$ with $x_0 = 1$ and $x_1 = 2$ then

Iteration 1: $x_2 = x_1 - f(x_1)(x_1 - x_0)/(f(x_1) - f(x_0)) = 2 - (2)(2 - 1)/(x - (-1))$
               $= 4/3$

Iteration 2: $x_3 = 4/3 - (-2/9)(4/3 - 2)/(-2/9 - 2) = 7/5$

Iteration 3: $x_4 = 7/5 - (-1/25)(7/5 - 4/3)/((-1/25) - (-2/9)) = 58/41 \approx 1.414634$

As before, the solution is $x^* = \sqrt{2} \approx 1.414214$.

The secant method converges superlinearly with rate $r = \frac{1}{2}(1 + \sqrt{2}) \approx 1.618$ and constant $C = |f''(x^*)/2f'(x^*)|^{1/r}$. This is somewhat difficult to prove and so we will not derive the result here. The proof can be found in the book by Ortega and Rheinbolt (1971). However, we can demonstrate the convergence for the example $f(x) = x^2 - 2 = 0$. We need two starting guesses for the secant method; we will use $x_0 = 1$ and $x_1 = 2$. These two points bracket the solution but this is not necessary.

x_i	e_i	e_{i+1}/e_i^r	C
1.333333333333333	8.1(−02)	0.1921	0.5259
1.400000000000000	1.4(−02)	0.8314	0.5259
1.414634146341463	4.2(−04)	0.4100	0.5259
1.414211438474870	2.1(−06)	0.6163	0.5259
1.414213562057320	3.1(−10)	0.4767	0.5259
1.414213562373095	4.4(−16)	1.0466	0.5259

The correspondence between theory and practice is not as exact here, but the essential convergence behavior is as predicted. The most important thing to notice is the column labelled e_i. Notice that the error is going to zero rapidly, at a rate that is much faster than linear, but not quite quadratic. This behavior is typical for the secant method. The numbers in the third column are not constant, but they are all of the same general magnitude, which is all that can be expected for most problems.

The secant method may not converge for much the same reasons that Newton's method may not converge. For example, the method is not defined if $f(x_i) = f(x_{i-1})$. However, as with Newton's method, safeguards can be incorporated in the algorithm to circumvent many of these problems. Generalizations of the secant method are among the most widely used methods for solving systems of nonlinear equations.

For both Newton's method and the secant method we have been implicitly assuming that x^* is a simple zero of $f(x)$, that is, that $f'(x^*) \neq 0$. If x^* is a multiple zero ($f'(x^*) = 0$) then the nonlinear equation is generally more difficult to solve. We saw in Chapter 2 that finding a zero of $f(x) = (x - a)^n$ is an ill-conditioned problem; this is also true for multiple zeroes of general nonlinear functions. Newton's method and the secant method can still be used, just as before, but the solution will in general be less accurately determined. Also, they will converge more slowly, at a linear rate. In the rare cases where the multiplicity of the root is known, the rapid convergence rates can be restored. These comments are not to imply that there is any deficiency in these methods; the difficulty is in the problem.

Further methods can be derived by considering quadratic approximations to the function $f(x)$. When $f(x)$ has been evaluated at more than two points, **inverse quadratic interpolation** can be used to improve later estimates of the zero. Let $g(y)$ be the quadratic in the variable y for which $x_j = g(f_j)$, $j = i - 2, i - 1, i$. As in Section 3 of Chapter 4,

$$g(y) = x_{i-2} \frac{(y - f_{i-1})(y - f_i)}{(f_{i-2} - f_{i-1})(f_{i-2} - f_i)} + x_{i-1} \frac{(y - f_{i-2})(y - f_i)}{(f_{i-1} - f_{i-2})(f_{i-1} - f_i)}$$
$$+ x_i \frac{(y - f_{i-2})(y - f_{i-1})}{(f_i - f_{i-2})(f_i - f_{i-1})}.$$

The next approximate zero is taken to be $x_{i+1} = g(0)$. This can be expressed directly in terms of the three x values and the three f values, but the actual formula is not important here. It is necessary for the three f values to be distinct. If they are not, a division by zero occurs in the formula.

The rate of convergence of inverse quadratic interpolation is 1.839, which is slightly faster than the secant method. However, three starting values are required, and the behavior of the algorithm can be erratic when the starting values are not close enough to a zero.

7.3 SUBROUTINE FZERO

One of the best computer algorithms that we have for finding a real zero of a single function combines the certainty of bisection with the ultimate speed of the secant method for smooth functions. It is called FZERO and was originated by van Wijngaarden, Dekker, and others at the Mathematical Center in Amsterdam in the 1960s. The algorithm was first published by Dekker (1969); the program here was written by Shampine and Watts. A description and analysis were given by Wilkinson (1967), especially pp. 8–12.

The prologue for FZERO appears at the end of this chapter. A typical call to FZERO might look like

CALL FZERO (F, B, C, B, RE, AE, IFLAG).

The second and third arguments B and C are end points of the interval in which the zero is sought; these are modified by FZERO. On return, B is the final estimate of the solution x^*. The parameter F is a real function subprogram having one real variable as an argument. AE and RE are the absolute and relative errors that can be tolerated in the result. The program expects that F(B) and F(C) have different signs.

FZERO iteratively improves the estimates B and C of the root. At all times B and C bracket the zero. Moreover $|F(B)| \le |F(C)|$. The algorithm terminates when the interval length $|B - C|$ has been reduced to satisfy

$$|B - C| \le 2(RE|B| + AE).$$

In addition, the convergence test involves the machine epsilon ϵ_{mach} to guard against the possibility that the tolerances RE and AE are too small.

The fourth argument to FZERO is your best guess at the location of the zero within the interval $[B, C]$. We suggest setting this to one of the endpoints B or C unless you have additional information. This argument is not altered on return from FZERO.

At each step, FZERO chooses the next iterate from two candidates—one obtained by the bisection algorithm and one obtained by the secant algorithm. If the point obtained by the secant method is "reasonable," it is chosen; otherwise the bisection point is chosen. The definition of "reasonable" is complicated, but essentially it means that the point is inside the current interval and not too close to the end points. Consequently the length of the interval is guaranteed to decrease at each step and to decrease by a large factor when the function is well behaved.

A few programming details in FZERO should be mentioned because they are applicable and important in other situations. The bisection point is computed by

XM = B + 0.5*(C-B)

rather than the conventional

XM = 0.5*(C+B).

To see why, take C = 0.982 and B = 0.984. Assume the calculations are being done on a machine with three-decimal-digit, rounded, floating-point arithmetic. Then C+B is computed as 1.97, and the conventional formula gives a midpoint 0.985 which is outside the interval. The general principle is that it is best to arrange formulas so that they express the desired quantity as a small correction to a good approximation.

Careful attention is paid to underflow and overflow problems. For example, the test to ensure that F(B) and F(C) have different signs is *not* the conventional

$$F(B)*F(C) < 0.0.$$

If F(B) = 10^{-30} and F(C) = -10^{-30}, then they have opposite signs, but on many computers the product will underflow and be set to zero, and the test will fail. The point obtained by the interpolation algorithms is expressed in the form

$$B + P/Q,$$

but the division is not done unless it is both necessary and safe to do so. When a bisection is to be done, then this quantity is not needed anyway.

In summary, what can be claimed for FZERO? First, it will always converge, even with floating-point arithmetic. Second, the number of function evaluations can be bounded in terms of ϵ_{mach} and the subroutine parameters. Third, the function F is guaranteed to change sign within the final interval [B, C] determined by the algorithm. Fourth, if F is smooth enough to have a continuous second derivative near a simple zero of F, then the algorithm FZERO (if started near enough to the zero) will eventually stop doing bisections and home in on the zero using the secant process, with a rate of convergence at least 1.618. Even though the algorithm is complex, it is possible to guarantee that it works well.

The following program illustrates the use of FZERO on a simple example, $f(x) = x^3 - 2x - 5$. It is easy to see from a rough graph of $f(x)$ that there is only one real zero and that it lies between 2 and 3. The output of the program is

```
INITIAL INTERVAL:    2.000000        3.000000
TOLERANCES:          1.0000000E-06  1.0000000E-06
ESTIMATE OF ZERO =   2.094552
FUNCTION VALUE =     1.4305115E-06
```

This example is a polynomial which could be treated by more specialized algorithms, but we have used it here because of its historical interest. The following excerpt from a letter written by de Morgan to Whewell in 1861 is contained in the book by Whittaker and Robinson (1924): "The reason I call $x^3 - 2x - 5 = 0$ a celebrated equation is because it was the one on which Wallis chanced to exhibit Newton's method when he first published it, in consequence of which *every* numerical solver has felt bound in duty to make it one of his examples. Invent a numerical method, neglect to show how it works on this equation, and you are a pilgrim who does not come in at the little wicket."

Here is the main program.

```
C SAMPLE PROGRAM FOR FZERO
C
      REAL    B, C, AE, RE
      EXTERNAL F
C
```

```
       B    = 2.0
       C    = 3.0
       AE   = 1.E-6
       RE   = 1.E-6
       WRITE (*,*) ' INITIAL INTERVAL: ', B, C
       WRITE (*,*) ' TOLERANCES:        ', AE, RE
C

       CALL FZERO (F, B, C, C, RE, AE, IFLAG)
C

       IF (IFLAG .NE. 1) WRITE (*,*) ' ERROR CODE =', IFLAG
       WRITE (*,*) ' ESTIMATE OF ZERO =', B
       WRITE (*,*) ' FUNCTION VALUE =  ', F(B)
C

       STOP
       END
C
C

       REAL FUNCTION F(X)
       REAL X
C

       F = X * (X*X-2.0) - 5.0
C

       RETURN
       END
```

Notice the Fortran statement

EXTERNAL F

at the beginning of the program. Such a statement is necessary when passing a function name to a subroutine. In a subroutine call, function names must be handled differently than other types of parameters, and the EXTERNAL statement warns Fortran to do this.

7.4 HISTORICAL PERSPECTIVE: ÉVARISTE GALOIS

It is to Évariste Galois that we owe the pessimistic result that it is impossible to compute solutions to general nonlinear equations. This result was not an isolated conclusion. Instead Galois's ideas and techniques initiated an important family of ideas in algebra, ideas which continue to be pursued by modern mathematicians.

Galois was born on October 25, 1811 in the village of Bourg-la-Reine on the edge of Paris. His father was to become mayor and his mother came from a family of distinguished jurists, but beyond this initial advantage of birth he had a life of monumental bad luck. He was so unfortunate that, were it not for the fact that we are reading history and not fiction, we might laugh at his story as if he were a character in a Dickensian melodrama.

Until the age of 12, Galois was taught at home by his mother. In 1823 he was sent to the Louis-le-Grand school in Paris, a strict school that emphasized the classics. He did well at first. But over the next two years, possibly out of boredom with his classes, his work deteriorated and he was demoted. When he was 14 he began to study mathematics, at first using a textbook written by Legendre, and later reading independently from the work of Lagrange and Abel. He became enthralled by mathematics and disinterested in everything else. He was not a success at school.

At the age of 16 he took the entrance examination to the École Polytechnique, the leading school for mathematics in France. He failed. (It has been speculated that the examiners did not recognize his genius, and that Galois was impatient with details and explanations of his ideas.) Nevertheless, he studied independently with a good teacher and published his first paper in 1829 (on continued fractions).

Shortly afterward, Galois submitted an important paper to Cauchy, the major French mathematician of the time, to be presented to the French Academy of Sciences. Cauchy forgot about the paper and lost the abstract. At 18, Galois tried again to pass the entrance examination to the École Polytechnique. In frustration, he threw a blackboard eraser at one of the examiners. He hit the examiner but he failed the exam.

Shortly afterward, as a consequence of a political scandal, Galois' father committed suicide. The funeral degenerated into a riot.

In 1830 at age 19, Galois submitted a paper on algebraic functions to the Academy of Sciences, to be considered for the Grand Prize in Mathematics. The Secretary of the Academy, Jean-Baptiste-Joseph Fourier, took the paper home and lost it.

Galois became politically active in the aftermath of the Revolution of 1830, joining the artillery of the National Guard. He was arrested on May 10, 1830 for his revolutionary actions, but was found not guilty by the jury at his trial. He was arrested again on July 14, 1831, but not charged until two months later for the silly crime of illegally wearing a uniform. He was sentenced to six months imprisonment. At about the same time, a third memoir that Galois sent to the Academy of Sciences was judged by Poisson to be "incomprehensible."

Galois was released on May 29, 1832, and the same day was challenged to a duel. The circumstances are not well documented. Overnight he wrote down the mathematical ideas that he had been developing, and entrusted them to his friend Auguste Chevalier. (They were not published until 1846, under the supervision of Joseph Liouville.) Then he wrote two final letters. The second, to two unnamed friends, reads in part

> I have been challenged by two patriots—it was impossible for me to refuse. I beg your pardon for having advised neither of you. But my opponents had put me on my honor not to warn any patriot. Your task is very simple: prove that I fought in spite of myself, that is to say after having exhausted every means of accommodation. ... Preserve my memory since fate has not given me life enough for my country to know my name. I die your friend
>
> E. Galois

(Taken from Bell (1975), p. 375.)

Early on May 30 Galois was shot through the intestines. He was found and taken to hospital some hours later, where he died of peritonitis as a result of his gunshot wounds. He was buried in an unmarked grave.

7.5 SYSTEMS OF NONLINEAR EQUATIONS

Now let us consider solving a set of n nonlinear equations in n unknowns

$$f_1(x_1, \ldots, x_n) = 0,$$
$$f_2(x_1, \ldots, x_n) = 0,$$
$$\vdots$$
$$f_n(x_1, \ldots, x_n) = 0.$$

To simplify our discussion we define $x = (x_1, \ldots, x_n)^T$ and $f(x) = (f_1(x), \ldots, f_n(x))^T$. In this notation our problem becomes

$$f(x) = 0.$$

There are many similarities between methods for one nonlinear equation and for a system of nonlinear equations; writing the problem in this way will help to make these similarities clearer.

We will consider analogs of Newton's method for nonlinear equations. In one dimension an iteration of Newton's method approximated the nonlinear equation by a linear equation obtained from the first two terms of the Taylor series. The same idea can be applied in n dimensions. Let $x^{(k)}$ be the current estimate of the solution x^*. Then the Taylor series about $x^{(k)}$ can be written in the form

$$f(x^{(k)} + p) = f(x^{(k)}) + J(x^{(k)})p + \cdots,$$

where $J(x^{(k)}) = \nabla f(x^{(k)})$ is the Jacobian matrix of first derivatives at $x^{(k)}$. That is,

$$(J(x^{(k)}))_{ij} = \left(\frac{\partial f_i(x)}{\partial x_j} \right)_{x = x^{(k)}}.$$

We will use the notation $J^{(k)} = J(x^{(k)})$ for the Jacobian and $f^{(k)} = f(x^{(k)})$ for the vector of function values.

Example 7.3 Computing the Jacobian.

Consider the moose and elk model from the introduction. Set the moose population equal to 1000, and let $x_1 = t$ and $x_2 = \text{elk}(t)$. Then the system of nonlinear equations becomes

$$f_1(x_1, x_2) = 1000 - 2e^{.1x_1} + .4x_2 = 0,$$

$$f_2(x_1, x_2) = x_2 - 5e^{.2x_1} + 600 = 0.$$

The Jacobian matrix is given by

$$J = \begin{pmatrix} -2(.1)e^{.1x_1} & .4 \\ -5(.2)e^{.2x_1} & 1 \end{pmatrix} . = \begin{pmatrix} -.2e^{.1x_1} & .4 \\ -e^{.2x_1} & 1 \end{pmatrix} . \quad \blacksquare$$

If we only consider the first two terms of the Taylor series, then we obtain a linear approximation to the nonlinear functions

$$f(x^{(k)} + p) \approx y = f^{(k)} + J^{(k)}p.$$

As in one dimension, we will use a zero of our linear approximation to determine the next estimate $x^{(k+1)}$ of the solution. Setting $y = 0$ we obtain the system of linear equations

$$J^{(k)}p = -f^{(k)}.$$

We can solve these equations using the methods of Chapter 3. The new estimate of the solution is given by $x^{(k+1)} = x^{(k)} + p$ or

$$x^{(k+1)} = x^{(k)} - (J^{(k)})^{-1}f^{(k)}.$$

This is the formula for Newton's method.

Example 7.4 Newton's Method.

Let's try out Newton's method on an example:

$$f_1(x_1, x_2) = x_1x_2 - x_2^3 - 1 = 0,$$

$$f_2(x_1, x_2) = x_1^2x_2 + x_2 - 5 = 0,$$

with $x^{(k)} = (2, 3)^T$. The Jacobian matrix is given by

$$J^{(k)} = \begin{pmatrix} x_2 & x_1 - 3x_2^2 \\ 2x_1x_2 & x_1^2 + 1 \end{pmatrix}_{x=(2,3)^T} = \begin{pmatrix} 3 & -25 \\ 12 & 5 \end{pmatrix}$$

and the Newton equations are

$$J^{(k)}p = \begin{pmatrix} 3 & -25 \\ 12 & 5 \end{pmatrix} \begin{pmatrix} p_1 \\ p_2 \end{pmatrix} = \begin{pmatrix} 22 \\ -10 \end{pmatrix} = -f^{(k)}.$$

Using Gaussian elimination, $p = (-0.4444, -0.9333)^T$ and thus the new estimate of the solution is $x^{(k+1)} = x^{(k)} + p = (1.5556, 2.0667)^T$. The solution of the nonlinear equations is $x^* = (2, 1)^T$. The complete iteration is illustrated below.

$x^{(k)}$	$\left\|x^{(k)} - x^*\right\|_2$	$\left\|f(x^{(k)})\right\|_2$
$(2.000000000000000, 3.000000000000000)$	$2.0(+00)$	$2.4(+01)$
$(1.555555555555556, 2.066666666666667)$	$1.2(+00)$	$6.9(+00)$
$(1.547205413881141, 1.477793334960415)$	$6.6(-01)$	$1.9(+00)$
$(1.780535029261150, 1.158864811005650)$	$2.7(-01)$	$5.2(-01)$
$(1.952843000499447, 1.028442688075807)$	$5.5(-02)$	$9.3(-02)$
$(1.997762973104506, 1.001240409084899)$	$2.6(-03)$	$4.4(-03)$
$(1.999995236211894, 1.000002599617500)$	$5.4(-06)$	$9.5(-06)$
$(1.999999999978874, 1.000000000011532)$	$2.4(-11)$	$4.2(-11)$
$(2.000000000000000, 1.000000000000000)$	$0.0(+00)$	$0.0(+00)$

Newton's method will converge quadratically, just as in Example 7.4, and this rapid convergence makes it a desirable method. On the other hand each iteration requires the solution of a system of linear equations—a relatively expensive task. Furthermore, Newton's method requires all the n^2 first partial derivatives of the nonlinear functions. If derivatives are difficult to compute, this can be a serious disadvantage. However, derivative values can be approximated using finite differencing (see Section 5 of Chapter 2), or generalizations of the secant method can be used that do not require derivative values.

As you might expect from the discussion in Section 2.2, when generalizations of the secant method are used, the method may not converge quadratically. In addition, as in one dimension, Newton's method and variations of it need not converge. This can happen if the Jacobian matrix is singular. For example, if we were to take $x^{(k)} = (0, 0)^T$ in Example 7.4 then both components of the first row of $J^{(k)}$ will be zero.

To improve the performance of Newton's method on general problems safeguards are included. Several approaches are possible. We discuss some of the details of one such method, called a **trust-region** strategy in the next section. Another approach based on a **line search** is discussed in Chapter 9.

*7.5.1 Safeguards for Newton's Method

The safeguards for Newton's method are based on two ideas. First we would like to guarantee that progress towards the solution is made at every iteration. Second we would like to prevent large and potentially disastrous steps from being taken. Progress towards the solution will be measured in terms of the values of the nonlinear functions; we will insist that

$$\left\|f(x^{(k+1)})\right\|_2 < \left\|f(x^{(k)})\right\|_2 .$$

To prevent large steps we will place a constraint on the step p:

$$\|p\|_2 \leq \delta$$

for some bound δ chosen by the algorithm. The bound δ represents our degree of *trust* in the model $f(x^{(k)} + p) \approx f(x^{(k)}) + J^{(k)}p$. If the model is good, then this approximation will be effective for large values of p and a large bound δ will be appropriate. If the model is poor then it will only be acceptable for small values of p and a small value of δ will be used. The set $\{\, p : \|p\|_2 \leq \delta \,\}$ is called the **trust region**.

Unless the bound δ is large, there is no reason to expect that the solution of the Newton equations will satisfy the constraint. Some compromise is necessary. One approach, having both theoretical and practical advantages, is not to insist that the step p actually solve $Jp = -f$, but instead to solve these equations as well as we can while still satisfying $\|p\|_2 \leq \delta$. In other words to solve the following optimization problem

$$\text{minimize}_p \quad \left\| J^{(k)}p + f^{(k)} \right\|_2 ,$$
$$\text{subject to} \quad \|p\|_2 \leq \delta.$$

If there were no constraint and if $J^{(k)}$ were nonsingular, then the setting $p = -(J^{(k)})^{-1} f^{(k)}$ would obviously minimize $\left\| J^{(k)}p + f^{(k)} \right\|_2$, since its value would be zero. Methods for solving this minimization problem will not be described here; for further information see Chapter 9, and the book by Dennis and Schnabel (1983).

A simple version of the overall algorithm proceeds as follows. At iteration k:

1. If $f(x_k) \approx 0$ then stop.
2. Compute the step p by solving the above constrained optimization problem.
3. If $\left\| f(x^{(k)} + p) \right\|_2 < \left\| f(x^{(k)}) \right\|_2$ then the step is accepted. Set $x^{(k+1)} \leftarrow x^{(k)} + p$, $k \leftarrow k + 1$. Go to step 1.
4. If the step is rejected then decrease the bound δ. Set $x^{(k+1)} \leftarrow x^{(k)}$, $k \leftarrow k + 1$. Go to step 1.

It may appear that the algorithm could cycle indefinitely, never accepting the step p and thus never making progress toward the solution. When the step is rejected, the algorithm is saying that the Taylor series is not a good approximation to the function at the point $x^{(k)}$. However, for points sufficiently close to $x^{(k)}$, the Taylor series model will be a good approximation to the function. Thus as the bound δ gets small and hence $\|p\|_2$ gets small, the step will eventually be accepted. For a well-written algorithm only one or two repetitions are likely to be required.

It is not possible to guarantee that our trust-region algorithms will converge to a solution of the nonlinear equations. For problems in more than one dimension, methods with guaranteed global convergence are still a topic of research. This is closely related to the problem of finding global solutions to optimization problems.

However, if a trust-region strategy is used it is possible to guarantee convergence to a *local* solution. A local solution $\bar{x}$ is a local minimizer of $\| f(x) \|_2^2$ (for more information about local solutions see Chapter 9). A local solution need not satisfy $f(\bar{x}) = 0$.

Example 7.5   Local Solution of a Nonlinear Equation.

As an illustration consider the one-dimensional problem

$$f(x) = x^3 - 3x + 18.$$

The solution of the nonlinear equation $f(x) = 0$ is $x^* = -3$. However the point $\bar{x} = 1$ is a local solution. To verify this, note that

$$\|f(x)\|_2^2 = f(x)^2 = x^6 - 6x^4 + 36x^3 + 9x^2 - 108x + 326.$$

At $x = 1$ the derivatives are

$$(d/dx)(\|f(x)\|_2^2) = 6x^5 - 24x^3 + 108x^2 + 18x - 108 = 0,$$
$$(d^2/dx^2)(\|f(x)\|_2^2) = 30x^4 - 72x^2 + 216x + 18 = 192 > 0,$$

and so $x = 1$ is a local minimizer of $\|f(x)\|_2^2$. However, since $f(1) = 16$, $x = 1$ does not solve the nonlinear equation. See Figure 7.5. ■

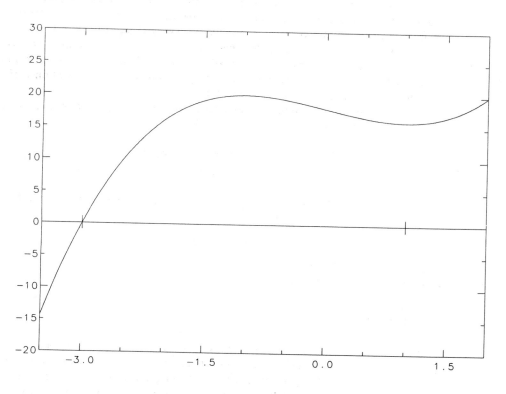

Figure 7.5 Local Solution of a Nonlinear Equation

More elaborate strategies can be used to avoid finding only local solutions to systems of nonlinear equations. One group of methods, called **homotopy methods**, is described in the survey article by Watson (1986). Related software is available in Hompack (see the Bibliography). These methods have higher computational costs than the strategies described above, since they require expensive auxiliary calculations, such as the solution of a differential equation or a sequence of systems of nonlinear equations. However, they have been successful at solving difficult problems that defy more traditional techniques.

A different approach based on interval analysis (see Section 3 of Chapter 2) is also possible. This is discussed in the book by Hansen (1969).

7.6 SUBROUTINE SNSQE

The routine SNSQE is designed for solving systems of nonlinear equations. It can be used for solving a single equation in one variable, but subroutine FZERO is likely to be more effective. However, FZERO requires that the user specify an interval containing a zero of the function. SNSQE is based on the ideas in the previous section; a more exact description of the algorithm can be found in the report by Moré et al. (1980). As with FZERO, the user must provide a subprogram to evaluate the nonlinear functions.

It is not possible to guarantee that SNSQE will find a solution of the system of nonlinear equations. It may only be possible to find a point that is a local minimum of $\|f(x)\|_2^2$; this is sometimes referred to as a local solution of the equations. Failure to converge is indicated using the variable INFO.

In the sample problem below, we solve a system of two equations in two unknowns, given by the formulas

$$f_1(x_1, x_2) = x_1 x_2 - x_2^3 - 1,$$
$$f_2(x_1, x_2) = x_1^2 x_2 + x_2 - 5.$$

The initial estimate of the solution is $x = (2, 3)^T$ and the solution is $x^* = (2, 1)^T$.

The output from the program is

```
INITIAL GUESS
    2.000000    3.000000
ESTIMATE OF SOLUTION
    1.999997    1.000001
VALUES OF NONLINEAR FUNCTIONS
-0.3457E-05 -0.7153E-05
```

Here is the calling program. The array W is a work vector of length LW. For a problem with n variables, the work vector must have length at least $n(3n + 13)/2$. TOL is a tolerance representing the accuracy desired in the solution. INFO indicates whether the equations were successfully solved. If INFO = 1 then the routine was successful. For the meaning of other values of INFO see the prologue for SNSQE.

```
C MAIN PROGRAM TO SOLVE A SYSTEM OF NONLINEAR EQUATIONS
C
      PARAMETER    (N = 2, LW = 19)
      REAL         X(N), FVEC(N), W(LW), TOL
      EXTERNAL     F
C
C SET UP PARAMETERS FOR SNSQE
C
      TOL  = 1.E-6
      X(1) = 2.0
      X(2) = 3.0
      WRITE (*,800) X(1), X(2)
      IOPT = 2
      NPRINT = 0
C
C SOLVE NONLINEAR EQUATIONS
C
      CALL SNSQE (F, F, IOPT, N, X, FVEC, TOL, NPRINT, INFO, W, LW)
C
C PRINT RESULTS
C
      IF (INFO .NE. 1) WRITE (*,810) INFO
      WRITE (*,820) X(1), X(2)
      WRITE (*,830) FVEC(1), FVEC(2)
C
      STOP
800   FORMAT (' INITIAL GUESS', /, 2F12.6)
810   FORMAT (' INFO =', I3)
820   FORMAT (' ESTIMATE OF SOLUTION ', /, 2F12.6)
830   FORMAT (' VALUES OF NONLINEAR FUNCTIONS', /, 2E12.4)
      END
C
C
      SUBROUTINE F (N, X, FVEC, IFLAG)
      REAL   X(N), FVEC(N)
C
C COMPUTE NONLINEAR FUNCTIONS
C
      FVEC(1) = X(1)*X(2) - X(2)**3 - 1.0
      FVEC(2) = X(1)**2*X(2) + X(2) - 5.0
C
      RETURN
      END
```

The second argument to SNSQE is the name of a user-written subroutine JAC to compute the Jacobian. SNSQE operates best when this routine is provided. But when IOPT = 2 the Jacobian is approximated by differences and JAC can be omitted. In that case we suggest that the formal name given to the first argument also be used for the second, as in the example problem above. For smaller problems, especially if they are only being solved once, we recommend using SNSQE with IOPT = 2. Use IOPT = 1 if you need to improve the performance of the subroutine. Writing a subroutine for the Jacobian can be time consuming, and there is a good chance of introducing an algebraic error in the derivatives. The subroutine CHKDER can be used to check your derivative subroutine; it compares your "exact" Jacobian with numerical approximations and attempts to determine if these are consistent. CHKDER is used independently of SNSQE.

input fzero.clam
fcnzero(f7,35,45)

7.7 PROBLEMS

P7–1.–In laying water mains, utilities must be concerned with the possibility of freezing. Although soil and weather conditions are complicated, reasonable approximations can be made on the basis of the assumption that soil is uniform in all directions. In that case the temperature in degrees Celsius $T(x, t)$ at a distance x (in meters) below the surface, t seconds after the beginning of a cold snap, is given approximately by

$$\frac{T(x, t) - T_s}{T_i - T_s} = \text{erf}\left(\frac{x}{2\sqrt{\alpha t}}\right),$$

R 0,l

where T_s is the constant surface temperature during the cold period, T_i is the initial soil temperature before the cold snap, and α is the thermal conductivity of the soil (in meters² per second). Assume that $T_i = 20°C$, $T_s = -15°C$, $\alpha = 0.138 \cdot 10^{-6} m^2/sec$. Use FZERO to determine how deep a water main should be buried so that it will only freeze after 60 days exposure at this constant surface temperature. Note that water freezes at $0°C$.

P7–2.–The purpose of this problem is to remind you that sometimes explicit solutions are not as useful as numerical methods. Kepler's equation in orbit determination is

$$M = E - e \sin E.$$

The symbols M, E, and e are referred to as the Mean anomaly, Eccentric anomaly and eccentricity of the orbit respectively.

(a) If $M = 24.851090$ and $e = 0.1$ use FZERO to find E.

(b) A formula for E is

$$E = M + 2\sum_{m=1}^{\infty} \frac{1}{m} J_m(me) \sin(mM),$$

where $J_m(x)$ is the Bessel function of the first kind of order m. Use this formula to find E for the values in (a). A routine BESJ to compute J_m is provided on the disk that accompanies this book.

(c) Find E from the formula in (b) using

$$J_m(me) = \sum_{n=0}^{\infty} \frac{(-1)^n (me/2)^{2n+m}}{n!(m+n)!}.$$

In (b) and (c) compare the effort and accuracy of your results with those in (a).

P7–3.–A long conducting rod of diameter D meters and electrical resistance R per unit length is in a large enclosure whose walls (far away from the rod) are kept at temperature $T_s^\circ C$. Air flows past the rod at temperature T_∞. If an electrical current I passes through the rod, the temperature of the rod eventually stabilizes to T, where T satisfies

$$Q(T) = \pi D h (T - T_\infty) + \pi D \epsilon \sigma (T^4 - T_s^4) - I^2 R = 0, \qquad 35\text{-}45$$

where

σ = Stefan-Boltzman constant = $5.67 \cdot 10^{-8}$ Watts/meter^{2}Kelvin^4,

ϵ = rod surface emissivity = 0.8,

h = heat transfer coefficient of air flow = 20 Watts/meter^{2}Kelvin,

$T_\infty = T_s = 25^\circ C,$ $273 + 25 = 298\ K$

$D = 0.1$ meter,

$I^2 R = 100.$

Use FZERO to find the steady state temperature T of the rod.

P7–4.–Use FZERO to find the ten smallest positive values of x for which the line $y = x$ crosses the graph of $y = \tan x$. The solution of this problem is important in determining the maximum load that a compressed rod will support without buckling. See, for example Timoshenko (1956) p 154.

P7–5.–Find the real roots of

$$x^4 - 3x^2 + 75x - 10000 = 0.$$

P7–6.–What output is produced by the following program? Explain why.

```
C MAIN PROGRAM
C
      REAL B, C, AE, RE, F
      EXTERNAL F
C
      B   = 0.0
      C   = 1.0
      AE  = 1.E-6
      RE  = 1.E-6
C
      CALL FZERO (F, B, C, B, RE, AE, IFLAG)
C
      WRITE (*,*) ' ESTIMATE OF ZERO =', B
```

```
        STOP
        END
C
C FUNCTION SUBPROGRAM
C
        FUNCTION F(X)
        REAL X, Y
C
        IF (X .EQ. 0.) Y = -1.0
        IF (X .GT. 0.) Y =  9.0
        WRITE (*,*) X, Y
        F = Y
C
        RETURN
        END
```

P7–7.–The following system of equations is to be solved for x_1, x_2, x_3,

$$(1 - s_1)^2 + (1.5 - s_2)^2 + (c_1 - c_2)^2 - (x_3 + 0.25)^2 = 0$$

$$(x_3 + 0.25)s_1 + 2x_3(c_2 s_1 - c_1) = 0$$

$$(x_3 + 0.25)s_2 + 2x_3(c_1 s_2 - 1.5c_2) = 0,$$

where s_i and c_i stand for $\sin(x_i)$ and $\cos(x_i)$. The equations occur in the following problem. A steel bar of unit length and mass is hinged at the point $u = 0, v = 1.5, w = 0$ so that it can swing freely in the uw plane. Another identical bar is hinged at $u = 1, v = 0, w = 0$ so that it can swing in the vw plane. The free ends of the bars are attached by a spring of natural length 0.25, and unit spring constant. When the bars stop vibrating the angle between the first bar and the vw plane is x_1, the angle between the second bar and the uw plane is x_2, and the distance between the free ends of the bars (along the spring) is $x_3 + 0.25$. Solve this system using SNSQE assuming that $0 \leq x_1, x_2 \leq \pi/2$, and that $x_3 > 0$. (Answer: $x_1 = 35.55°, x_2 = 44.91°, x_3 = 0.65$).

P7–8.–The Cray 1 supercomputer does not have a divide unit. Instead, to compute $1/R$, $R > 0$ it forms a "reciprocal approximation," accurate to about half of a floating point word and then uses that as the initial guess for one Newton iteration. Let $f(x) = (1/x) - R$. Show that the formula for Newton's method is

$$x_{k+1} = x_k(2 - Rx_k).$$

(a) Using this Newton iteration compute $1/R$ for $R = 1, 2, \ldots, 10$. Tabulate the number of iterations needed to generate a result that is accurate to six digits. Use $x_0 = 0.01$.

(b) Using the same function $f(x)$ repeat the computations with the bisection algorithm and compare the amount of work to that in (a). Use the initial interval $[.01, 2]$.

P7–9.–Consider using Newton's method on the equation $f(x) = (x - 1)^2$ that has a double zero. Show that for this equation the iterations converge linearly rather than quadratically. Verify numerically that the convergence is linear at a multiple zero by trying to solve $f(x) = \exp(2x) - 2\exp(x) + 1 = 0$ and printing out the error at each iterate.

P7–10.–We have used Newton's method to find real-values solutions of nonlinear equations. It can also be used to find complex roots merely by choosing an initial guess x_0 that is complex. Use Newton's method to find a complex root of our historical example,

$$x^3 - 2x - 5 = 0.$$

P7–11.–One classical method for solving cubics is Cardano's solution. The cubic equation

$$x^3 + ax^2 + bx + c = 0$$

is transformed to a reduced form

$$y^3 + py + q = 0$$

by the substitution $x = y - a/3$. The coefficients in the reduced form are

$$p = b - \frac{a^2}{3},$$

$$q = c - \frac{ab}{3} + 2\left(\frac{a}{3}\right)^3.$$

One real root of the reduced form can be found by

$$s = \left[\left(\frac{p}{3}\right)^3 + \left(\frac{q}{2}\right)^2\right]^{\frac{1}{2}},$$

$$y_1 = \left[-\frac{q}{2} + s\right]^{\frac{1}{3}} + \left[-\frac{q}{2} - s\right]^{\frac{1}{3}}$$

and then a real root of the original equation by

$$x_1 = y_1 - \frac{a}{3}.$$

The other two roots can be found by similar formulas or by dividing out x_1 and solving the resulting quadratic equation.

(a) Apply Cardano's method to find the real root of

$$x^3 + 3x^2 + \alpha^2 x + 3\alpha^2 = 0$$

for various values of α. Investigate the loss of accuracy from roundoff for large α, say α about the reciprocal of machine epsilon.

(b) Apply Newton's method to the same equation for the same values of α. Investigate the effects of roundoff error and the choice of starting value.

P7–12.–In Section 2.1 of Chapter 5, the 2-point Gauss quadrature rule was derived by solving the following system of nonlinear equations

$$(b - a) = w_1 + w_2,$$

$$0 = w_1(x_1 - m) + w_2(x_2 - m),$$

$$(b - a)^3/12 = w_1(x_1 - m)^2 + w_2(x_2 - m)^2,$$

$$0 = w_1(x_1 - m)^3 + w_2(x_2 - m)^3,$$

where $m = (a+b)/2$. Set $[a,b] = [-1,1]$, and solve this system using SNSQE. Compare your results with the true solution $x_1 = 1/\sqrt{3}$, $x_2 = -1/\sqrt{3}$, $w_1 = w_2 = 1$.

7.8 PROLOGUES: FZERO AND SNSQE

```
      SUBROUTINE FZERO(F,B,C,R,RE,AE,IFLAG)
C***BEGIN PROLOGUE  FZERO
C***DATE WRITTEN    700901    (YYMMDD)
C***REVISION DATE   860411    (YYMMDD)
C***CATEGORY NO.  F1B
C***KEYWORDS  BISECTION,NONLINEAR,ROOTS,ZEROS
C***AUTHOR  SHAMPINE,L.F.,SNLA
C           WATTS,H.A.,SNLA
C***PURPOSE  FZERO searches for a zero of a function F(X) in a given
C            interval (B,C).  It is designed primarily for problems
C            where F(B) and F(C) have opposite signs.
C***DESCRIPTION
C
C    From the book "Numerical Methods and Software"
C       by  D. Kahaner, C. Moler, S. Nash
C          Prentice Hall 1988
C
C    Based on a method by T J Dekker
C    written by L F Shampine and H A Watts
C
C            FZERO searches for a zero of a function F(X) between
C            the given values B and C until the width of the interval
C            (B,C) has collapsed to within a tolerance specified by
C            the stopping criterion, ABS(B-C) .LE. 2.*(RW*ABS(B)+AE).
C            The method used is an efficient combination of bisection
C            and the secant rule.
C
C    Description Of Arguments
C
C    F,B,C,R,RE and AE are input parameters
C    B,C and IFLAG are output parameters (flagged by an * below)
C
C       F     - Name of the real valued external function.  This name
C               must be in an EXTERNAL statement in the calling
C               program.  F must be a function of one real argument.
C
C      *B     - One end of the interval (B,C).  The value returned for
C               B usually is the better approximation to a zero of F.
C
C      *C     - The other end of the interval (B,C)
C
C       R     - A (better) guess of a zero of F which could help in
```

```
C               speeding up convergence.  If F(B) and F(R) have
C               opposite signs, a root will be found in the interval
C               (B,R); if not, but F(R) and F(C) have opposite
C               signs, a root will be found in the interval (R,C);
C               otherwise, the interval (B,C) will be searched for a
C               possible root.  When no better guess is known, it is
C               recommended that R be set to B or C; because if R is
C               not interior to the interval (B,C), it will be ignored.
C
C        RE    - Relative error used for RW in the stopping criterion.
C               If the requested RE is less than machine precision,
C               then RW is set to approximately machine precision.
C
C        AE    - Absolute error used in the stopping criterion.  If the
C               given interval (B,C) contains the origin, then a
C               nonzero value should be chosen for AE.
C
C       *IFLAG - A status code.  User must check IFLAG after each call.
C               Control returns to the user from FZERO in all cases.
C
C               1  B is within the requested tolerance of a zero.
C                  The interval (B,C) collapsed to the requested
C                  tolerance, the function changes sign in (B,C), and
C                  F(X) decreased in magnitude as (B,C) collapsed.
C
C               2  F(B) = 0.  However, the interval (B,C) may not have
C                  collapsed to the requested tolerance.
C
C               3  B may be near a singular point of F(X).
C                  The interval (B,C) collapsed to the requested tol-
C                  erance and the function changes sign in (B,C), but
C                  F(X) increased in magnitude as (B,C) collapsed,i.e.
C                    abs(F(B out)) .GT. max(abs(F(B in)),abs(F(C in)))
C
C               4  No change in sign of F(X) was found although the
C                  interval (B,C) collapsed to the requested tolerance.
C                  The user must examine this case and decide whether
C                  B is near a local minimum of F(X), or B is near a
C                  zero of even multiplicity, or neither of these.
C
C               5  Too many (.GT. 500) function evaluations used.
C***REFERENCES  L. F. SHAMPINE AND H. A. WATTS, *FZERO, A ROOT-SOLVING
C                  CODE*, SC-TM-70-631, SEPTEMBER 1970.
C               T. J. DEKKER, *FINDING A ZERO BY MEANS OF SUCCESSIVE
C                  LINEAR INTERPOLATION*, 'CONSTRUCTIVE ASPECTS OF THE
C                  FUNDAMENTAL THEOREM OF ALGEBRA', EDITED BY B. DEJON
C                  P. HENRICI, 1969.
C***ROUTINES CALLED  R1MACH
```

```
C***END PROLOGUE  FZERO



      SUBROUTINE SNSQE(FCN,JAC,IOPT,N,X,FVEC,TOL,NPRINT,INFO,WA,LWA)
C***BEGIN PROLOGUE  SNSQE
C***DATE WRITTEN    800301   (YYMMDD)
C***REVISION DATE   840405   (YYMMDD)
C***CATEGORY NO.  F2A
C***KEYWORDS  EASY-TO-USE,NONLINEAR SQUARE SYSTEM,POWELL HYBRID METHOD,
C             ZERO
C***AUTHOR  HIEBERT, K. L., (SNLA)
C***PURPOSE  SNSQE is the easy-to-use version of SNSQ which finds a zero
C            of a system of N nonlinear functions in N variables by a
C            modification of Powell hybrid method.  This code is the
C            combination of the MINPACK codes(Argonne) HYBRD1 and HYBRJ1
C***DESCRIPTION
C
C    From the book "Numerical Methods and Software"
C        by  D. Kahaner, C. Moler, S. Nash
C            Prentice Hall 1988
C
C 1. Purpose.
C
C
C        The purpose of SNSQE is to find a zero of a system of N non-
C        linear functions in N variables by a modification of the Powell
C        hybrid method.  This is done by using the more general nonlinear
C        equation solver SNSQ.  The user must provide a subroutine which
C        calculates the functions.  The user has the option of either to
C        provide a subroutine which calculates the Jacobian or to let the
C        code calculate it by a forward-difference approximation.  This
C        code is the combination of the MINPACK codes (Argonne) HYBRD1
C        and HYBRJ1.
C
C
C 2. Subroutine and Type Statements.
C
C        SUBROUTINE SNSQE(FCN,JAC,IOPT,N,X,FVEC,TOL,NPRINT,INFO,WA,LWA)
C        INTEGER IOPT,N,NPRINT,INFO,LWA
C        REAL TOL
C        REAL X(N),FVEC(N),WA(LWA)
C        EXTERNAL FCN,JAC
C
C
C 3. Parameters.
C
C        Parameters designated as input parameters must be specified on
```

```
C        entry to SNSQE and are not changed on exit, while parameters
C        designated as output parameters need not be specified on entry
C        and are set to appropriate values on exit from SNSQE.
C
C        FCN is the name of the user-supplied subroutine which calculates
C           the functions.  FCN must be declared in an EXTERNAL statement
C           in the user calling program, and should be written as follows.
C
C           SUBROUTINE FCN(N,X,FVEC,IFLAG)
C           INTEGER N,IFLAG
C           REAL X(N),FVEC(N)
C           - - - - - - - - - - - - - - - - - - - - - - - - - - - - - - -
C           Calculate the functions at X and
C           return this vector in FVEC.
C           - - - - - - - - - - - - - - - - - - - - - - - - - - - - - - -
C           RETURN
C           END
C
C           The value of IFLAG should not be changed by FCN unless the
C           user wants to terminate execution of SNSQE.  In this case, set
C           IFLAG to a negative integer.
C
C        JAC is the name of the user-supplied subroutine which calculates
C           the Jacobian.  If IOPT=1, then JAC must be declared in an
C           EXTERNAL statement in the user calling program, and should be
C           written as follows.
C
C           SUBROUTINE JAC(N,X,FVEC,FJAC,LDFJAC,IFLAG)
C           INTEGER N,LDFJAC,IFLAG
C           REAL X(N),FVEC(N),FJAC(LDFJAC,N)
C           - - - - - - - - - - - - - - - - - - - - - - - - - - - - - - -
C           Calculate the Jacobian at X and return this
C           matrix in FJAC.  FVEC contains the function
C           values at X and should not be altered.
C           - - - - - - - - - - - - - - - - - - - - - - - - - - - - - - -
C           RETURN
C           END
C
C           The value of IFLAG should not be changed by JAC unless the
C           user wants to terminate execution of SNSQE.  In this case, set
C           IFLAG to a negative integer.
C
C           If IOPT=2, JAC can be ignored (treat it as a dummy argument).
C
C        IOPT is an input variable which specifies how the Jacobian will
C           be calculated.  If IOPT=1, then the user must supply the
C           Jacobian through the subroutine JAC.  If IOPT=2, then the
C           code will approximate the Jacobian by forward-differencing.
```

```
C
C       N is a positive integer input variable set to the number of
C          functions and variables.
C
C       X is an array of length N.  On input, X must contain an initial
C          estimate of the solution vector.  On output, X contains the
C          final estimate of the solution vector.
C
C       FVEC is an output array of length N which contains the functions
C          evaluated at the output X.
C
C       TOL is a non-negative input variable.  Termination occurs when
C          the algorithm estimates that the relative error between X and
C          the solution is at most TOL.  Section 4 contains more details
C          about TOL.
C
C       NPRINT is an integer input variable that enables controlled
C          printing of iterates if it is positive.  In this case, FCN is
C          called with IFLAG = 0 at the beginning of the first iteration
C          and every NPRINT iteration thereafter and immediately prior
C          to return, with X and FVEC available for printing. Appropriate
C          print statements must be added to FCN (see example). If NPRINT
C          is not positive, no special calls of FCN with IFLAG = 0 are
C          made.
C
C       INFO is an integer output variable.  If the user has terminated
C          execution, INFO is set to the (negative) value of IFLAG.  See
C          description of FCN and JAC.  Otherwise, INFO is set as follows.
C
C          INFO = 0  improper input parameters.
C
C          INFO = 1  algorithm estimates that the relative error between
C                    X and the solution is at most TOL.
C
C          INFO = 2  number of calls to FCN has reached or exceeded
C                    100*(N+1) for IOPT=1 or 200*(N+1) for IOPT=2.
C
C          INFO = 3  TOL is too small.  No further improvement in the
C                    approximate solution X is possible.
C
C          INFO = 4  iteration is not making good progress.
C
C          Sections 4 and 5 contain more details about INFO.
C
C       WA is a work array of length LWA.
C
C       LWA is a positive integer input variable not less than
C          (3*N**2+13*N))/2.
C
```

```
C
C
C 4. Successful Completion.
C
C       The accuracy of SNSQE is controlled by the convergence parame-
C       ter TOL.  This parameter is used in a test which makes a compar-
C       ison between the approximation X and a solution XSOL.  SNSQE
C       terminates when the test is satisfied.  If TOL is less than the
C       machine precision (as defined by the function R1MACH(4)), then
C       SNSQE attemps only to satisfy the test defined by the machine
C       precision.  Further progress is not usually possible.  Unless
C       high precision solutions are required, the recommended value
C       for TOL is the square root of the machine precision.
C
C       The test assumes that the functions are reasonably well behaved,
C       and, if the Jacobian is supplied by the user, that the functions
C       and the Jacobian  coded consistently.  If these conditions
C       are not satisfied, SNSQE may incorrectly indicate convergence.
C       The coding of the Jacobian can be checked by the subroutine
C       CHKDER.  If the Jacobian is coded correctly or IOPT=2, then
C       the validity of the answer can be checked, for example, by
C       rerunning SNSQE with a tighter tolerance.
C
C       Convergence Test.  If ENORM(Z) denotes the Euclidean norm of a
C          vector Z, then this test attempts to guarantee that
C
C               ENORM(X-XSOL) .LE.  TOL*ENORM(XSOL).
C
C       If this condition is satisfied with TOL = 10**(-K), then the
C       larger components of X have K significant decimal digits and
C       INFO is set to 1.  There is a danger that the smaller compo-
C       nents of X may have large relative errors, but the fast rate
C       of convergence of SNSQE usually avoids this possibility.
C
C
C 5. Unsuccessful Completion.
C
C       Unsuccessful termination of SNSQE can be due to improper input
C       parameters, arithmetic interrupts, an excessive number of func-
C       tion evaluations, errors in the functions, or lack of good prog-
C       ress.
C
C       Improper Input Parameters.  INFO is set to 0 if IOPT .LT. 1, or
C          IOPT .GT. 2, or N .LE. 0, or TOL .LT. 0.E0, or
C          LWA .LT. (3*N**2+13*N)/2.
C
C       Arithmetic Interrupts.  If these interrupts occur in the FCN
C          subroutine during an early stage of the computation, they may
```

```
C          be caused by an unacceptable choice of X by SNSQE.  In this
C          case, it may be possible to remedy the situation by not evalu-
C          ating the functions here, but instead setting the components
C          of FVEC to numbers that exceed those in the initial FVEC.
C
C          Excessive Number of Function Evaluations.  If the number of
C             calls to FCN reaches 100*(N+1) for IOPT=1 or 200*(N+1) for
C             IOPT=2, then this indicates that the routine is converging
C             very slowly as measured by the progress of FVEC, and INFO is
C             set to 2.  This situation should be unusual because, as
C             indicated below, lack of good progress is usually diagnosed
C             earlier by SNSQE, causing termination with INFO = 4.
C
C          Errors in the Functions.  When IOPT=2, the choice of step length
C             in the forward-difference approximation to the Jacobian
C             assumes that the relative errors in the functions are of the
C             order of the machine precision.  If this is not the case,
C             SNSQE may fail (usually with INFO = 4).  The user should
C             then either use SNSQ and set the step length or use IOPT=1
C             and supply the Jacobian.
C
C          Lack of Good Progress.  SNSQE searches for a zero of the system
C             by minimizing the sum of the squares of the functions.  In so
C             doing, it can become trapped in a region where the minimum
C             does not correspond to a zero of the system and, in this situ-
C             ation, the iteration eventually fails to make good progress.
C             In particular, this will happen if the system does not have a
C             zero.  If the system has a zero, rerunning SNSQE from a dif-
C             ferent starting point may be helpful.
C
C
C 6. Characteristics of the Algorithm.
C
C          SNSQE is a modification of the Powell hybrid method.  Two of
C          its main characteristics involve the choice of the correction as
C          a convex combination of the Newton and scaled gradient direc-
C          tions, and the updating of the Jacobian by the rank-1 method of
C          Broyden.  The choice of the correction guarantees (under reason-
C          able conditions) global convergence for starting points far from
C          the solution and a fast rate of convergence.  The Jacobian is
C          calculated at the starting point by either the user-supplied
C          subroutine or a forward-difference approximation, but it is not
C          recalculated until the rank-1 method fails to produce satis-
C          factory progress.
C
C          Timing.  The time required by SNSQE to solve a given problem
C             depends on N, the behavior of the functions, the accuracy
C             requested, and the starting point.  The number of arithmetic
```

```
C          operations needed by SNSQE is about 11.5*(N**2) to process
C          each evaluation of the functions (call to FCN) and 1.3*(N**3)
C          to process each evaluation of the Jacobian (call to JAC,
C          if IOPT = 1).  Unless FCN and JAC can be evaluated quickly,
C          the timing of SNSQE will be strongly influenced by the time
C          spent in FCN and JAC.
C
C      Storage.  SNSQE requires (3*N**2 + 17*N)/2 single precision
C          storage locations, in addition to the storage required by the
C          program.  There are no internally declared storage arrays.
C
C***REFERENCES  POWELL, M. J. D.
C                    A HYBRID METHOD FOR NONLINEAR EQUATIONS.
C                    NUMERICAL METHODS FOR NONLINEAR ALGEBRAIC EQUATIONS,
C                    P. RABINOWITZ, EDITOR.  GORDON AND BREACH, 1970.
C***ROUTINES CALLED  SNSQ,XERROR
C***END PROLOGUE  SNSQE
```

8

Ordinary Differential Equations

8.1 INTRODUCTION

A differential equation describes how a function changes. It allows us to model phenomena that are in motion or that change continuously over time. As a simple example, suppose that a ball is dropped from a building. Then the height of the ball satisfies the differential equation

$$\frac{d}{dt}\text{height}(t) = -gt$$

where g is a constant representing the force of gravity, and t is a variable representing the elapsed time since the ball was dropped. This equation can be solved by integrating the right-hand side, giving

$$\text{height}(t) = \text{height}(0) - \tfrac{1}{2}gt^2.$$

There are several things to notice. First, the solution depends on the height of the building, height(0), and hence is not uniquely determined by the differential equation. Second, the solution is not a number but a function.

Models based on differential equations are used extensively in the sciences to describe the world around us. From the atomic level in the analysis of chemical reactions, to global level weather prediction, to the cosmic level in the motion of planets, comets, and investigation of the galaxies, they are essential tools. Even systems which are inherently discrete, such as the number of rabbits in a biological ecosystem, can often be accurately approximated by differential equations. This section introduces some of the major concepts which are needed to understand ordinary differential equations. In the sections that follow we return to each of these ideas and develop them much more fully.

An **ordinary differential equation** (ODE, pronounced oh-dee-ee) is an equation involving a function and its derivatives. The simplest ODE is

$$\frac{dy}{dt} = f(t).$$

For the example above, y is the height of the ball, and $f(t) = -gt$. Suppose that we are interested in the solution starting at some time $t = t_0$. If $f(t)$ is continuous on the interval $[t_0, T]$ then this equation has solution

$$y(t) = c + \int_{t_0}^{t} f(\tau)\, d\tau, \qquad t_0 \le t \le T.$$

The function $y(t)$ is only unique up to the constant c. In the above example, c corresponds to height(0), the initial height. Consequently, the ODE has a **family** of solutions each of which differs from the others by an additive constant. The form of this equation implies that the slope dy/dt of the solution at a fixed t does not depend upon which member of the solution family we examine. One particular member of this family will be selected by asking for the unique solution taking on a specified value A at $t = t_0$. The value of A is determined by the application; above it represents the height of the building.

This equation illustrates the close ties between quadrature and the solution of differential equations. However there are some differences. One of the most important is that in the quadrature problem the solution is a single number, the value of the integral. Here, the solution is a function defined on an interval. Sometimes the solution of a differential equation is only needed at one particular point, for example the height of the ball after 3 seconds. In that case the solution of the problem is a number just as for quadrature. But most often the solution is needed for an interval of values; we want a graph of the height of the ball as a function of time, and thus it is best to think about the solution as a function. In simple cases, an explicit formula for the solution can be obtained; but in general this is impossible, or maybe it is just too complicated or time consuming to be of practical value. Another difference between ordinary differential equations and quadrature is that the former are univariate, e.g., involve one independent variable; multidimensional quadrature is common.

Because the solution of a differential equation is a function, if we expect to "solve" this on a computer we have to agree in advance on what we mean by this term. A common approach is to **discretize** the problem. This means that we will only compute the solution $y(t)$ at a finite set of *discrete* points, as opposed to finding the solution on a *continuous* interval. For example, if $[t_0, T] = [0, 1]$ the solution might be determined at the points 0.0, .01, .02, ..., .99, 1.0. To estimate the value of the solution at a general point, some form of interpolation is used (see Chapter 4). The set of discrete points need not be equally spaced. Sometimes the points are specified in advance, since the value of the solution at certain points may have special significance, or we need a particular resolution of output points for a plot or a table. It is better to let the software determine the discretization points, since the solution can often be obtained more accurately and efficiently by choosing the points based on the properties of the differential equation. For example, if the solution is varying rapidly in some region, it will usually be necessary to put many discretization points in that region in order to effectively follow the changes in the solution. In this sense too, differential equations and quadrature are similar.

There are several ways in which errors can arise when solving differential equations. First, discretizing the problem introduces errors just as discretizing the derivative did (see Example 2, Chapter 2). In many, but unfortunately not all cases, if we rerun the problem using more finely spaced discretization points, the discrete solution converges to the solution of the original problem. Sometimes though, small errors introduced at the beginning of the calculation can overwhelm the solution unless the numerical method is selected carefully. This is sometimes called an "instability." Also, there are the usual rounding errors that arise in the course of the computations. Methods must often compromise when trying to control these forms of errors while at the same time trying to solve the equation efficiently.

Frequently it is necessary to solve systems of differential equations involving a set of functions. Many of the methods for a single equation can be immediately applied to a system of equations although the programming details become more tedious and geometric intuition is more difficult to apply. Higher-order differential equations—those involving second, third, and higher derivatives—can be transformed to systems of first-order equations (see Section 1.2), so there is no reason to consider anything more than first-order equations.

When a differential equation is modeling two interrelated phenomena, one changing rapidly and one changing slowly, difficulties can arise. An example might be a model of a drum beat where there are rapid vibrations corresponding to the tone of the drum, along with a slow decay in the volume of the sound. Many discretization points would have to be used to model the rapid vibrations, leading to a great many calculations. However, a short time after the drum is struck, these rapid vibrations will be less important than the general decay, which can be approximated well by using only a few discretization points. Such a problem is called **stiff**. Some numerical methods have difficulty with stiff problems since they slavishly follow the rapid motions even when they are less important than the general trend in the solution. Special algorithms have been devised for solving stiff problems, and some of these are developed below.

In this book we discuss only initial-value problems for ordinary differential equations. By an initial-value problem (IVP) we mean a differential equation where the function value is specified at some initial point (labelled t_0 above), and then the behavior of the solution is followed as t is increased or decreased. There are other kinds of problems which occur in connection with ordinary differential equations. Section 20 mentions a few of them.

The numerical methods for initial value problems begin at the initial point and try to follow or "sniff out" the direction of the solution curve (function). They are often called **marching** methods. To compute a solution any numerical method must evaluate the right hand side, f, of the ODE. If f is infinite somewhere the methods can break down. For example, this might happen near $t = 0$ when trying to solve the problem

$$y'(t) = \frac{1}{\sin \sqrt{|t|}}, \quad y(-1) = 0.$$

Nevertheless, the quadrature problem

$$\int_{-1}^{2} \frac{dt}{\sin \sqrt{|t|}} = I = 5.3141\ldots,$$

is not difficult to solve by most modern quadrature software such as Q1DA; this is yet another difference between differential equations and quadrature. Problem 8–1 gives you an opportunity to further compare initial value problems and quadrature.

8.1.1 Review of Basic Terminology

As mentioned above, the simplest ordinary differential equation is

$$\frac{dy}{dt} = f(t).$$

This ODE is called **first-order** because the highest appearing derivative is the first. Its solutions can be found by integrating the right-hand side, and they all differ by an additive constant. A more general first-order ODE is

$$\frac{dy}{dt} = f(t, y).$$

This equation also has a family of solutions (under modest assumptions on f), but distinct members of the family do not differ by an additive constant, because for fixed t, each family member has a different slope. We ignore further questions of the existence of solutions, which are covered thoroughly in the book by Shampine and Gordon (1975).[1]

Example 8.1   Solution Families.

Writing y' for $\frac{dy}{dt}$ consider the two different equations

$$y' = e^{-t} \quad \text{and} \quad y' = -y,$$

which have solutions

$$y(t) = c - e^{-t} \quad \text{and} \quad y(t) = ce^{-t},$$

respectively. Figure 8.1 illustrates the solution families of these equations. For a fixed t, notice that y' varies with y in the second but not in the first. ■

---

[1] L. F. Shampine, formerly at Sandia Laboratories and now teaching at Southern Methodist University, is one of the most active researchers into numerical methods for the solution of ODEs. His subroutines DEABM and DERKF are considered the best implementations of Adams Bashforth and Runge Kutta algorithms (see Section 16 and 19) for the solution of nonstiff problems.

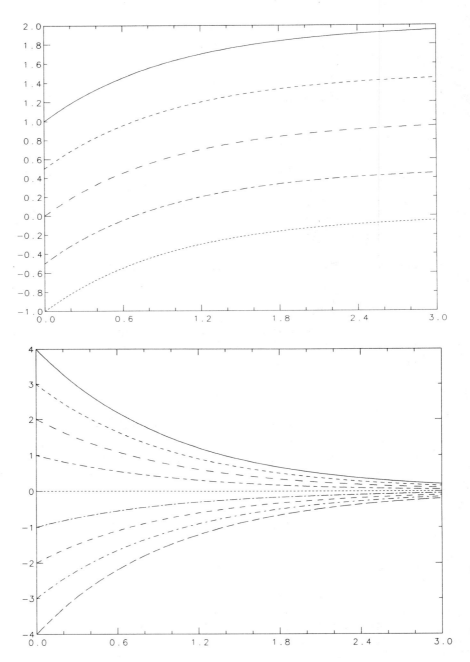

Figure 8.1   Solution Families

The independent variable t is often time, and then the ODE can be thought of as describing the evolution of a physical system. The starting state of such a system is usually known, and accounts for the term **initial value problem**. If t_0 and A are given numbers, this is written

$$y' = f(t, y), \qquad y(t_0) = A$$

and (t_0, A) is called the **initial point**.

If t represents time the solution $y(t)$ is desired on a fixed interval, $t_0 \leq t \leq T$. If t represents distance the solution might be needed for $T \leq t \leq t_0$. The theory and algorithms work in either case, but we will consistently describe only the former.

The distinction between the solution families of $y' = f(t)$ and $y' = f(t, y)$ also uncovers another difference between differential equations and quadrature. In estimating a definite integral, say from 0 to 1, we are free to ask for the evaluation of the integrand at points on [0,1] in any order which will be helpful. Indeed, the adaptive quadrature algorithm in Chapter 5 selects subintervals based on their error estimates and not their position within the original interval. This is correct for quadrature because the integrand can be explicitly evaluated anywhere. This is not possible for $y' = f(t, y)$; the value of the derivative of the solution at $t = 0.5$ cannot be evaluated until the solution is known there too, e.g., $y'(0.5) = f(0.5, y(0.5))$.

8.1.2 Higher Order Equations and Systems

Most physical problems do not lead to a single first-order ODE. Systems of first-order equations also occur directly if there is more than one unknown. For example, a simple model of a "predator-prey" problem composed of rabbits $r(t)$, and foxes $f(t)$ who eat only the rabbits, is governed by (see P8–5) the system of two first-order equations

$$r' = 2r - \alpha rf,$$
$$f' = -f + \alpha rf.$$

The interaction of the two species is proportional to the product of the values of each, with proportionality factor α. The natural physical problem is to simultaneously solve for $r(t)$ and $f(t)$ given their values at an initial time t_0.

A more general situation is to have n unknown functions $y_i(t)$, $i = 1, \ldots, n$ which are governed by the same number of differential equations.

$$y_1' = f_1(t, y_1, y_2, \ldots, y_n)$$
$$y_2' = f_2(t, y_1, y_2, \ldots, y_n)$$
$$\vdots$$
$$y_n' = f_n(t, y_1, y_2, \ldots, y_n).$$

In the predator-prey example, $n = 2, y_1(t) = r(t), y_2(t) = f(t)$, and

$$f_1(t, y_1, y_2) = 2y_1 - \alpha y_1 y_2, \qquad f_2(t, y_1, y_2) = -y_2 + \alpha y_1 y_2.$$

The initial conditions for a first-order system are

$$y_i(t_0) = A_i = \text{given}, \qquad i = 1, \ldots, n.$$

A standard notational simplification rewrites a system in vector form. Define

$$\mathbf{y}(t) = \begin{pmatrix} y_1(t) \\ y_2(t) \\ \vdots \\ y_n(t) \end{pmatrix}, \qquad \mathbf{f}(t, \mathbf{y}) = \begin{pmatrix} f_1(t, y_1, \ldots, y_n) \\ f_2(t, y_1, \ldots, y_n) \\ \vdots \\ f_n(t, y_1, \ldots, y_n) \end{pmatrix}, \qquad \mathbf{y}_0 = \begin{pmatrix} A_1 \\ A_2 \\ \vdots \\ A_n \end{pmatrix}.$$

Then the initial value problem for a system is

$$\mathbf{y}' = \mathbf{f}(t, \mathbf{y}), \qquad \mathbf{y}(t_0) = \mathbf{y}_0.$$

If the initial condition is not specified, the system will have an n-parameter family of solutions; each solution is a set of n functions.

Remarkably, most of the theory and the numerical algorithms for a single equation carry over to a system. Problem P8–10 gives you an opportunity to convince yourself by writing such a program for two equations using a simple algorithm. We will henceforth use a single equation as our model for most of the remainder of this chapter.

A **second-order** ODE is an equation of the form

$$u'' = f(t, u, u').$$

For example, the inverted pendulum (P8–13) is described by the nonlinear, second-order equation

$$\theta'' = \frac{3}{2L}(g - A\omega^2 \sin \omega t) \sin \theta.$$

Here $\theta(t)$ is the angular position of the pendulum at time t; A, ω and L are physical parameters and g is the acceleration of gravity. An initial value problem for this equation specifies both position θ, and velocity θ', at time $t = t_0$.

Second, higher order equations are usually converted to a system of first-order equations by a standard change of variable. Define two new unknowns $y_1(t)$ and $y_2(t)$ by

$$y_1(t) = \theta(t) \qquad \text{and} \qquad y_2(t) = \theta'(t).$$

Then the resulting system of two first-order equations for these new unknowns is

$$\begin{aligned} y_1' &= y_2 \\ y_2' &= \frac{3}{2L}(g - A\omega^2 \sin \omega t) \sin y_1, \end{aligned}$$

with initial condition

$$y_1(t_0) = \theta(t_0) = \text{given, and } y_2(t_0) = \theta'(t_0) = \text{given.}$$

For an n-th order equation

$$u^{(n)} = f(t, u, u', \ldots, u^{(n-1)}),$$

define the n new unknowns $y_1(t), y_2(t), \ldots, y_n(t)$ by

$$y_1(t) = u,$$
$$y_2(t) = u',$$
$$\vdots$$
$$y_n(t) = u^{(n-1)}.$$

Then the system of n first-order equations for these new unknowns is

$$y_1' = y_2,$$
$$y_2' = y_3,$$
$$\vdots$$
$$y_{n-1}' = y_n,$$
$$y_n' = f(t, y_1, y_2, \ldots, y_n).$$

For the pendulum $f_1(t, y_1, y_2) = y_2$ and $f_2(t, y_1, y_2) = 3(g - A\omega^2 \sin \omega t)(\sin y_1)/(2L)$. Note that in the system for foxes/rabbits the variable t does not appear explicitly in any of the f_i's whereas for the pendulum it does (in f_2). The former case is called **autonomous**, the latter **non-autonomous**. When a second-order ODE is being solved it is often insightful to plot one dependent variable against another, with the independent variable acting as a parameter. When the dependent variables are position and velocity this plot is called the **phase plane** since position and velocity are referred to as the two phases of the system. The curve sketching the path of the solution is called the trajectory in the phase plane. The term is also widely used when solving any system of ODEs to mean the curve obtained by plotting one dependent variable against another. For example if the dependent variables represent the x and y coordinates of a satellite in planar orbit (see problem P8–6), the phase plane is the orbit seen from above; thus it has much more physical significance than two separate plots of x versus t and y versus t.

Almost all library software for solving initial value problems solves a system of n first-order equations. If a user has a second or higher order equation, the equation must first be transformed to a system (by hand). Being able to perform this transformation is a *must* for anyone who wants to use most current software. The system resulting from a converted n-th order equation is much simpler than the general system, but few of the current popular routines take advantage of this.

The transformation of an n-th order equation into a system of n first-order equations has a benefit. While the original variable $y_1(t)$ is of primary interest, its derivatives, especially the first $y_1'(t)$, also have physical significance. These values are available at "no charge" as output of the system.

8.2 STABLE AND UNSTABLE EQUATIONS, NUMERICAL METHODS

Most initial value problems which are encountered in practice cannot be solved analytically. From the fifteenth century through the mid 1950's they were solved, approximately, by clever mechanical devices. That these held on for so long was perhaps due to the fact that their output was a function rather than a sequence of numbers. Numerical methods date back to Newton in the seventeenth century who proposed using them to describe the motion of a comet. In 1748 French scientists predicted the return of Halley's comet with six months of hand computation "from morning to night, sometimes even at meals." Hand computation continued to be used until the 1960's. Currently, the most popular technique for numerical integration works as follows: Rather than try to compute an approximate solution for all t in $[t_0, T]$ we content ourselves with finding approximate values of the solution at a sequence of points $t_0, t_1, \ldots, T$. If the true solution at t_k is $y(t_k)$ we denote the approximate solution by y_k. We also adopt the notation $y_k' \equiv f(t_k, y_k)$. Note that y_k' is *not* the same as $y'(t_k)$. (In Section 1.2 we also used y_k to denote the k-th solution component of a system. Both uses are standard but rarely give rise to confusion.)

The numerical estimate y_k is computed from knowledge of the solution at earlier times $y_{k-1}, \ldots, y_1, y_0$. If the formula to compute y_k depends explicitly only on y_{k-1} the method is called **single-step**. If y_k depends on both y_{k-1} and y_{k-2} the method is called **two-step**. *Multi-step* methods present obvious difficulties, for example at the onset of the integration.

The sequence of points t_k might be selected by the user or automatically by the program. In either case the point spacing $h_k \equiv t_{k+1} - t_k$ depends upon the problem, the numerical method, and the accuracy we wish to achieve in the results y_k. We know that $y_0 = y(t_0)$, and for practical problems, $y_1 \approx y(t_1)$ but $y_1 \neq y(t_1)$. Thus the point (t_1, y_1) does *not* lie on the solution curve which passes through the initial point (t_0, y_0). Rather, it lies on another solution family member. If we visualize following this new curve *back* to t_0, its value there will not be y_0. The point (t_1, y_1) can thus be thought of as the *exact* solution to the given differential equation but with a *different* initial condition. Figure 8.2 illustrates these concepts.

Now in going from t_1 to t_2 the best we can hope for is that y_2 will be on the curve through (t_1, y_1). Actually, this is unlikely. It is even more unlikely to expect to get back on the original curve through (t_0, y_0). So if we make an error on the first step, even if we make no further errors, the numbers $y_2, y_3, \ldots$ will also be in error. The size of these errors is controlled not only by the numerical method and its step sizes but also by the solution family.

Refer to the solution family ce^{-t} in Figure 8.1. Notice that members of the family move toward each other as t increases. In general, if the family of curves of the equation

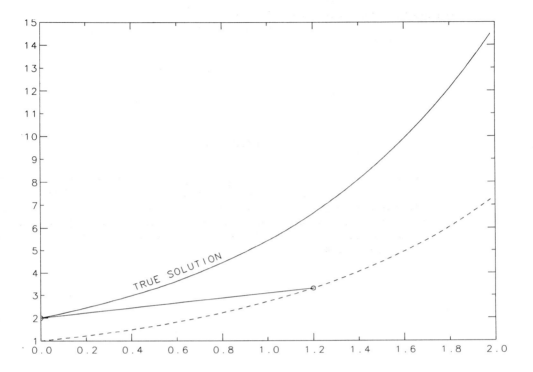

Figure 8.2 Stepping off the Solution Curve

$y' = f(t, y)$ separate as t moves away from the initial point the equation is called **unstable**. If the curves come together the equation is said to be **stable**. For a general equation either type of behavior can occur in different portions of the domain of interest. Systems of equations have a similar characterization although it is more difficult to visualize. If the equation is unstable, errors generally will tend to be magnified.[2] Stable equations reduce errors. What matters is not the particular solution but the entire solution family. The two problems, $y' = y, y(0) = 1$, and $y' = \exp(t), y(0) = 1$ have identical solutions but different solution families. Which do you suppose is more difficult to solve numerically?

At a fixed t the slopes of the members of a general solution family vary as a function of y. This variation is governed by f_y and this quantity measures the stability of the equation. The quantity f_y is called the **Jacobian** of the equation and denoted by J. Stable problems have negative values of J. You can see an example of this by looking at Figure 8.1.

To conclude, do not expect to accurately solve an unstable ODE with a numerical method.

---

[2] When we use the term error we mean *absolute error*. Relative error (error scaled by the magnitude of the solution) might not behave the same way. In practice this latter is the quantity we want the software to control (see for example Section 5.1) but absolute error is easier to treat and we concentrate on it.

Example 8.2   Stable and Unstable Equation.

The equation

$$y' = -2\alpha(t-1)y$$

has solution family

$$y = c\exp(-\alpha(t-1)^2).$$

Figure 8.3 shows a few members of this family for the particular case of $\alpha = 5$ on the interval $[0, 2]$. We see that the family members are at first separating and then converging. This corresponds to the fact that $J = -2\alpha(t-1)$, which is positive for $t < 1$ and negative for $t > 1$. These are regions of mild instability and stability respectively. A change in the initial value at $t = 0$ from 0.01 to 0.1 changes the value of the solution at $t = 1$ from about 1.0 to about 15.0. ■

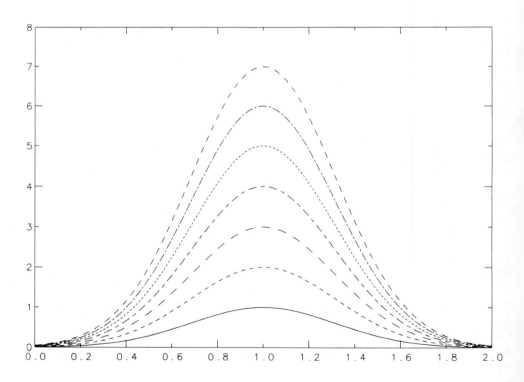

Figure 8.3   Equation with Stable and Unstable Regions

*8.2.1 Stability for Systems

The concepts of stability also apply to systems. The one new idea is related to the **Jacobian** matrix J, defined in Section 5 of Chapter 7. Recall that J is the matrix whose i, j component is $J_{i,j} = \partial f_i / \partial y_j$.

For the pendulum problem J is the 2×2 matrix

$$J = \begin{pmatrix} 0 & 1 \\ 3(g - A\omega^2 \sin t)(\cos y_1)/(2L) & 0 \end{pmatrix}.$$

For the rabbits/foxes problem J is also 2×2,

$$J = \begin{pmatrix} 2 - \alpha y_2 & -\alpha y_1 \\ \alpha y_2 & -1 + \alpha y_1 \end{pmatrix}.$$

For a system of n equations J is an $n \times n$ matrix. For a single equation, J is a 1×1 matrix whose sole element is the scalar partial derivative f_y. Thus J refers to the Jacobian even when we are discussing a single equation. As we will see, the Jacobian occurs in several contexts. Some programs even require that the user write a subroutine to compute its elements.

The stability of the system turns out to be related to the *eigenvalues* of J. Recall that these are n numbers $\lambda_1, \lambda_2, \ldots, \lambda_n$ which make the matrix $J - \lambda_i I$ singular, i.e., make the determinant of this matrix zero. For a single equation $\lambda_1 = J$. For the pendulum problem they are the roots of the quadratic

$$\lambda^2 - \frac{3}{2L}(g - A\omega^2 \sin t) \cos y_1 = 0.$$

In general, the eigenvalues are complex numbers whose values are not constant but depend on t and the solution $\mathbf{y}$. It can be shown that the eigenvalue with the largest real part governs stability. Positive real parts usually correspond to regions of separating (unstable) solutions. For most problems of interest the presence of negative real parts in all the eigenvalues corresponds to stable solutions. At a fixed t there can be eigenvalues with positive and negative real parts.

That stability is governed by the eigenvalues can be made more plausible by studying a linear problem, with the expectation that the results will provide a useful guide to the more general situation. Thus consider an object of mass $m = 1/2$ lying on a frictionless table and attached to an adjacent wall by means of a spring and a shock absorber (dashpot).[3]

The spring and dashpot are assumed to be linear in the sense that the resistive spring force is $-x/2$, and the dashpot force is $-dx'$, with x measuring the distance from the wall to the mass. If gravity is ignored the total force on the mass is mx'', which also equals the sum of the spring and dashpot forces

$$x'' = -2dx' - x.$$

---

[3] This example is extracted (with permission) from notes originally given by A. Hindmarsh in 1974.

Let us try to find a solution of the form $x(t) = \exp(\lambda t)$. Substituting this expression into the equation shows that such an $x(t)$ will be a solution if λ satisfies the quadratic equation

$$\lambda^2 + 2d\lambda + 1 = 0.$$

Hence

$$\lambda_i = -d \pm \sqrt{d^2 - 1}, \qquad i = 1, 2.$$

Using standard theory it is known that every solution of this differential equation can be formed from a linear combination of the form

$$x(t) = A \exp(\lambda_1 t) + B \exp(\lambda_2 t).$$

(The physical time constants of this problem are $1/\lambda_1$ and $1/\lambda_2$.) Since λ_i is always negative every member of the solution family of this equation is the sum of two decaying exponentials and any two different solution curves will come together, i.e. the equation is stable in the sense of this section. Our experience tells us that the problem is physically stable, i.e., the movement of the mass will diminish with time, and we expect that the ODE will have the same property.

Finally, if we write the second order ODE as a a system of two first order equations,

$$y_1' = y_2 \qquad\qquad J = \begin{pmatrix} 0 & 1 \\ -1 & -2d \end{pmatrix}$$
$$y_2' = -2dy_2 - y_1,$$

then the 2×2 Jacobian matrix J has eigenvalues λ_1 and λ_2.

Now consider what will happen if the dashpot is stiff ($d \gg 1$). Then the eigenvalues are

$$\lambda_i = d\left(-1 \pm (1 - 1/d^2)^{1/2}\right) \approx d\left(-1 \pm (1 - 1/(2d^2))\right), \qquad \lambda_1 \approx -2d, \qquad \lambda_2 \approx -\frac{1}{2d}.$$

The stiffness of the physical system is associated with the large negative eigenvalue.

An initial value problem for a system of ODEs is **stiff** if at least one eigenvalue of J has a large negative real part, and the solution is slowly varying on most of the integration interval of interest.

8.3   STIFF DIFFERENTIAL EQUATIONS

A single ODE is said to be super stable if $J \ll 0$. We expect that initial value problems for these equations will be easier to solve numerically than unstable ones, because getting off the correct solution curve should not be too devastating. We will see though, that this depends on the numerical method. Super stable equations are usually called **stiff**, although the term is more applicable to a system than to one equation. Practical problems are usually stable, stiff or only mildly unstable.

Stiffness is a complicated idea that we have not precisely defined. A working definition of a stiff problem is one in which the underlying physical process contains components operating on disparate time scales, or one in which the time scale of the process is small compared to the interval over which it is to be studied. (A better description was given in Section 2.1 and depends on the size of the eigenvalues of the Jacobian J.) Many practicing scientists equate stiff with "difficult," although this is misleading because the stiffer a system the better a good stiff solver likes it. The term really should refer to a combination of the problem, the interval on which we want the solution, the numerical method employed, and seemingly minor details of implementation. The reader who wishes to pursue this matter further can find many useful expository articles and references in the book edited by Aiken (1985) or the survey paper by Byrne and Hindmarsh (1987). Stiff problems occur in almost every field of science but are extremely common in chemical kinetics.

Example 8.3   A Stiff Equation.

Consider the problem $y' = -\alpha(y - \sin t) + \cos t$, $y(0) = 1$, which has the solution

$$y(t) = e^{-\alpha t} + \sin t.$$

If we want to solve this on the interval $0 \leq t \leq 1$, and α is large, $\alpha \approx 1000$, then $J = -1000$ and this is a stiff problem. (On another interval of integration, say $t \leq 0.002$ the problem would not be considered stiff.) For t near zero, y rapidly decreases almost to zero. From then on it is almost indistinguishable from $\sin t$. The solution near $t = 0$ is qualitatively different from that for larger values of t. The early portion of the solution is called the *transient* and the remaining portion is called *slowly varying*. Figure 8.4 illustrates the solution family for this problem. An important fact to note is that for points in the plane just above (or below) the slowly varying $\sin t$, the solution family member through that point is tending rapidly toward $\sin t$. In this figure we have replaced the constant 1000 by 20 to allow better visualization. This new problem is much less stiff. ∎

In practice stiff equations are often characterized by both transient and slowly varying behavior because they model physical phenomena which are occuring on several different time scales. Often the transient does not occur at the beginning of the integration, but only when some "driving term" changes abruptly. We will see an example of this later when we discuss a model for ozone in the atmosphere. For some applications the transient is crucial; for others only the long time (slowly varying) behavior matters. This distinction is relevant when thinking about solving the equation. We expect that it will be difficult to follow the transient and easy to track the slowly varying portion. When we consider applying some of our numerical methods to such a problem we will see that our expectations are not always born out in practice unless we are careful in the selection of methods.

8.4   EULER'S METHOD

Taylor's theorem is the fundamental tool used to derive most ODE methods. Of course, the goal of this chapter is not derivation but to provide a practical understanding of

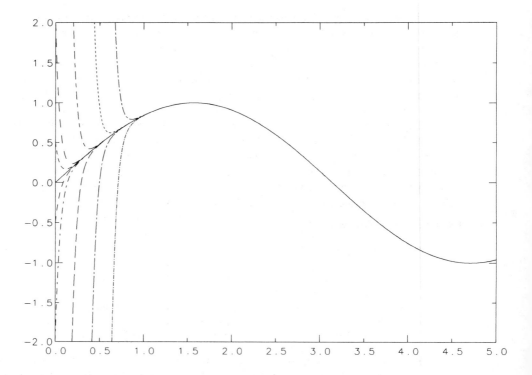

Figure 8.4 A Stiff Equation

how one particular program works. To that end we *will* derive a few simple methods including Euler's method, the backward Euler method, and the trapezoidal rule. These will be used to illustrate the ideas of order, accuracy, and stability.

Suppose that we wish to solve $y' = f(t, y)$, $y(t_0) = y_0$ on the interval $[t_0, T]$. We assume that the unknown $y(t)$ has at least two continuous derivatives. Then

$$y(t + h) = y(t) + hy'(t) + h^2 y''(\xi)/2 = y(t) + h f(t, y(t)) + h^2 y''(\xi)/2.$$

If we already have an approximate solution y_k at t_k, and have selected t_{k+1}, the **Euler's method** (EM), one step approximation, y_{k+1}^{EM} is obtained from above by dropping the remainder term

$$y_{k+1}^{EM} \equiv y_k + (t_{k+1} - t_k) f(t_k, y_k) = y_k + h_k y_k'.$$

Example 8.4   Use of Euler's Method.

Use Euler's method to solve $y' = -10(t - 1)y$, $y(0) = \exp(-5)$ on the interval $[0, 2]$.

We select *a priori* 21 points, equally spaced on the integration interval, $t_0 = 0$, $t_1 = 0.1, \ldots, t_{20} = 2.0$. The computation begins

$$y_0 = \exp(-5)$$

$$y_1 = y_0 + hy_0' = y_0 + (.1)[-10(t_0 - 1)y_0]$$

$$y_2 = y_1 + (.1)[-10(t_1 - 1)y_1].$$

The table below lists the values y_k obtained and the actual error at the points t_k. The exact solution is $y(t) = \exp(-5(t - 1)^2)$.

t_n	y_n	Error	t_n	y_n	Error
0.1	0.0134	0.004	1.1	0.452	0.499
0.2	0.0256	0.015	1.2	0.406	0.412
0.3	0.0461	0.040	1.3	0.325	0.312
0.4	0.0784	0.087	1.4	0.223	0.222
0.5	0.1254	0.161	1.5	0.137	0.150
0.6	0.1880	0.261	1.6	0.068	0.097
0.7	0.263	0.374	1.7	0.003	0.059
0.8	0.342	0.477	1.8	0.0082	0.033
0.9	0.411	0.541	1.9	0.0016	0.016
1.0	0.452	0.548	2.0	0.0002	0.007

We know from Example 8.2 that this equation is unstable on $[0, 1]$ and thus we are not surprised that the error at $t = 1$ is so large. If we repeat this calculation with a smaller step h, the error at 1.0 will decrease.

h	Error
0.1	0.548
0.05	0.375
0.025	0.229
0.0125	0.128
0.01	0.105
0.001	0.01154
0.0001	0.001165
0.00001	0.0001167
0.000001	0.00001167

Error at $t = 1$ with step size h   ■

8.5 ACCURACY AND STABILITY OF NUMERICAL METHODS

When we solve an initial value problem numerically with Euler's method the one parameter we have at our disposal is the step size h. Typical numerical integration programs, such as the one in this chapter, require the user to input a tolerance ϵ. This is then used to adjust the step size during the course of the integration. By taking h small enough we hope reduce the error to below ϵ. To do this we must have a way of assessing the accuracy of our computations. This section explains how this is done.

We return to the Taylor series and set $t = t_k$, and $h = h_k$,

$$y(t_{k+1}) = y(t_k) + h_k f(t_k, y(t_k)) + h_k^2 y''(\xi_k)/2, \quad t_k < \xi_k < t_{k+1}.$$

Normally $y_k \neq y(t_k)$. The sole exception is at t_0 where $y_0 = y(t_0)$. If we subtract Euler's method from this we get

$$y(t_{k+1}) - y_{k+1}^{EM} = y(t_k) - y_k^{EM} + h_k[f(t_k, y(t_k)) - f(t_k, y_k^{EM})] + h_k^2 y''(\xi_k)/2.$$

The left hand side is the error at the point t_{k+1}. This is the **global** error, and it is the quantity which we are interested in keeping under control. The global error is the difference between the computed and theoretical solution. We see that this arises from two sources.

(1) Local error. If at t_k we were fortunate and had $y_k^{EM} = y(t_k)$ the two differences on the right hand side would be zero. The global error is not zero though. The term remaining is called the **local**, or **truncation** error. It is exactly the Taylor series remainder $y''(\xi)h^2/2$. The local error is the error in going from t_k to t_{k+1} when all the data at t_k (and to the left) are exact; in that sense it is artificial.

(2) Propagation error. The second term on the right hand side above can be rewritten using the Mean Value Theorem as

$$h_k[f(t_k, y(t_k)) - f(t_k, y_k^{EM})] = h_k J(\xi)(y(t_k) - y_k^{EM}).$$

The notation $J(\xi)$ means that the Jacobian is evaluated at a specific, but unknown, point. So (1) and (2) can be summarized as

$$\text{Global error at } t_{k+1} = (1 + h_k J)(\text{Global error at } t_k) + \text{Local error at } t_{k+1}.$$

The global error at t_k, $y(t_k) - y_k^{EM}$, is multiplied by $(1 + h_k J)$, which is called the **amplification factor**. We see that the Jacobian J which occurs in studying the stability of the ODE, also occurs, via the amplification factor in studying the behavior of the numerical scheme. As long as $|1 + h_k J| < 1$ errors in Euler's method do not grow, whereas if this term is greater than one they do. In the latter case Euler's method is said to be **unstable**. Thus the term *stability* can occur in two contexts, the equation and the numerical method.

The inequality $|1 + hJ| < 1$ is equivalent to $-2 < hJ < 0$, and this is called the **stability interval** for Euler's method. If the ODE is unstable these inequalities can never be satisfied and instability in EM always occurs. If the ODE is stable ($J < 0$) the inequalities are equivalent to $h < -2/J$. Thus even for a stable problem EM will be unstable unless the step size h is restricted to be less than $|2/J|$.[4]

---

[4] If these ideas are confusing, you are in good company. Until the early 1950's most scientists did not think much about stability. Indeed, in 1953, W. E. Milne analyzed a then-popular class of integration methods and showed that one of them was unstable for *any* positive value of h.

Example 8.5   Stability for Euler's Method.

Consider again the problem of Examples 8.2 and 8.4. For $t < 1$ the equation is unstable and this leads to error growth using EM because $|1 + hJ| > 1$ for any value of h. For $t > 1$ the equation is stable. But EM will become unstable unless $h < 1/[\alpha(t - 1)]$ even though the equation is stable. For $\alpha = 5$, and $h = .1$ in Example 8.4, we should expect to see error growth once $t > 3$. Figure 8.5a shows a plot of the points (t_n, y_n) joined together with straight lines. The instability takes a while to develop but from $t = 4$ on grows rapidly. For this problem the amplification factor is negative and hence the global error changes sign at each successive integration point. Output that looks similar, i.e., a growing oscillation (whatever its original source), is almost always the result of an instability in the numerical method. Readers should be alert for this and never associate the values with those of the physical process being studied. ∎

It is important to realize that the numerical method can be unstable even when the ODE is stable. In Example 8.4 we showed a problem which has a slowly varying portion for almost all $t > 0$. If $\alpha = 1000$, the step size condition for EM is $h < 10^{-3}$. If we are interested in the solution during the transient portion this small step size makes physical sense, but during the slowly varying portion it does not. In the first case we say that the numerical method has a step size determined by accuracy considerations, in the second, by stability. A view of the difficulties in using EM on a stiff problem can be seen by examining Figure 8.5b. The further off the solution we go, the more the computed slope differs from its correct value. EM is rarely used for the solution of stiff problems, unless we intend to focus entirely on the transient portion.

The step size h appears both in the local error $h^2 y''$ and in the amplification factor $1 + hJ$. By taking h small enough we can make the local error as small as we like, and for a stable problem make the amplification less than one. But neither y'' nor J are known explicitly. Most programs try to estimate y'' but ignore J. Sometimes this is described by saying that the program has "only local error control."

The error control process tries to make the step size roughly inversely proportional to the square root of the second derivative of the solution, and hence deals with the local error. But we have already explained that local error is only one part of the whole picture; the amplification factor is also significant in determining global error. Thus, what happens when we use such a step selection algorithm? For Examples 8.2 or 8.3 we know the exact solution and hence could compute $y''(t)$ directly; this is (approximately), when t is large) $t^2 \exp(-\alpha t^2)$ and $-\sin t$ respectively. The first is decreasing with t and the second is never more than 1. Thus a step size strategy based on these values would be continually increasing h in Example 8.2, and keeping it more or less constant in Example 8.3. Both contradict the stability requirement for EM and would result in a rapid and pronounced instability. But if we program a step selection algorithm, such as the one in Section 5.1 below, without knowing *a priori* the second derivative but only estimating it, we find that neither problem "blows up" nor produces monstrously large errors. Instead, h will be kept roughly proportional to $1/|J|$ for both problems and errors will be controlled. Note that this means a small step even during the slowly varying portion of the stiff problem. How does this come about? The answer, again, depends upon the solution family.

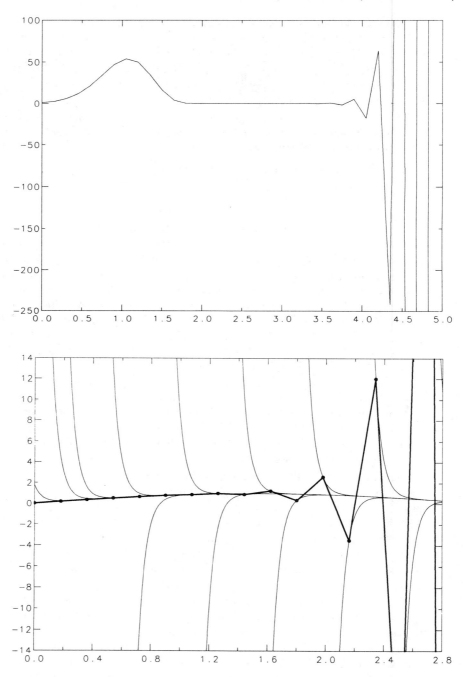

Figure 8.5 Euler's Method: (a) Becomes Unstable, (b) On A Stiff Problem

The estimate of the second derivative is based on differences of derivatives, that is y'_{n-1} and y'_n. If we are on the correct solution curve, then these derivatives are exact, and the second derivative estimate is small because all values of $y'(t)$ are small. But, $y'_{n-1} = f(t_{n-1}, y_{n-1})$ and $y'_n = f(t_n, y_n)$. Neither y_{n-1} nor y_n are on the same, or correct, solution curve. Thus these two numbers can be thought of as exact values of the derivative on different solutions. But for a stiff problem, changing solution curves even a little causes a large change in y' (because J is large). So the difference between them is likely to be large, forcing a small integration step. Example 8.2 acts stiff for large t and we see the same behavior.

Example 8.6   Variable Step EM on a Stiff Problem.

We consider the stiff problem

$$y' = -500(y - \sin t) + \cos t, \qquad 0.5 \le t \le 1.$$

Beginning with the exact initial condition at $t = 0.5$, we attempt to integrate to $t = 1.0$ using EM along with the simple step selection strategy described in Section 5.1 below. The initial step is 0.05 and the input error tolerance is $\epsilon = .001$. If the step from t to $t + h$ is unsuccessful, e.g., the error estimate is too large, we try again with $h/2$. The computation is easy to program. We find that 669 evaluations of $f(t, y)$ are required to reach $t = 1$ and that the relative error in the final estimate is 0.005. Thus the average step size is $1/(2 \cdot 669)$, about 0.0007, i.e., using fixed step EM with $h = .0007$ requires about the same number of f evaluations. This is much smaller than intuition suggests to track such a slowly varying function and is due to the stiffness of the equation. ∎

The concept of stability of a numerical method is fundamental to our understanding. Fixed step size programs can exhibit instability as illustrated in Example 8.5. Variable step size programs rarely show the same unstable behavior, but they can be inefficient, as this example demonstrates.

Summarizing:

An ODE is stable or unstable independent of how it is to be solved. A numerical integration method can be stable or unstable for a particular problem depending on the stepsize.

*8.5.1 A Simple Step Size Control Strategy

Many programs which provide automatic integration are based upon the following general algorithm. Given an approximate solution y_n at t_n (as well as other information):

(1) Estimate the next step size h_n using known information about the current local error for EM. There are various ways to estimate y''. Here is one. Suppose we have computed y_{n-1}, y_n and consequently y'_{n-1}, y'_n, and wish to integrate forward to a new point t_{n+1} while at the same time keeping the local error less than a given ϵ. An estimate of y'' is

$$y'' \approx \frac{y'_n - y'_{n-1}}{t_n - t_{n-1}} \equiv y''_{n-1,n}.$$

In going from t_n to t_{n+1}, to force the local error to be less than ϵ, we pick $h = t_{n+1} - t_n$ so that

$$|y''|h^2/2 \approx |y''_{n-1,n}|h^2/2 \le \epsilon, \quad \text{or} \quad h \le \sqrt{\frac{2\epsilon}{|y''_{n-1,n}|}}.$$

As a practical matter h is usually selected less than this value, say about 0.9 of it, and also never allowed to be too large, even if $y''_{n-1,n}$ is small. Finally, most scientists want results which are accurate to a fixed number of digits, usually three to five, regardless of the absolute size of the solution. Thus relative error control is desired. For this we replace ϵ by $\epsilon|y_n|$, i.e.,

$$\text{pick} \quad h \le \min\left(0.9\sqrt{\frac{2\epsilon|y_n|}{|y''_{n-1,n}|}}, \quad h_{max}\right).$$

This is a typical step size selection strategy. It can be implemented to allow a step-size change on each integration step, but there is theoretical evidence to suggest that this is not a good idea. Most programs are designed to use several equal length steps before a change is permitted. Also, notice that there is no prescription given for computing the initial step $t_1 - t_0$. Further, if y_n is near zero the formula above will force h to be small. Most production programs have a *minimum* step size that they will allow.

(2) Take the step to obtain values of y_{n+1} at t_{n+1} using EM or some other integration formula.

(3) Using updated information at t_{n+1} verify that the step was not too large. If it appears to be okay then proceed; otherwise go to (1) and try again. The estimate of y'' uses information to the left of t_{n+1} in order to select t_{n+1}. Once the step has been determined we then use EM to compute y_{n+1} and from it y'_{n+1}. It is possible that something untoward happened between t_n and t_{n+1} which will make itself known in an unusual value for y'_{n+1}. In that case our step might have been too large. To guard against that we estimate y'' again but this time based on data at t_{n+1} and t_n. That is

$$y''_{n,n+1} = \frac{y'_{n+1} - y'_n}{t_{n+1} - t_n}$$

and check that $|y''_{n,n+1}|h^2/2$ is still small enough. If it is, we accept the step and attempt to integrate forward from t_{n+1}. If it is not, then y'' or y has changed enough that our step size strategy was unjustified and we must go back to t_n and use a smaller step.

We conclude this section with a reminder that most step size strategies are only partially based on a solid theoretical foundation. More often they are a combination of theory, experiment and intuition.

*8.5.2 Stability Analysis for Systems of ODEs

When we are dealing with a system it is possible to perform a stability analysis that is similar to the one in Section 5. The results are much the same: EM will become unstable on a stable problem unless $\|I + hJ\| < 1$, where I is the identity matrix and $\| \cdot \|$ is some matrix norm (see Section 7 of Chapter 3). This will be satisfied if $|1 + h\lambda| < 1$, where λ is the largest eigenvalue of J. For a single real equation this means that hJ has to be in the interval $(-2, 0)$; but eigenvalues are usually complex numbers, and then the inequality $|1 + h\lambda| < 1$ no longer describes an interval. Instead it forces $h\lambda$ to be no further from the point -1 than 1 unit (note that $|1 + h\lambda| = |h\lambda - (-1)|$). The set of such points satisfying this is the interior of a circle centered at $z = -1$ of radius 1. This circle is called the **stability region**; if $h\lambda$ is inside it then EM is stable, otherwise it is not. Notice that the stability region includes the interval $(-2, 0)$ so that the case of one real equation is covered too. The quantity $I - hJ$ is more properly called the amplification *matrix*, but we will continue to call it the amplification factor.

8.5.3 Historical Perspective: Euler

Leonhard Euler (1707–1783), born and educated in Basel, Switzerland, had Johann Bernoulli as one of his teachers. The latter eventually referred to him as the "most famous and wisest mathematician." He wrote and published prodigiously; almost 600 books and articles appeared during his lifetime. His impact was so great that, in mathematics at least, the eighteenth century can fairly be called the age of Euler. Euler contributed not only to mathematics, but to physics, astronomy, hydrodynamics, optics, and electricity and magnetism. In 1768 he proposed the method for solving initial value problems which we described in Section 4. He also introduced many of the present conventions of mathematical notation including the symbols e for the base of natural logarithims, i for $\sqrt{-1}$, Δy for finite differences, $\sum$ for sum, the use of the letter f and of parentheses for a function, and the modern names for trigonometric functions, among others.

Euler worked fourteen years at the Russian St. Petersburg Academy of Sciences. He then moved to the Berlin Academy where he worked until differences with King Frederick the Great caused him to return to Russia where he spent the remainder of his life. He never returned to Basel after leaving in 1727.

Brilliant and successful people are often called upon to perform a variety of tasks; Euler was no exception. He worked on problems of ship building, joined various technical committees, engaged in testing scales and fire pumps, managed the publication of calendars and geographical maps, and supervised the botanical gardens. His original plans of becoming a minister were abandoned early, but his interest in religion and humanities gave him a perspective to view the other activities of his life. Euler had a remarkable memory. He was blinded by illness in 1771 but almost half of his published works were produced after that date, and at age seventy he could recall precisely the lines printed at the top and bottom of each page of Vergil's *Aeneid* which he had memorized when he was young. After he died a massive monument was erected at his grave in the Lutheran cemetery of St. Petersburg (now Leningrad), but in 1956 his remains and the monument were transferred to the city's necropolis.

8.6 ORDER OF AN INTEGRATION METHOD

Euler's method is not the only one-step method. Other important examples will be found in Sections 8 and 19. For all of these the concepts of local and global error, order, and stability region are relevant and can be defined, see for example the book by Ortega and Poole (1981). For our use it is enough that it is usually possible to derive an expression of the form

$$y(t_{k+1}) - y_{k+1} = a_k \cdot (y(t_k) - y_k) + L_k,$$

which is analogous to the global error formula in Section 5. The amplification factor is a_k; this is $1 + hJ$ for Euler's method. L_k is called the **local error**; it is $h^2 y''/2$ for Euler's method. As before, the local error is the error at t_{k+1} if all the values used to compute y_{k+1} are exact. Using the O notation (see Section 4 of Chapter 2), we say that the local error for Euler's method is $O(h^2)$. More generally if the local error can be written as $O(h^{p+1})$ we say that the method is p-th **order**. (Don't confuse this with the order of the differential equation.) Euler's method is first-order. If the local error involves $p+1$ why isn't it called a $(p+1)$-st order method? Often applying a p-th order method to an equation will lead to a *global* error that is proportional to h^p if we use a sufficiently small constant step size h.

Today most programs do not use a constant h, but the idea of local error and order are still useful.

(1) They suggest that a higher-order method is likely to be more accurate than a lower-order one which uses the same h.

(2) Repeating a computation from t_0 to T with a new step $h' < h$ ought to reduce the error at T by a factor of about $(h'/h)^p$. This can be the basis for an effective error estimation and step size selection strategy.

Example 8.7   Euler's Method is First-Order.

Refer to the table in Example 8.4 which displays the error in EM at $t = 1$ with various step sizes h. It is clear that by the time $h \le 0.001$ each tenfold reduction in h results in a comparable tenfold reduction in the global error. ∎

8.7 **SUBROUTINE** SDRIV2

is the generic name of a suite of routines by Kahaner and Sutherland, described in the paper by Boisvert et al. (1984) for solving a variety of initial value problems for ordinary differential equations. Actually, there are three related packages, SDRIV, DDRIV, and CDRIV associated with real single precision, real double precision and complex-valued single precision problems. Here we present one subroutine, SDRIV2, from the first package. SDRIV2 solves problems of the form $y' = f(t, y)$ using the multi-value methods described later. It is specifically designed to solve stiff problems but also is appropriate for non-stiff problems. Within the package, SDRIV2 is intermediate in flexibility and complexity. For SDRIV2 the user's job is to write a main program, and a subroutine that evaluates the right hand sides of the equations. Here we call the subroutine F although any other name can be selected.

Some of the parameters of SDRIV2 such as WORK, LENW and the use of EX-TERNAL statements will be familiar from their analagous use in earlier chapters. Some other input values require particular attention.

MINT —Method flag. In Section 16 we discuss the different methods that are available. For routine problems we suggest a value of MINT=2, enabling SDRIV2 to integrate stiff as well as nonstiff problems.

EPS, EWT—Tolerance and problem zero. The first is the requested error tolerance for the integration. By setting EWT > 0.0 the user declares that it is unnecessary to accurately follow solution components while their magnitudes are below EWT. We call this "problem zero." Since asking SDRIV2 to track solution components when they are below physically meaningful levels can be expensive, it is always worthwhile for the user to give this value some thought. Setting EWT = 0.0 asks for a relative accuracy of EPS on all the solution components.

NROOT, G—Number of roots and root function. SDRIV2 requires the user to specify the interval [T, TOUT] on which the problem is to be solved, and it will return the solution values at TOUT. In some cases the solution is needed at a specific value of $t <$ TOUT that is not known in advance. In fact the determination of this point can be an important part of the problem. To this end the user can invoke a "root finding" option in SDRIV2, asking that the integration terminate before TOUT whenever a user written function that we call G changes sign. The user's name for this function appears as the last argument to SDRIV2. If this option is not used (NROOT = 0), we suggest that in your call to SDRIV2 the last argument be the name you have selected for the subroutine to evaluate the right hand sides of the ODE. We have called it F.

In most physical problems the equations involve parameters. When these are known in advance their values can be included in F by an assignment (=), DATA, or PARAMETER statement. If they vary from one problem to another within the same run their values must be communicated to F from the main program. It is always possible to utilize the

Fortran COMMON, but most packages provide some other mechanism. SDRIV2 deals with this as follows. Normally, an array Y is initialized by the main program to contain the N initial values of the system to be solved, and this array is passed by SDRIV2 to F. If Y is dimensioned larger than N then the routine that calls SDRIV2 can store parameters in Y beyond the N-th element and then these can be used within F. One disadvantage of this approach is that in F, the parameters have to be referred to as subscripted values of Y, for example Y(N+1), rather than by more easily remembered names.

To illustrate, consider the motion of a mass m that falls to earth from a height $H > 0$. An ODE giving the height as a function of time is

$$y'' = -32 + \frac{a(t)}{m}, \qquad 0 \le t \le H, \qquad y(0) = H, \ y'(0) = 0.$$

The function $a(t)$ represents the air resistance. For this example we will assume that $a(t) = -y'(t)$, i.e. that it is equal to velocity, and we take $H = 10$. First we convert the equation to a system with $y \equiv y_1$

$$\begin{array}{lll} y_1' = y_2 & y_1(0) = H & \\ y_2' = -32 - y_2/m & y_2(0) = 0 & \end{array} \qquad 0 \le t \le H.$$

The task is to determine the time at which the mass reaches the ground. We use the "root function" $g(t, y_1, y_2) = y_1$; the integration will stop when this function, which is initially 10.0, becomes zero.

```
C An example of the use of SDRIV2
C
C Notice that NROOT is set to 1
C
      PARAMETER (N=2, H=10.0, NROOT=1, MINT=2,
     *           LW=N*N+10*N+2*NROOT+204, LIW=23)
      REAL       Y(N+1),  W(LW), MASS
      INTEGER    IW(LIW)
      EXTERNAL   FSUB,GFUN
      DATA       MASS /0.125/
C
      EPS = 1.E-5
C Set initial point
      T = 0.
      TOUT = T
C Set for pure relative error
      EWT = 0.0
C Set initial conditions
      Y(1) = H
      Y(2) = 0.0
C Set parameter value
      Y(3) = MASS
      MS = 1
```

```
          WRITE (*,*) '    T,           Y(1),        Y(2),       MS '
C
      10 CALL SDRIV2
       *    (N,T,Y,FSUB,TOUT,MS,NROOT,EPS,EWT,MINT,W,LW,IW,LIW,GFUN)
          TOUT = TOUT+0.1
          IF (MS .EQ. 5) THEN
             WRITE (*,'(3F11.5,I4,A,F11.5)')
       *          T,Y(1),Y(2),MS, ' <- - Y=0 AT T= ', T
             STOP
          ELSE
             WRITE (*,'(3F11.5,I4)') T,Y(1),Y(2),MS
C Stop if any output code but 1 or 2.
             IF (MS .GT. 2) STOP
          END IF
          GOTO 10
          END
C
          SUBROUTINE FSUB (N,T,Y,YDOT)
C Routine for evaluating right hand sides of equations.
          REAL T, Y(*), YDOT(*), MASS, G
          DATA G/32.0/
C Retrieve parameter
          MASS = Y(3)
C
          YDOT(1) = Y(2)
          YDOT(2) = -(G+1.0/MASS*Y(2))
          RETURN
          END
C
          REAL FUNCTION GFUN(N,T,Y,IROOT)
C Routine for root finding.
C Integration will stop when GFUN changes sign.
          REAL Y(*)
          GFUN = Y(1)
          RETURN
          END
C
C      Typical output
C
C   T,           Y(1),        Y(2),        MS
C   0.00000   10.00000     0.00000     1
C   0.10000    9.87534    -2.20270     2
C               .
C               .
C               .
C   2.60000    0.10000    -4.00000     2   (Terminal velocity)
C   2.62500    0.00000    -4.00000     5 <- - Y=0 AT T=      2.62500
```

8.8 IMPLICIT METHODS

Recall from Section 5 that the stability interval for Euler's method was $(-2, 0)$. This is simply not large enough to permit stiff problems to be integrated with realistic step sizes. Many other methods have similar restrictions. To understand how this difficulty is avoided we must introduce **implicit** methods, which can have infinite stability regions.

An alternative derivation of EM formally integrates $y' = f(t, y)$,

$$y(t_{n+1}) = y(t_n) + \int_{t_n}^{t_{n+1}} f(t, y(t))\, dt,$$

and approximates the integral by $(t_{n+1} - t_n) f(t_n, y(t_n))$. Other approximate quadratures are possible. One way is to use the right endpoint, another is to use the average of both endpoints. These lead to the integration formulas

$$y_{n+1}^{BE} \equiv y_n + h_n f(t_{n+1}, y_{n+1}^{BE}), \quad \text{and} \quad y_{n+1}^{TR} \equiv y_n + \frac{h_n}{2}\left[f(t_n, y_n) + f(t_{n+1}, y_{n+1}^{TR})\right],$$

or equivalently,

$$y_{n+1}^{BE} \equiv y_n + h_n y_{n+1}', \quad \text{and} \quad y_{n+1}^{TR} \equiv y_n + \frac{h_n}{2}\left[y_n' + y_{n+1}'\right],$$

which are called the **backward Euler** method (BE) and **trapezoidal rule** (TR) respectively. A more geometric derivation can be obtained by writing the formula for the line $l(t)$ of slope m through the point $(t_n, y(t_n))$. This is

$$l(t) - y(t_n) = m(t - t_n).$$

Setting $t = t_{n+1}$, and writing y_{n+1} for $l(t_{n+1})$ we have

$$y_{n+1} = y(t_n) + mh.$$

Different m's will produce different y_{n+1}'s. EM uses $y'(t_n)$ for m, BE uses $y'(t_{n+1})$, and TR uses the average of these values. Figure 8.6 illustrates these ideas.

Example 8.8 Backward Euler.

Use BE to repeat Example 8.4.

We have

$$y_{n+1}^{BE} = y_n + h_n y_{n+1}'^{BE} = y_n + h_n[-10(t_{n+1} - 1)y_{n+1}^{BE}],$$

or

$$y_{n+1}^{BE} = \frac{y_n}{1 + 10 h_n(t_{n+1} - 1)}.$$

With the same step size, 0.1, the corresponding errors are given below.

t_n	Error	t_n	Error
0.1	0.050	1.1	15.93
0.2	0.337	1.2	13.25
0.3	1.123	1.3	10.18
0.4	2.642	1.4	7.28
0.5	5.328	1.5	4.87
0.6	8.909	1.6	3.06
0.7	12.731	1.7	1.81
0.8	15.899	1.8	1.01
0.9	17.617	1.9	0.54
1.0	17.570	2.0	0.27

Referring to Figure 8.3 we see that when $t \leq 1$, being above the correct solution is more serious than being below. By studying Figure 8.6 convince yourself that BE will place y_1 above $y(t_1)$ thus moving above the correct solution curve, but EM will be below it. Hence we expect BE to be less accurate than EM on this particular problem. These expectations are reflected in the errors listed for this example. ■

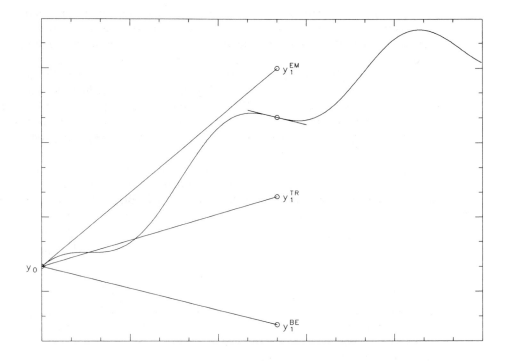

Figure 8.6   Graphical Interpretation of EM, BE and TR

A numerical method, such as BE or TR, where the quantity to be computed, y_{n+1}, appears inside the f term is called **implicit**; otherwise it is called **explicit**. We were able to solve for y_{n+1}^{BE} in Example 8.8. In the more usual situation, it is not possible to solve explicitly for y_{n+1}^{BE}. For example, suppose we wanted to solve $y' = \sin y$, $y(0) = 1$ using BE. To find the value of y_1^{BE} requires solving

$$y_1^{BE} = 1 + h \sin y_1^{BE} = y_0 + h \sin y_1^{BE}.$$

If we elected to use TR instead, the corresponding equation becomes

$$y_1^{TR} = y_0 + \frac{h}{2} \left(\sin y_1^{TR} + \sin y_0 \right).$$

In both cases this has the general form

$$y_{n+1} = y_n + \beta h f(t_{n+1}, y_{n+1}) + g,$$

where the last term g does not depend on y_{n+1}. This is a nonlinear equation; it is not possible to solve explicitly for y_{n+1} and techniques such as those described in Chapter 7 are required. Consequently, relative to EM, an implicit method always leads to a more complicated algorithm which is more expensive to advance from t_n to t_{n+1}. Why bother?

The answer is that implicit methods are often much more stable than explicit methods. By that we mean that the stability region of the method can be large, even infinite. In fact for both BE and TR the stability interval is $(-\infty, 0)$. This means that for these methods the amplification factor will be less than one for any positive h (for a stable problem), and thus errors will not be magnified. We call such methods **absolutely stable**. You should realize this does not imply there will not be errors, or that we may use any large step size we like. Like Euler's method, implicit methods have a global error which is the sum of two terms, the amplified old global error and the local error. For example, here is the formula for the global error in BE,

$$y_{n+1}^{BE} - y(t_{n+1}) = (1 - hJ)^{-1} \left([y_n - y(t_n)] + \frac{h^2}{2} y''(\xi_n) \right)$$

$$= (1 - hJ)^{-1}[y_n - y(t_n)] + O(h^2).$$

The amplification factor is $(1 - hJ)^{-1}$ which is less than one in magnitude as long as $J < 0$. Adjusting h also affects the local error. Stability only means that errors are not amplified. We sometimes say that for a stable method, step size selection will be governed entirely by considerations of local accuracy.

Examining the formula for the global error in the backward Euler method we see that it is a first order method just like Euler's method.

Example 8.9   BE is First-Order and Stable.

We list below the error at $t = 2$ and also at $t = 6$ using BE on the problem of Example 8.4, using step sizes h from 10^{-1} to 10^{-6}. These numbers support the conclusion that BE is first-order.

h	$|y_n - y(2)|$	$|y_n - y(6)|$
10^{-1}	$2.7 \cdot 10^{-1}$	$8 \cdot 10^{-25}$
10^{-2}	$1.8 \cdot 10^{-3}$	$9 \cdot 10^{-48}$
10^{-3}	$1.6 \cdot 10^{-4}$	$3 \cdot 10^{-54}$
10^{-4}	$1.6 \cdot 10^{-5}$	$1 \cdot 10^{-55}$
10^{-5}	$1.6 \cdot 10^{-6}$	$1 \cdot 10^{-56}$
10^{-6}	$1.6 \cdot 10^{-7}$	$1 \cdot 10^{-57}$

For the trapezoidal rule the global error can be shown to be

$$y_{n+1}^{TR} - y(t_{n+1}) = (1 - 0.5hJ)^{-1}(1 + 0.5hJ)(y_n - y(t_n)) + O(h^3).$$

This has amplification factor

$$(1 - 0.5hJ)^{-1}(1 + 0.5hJ)$$

which has magnitude less than one as long as $J < 0$. Thus TR is also absolutely stable. If $y_n = y(t_n)$ then $y_{n+1}^{TR} - y(t_{n+1}) = O(h^3)$, hence the trapezoidal rule is second order.

We explained in Section 6 that higher order is usually better than lower order *for small enough values of h*. For nonstiff problems this implies that the trapezoidal rule is a better choice than either backward Euler or (forward) Euler's method. You can confirm this yourself by working Problem P8–10. For a stiff problem EM cannot be used with reasonable step sizes because it is unstable. Both TR and BE are stable, i.e., errors are not amplified, but in this case the order of the method is not a good indicator of which will be more accurate. TR is of higher order, but this only matters for vanishingly small h, much too small to be of interest for a stiff problem. More relevant is that when J is large and negative the amplification factor of TR is nearly equal to -1. Thus errors in successive steps are changed in sign and only slightly reduced in magnitude. On the other hand the amplification factor for BE is much smaller than 1 and positive. Thus errors are rapidly damped and do not oscillate. We conclude that TR although stable and of second-order is not accurate enough for the integration of stiff equations.

Figure 8.7 illustrates BE when it is applied to a stiff problem. Note what helps BE, even though we are off the correct solution, the slope is computed at a later time, when its value is less. Based on this figure, and the comparable one for EM (Figure 8.5b), you can see why TR would be less useful for stiff problems.

The sad fact is that all well-known explicit methods have some stability step size restriction of the form $|hJ| <$ small constant, whereas many implicit methods do not. This accounts for the extensive interest in the latter. Do not be misled, though, into

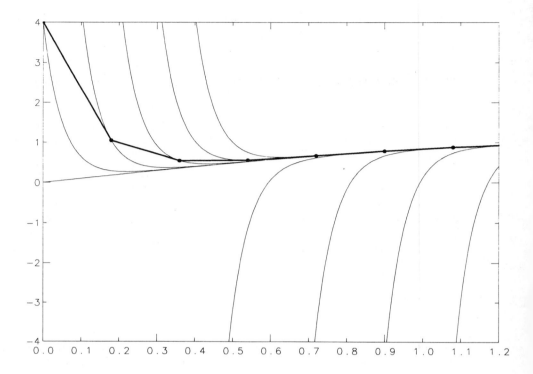

Figure 8.7   Backward Euler for a Stiff Problem

thinking that all implicit methods have good stability properties. Implicitness is necessary but definitely not sufficient. Conversely, a stable method is not always accurate, as our discussion above of BE and TR indicates.

Returning to the nonlinear equation which must be solved for either y_{n+1}^{BE}, or y_{n+1}^{TR}, and taking a cue from Chapter 7, we write

$$0 = F(y) \equiv y - y_n - \beta h f(t_{n+1}, y) - g.$$

As we have seen, such equations are usually solved by an iteration, such as bisection, Newton's method, etc. Select a starting estimate of y_{n+1}, call it $y_{n+1}^{(0)}$. Then repetitively compute estimates $y_{n+1}^{(1)}, \ldots, y_{n+1}^{(m)}$. We hope m need not be too large but this is obviously determined by the problem and the quality of $y_{n+1}^{(0)}$.

Of the general methods described in Chapter 7, Newton's method is one of the most effective for this equation. But our $F(y)$ is not the arbitrary function described there. It is special in the sense that the unknown y not only appears within f, but also in a term by itself. This suggests a different method, called **functional iteration**,

$$y_{n+1}^{(k+1)} = y_n + \beta h f(t_{n+1}, y_{n+1}^{(k)}) + g.$$

Both Newton's method and functional iteration have been analyzed in terms of how rapidly they converge, and the difficulties associated with their implementation. Some of these details are presented in Section 12.

*8.8.1 Stability of Backward Euler and Trapezoidal Rule for Systems of ODEs

All the stability results from the previous section carry over to a system of ODEs in the same way that the corresponding statements were generalized for Euler's method in Section 5.2. For example, for backward Euler to be stable we must have $|1 - h\lambda|^{-1} < 1$, with λ representing the largest complex eigenvalue of the Jacobian. As long as the complex number $h\lambda$ is more than one unit away from the point $z = 1$ the amplification factor will be less than one. Thus the stability region for BE is the entire complex plane except for a circle of radius one centered about $z = 1$. In particular, for a stable problem (eigenvalues in left half plane) the backward Euler method is stable.

The trapezoidal rule will be stable when its amplification factor has magnitude less than one, i.e.,

$$\left| \frac{1 + 0.5h\lambda}{1 - 0.5h\lambda} \right| < 1.$$

The stability region for TR is the left half plane. You can see this geometrically by interpreting the magnitude of the amplification factor as the ratio of two distances, in the numerator from the point $z = 0.5h\lambda$ to -1 and in the denominator from the same z to $+1$. As long as z is closer to -1 than to $+1$ the ratio is less than one.

8.9   MULTI-STEP METHODS

Recall that in Section 2 we explained that a multi-step (MS) method uses previously computed values of the solution (and perhaps its derivative) at several preceding points in order to advance from t_n to t_{n+1}. A typical three-step method uses solution and derivative values at t_n, t_{n-1}, and t_{n-2} to estimate the solution at t_{n+1} and has the form

$$y_{n+1} = \alpha_1 y_n + \alpha_2 y_{n-1} + \alpha_3 y_{n-2}$$
$$+ h[\beta_0 f(t_{n+1}, y_{n+1}) + \beta_1 f(t_n, y_n) + \beta_2 f(t_{n-1}, y_{n-1}) + \beta_3 f(t_{n-2}, y_{n-2})].$$

The constants α_i and β_i are determined in advance. If $\beta_0 = 0$ this method is explicit; a nonzero β_0 defines an implicit method. General k-step methods are analogously defined. The coefficients for two standard three-step methods are listed in Example 8.10 below.

An explicit multi-step method only requires 1 new evaluation, which is $f(t_n, y_n)$, to estimate the solution at t_{n+1} no matter how many steps are involved. An implicit method requires one evaluation for each iteration, but if convergence is rapid only a few evaluations are needed. Compared to one-step methods MS can be much more efficient for the same accuracy, or more accurate for the same amount of work.

Most implicit MS methods are implemented with either functional iteration or Newton's method. At each step an explicit MS formula is used to generate an initial guess and then an implicit one is iterated to get the final estimate at t_{n+1}. For reasons of accuracy as well as ease of implementation the two formulas are designed to be compatible, that is, use the same amount of data and be comparably accurate.

For years, MS methods were the standard tools used to solve initial value problems. Nevertheless, they have always had serious implementation difficulties.

(1) A three-step method requires knowledge of the solution at two preceding times as well as at the current time. At the beginning of the integration only t_0 and y_0 are available. One approach was to use a one-step method to generate y_1 and y_2 and then switch over to the three-step MS integrator.

(2) As written, MS can only be applied to march with a fixed step size h from t_0 to T; a one-step method can use a completely variable step size.

The inherent advantages of MS have led to programs dealing with (1) and (2) in various ways, some of them quite ad hoc. Almost every text on numerical methods has a section on this topic. In 1962 Nordsieck described another approach, now known as **Multi-value** (MV), which elegantly addresses each MS difficulty. Also important, was the related work of Krogh (1970). Today most production programs such as SDRIV implement Nordsieck's formulation or variants of it. However, it has been shown that although they look different, MS and MV are essentially the same process. See for example Osborne (1966) or Skeel (1979). An overview of some of this material is presented in Sections 14–19. If your goal is only to use SDRIV2 or related programs, these sections can be safely skipped.

*8.10 ORDER AND ERROR OF A MULTI-STEP METHOD

Recall the remarks in Section 6 about the order, stability and error of a one-step method. The same terms can be defined for multi-step methods. A multi-step method is p-th order if it produces exact results when applied to a problem whose solution is a p-th degree polynomial. As you might expect, using more steps allows for higher order. If a p-th order method is used on a problem with a nonpolynomial solution there will be an error. Even if all the preceding values of the solution are exact, the multi-step method will introduce some error, called the local error in analogy with one-step methods. For a p-th order method, the local error will be

$$L_k = ch^{p+1}y^{(p+1)}(\xi_k) = O(h^{p+1}).$$

Of course, the global error results from the local error and possible amplification of preceding errors.

The definition of order also provides a way to determine the coefficients. For example, when $y(t) \equiv 1$, e.g., when we integrate $y' = 0$, with $y(0) = 1$, the error will be zero for a p-th order method ($p > 0$). For the three-step method of Section 9 this implies

$$\alpha_1 + \alpha_2 + \alpha_3 = 1.$$

Similarly when $y'(t) = (t^k)' = kt^{k-1}$, $k = 1, 2, \ldots, p - 1$, the error should also be zero. This leads to a set of p linear equations for the α's and β's, and if we select p equal to the number of these constants we can hope to solve this system uniquely. Often some of the constants are prescribed *a priori*, but we still solve for the remaining ones.

Example 8.10   Derivation of Two MS Methods.

We will derive the highest order three-step method a) when $\alpha_2 = \alpha_3 = 0$, and b) when $\beta_1 = \beta_2 = \beta_3 = 0$.

For a) there are five unknown constants. These are determined by requiring that the multi-step method integrate five equations, $y' = kt^{k-1}, k = 0, 1, \ldots, 4$. With $n = 0$ we get the following set of five equations

$$
\begin{aligned}
k = 0, \quad & y(t) = 1, \quad && 1 = \alpha_1 \\
k = 1, \quad & y(t) = t, \quad && 1 = \beta_0 + \beta_1 + \beta_2 + \beta_3 \\
k = 2, \quad & y(t) = t^2, \quad && 1 = 2[\beta_0 + 0 - \beta_2 - 2\beta_3] \\
k = 3, \quad & y(t) = t^3, \quad && 1 = 3[\beta_0 + 0 + \beta_2 + 4\beta_3] \\
k = 4, \quad & y(t) = t^4, \quad && 1 = 4[\beta_0 + 0 - \beta_2 - 8\beta_3]
\end{aligned}
$$

which has the solution

$$\alpha_1 = 1, \qquad \beta_0 = \frac{9}{24}, \qquad \beta_1 = \frac{19}{24}, \qquad \beta_2 = -\frac{5}{24}, \qquad \beta_3 = \frac{1}{24}.$$

This becomes an implicit fourth-order method. In a similar manner, if we also set $\beta_0 = 0$, we can find the constants to generate an explicit, third-order three-step method.

For b) there are only four unknowns and we can hope to generate an implicit third order 3-step method. The results are

$$\beta_0 = \frac{6}{11}, \qquad \alpha_1 = \frac{18}{11}, \qquad \alpha_2 = -\frac{9}{11}, \qquad \alpha_3 = \frac{2}{11}. \quad \blacksquare$$

Once we know the order of a MS method we can also derive some information about its local error. For example, for Example 8.10 a) above, we know the method is fourth-order. That means it will not be able to integrate exactly the equation $y' = 5t^4$. Using this fact, we can show that for this method

$$L_k = -\frac{19}{720}h^5 y^{(5)}(\xi_n).$$

8.11 STABILITY FOR MULTI-STEP METHODS

MS methods are exciting because they can be of arbitrarily high order. But, if we implement an explicit MS method we will discover that for many stable problems an instability will develop unless h is small. This shouldn't surprise us in light of the remarks in Section 8 that explicit methods are not always stable. What is surprising is that the same behavior can be exhibited by *implicit* methods. In 1963, Dahlquist proved that no MS method can be absolutely stable (see Section 8) if it is above second order; even being implicit doesn't help.

While this is certainly disappointing, it's not quite the setback you might imagine. Luckily, most physical problems do not have time constants, i.e., values of J, that are completely arbitrary. Recall that these are what determine how stable the system is. Stiff problems have $J \ll 0$, non-stiff problems have more modest values. Thus it seems practical to look for methods which would be stable for certain classes of problems, although not for all. There has been a tremendous amount of research in this direction and one of the most fruitful results is the set of *stiffly-stable* formulas of Gear.[5] We display one of these formulas in Section 16. Gear's formulas are efficient for the integration of stiff problems as long as these do not have any oscillating solutions. For problems with highly oscillating components Gear's formulas can still be used, but with smaller step size.

*8.12 FUNCTIONAL ITERATION AND NEWTON'S METHOD FOR SOLVING THE IMPLICIT EQUATIONS

Implicit ODE methods (such as BE or TR) lead to an equation of the form

$$y_{n+1} = y_n + \beta h f(t_{n+1}, y_{n+1}) + g, \qquad \text{or} \qquad 0 = F(y) \equiv y - y_n - \beta h f(t_{n+1}, y) - g.$$

The two most popular algorithms for solving this are functional iteration,

$$y_{n+1}^{(k+1)} = y_n + \beta h f(t_{n+1}, y_{n+1}^{(k)}) + g,$$

and Newton's method,

$$y_{n+1}^{(k+1)} = y_{n+1}^{(k)} - [F'(y_{n+1}^{(k)})]^{-1} F(y_{n+1}^{(k)}), \qquad F'(y) = I - \beta h \frac{\partial f(t_{n+1}, y)}{\partial y}.$$

We now discuss the important characteristics of each of these iteration methods.

---

[5] Charles William (Bill) Gear, (1935–) is a Cambridge educated mathematician (now at University of Illinois) whose name became a household word to engineers when his 1971 book not only described a new class of methods for solving stiff equations but included a program which implemented them too. Most of the stiff solvers currently used in production applications (including SDRIV) are traceable to that one. Almost overnight, a previously intractable set of problems could be dealt with on a routine basis. Gear received so much attention from this work that it obscured his other important contributions to automatic problem solving systems, simulation, and network analysis. In 1986 though, the scientific computing community has recognized his broader achievements by electing him president of its most important organization, the Society for Industrial and Applied Mathematics (SIAM).

*8.12.1 Functional Iteration

Each iteration requires one evaluation of f. Early strategies continued to iterate until successive iterates agreed to within machine accuracy at which time convergence was said to have occurred. But even after complete convergence ($m \approx \infty$) the converged result is y_{n+1} not $y(t_{n+1})$. Since y_{n+1} is not expected to be the exact solution there is no point in computing it more accurately than necessary. For nonstiff problems, the current approach is to perform a fixed number of iterations, usually only one or two, and stop regardless of how close this is to convergence. The estimate $y_{n+1}^{(0)}$ is called a **prediction** (P), the evaluation of f is denoted by (E), and the computation of $y_{n+1}^{(1)}$ a **correction** (C). By convention after the last correction, an evaluation is performed to get an estimate of $y'(t_{n+1}) \approx f(t_{n+1}, y_{n+1}^{(m)})$. Thus a single iteration can be described by the initials PECE. Two iterations would be represented as PECECE, etc. The general technique is usually called **predictor-corrector**.

Example 8.11   BE with Functional Iteration.

Use EM to obtain a prediction for each iteration, e.g., $y_{n+1}^{(0)} \equiv y_n + h f(t_n, y_n)$ and then apply BE with functional iteration on

$$y' = \sin(y), \qquad y(0) = 1, \qquad 0 \le t \le 2.$$

For this equation BE is

$$y_{n+1} = y_n + h \sin y_{n+1}.$$

We select a step size of $h = 0.5$. Below we list the prediction and corrections at each time step. We continue to correct until

$$|y_{n+1}^{(k+1)} - y_{n+1}^{(k)}| \le 10^{-6}|y_{n+1}^{(k+1)}|.$$

The first figure is the prediction and this is followed by all the functional iteration corrections which occur until convergence. ■

Prediction	Iterates	Prediction	Iterates
$t = 0.0$	1.420735492	$t = 0.5$	1.997402250
	1.494380992		1.953888825
	1.498540884		1.962457688
	1.498695355		1.960839202
	1.498700925		1.961147505
	1.498701126		1.961088870
			1.961100025
			1.961097903
			1.961098306

Prediction	Iterates	Prediction	Iterates
$t = 1.0$	2.423495411	$t = 1.5$	2.689656949
	2.290074794		2.543731664
	2.337238957		2.606815696
	2.321289573		2.580202380
	2.326774589		2.591559605
	2.324898708		2.586735507
	2.325541495		2.588788747
	2.325321383		2.587915590
	2.325396774		2.588287043
	2.325370953		2.588129046
	2.325379797		2.588196254
	2.325376768		2.588167666
	2.325377805		2.588179827
			2.588174654
			2.588176854

Predictor-corrector methods were described in 1953 by Milne and have been in wide use since the 1960's.[6] Convergence to completion has been replaced by a fixed number of PECE iterations since the mid 1970's. The latter are effective for solving most standard problems but suffer from one major shortcoming. A fixed iteration PECE method is an explicit method! When we fix the number of iterations m in advance, the best estimate of y_{n+1}, given by $y_{n+1}^{(m)}$, can be written as an explicit (although complicated) formula in terms of known quantities. For example, the one iteration PECE composed of EM and BE gives

$$y_{n+1} \approx y_{n+1}^{(1)} = y_n + h f(t_{n+1}, y_n + h f(t_n, y_n)).$$

The absolute stability of BE does *not* apply to $y_{n+1}^{(1)}$. All fixed iteration PECE methods are governed by the same sort of stability condition, that

$$|hJ| < \text{constant}$$

---

[6] An early popularizer of predictor-corrector methods was Richard W. Hamming (1915–), a mathematician who combined a quick creative mind with often irreverent and unorthodox views. After contributing to the Manhattan Project in Los Alamos during 1944–45, he spent most of his career at Bell Telephone Laboratories, eventually heading the numerical methods and computer science research departments. He made numerous contributions to problems important to the communications industry, including the Hamming window for digital filtering, the Hamming code for error detecting/correcting in data transmission, and the Hamming predictor-corrector method He was an active proponent of trigonometric approximations for numerical problems. The preface of his 1962 text "Numerical Methods for Scientists and Engineers" forthrightly sums up his philosophy that "The purpose of computing is insight, not numbers." Well before 1970 he was astonishing big-time computer users at the national labs by claiming that by the 1980's one of the most significant uses of computers would be to entertain people in their homes. We'll have to wait though, to see if his prediction of artwork, computer controlled to respond to environmental stimuli such as weather or time of day, will come to pass.

and *should not* be used for stiff problems. For stiff problems a fixed number of iterations will not suffice. The iteration scheme must continue to convergence, or at least until we are close enough to get the full benefit of an implicit method.

We know, however, that iterative schemes do not always converge, and unfortunately, for functional iteration that is often the case. In fact it can be shown that unless

$$|\beta h J| < 1$$

functional iteration will not converge *no matter how good our prediction*. As β depends on the method but usually is a number near one (for BE, $\beta=1$), we conclude that the functional iteration will converge if $|hJ|$ is less than a small constant and diverge otherwise. For stiff problems this would excessively restrict the step size h. In fact it was this condition which we obtained in our discussion of stability for explicit methods, and this step size restriction was the motivation for our interest in implicit ones. But when using an implicit method we have to solve some nonlinear equations. To do that, functional iteration forces the same restriction on the step size. The solution is to pick another iteration scheme.

*8.12.2 Newton's method

Each Newton iteration to solve $F(y) = 0$ requires solving a system of linear equations with matrix $F'(y)$. Of course Newton's method doesn't always converge either. However, it can be shown that given any step size h, Newton's method will converge if the initial prediction $y_{n+1}^{(0)}$ is good enough. Compare this with functional iteration where convergence has nothing to do with the prediction and depends only on the step size. The step size obviously affects the accuracy of the prediction. For functional iteration it also determines if the iterations will converge, but for Newton's method it does not. Intuitively, we want the step size to be used to control errors, not to be necessary to force other computations to work properly. We can select h to get an accurate prediction and usually only a few Newton iterations are necessary for convergence, although the number is not fixed in advance. As a result, Newton's method or some variant of it is always used as the iteration scheme for implicit problems. This requires much more work. We have discussed the role the Jacobian plays but we never actually needed its values for any numerical scheme. But to implement Newton's method the matrix of partial derivatives is definitely required. It must either be provided by the user who will have to differentiate the system on paper, or must be approximated within the program. (Some new programs attempt to analyze the user's Fortran subroutine for f, the right hand side of the system, and then generate an analytic Jacobian subroutine automatically. These ideas have not yet been incorporated into production software.) In either case storage must be allocated for this matrix. Solving the linear system takes time too, and this must be done as often as necessary until the iteration converges.

Note that $F'(y_{n+1}^{(k)})$ depends upon both h and the current iterate $y_{n+1}^{(k)}$. Thus it ought to be updated every iteration. Packages such as SDRIV do not automatically update F' unless it appears to be necessary. Experience has shown that this does not reduce reliability and can substantially improve performance. Clever implementation details notwithstanding, using an implict method method entails substantially more computation than using an explicit one. Nevertheless implicit methods are effective, and currently there does not seem to be a better alternative; almost all programs for solving stiff equations use implicit integrators.

Example 8.12   BE with Newton's Method.

We repeat Example 8.11 using Newton's method. Our preceding analysis allows us to determine that for this problem functional iteration will converge as long as

$$|hJ| = |h\cos y| < 1.$$

This will certainly be satisfied for $h < 1$, thus there is no need to use Newton's method. Nevertheless, for illustration we perform the computations anyway. The iteration is

$$y_{n+1}^{(k+1)} = y_{n+1}^{(k)} - \frac{y_{n+1}^{(k)} - y_n - h\sin y_{n+1}^{(k)}}{1 - h\cos y_{n+1}^{(k)}}.$$

Below we list the prediction and Newton correction at each time step from $t = 0$ to $t = 4.5$. At the conclusion the solution is given at $t = 5.0$. Notice that fewer iterations are required with Newton's method than with functional iteration (Example 8.11).

Prediction	Iterates	Prediction	Iterates
$t = 0.0$	1.420735492404	$t = 0.5$	1.997402267036
	1.500330663274		1.961348131480
	1.498701819848		1.961098260881
	1.498701133518		1.961098248754
$t = 1.0$	2.423495363990	$t = 1.5$	2.689656746824
	2.326570008884		2.589004843113
	2.325377690520		2.588176045381
	2.325377497789		2.588175982114
$t = 2.0$	2.850974466440	$t = 2.5$	2.904679600000
	2.770160735513		2.893007237193
	2.769813021941		2.892887564025
	2.769813014455		2.892887563431
$t = 3.0$	3.015962112407	$t = 3.5$	3.058182875049
	2.975573085881		3.030823764432
	2.975535219280		3.030812214168
	2.975535219240		3.030812214166
$t = 4.0$	3.086089209091	$t = 4.5$	3.104621053709
	3.067720100573		3.092336368097
	3.067716633938		3.092335335059
	3.067716633937		

∎

Summarizing:

(1) All explicit methods have a step size restriction $h \approx |J|^{-1}$.

(2) Implicit methods must be iterated to convergence else they have a similar restriction for stability.

(3) Implicit methods can be iterated by various techniques; functional iteration and Newton's method are the two most popular.

(4) Functional iteration has a step size restriction $h \approx |J|^{-1}$ for convergence and is unsuitable for stiff problems.

(5) Newton's method has no step size restriction for convergence but requires an accurate prediction.

*8.13 OZONE IN THE ATMOSPHERE—A STIFF SYSTEM

There is no "typical" stiff initial value problem. But several important numerical methods have been developed while studying problems in chemical kinetics. This section describes one such problem, that of modeling the amount of ozone in the atmosphere. High altitude ozone protects us from most of the sun's harmful ultraviolet radiation, and some scientists feel that the increase in fossil fuel use and associated pollution are causing permanent changes in its level. For us, the model illustrates the kind of stiff initial value problems which are routinely being solved by current software. An excellent description of this problem, as well as detailed remarks on the difficulties involved in solving it, can be found in the paper by Hindmarsh (1980).

When we combine chemicals, such as hydrogen and oxygen in the generation of water, we normally think that the reaction occurs instantaneously. Actually, of course, the chemicals react at different rates depending on certain specific properties, including the actual reactants, temperature, pressure, as well as what other chemicals are around. In a complex system the amount or concentration of any particular chemical may be increasing due to some reactions and decreasing due to others. Each individual reaction is governed by a "rate constant," denoted k_i, which is determined empirically. Sometimes the rate constants are functions of time and other variables too. An important problem in chemical kinetics is to describe the time evolution of the concentrations of a set of chemicals, given their initial concentrations and rate constants.

A bimolecular reaction has the form

$$A + B \xrightarrow{\;k\;} C + D,$$

meaning that A and B combine to form C and D at a rate k per unit time. Presumably the actual physical process also has other reactions in which these chemicals are involved. If N chemicals interact this will determine N ordinary differential equations regardless of the number of reactions each chemical participates in. Thus from the single reaction above we know that there will be at least four ODEs, one each for A, B, C, D. The reaction tells us that the equations for A and B must include "loss" terms, because the concentrations of these are being reduced by the reaction. Similarly the equations for C and D involve "production" or "gain." The loss *or* gain is defined as the product of the concentration of the two *losing* chemicals and the rate constant. If we let $[A]$ denote the concentration of A, etc., then this one reaction affects four equations as follows,

$$[A]' = \cdots - k[A][B] + \cdots$$
$$[B]' = \cdots - k[A][B] + \cdots$$
$$[C]' = \cdots + k[A][B] + \cdots$$
$$[D]' = \cdots + k[A][B] + \cdots$$

If there are no other reactions the "$\cdots$" do not appear. Thus the process of generating equations from reactions can be automated, and programs which model complicated systems can easily require the solution of several hundred equations. These equations are notoriously stiff, because the rate constants can differ by many orders of magnitude.

As an example, we study a simple model for ozone in the atmosphere. Assume that the earth's atmosphere is a closed system held at constant temperature and volume and consider the simultaneous interaction of three chemicals, free oxygen O, ozone O_3, and molecular oxygen O_2. One reaction mechanism for these chemicals is

$$O + O_2 \xrightarrow{k_1} O_3$$
$$O + O_3 \xrightarrow{k_2} 2O_2$$
$$O_2 \xrightarrow{k_3(t)} 2O$$
$$O_3 \xrightarrow{k_4(t)} O + O_2.$$

The notation $k_3(t)$ and $k_4(t)$ means that the rate constants change with time. This is because the last two reactions describe the effect of sunlight, which causes molecular oxygen and ozone to "photodissociate."

This model is built upon some questionable assumptions, in particular that the concentrations do not vary with altitude. Nevertheless by the process described above we can deduce the differential equations

$$[O]' = -k_1[O][O_2] - k_2[O][O_3] + 2k_3(t)[O_2] + k_4(t)[O_3],$$
$$[O_3]' = \quad k_1[O][O_2] - k_2[O][O_3] - \quad k_4(t)[O_3].$$

Since $[O_2]$ is many orders of magnitude larger than $[O]$ and $[O_3]$ we assume that it is essentially unaffected by the other two and hence is constant for all time. Thus we can ignore its differential equation. The rate constants k_1, k_2 are known to be

$$k_1 = 1.63 \cdot 10^{-16}, \qquad k_2 = 4.66 \cdot 10^{-16},$$

whereas the other two rate constants vary twice a day, and are modelled by

$$k_i(t) = \begin{cases} \exp(-c_i/\sin \omega t), & \sin \omega t > 0, \\ 0, & \sin \omega t \leq 0, \end{cases} \qquad i = 3, 4$$

with $\omega = \pi/43200$ second^{-1} ($= \pi/12$ hour^{-1}), $c_3 = 22.62, c_4 = 7.601$. The values of k_3 and k_4 rise rapidly beginning at dawn ($t = 0$), reach a peak at noon ($t = 6 \times 3600$ seconds), and decrease to zero at sunset ($t = 12 \times 3600$sec). The units of t are in seconds. Reasonable initial conditions are

$$[O](0) = 10^6 \text{ cm}^{-3}, \qquad [O_3](0) = 10^{12} \text{ cm}^{-3}, \qquad [O_2](0) = 3.7 \cdot 10^{16} \text{ cm}^{-3}.$$

When t is in an interval corresponding to daylight there are rapid changes in the solution. For example, $[O]$ goes from its initial value (10^6) to virtually zero ($< 10^{-30}$) by the time t reaches 60 seconds, increases to almost 10^8 around noon but then falls off again at night. At night, both $[O]$ and $[O_3]$ are essentially constant; the process begins again the next morning when the sun rises. Figure 8.8 plots $[O]$ for a day and a half. The initial drop occurs so fast that it cannot be seen on the plot. Notice that the vertical scale is logarithmic and that time is in seconds.

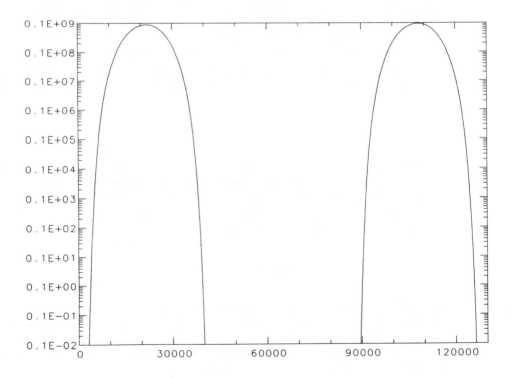

Figure 8.8   Atomic Oxygen in a Simple Ozone Model

This system is stiff during simulated night, for example when $12 \leq t/3600 \leq 24$ hours. When the integration is simulating daylight small integration steps are required for resolution and the equations could be integrated with conventional nonstiff methods.

Nevertheless the equations are still moderately stiff and it is more efficient to use a stiff solver. There are two eigenvalues of the system; $\lambda_1 \approx -6.03$ always, and λ_2 which reflects the rate of change of the system. $|\lambda_2|$ increases toward noon as the system accelerates and decreases toward sundown as it decelerates. At night λ_2 is computationally zero, but during the day it becomes as negative as $-1.6 \cdot 10^{-7}$. Although this problem represents a simple model it would have been impossible to solve before the advent of stiff solvers.

8.14 MULTI-VALUE METHODS

The key problem in implementing an efficient MS method is how to deal with a step h' which is not the same as the h used in the preceding few steps. The way to do this is to realize that the number y_{n+1} which is the output of the method at t_{n+1} (for example the three-step one described in Sections 9–10) can be viewed in two ways:

(1) as obtained via the MS method itself—this is how we have described it up to now,

(2) as obtained as the value at t_{n+1} of a certain polynomial $P(t)$.

The advantage of the second approach is that if we know $P(t)$ we can evaluate it anywhere we like, not only at t_{n+1}. For example, let us integrate $y' = f(t, y)$ between t_n and t,

$$y(t) = y(t_n) + \int_{t_n}^{t} f(\tau, y(\tau)) \, d\tau,$$

and approximate f by a cubic P_3. We can evaluate the integral of P_3 over *any* interval. The cubic we use is the unique one which interpolates to f at the four points, $t_{n-2}, t_{n-1}, t_n, t_{n+1}$. If we call the four ordinate values f_{n-2}, f_{n-1}, f_n, and f_{n+1} then using any of the representations given in Chapter 4 we can write P_3 explicitly. Further, we can compute its integral and this can be shown to be

$$\int_{t_n}^{t_{n+1}} P_3(t) \, dt = \frac{h}{24}[f_{n-2} - 5f_{n-1} + 19f_n + 9f_{n+1}],$$

which when y_n is added to it gives an estimate of y_{n+1} exactly equivalent to the implicit fourth-order one derived in Example 8.10. To repeat, the MS formula should be thought of as the *value* at t_{n+1} of a more general formula. All the other MS formulas can be obtained in a similar way and this point of view gives us the ability to evaluate an estimate of the solution at an entirely arbitrary point.

The final step in this approach is to think about how to *represent* $P_3(t)$. The MS approach focuses on the Lagrange representation, see Section 3 of Chapter 4. That is, the data that define $P_3(t)$ are the calculated values of y (and y') at a set of points. As the integration advances, the set of points in use is shifted to the right, somewhat like sliding a template. But we know that there are other representations which are mathematically equivalent although they appear different. The integral of P_3 from t_n to an arbitrary value of t could be worked out from any of its representations, but it might be easier to do this for some than for others. If we represented P_3 in the form

$$P_3(t) = a_0 + a_1(t - t_n) + a_2(t - t_n)^2 + a_3(t - t_n)^3$$

then evaluating its integral is immediate. We know a_0 trivially, because $P_3(t_n) = f_n = a_0$. Since $P_3(t) \approx f(t, y) \equiv y'$, the a_i are approximations to the derivatives,

$$a_i = P_3^{(i)}(t_n)/i! \approx y^{(i+1)}(t_n)/(i+1)!$$

Thus the a_i represent approximations to successive derivatives of the solution at the point t_n. This is the heart of the Multi-Value (MV) approach. Up to now we have been thinking of the data available to the integrator as past values of the solution (and its derivative), but MV treats as data the first few coefficients of a Taylor series centered about the current point. The goal of the integration is to generate the same Taylor coefficients centered around the next integration point. To explain this in more detail, we work our way through the derivation of a specific example.

Assume at the point t_n we know the exact solution $y(t_n)$ and its first three deriva-tives. We wish to march to $t_n + h \equiv t_{n+1}$, i.e., to compute $y(t_{n+1})$, $y'(t_{n+1})$, $y''(t_{n+1})$, and $y'''(t_{n+1})$. Write this symbolically as

$$\tilde{\mathbf{y}}(t_n) \equiv \begin{pmatrix} y(t_n) \\ hy'(t_n) \\ \frac{h^2}{2}y''(t_n) \\ \frac{h^3}{6}y'''(t_n) \end{pmatrix} \quad \rightarrow \quad \tilde{\mathbf{y}}(t_{n+1}) \equiv \begin{pmatrix} y(t_{n+1}) \\ hy'(t_{n+1}) \\ \frac{h^2}{2}y''(t_{n+1}) \\ \frac{h^3}{6}y'''(t_{n+1}) \end{pmatrix}.$$

We call a method which tries to compute $\tilde{\mathbf{y}}(t_{n+1})$ a **four-value** method. In practice the data are not exact, but are the results of earlier computation. We denote by $\tilde{\mathbf{y}}_n$ and $\tilde{\mathbf{y}}_{n+1}$ the corresponding history arrays with inexact values. The constants $h^2/2$ etc., are only included for scaling, but they always appear in implementations and we use the same notation.

Notice that the components of $\tilde{\mathbf{y}}(t_n)$ are not independent numbers; for example, once we know the first we can compute the second by evaluating the differential equation. For our approximations, $\tilde{\mathbf{y}}_n$ we will *require* the same property.

We proceed from t_n to t_{n+1} according to the following scheme. Expand the solution $y(t)$ in a Taylor series about the point t_n, differentiate this three times and evaluate these series at the point t_{n+1}. We get

$$y(t_{n+1}) = y(t_n) + hy'(t_n) + \frac{h^2}{2}y''(t_n) + \frac{h^3}{6}y'''(t_n) + \frac{h^4}{24}y''''(t_n) + \cdots,$$

$$hy'(t_{n+1}) = hy'(t_n) + h^2 y''(t_n) + \frac{h^3}{2}y'''(t_n) + 4\frac{h^4}{24}y''''(t_n) + \cdots,$$

$$\frac{h^2}{2}y''(t_{n+1}) = \frac{h^2}{2}y''(t_n) + \frac{h^3}{2}y'''(t_n) + 6\frac{h^4}{24}y''''(t_n) + \cdots,$$

$$\frac{h^3}{6}y'''(t_{n+1}) = \frac{h^3}{6}y'''(t_n) + 6\frac{h^4}{24}y''''(t_n) + \cdots.$$

If we drop the terms in these equations involving the fourth and higher derivatives, we can write this as

$$\tilde{\mathbf{y}}_{n+1}^P \equiv \begin{pmatrix} y_{n+1}^P \\ hy_{n+1}^{P\prime} \\ \frac{h^2}{2} y_{n+1}^{P\prime\prime} \\ \frac{h^3}{6} y_{n+1}^{P\prime\prime\prime} \end{pmatrix} \equiv B\tilde{\mathbf{y}}(t_n), \qquad B = \begin{pmatrix} 1 & 1 & 1 & 1 \\ 0 & 1 & 2 & 3 \\ 0 & 0 & 1 & 3 \\ 0 & 0 & 0 & 1 \end{pmatrix}.$$

We can think of $\tilde{\mathbf{y}}_{n+1}^P$ as a polynomial "prediction" of the solution at t_{n+1}. Note that the prediction is defined as the product $B\tilde{\mathbf{y}}(t_n)$. It does not make use of the differential equation in any way. Now we have

$$\tilde{\mathbf{y}}(t_{n+1}) = \tilde{\mathbf{y}}_{n+1}^P + \begin{pmatrix} \frac{h^4}{24} y^{(4)}(t_n) + \cdots \\ 4\frac{h^4}{24} y^{(4)}(t_n) + \cdots \\ 6\frac{h^4}{24} y^{(4)}(t_n) + \cdots \\ 6\frac{h^4}{24} y^{(4)}(t_n) + \cdots \end{pmatrix}.$$

These four equations are exact. We now drop this requirement by ignoring all terms in the right array which are not explicitly written out. Then the right hand side becomes

$$B\tilde{\mathbf{y}}(t_n) + \begin{pmatrix} 1 \\ 4 \\ 6 \\ 6 \end{pmatrix} y^{(4)}(t_n)h^4/24$$

which no longer equals $\tilde{\mathbf{y}}(t_{n+1})$ but is some approximation $\tilde{\mathbf{y}}_{n+1}$ to it. This suffers in two ways:

(1) The $y^{(4)}$ term is unknown. Recall that our data at t_n only include the first three derivatives.

(2) If we knew $y^{(4)}(t_n)$ exactly that would completely determine $\tilde{\mathbf{y}}_{n+1}$ without using the differential equation. We cannot expect much of such an approximation and in particular if we call y_{n+1} the first component of $\tilde{\mathbf{y}}_{n+1}$ we would have no reason to believe that $hf(t_{n+1}, y_{n+1})$ will agree with the second component. The two components will not be consistent.

Of these (2) is more serious; we have to use the differential equation. A fruitful way around this problem is to abandon $y^{(4)}(t_n)$ on the right and replace it by some approximation a_n, chosen to utilize the differential equation in some manner. The criteria we use is that the first two components of $\tilde{\mathbf{y}}_{n+1}$ must satisfy $y' = f(t, y)$. The right hand side of the second equation now reads

$$hy_{n+1}^{P\prime} + 4a_n$$

and if we define

$$4a_n \equiv hf(t_{n+1}, y_{n+1}) - hy^{P\prime}_{n+1}, \qquad \tilde{\mathbf{y}}_{n+1} \equiv \begin{pmatrix} y_{n+1} \\ hy'_{n+1} \\ \frac{h^2}{2}y''_{n+1} \\ \frac{h^3}{6}y'''_{n+1} \end{pmatrix},$$

then we will be assured that $y'_{n+1} = f(t_{n+1}, y_{n+1})$.

In this way the set of equations reads

$$\tilde{\mathbf{y}}_{n+1} = B\tilde{\mathbf{y}}_n + \mathbf{r}[hf(t_{n+1}, y_{n+1}) - hy^{P\prime}_{n+1}], \qquad \mathbf{r} = \begin{pmatrix} 1/4 \\ 1 \\ 3/2 \\ 3/2 \end{pmatrix}.$$

The entire process is thus defined by the matrix B and the vector $\mathbf{r}$.

We repeat that the elements in the $\tilde{\mathbf{y}}_{n+1}$ array are *approximations* to the values of the solution and its derivatives. The process has been set up so that the first two components satisfy $f(t_{n+1}, y_{n+1}) = y'_{n+1}$, but no such relations are associated with the last two components; the double and triple primes are suggestive and notational, but we would have to prove that these approximate y'' and y'''. Also note that a_n is precisely equal to the difference between the second component of the final result $\tilde{\mathbf{y}}_{n+1}$ and the second component of the prediction $\tilde{\mathbf{y}}^P_{n+1}$. This is an approximation to $h^4 y^{(4)}/24$; MV programs make use of this fact.

8.15   AN EXAMPLE OF MULTI-VALUE

Let us use MV to solve

$$y' = -10(t-1)y, \qquad y(0) = 1.$$

We will use a step size of $h = 0.1$. To begin we need

$$\tilde{\mathbf{y}}(0) = \begin{pmatrix} y(0) \\ hy'(0) \\ (h^2/2)y''(0) \\ (h^3/6)y'''(0) \end{pmatrix}.$$

The first component of this array is given as initial data; the second is obtained directly from the differential equation, $hy'(0) = y(0)$. The remaining components are not known except by differentiating the equation. Modern implementations get around this but for the example we assume that these numbers are given. We have

$$\tilde{\mathbf{y}}(0) = \begin{pmatrix} 1 \\ 1 \\ .45 \\ 7/60 \end{pmatrix}.$$

The first prediction is

$$\tilde{\mathbf{y}}_1^p = \begin{pmatrix} 2.56667 \\ 2.25 \\ 0.8 \\ 0.11667 \end{pmatrix},$$

and then

$$\tilde{\mathbf{y}}_1 = \begin{pmatrix} y_1 \\ .1y_1' \\ (.01/2)y_1'' \\ (.001/6)y_1''' \end{pmatrix} = \tilde{\mathbf{y}}_1^p + \mathbf{r}[0.9y_1 - 2.25].$$

There are four unknowns, but only one, y_1, appears on both sides of the system. Thus MV is implicit, but only in this variable. Once y_1 is known, the remaining components of $\tilde{\mathbf{y}}_1$ are directly computable. The first equation is

$$y_1 = 2.56667 + 0.25[0.9y_1 - 2.25].$$

If the original equation were nonlinear, $0.9y_1$ on the right would be replaced by $0.1f(0.1, y_1)$ and we would have to solve this nonlinear equation iteratively. Our present example is linear and we can solve for y_1 explicitly as

$$y_1 = \frac{1}{1 - 0.25 \cdot 0.9}[2.56667 - 0.25 \cdot 2.25] = 2.58603.$$

Then

$$\tilde{\mathbf{y}}_1 = \begin{pmatrix} 2.5860 \\ 2.3274 \\ 0.9161 \\ 0.2328 \end{pmatrix}.$$

Recall that $\tilde{\mathbf{y}}_1$ should approximate $\tilde{\mathbf{y}}(0.1)$ which is

$$\tilde{\mathbf{y}}(0.1) = \begin{pmatrix} 2.5857 \\ 2.3272 \\ 0.9179 \\ 0.1978 \end{pmatrix}.$$

The computation of $\tilde{\mathbf{y}}_2, \tilde{\mathbf{y}}_2, \dots, \tilde{\mathbf{y}}_{10} \approx \tilde{\mathbf{y}}(1.0)$, is analogous. The table below lists the computed and exact solution at $t = 0.1, \dots, 1.0$. We see that initially the error goes down, but soon begins to rise rapidly.

t	MV	Exact
0.1	2.58062	2.58571
0.2	6.05981	6.04965
0.3	12.78747	12.80710
0.4	24.66495	24.53253
0.5	42.02347	42.52108
0.6	69.06068	66.68633
0.7	83.93899	94.63241
0.8	171.92620	121.51042
0.9	−99.60609	141.17496
1.0	1323.69770	148.41316

Approximate and Exact Solution of $y' = -10(t - 1)y$, $y(0) = 1$ **Using MV**

Although the early MV estimates are acceptable they soon are significantly in error and even begin to oscillate near $t = 1$. An unsuspecting scientist might think this is because the interval $[0, 1]$ is unstable for this problem. However, repeating the computation with a smaller step size only makes things worse. As mentioned in Example 8.5 unphysical, growing oscillations in the computed solution which are exacerbated by reducing the step size almost always signify an unstable method. In the next section we will see that this is indeed the case.

8.16 OTHER MULTI-VALUE METHODS

The failure of MV to work on our simple example is disappointing, and at first there appears to be no flexibility in the choice of method except for the number of values we retain, e.g., the length of the vector $\tilde{\mathbf{y}}_n$. Let us consider this in more detail. If the true solution to the differential equation is a polynomial of degree three or less and if $\tilde{\mathbf{y}}_0 = \tilde{\mathbf{y}}(t_0)$, i.e., the starting values are exact, then all subsequent $\tilde{\mathbf{y}}_n$ are equal to $\tilde{\mathbf{y}}(t_n)$ for any $\mathbf{r}$. This suggests that a value of $\mathbf{r}$ other $\mathbf{r} = (1/4, 1, 3/2, 3/2)^T$ might be useful too. It is necessary to require that $r_2 = 1$ in order to maintain consistency of the first two components through the differential equation. A stability analysis can be done for MV methods. It can be shown that for the value of $\mathbf{r}$ above the method is *never* stable, and that the numerical results in the preceding section are typical. Fortunately, there are other MV methods with better properties. Two of the most famous are $\mathbf{r} = (3/8, 1, 3/4, 1/6)^T$, called Adams-Moulton MV, and $\mathbf{r} = (6/11, 1, 6/11, 1/11)^T$, called Gear MV. Adams-Moulton MV has a finite stability region and it is appropriate for the solution of nonstiff problems. Gear MV has an infinite stability region, although the method is not absolutely stable. It is suitable for the integration of both stiff and nonstiff problems. It is slightly less accurate than Adams-Moulton MV and is not particularly effective on problems with oscillating solutions. SDRIV2 gives users the opportunity to select either Gear's methods, MINT=2, or Adams-Moulton methods, MINT=1.

Example 8.13   Adams-Moulton MV.

Repeat the calculation of Section 15 using Adams-Moulton MV.

The first step, to $t = 0.1$, gives

$$\tilde{\mathbf{y}}_1 = \begin{pmatrix} 2.60063 \\ 2.34057 \\ 0.86792 \\ 0.13176 \end{pmatrix}.$$

The table below lists the relative error at $t = 1$ and is directly comparable to the table in the preceding section. In the present case, we see that initially there is some error growth as h decreases but as soon as h becomes small enough the error also begins to decrease. The largest magnitude of hJ is $10h$ and for $h \leq\approx 0.25$ the method is stable.

Error $\cdot e^5$	Step size h
0.026390	1.0000
0.093329	0.5000
0.190202	0.3333
0.094691	0.2500
0.048559	0.2000
0.004978	0.1000
0.000425	0.0500
0.000032	0.0250
0.000002	0.0125

Error at $t = 1$ Using Adams-Moulton MV on $y' = -10(t-1)y,\ y(0) = 1$ ■

Example 8.14   Gear MV.

Repeat the calculation in the preceding example using Gear's method. The first step, to $t = 0.1$, gives

$$\tilde{\mathbf{y}}_1 = \begin{pmatrix} 2.63095 \\ 2.36786 \\ 0.86429 \\ 0.12738 \end{pmatrix}.$$

The table below lists the relative error at $t = 1$ and is directly comparable to the table above. As we mentioned the method is stable but not as accurate as Adams-Moulton MV.

Error $\cdot e^5$	Step size h
0.028380	0.1000
0.005067	0.0500
0.000841	0.0250
0.000124	0.0125

Error at $t = 1$ Using Gear's method on $y' = -10(t-1)y,\ y(0) = 1$ ■

Summarizing:

(1) MV methods are determined by the array **r**.

(2) The most "accurate" MV method is not stable.

(3) Adams-Moulton MV methods are accurate and suitable for nonstiff problems.

(4) Gear's method is less accurate than Adams-Moulton for nonstiff problems, but works well on stiff problems.

8.17 RELATION OF MULTI-STEP AND MULTI-VALUE METHODS

MS methods are easy to think about but difficult to implement efficiently; MV methods are the reverse. In 1968, Skeel showed that these two are essentially equivalent. By equivalent we mean that the first component of $\tilde{\mathbf{y}}_{n+1}$, which is the number y_{n+1} approximating $y(t_{n+1})$ is identical to that which we would get using a MS method with appropriate α's and β's. This is an important result, because it says we can adopt whichever point of view is most appropriate. In practice we almost always implement a method using MV but often think about it as a MS which has been organized to carry the same information, but in a different form. We illustrate a few of the equivalent MS-MV methods below.

Example 8.15   Equivalent MS and MV Methods.

a) The Adams-Moulton four-value method, $r_1 = 3/8$, $r_3 = 3/4$, $r_4 = 1/6$ is equivalent to

$$y_{n+1} = y_n + \frac{h}{24}(9y'_{n+1} + 19y'_n - 5y'_{n-1} + y'_{n-2}),$$

called the three-step Adams-Moulton method, first seen in Example 8.10 and shown there to be fourth-order.

b) The four-value method $r_1 = 5/12$, $r_3 = 3/4$, $r_4 = 1/6$ is equivalent to

$$y_{n+1} = y_n + \frac{h}{12}(5y'_{n+1} + 8y'_n - y'_{n-1}),$$

which is called the two-step Adams-Moulton method.

c) The four-value method $r_1 = 0$, $r_3 = 3/4$, $r_4 = 1/6$ is explicit (because $r_1 = 0$) and is equivalent to

$$y_{n+1} = y_n + \frac{h}{12}(23y'_n - 16y'_{n-1} + 5y'_{n-2}).$$

Multi-step formulas with $\alpha_1 = 1$ and the other α's equal zero (see Section 9) are called Adams formulas; implicit ones are Adams-Moulton, explicit rules such as this one are called Adams-Bashforth.

d) The Gear four-value method $r_1 = 6/11$, $r_3 = 6/11$, $r_4 = 1/11$ is equivalent to the *three-step* Gear integrator.

$$y_{n+1} = \frac{1}{11}(18y_n - 9y_{n-1} + 2y_{n-2}) + h\frac{6}{11}y'_{n+1}.$$

Notice that for the last formula the coefficients of all the derivative terms are zero except for y'_{n+1} (this is a characteristic of these methods). If we isolate that term on one side of the equation we can interpret the result as a formula for approximating a derivative at the new point based on function values at the new, current and two past points. For this reason, Gear formulas are sometimes referred to as **backward differentiation** formulas. ∎

*8.18 ATTRACTIVE CHARACTERISTICS OF MULTI-VALUE: STEP SIZE AND ORDER CHANGING

A MS algorithm uses past values of the solution and its derivative. A MV algorithm uses current values of the solution and several of its derivatives, the first few coefficients of an approximate Taylor series. Taylor series are easy to manipulate, so it is not surprising that MV algorithms permit the integrator to easily modify the step size. This section discusses a few of the implementation details of such algorithms.

*8.18.1 Changing Step Size

Suppose that we have been integrating with step size h, and that at time t_n we decide to advance the solution to $t_n + h'$. That is, we wish to alter the step size from h to h'. MV allows this to be accomplished in a simple manner. The array $\tilde{\mathbf{y}}_n$ has already been calculated. The step size h is "built in" to its values. Before we begin the computation to march forward from t_n, we only need to multiply the components of $\tilde{\mathbf{y}}_n$ by the correct power of h'/h. For example, referring to Sections 15–16, if after reaching $t_1 = 0.1$ we elect to integrate to 0.15 rather than 0.20, i.e., to use a step size of 0.05 instead of 0.10 we simply compute

$$\text{new } \tilde{\mathbf{y}}_1 = \begin{pmatrix} 1 & 0 & 0 & 0 \\ 0 & 0.5 & 0 & 0 \\ 0 & 0 & 0.25 & 0 \\ 0 & 0 & 0 & 0.125 \end{pmatrix} \cdot \tilde{\mathbf{y}}_1 = \begin{pmatrix} 2.5860 \\ 1.1637 \\ 0.2129 \\ 0.0210 \end{pmatrix}.$$

The computation now proceeds from t_1 just as before, except using the new value of $\tilde{\mathbf{y}}_1$ and step size h'.

There is a small amount of extra overhead in changing the step size but as a practical matter, step changes with MV are as easy as with a one-step method. The same general technique is used to evaluate the solution at a point not equal to an integration output point. We know that the components of the array $\tilde{\mathbf{y}}_n$ are scaled approximations to the first terms in the Taylor series for $y(t)$ centered around t_n. So a good estimate of the solution at the point $t_n + h'$ is given by the dot product

$$\left(1, \frac{h'}{h}, \left(\frac{h'}{h}\right)^2, \left(\frac{h'}{h}\right)^3\right) \cdot \tilde{\mathbf{y}}_n.$$

We say that the solution has been *interpolated* to the new point. SDRIV and similar programs work in exactly this way. The user specifies a point at which output is needed, but SDRIV uses its own internal steps until just past the output point, and then interpolates back. In this case h' is negative but the procedure is exactly the same. Consequently, integration and output are uncoupled. In fact, if the solution is smooth enough the integrator can be far ahead of the output values.

*8.18.2 Error Estimation

We have described several techniques for estimating errors in numerical integration methods. Recall that we can only deal with local or per step error. Any method capitalizes on the fact that the local error is proportional to the product of a known power of the step size and a derivative of the solution,

$$\text{local error} = ch^{p+1}y^{(p+1)}(\xi).$$

The value of p is the order of the method, where p is the highest power of t which the numerical method can integrate exactly.

We've seen that MV and MS can produce identical results if we select the correct values of **r**. This is useful when finding the error constant for MS formulas. Thus, when $\mathbf{r} = (3/8, 1, 3/4, 1/6)^T$, from the remark following Example 8.10 the local error estimate is

$$y_{n+1} - y(t_{n+1}) = -\frac{19}{720}h^5 y^{(5)}(\xi).$$

If we had an estimate of $y^{(5)}$ we could develop a step size strategy. Such can be obtained by differencing two estimates for $y^{(4)}$, which are available to us. Recall from the end of Section 14 that a_n represented an approximation to $y^{(4)}(t_n)h^4/4!$ and could be computed by subtracting the second component of the prediction from the second component of the final estimate or equivalently, the amount by which the prediction is altered. Most MV programs add this as a fifth component to the $\tilde{\mathbf{y}}_{n+1}$ array which is updated along with the first four. The difference between two successive estimates is about

$$\frac{h^4 y_{n+1}^{(4)}}{4!} - \frac{h^4 y_n^{(4)}}{4!} \approx \frac{h^5 y^{(5)}}{4!} \equiv e_5,$$

so

$$|y_{n+1} - y(t_{n+1})| \approx \left| \frac{19}{720}4!e_5 \right|$$

which can be tested against the input quantity ϵ to see if the step is successful.

The step size to be selected for the next integration step, or the reduced step to be tried in case the above test fails, is αh, where α is chosen so that the local error ought to be less than ϵ. That is,

$$\epsilon \geq \left| \frac{19}{720}(\alpha h)^5 y^{(5)}(\xi) \right| \approx \alpha^5 \left| \frac{19}{720}4!e_5 \right|.$$

Conservatively, α is picked to be about 80% of this,

$$\alpha = \frac{1}{1.2}\left[\frac{\epsilon}{19/720 \cdot 4!e_5}\right]^{1/5}.$$

Also, based on experimentation and some analysis for programs such as SDRIV, it has been found that step size increases should not be allowed every step, even if apparently warranted; if we are using a p-th order method, step increases are allowed no more than once every $p + 1$ steps in order to allow estimates to "stabilize."

*8.18.3 Order Changing

The ultimate goal is to get from t_0 to T with as little work as possible consistent with our accuracy request. The total work is a complicated function of the step size because the latter affects how many iterations are required to solve for y_{n+1}, how often the Jacobian needs to be recomputed if a Newton-type iteration is used, etc. For a fixed method, for example, the four-value Adams method, the step selection algorithm outlined above does a good job. But there is nothing special about that method and programs like SDRIV also dynamically select the method from a fixed family.

If we know $\tilde{\mathbf{y}}_n$ and have estimated αh from above, at the next step we might wonder if choosing a different method (different $\mathbf{r}$'s) might result in a larger α. We already know that a number of different methods can be used beginning with the same array $\tilde{\mathbf{y}}_n$. A glance at the formula above for α shows that using a method of the same order, four, but different coefficients will yield a different step selection. Similarly if we consider a formula which also has a different order, the derivative changes too. For example, if we want to try a third-order method to advance from t_n its local error is

$$c_3(\alpha h)^4 y^{(4)}(\xi).$$

We already have an estimate of $h^4 y^{(4)}/24$; this is the extra component we added to the $\tilde{\mathbf{y}}_n$ array. If we know c_3 we can estimate the proposed step αh from t_n and if it is larger than our first estimate of αh, use that instead. Of course this requires using the $\mathbf{r}$ associated with the new method.

In the same way we estimate the step proposed by a method of one higher order, i.e., fifth-order. This requires $y^{(6)}(t_n)$ which can be estimated by taking the fifth component of $\tilde{\mathbf{y}}_n - \tilde{\mathbf{y}}_{n-1}$. Of course, there are a great many defaults and "tuning" parameters, set by experience and judgment. But overall the algorithm marches forward by selecting from the current formula or one of one higher or lower order depending on which estimates the largest next step size. The formulas are all from within a single family, for example, all Gear MV, or alternatively, all Adams MV. There has been some limited success at "cross formula" methods. SDRIV2 can switch between Adams and Gear, (MINT=3) and for some problems this results in a faster integration.

One of the most important aspects of dynamic order selection is that such programs are self starting; no special methods are needed at the beginning of the integration. From the initial condition and the differential equation, hy_0' can be calculated. This is enough to get a first-order method such as BE going, and then the order control mechanism can increase the order to an appropriate level.

Even the inital step size can be estimated. The first prediction is $y_1^P = y_0 + h y_0'$, which is EM and hence should have an error of $y'' h^2 / 2$. Thus

$$\frac{h^2}{2} y'' = y_1^P - y(t_1) = y(t_0) + h y'(t_0) - y(t_1) \approx y(t_0) + h y'(t_0) - y(t_0) = h y'(t_0).$$

Thus as long as $y'(t_0) \neq 0$, a crude estimate for the initial step is $\epsilon / |y'(t_0)|$.

*8.19   TAYLOR SERIES AND RUNGE KUTTA METHODS

One-step methods have continued to be popular because of their simplicity. But Euler's method is of too low an order to be useful. In this section we describe some approaches to getting higher order one-step methods.

Our derivation of EM expanded the solution $y(t)$ in a Taylor series. Taking one additional term

$$y(t_{k+1}) = y(t_k) + y'(t_k) h_k + y''(t_k) h_k^2 / 2 + y'''(\xi) h_k^3 / 6,$$

suggests a numerical scheme

$$y_{k+1} = y_k + h_k y_k' + y_k'' h_k^2 / 2.$$

This has local error $y'''(\xi) h_k^3 / 6$, hence this method is second-order. It is possible to consider Taylor series with any number of terms.

What should we make of the term y_k'', whose value is needed to compute y_{k+1}? In Section 5.1 we used

$$y_k'' \approx \frac{y_k' - y_{k-1}'}{t_k - t_{k-1}},$$

but this leads to a two-step method. Alternatively, differentiating $y' = f(t, y)$ by the chain rule we know that

$$y''(t) = f_t(t, y) + f(t, y) f_y(t, y), \qquad \text{or} \qquad y_k'' = f_t(t_k, y_k) + y_k' f_y(t_k, y_k),$$

giving the one-step method

$$y_{k+1} = y_k + h_k y_k' + \frac{h_k^2}{2} \left(f_t(t_k, y_k) + y_k' f_y(t_k, y_k) \right).$$

If $f(t, y)$ is not too complicated, these partial derivatives can be obtained algebraically. For example, for $y' = -10(t-1)y$, we have $f_t = -10y$ and $f_y = -10(t-1)$.

Example 8.16   Second-Order Taylor Series Method.

We use the one-step second-order Taylor series method to solve the problem in Example 8.4. With the same step size, 0.1 the corresponding errors are

t_n	Error	t_n	Error
0.1	0.0022283	1.1	0.6497108
0.2	0.0093105	1.2	0.5865614
0.3	0.0266926	1.3	0.4762708
0.4	0.0621891	1.4	0.3476736
0.5	0.1241069	1.5	0.2280530
0.6	0.2171000	1.6	0.1343194
0.7	0.3368916	1.7	0.0709585
0.8	0.4668690	1.8	0.0335548
0.9	0.5800153	1.9	0.0141430
1.0	0.6473466	2.0	0.0052622

For the first few steps of the integration the local error is the dominant contribution to the global error. Comparing this table with the corresponding errors in Example 8.4 shows that this second-order method has much smaller local errors than EM. Further along in the integration there isn't much difference between these two. We know that the equation is unstable on $[0, 1]$, and because of this, error amplification is much more significant than differences in the local errors. ∎

For general differential equations there have been attempts to utilize "symbolic differentiation" to compute the partial derivatives which are required in second and higher-order Taylor series methods. However, Runge Kutta, multi-step and multi-value methods have most of the same good properties of these Taylor series methods and are not nearly so clumsy to implement. As a result Taylor series methods have not proved to be practical and are not used much at present.

Runge Kutta methods are one-step integrators which have the accuracy of higher-order Taylor series methods without the problems associated with symbolic differentiation. They are popular and often presented in engineering texts. Because they are one-step, Runge Kutta methods appear easy to program. To compute y_{n+1} we need only y_n; the preceding solution estimates $y_{n-1}, \ldots$ do not enter. This is particularly important at the beginning of the integration when y_0 is the only solution value known.

Runge Kutta methods are tedious to derive, requiring careful expansions in Taylor series of two variables. They are described by saying how many **stages** are used. For example, a **two-stage** Runge Kutta is

$$y_{n+1} = y_n + \frac{1}{2}k_1 + \frac{1}{2}k_2 \quad \text{with} \quad k_1 = hf(t_n, y_n), \quad k_2 = hf(t_n + h, y_n + k_1).$$

Note that if t_n, y_n and h are known then k_1 can be evaluated, then k_2, and then y_{n+1}. Thus this Runge Kutta requires two evaluations of the right hand side of the equation, hence the term two-stage, but it is a one-step method.

This method is far easier to implement than the second-order Taylor series method because only values of f are required. But it can be shown that, like the Taylor series method above, it is also second-order.

Example 8.17   Two-stage Second-Order Runge Kutta.

Use the two-stage second-order Runge Kutta to solve the problem in Example 8.4. With the same step size 0.1, the corresponding errors are given below. Comparing this table with the error for the second-order Taylor series method used in Example 8.16, shows that this Runge Kutta method has comparable, and even slightly smaller errors.

t_n	Error	t_n	Error
0.1	0.0012513	1.1	0.2086880
0.2	0.0050241	1.2	0.1801451
0.3	0.0137453	1.3	0.1395314
0.4	0.0303591	1.4	0.0956802
0.5	0.0571071	1.5	0.0566331
0.6	0.0937626	1.6	0.0273759
0.7	0.1362795	1.7	0.0090567
0.8	0.1770045	1.8	0.0001733
0.9	0.2068270	1.9	0.0034548
1.0	0.2183775	2.0	0.0037006

Incidentally, if we were to continue the integration to larger values of t both the Runge Kutta and the second-order Taylor series method will become unstable, just as Euler's Method does. The instability manifests itself at earlier values of time, near $t = 4$. As with EM we can delay the onset of the instability by using smaller steps. If we are only interested in the integration to some T which is known *a priori*, we can select h small enough to allow us to integrate accurately to this point, although this might require more steps than we are willing to use. ∎

The second-order Runge Kutta method is not of much practical use. Most programs incorporate a higher order scheme. The best known, or "classical"[7], Runge Kutta method is a four-stage, fourth-order method:

$$y_{n+1} = y_n + \tfrac{1}{6}(k_1 + 2k_2 + 2k_3 + k_4)$$
$$k_1 = hf(t_n, y_n)$$
$$k_2 = hf(t_n + \tfrac{1}{2}h, y_n + \tfrac{1}{2}k_1)$$
$$k_3 = hf(t_n + \tfrac{1}{2}h, y_n + \tfrac{1}{2}k_2)$$
$$k_4 = hf(t_n + h, y_n + k_3).$$

---

[7] Runge (see Chapter 5 Section 6) published several specific methods in 1895. Wilhelm Kutta independently introduced the general method in 1901 and derived this one.

This requires four evaluations of $f(t, y)$ to get from t_n to t_{n+1} and we still have no prescription for estimating the local error or adjusting the step size. The standard technique has been to select a step h and then integrate from t_n to t_{n+1} twice, the first time using step size h to get y_{n+1} and then again using two steps of size $h/2$. Knowing theoretically that classical Runge Kutta is fourth-order, it is possible to compare the two estimates at t_{n+1} and estimate the local error from the difference.

This process works; the details are presented in many older references, but it is expensive. Even a successful step requires 11 evaluations of f. In the 1960's Fehlberg devised a Runge Kutta method as accurate as the classical one, but which integrates *and* estimates the local error with only 6 evaluations. He began with a six-stage Runge Kutta method of the following form

$$y_{n+1} = y_n + \gamma_1 k_1 + \gamma_2 k_2 + \cdots + \gamma_6 k_6$$
$$k_1 = hf(t_n, y_n)$$
$$k_2 = hf(t_n + \alpha_2 h, y_n + \beta_{21} k_1)$$
$$\vdots$$
$$k_6 = hf(t_n + \alpha_6 h, y_n + \beta_{61} k_1 + \cdots + \beta_{65} k_5).$$

There are a total of 26 coefficients, i.e., the α's, β's and γ's. Fehlberg found a set so that the method is fifth-order. It might be a little surprising but it can be shown that six evaluations of f are always needed for any fifth-order method whereas only four evaluations are needed for a fourth-order method. The "extra" evaluation is not wasted, however. Fehlberg also showed that four of the k's, hence four of the already computed f values, can be combined with different γ's to produce a four-stage, fourth-order Runge Kutta, which is *not* the classical one. Thus six f evaluations provide two estimates of the solution at t_{n+1}, and their difference can be used to estimate the local error and the next step size h.

As in the case of quadrature, one measure of efficiency is the number of these evaluations to complete the integration. The total number of evaluations of f in going from t_0 to T depends upon both the number of evaluations per step and the number of steps. These are not independent quantities.

As we mentioned earlier in this section, the formulas for classical Runge Kutta are simple, encouraging a generation of scientists and engineers to write their own "quick and dirty" versions. Not surprisingly, the development of a Runge Kutta program that deals reliably and efficiently with step size, output, and error control is *not* easy. An excellent implementation which also includes Fehlberg's technique is given in the text by Forsythe, Malcolm, and Moler (1977) in the program RKF45. Programs such as this have done a great deal to improve the already substantial popularity of Runge Kutta methods. Furthermore a number of experimental studies have shown that for many non-stiff problems with standard accuracy requirements ($\approx 10^{-5}$) these programs are as efficient as any others now available. They are not as efficient for high accuracy requirements, but this is not normally a serious issue. Most practical modeling problems rarely need to obtain more than five or six digits. On the other hand, for problems which are not stiff the differences between the best routines of any type are often a matter of no more than 50 percent, and usually much less. Thus considerations such as reliability and applicability become more important when selecting between different programs.

A major shortcoming of the current Runge Kutta programs is their inability to deal effectively with stiff problems which are becoming increasingly common in practical applications. Another problem is related to output. Sometimes the automatic step selection mechanism is so efficient that it will try to take steps which are larger than the user wishes. This is particularly true when the solution is to be plotted; a smooth plot requires many data points but the program might be able to use much larger steps and still keep the error under control. Asking for closely spaced output with a program like RKF45 has the effect of forcing the integrator to use a smaller step than the error dictates. Since each step requires six derivative evaluations this can be quite inefficient; doubling the number of output points on the same integration interval $[t_0, T]$ makes the program run twice as long. Multi-value programs do not suffer from this limitation. For them there is essentially no relationship between internal integration steps and output points. Doubling the number of output points might only increase the running time of a multi-value program by a few percent. Multi-step and multi-value methods have the further advantage, in principle, of only requiring one evaluation per internal integration step. However, there is still a great deal of active research into Runge Kutta methods and perhaps in a few years a new generation of Runge Kutta programs will again be the best available.

*8.20   SOME TOPICS OMITTED

In this section we mention briefly a few topics which have been omitted for reasons of space or complexity.

An ODE is **linear** if $f(t, y) = \alpha(t)y + \beta(t)$, otherwise it is nonlinear. The equation

$$y' = t^2 e^{-t^2} y + \sin t$$

is linear, whereas the equations

$$y' = \sin y, \quad y' = y^2, \quad y' = \frac{1}{t - y}$$

are nonlinear.

Linear equations and the even more special subcase of linear constant-coefficient equations (where α and β do not depend on t) form an important, insightful part of the theory and suggest numerical techniques, but are rarely encountered in practice.

Linear differential equations (which often can be solved analytically) are sometimes used as approximations. This trend was popular before the advent of reliable numerical methods. Today most scientists resort to numerical solution of the nonlinear equations. Analytic methods are now used mostly to obtain qualitative information about the solution, or asymptotic estimates of its long term behavior, i.e., for $t \gg t_0$.

The **mixed**, or **hybrid, algebraic-differential** system occurs when one or more of the left hand sides is zero. There are still n unknown functions and n equations, but only $m < n$ differential equations. For example, if the last equation of the system were replaced by

$$0 = f_n(t, y_1, \ldots, y_n).$$

The function y_n still varies with time, $y_n(t)$, and hence its values are needed to compute all the f_i. In simple cases it is possible to solve this equation for y_n and the resulting expression can be substituted into the preceding equations. This eliminates y_n and we can solve the reduced system of equations of size $n - 1$. Usually though, we cannot solve for y_n and the algebraic equation has to be solved simultaneously with the differential equations. Often the algebraic equations are included to force some kind of balance or conservation principle to apply, or to impose a physical constraint.

A **semi-linear implicit** system is of the form

$$A\mathbf{y}' = \mathbf{f}(t, \mathbf{y})$$

where A is an n by n matrix whose elements may depend upon t and y_i. If A is nonsingular (this is often, but not always the case) multiplying by A^{-1} recovers the form of an ordinary system, but this isn't always practical. For example, in solving partial differential equations by the finite element method large implicit systems arise. The matrix A might be tridiagonal in which case we could easily store its nonzero elements. On the other hand A^{-1} is usually a full matrix and too large to store.

The most general first-order ODE is

$$f(t, y, y') = 0.$$

It is more difficult to treat both theoretically and numerically.

In some problems, $\mathbf{y}' = \mathbf{f}(t, \mathbf{y})$, $\mathbf{y}(t_0) = \mathbf{A}$, the given values A_i are not all specified at the same point t_0. That is we may have

$$y_i(t_i) = A_i = \text{given} \qquad i = 1, \ldots, n.$$

Such a problem is not an initial value problem, but is called a **boundary value problem**. Theoretical and numerical techniques for boundary value problems are more sophisticated than those for initial value problems. Initial value problems almost always have a solution, but this is not true for boundary value problems. Thus the facts concerning them are, perforce, more complicated. Some methods for initial-value problems can be adapted to solve boundary-value problems. For other approaches, see the reference list in Chapter 1. We will also not discuss partial differential equations, where the solution is a function of two or more variables. Again, even though this is an important practical problem, it is beyond our scope.

8.21 PROBLEMS

P8–1.–The error function is usually defined by an integral,

$$\text{erf}(x) = \frac{2}{\sqrt{\pi}} \int_0^x e^{-t^2}\, dt,$$

but it can also be defined as the solution of the problem

$$y'(x) = \frac{2}{\sqrt{\pi}} e^{-x^2}, \qquad y(0) = 0.$$

(a) Write a program using SDRIV2 to print a table of erf(x) for $x = 0.0, 0.1, 0.2, \ldots, 1.9, 2.0$. Compare your table with the one produced for Problem P5–1. If you have access to precise execution times for programs run on your computer, measure the amount of time required by the two methods for generating the same table.

(b) In the subroutine which you must write to evaluate y' above, insert a statement to print the input values of x and y, and the output y'. Why do these numbers indicate that, when used to do a quadrature, SDRIV2 uses about twice as many evaluations as necessary? What is the source of this inefficiency?

(c) Try to compute $y(2) = 5.3141\ldots$, where

$$y' = \frac{1}{\sin\sqrt{|t|}}, \qquad y(-1) = 0,$$

using SDRIV2. Then use Q1DA on the same problem. Comment on the difficulties, if any, you encounter. Hint: Be careful that the subroutine you use for SDRIV2 is not called F. Also, better results will be obtained if you split the interval at the origin.

P8–2.–Use SDRIV2 to solve the stiff initial value problem $y' = -1000(y - \sin t) + \cos t$, $y(0) = 1$, $0 \le t \le 1$. Take EPS=0.001. Run your program twice, first with MINT=2 and then with MINT=1. In each case print the solution at $0.1, 0.2, \ldots, 1.0$. Also print the number of integration steps required, IWORK(3). Which method requires fewer steps?

P8–3.–A 200 pound paratrooper is dropped out of an airplane at a height of 1000 feet. After 5 seconds the parachute opens. Assume that the air resistance is $a(t) = k_1 y'^2(t)$ in free fall and $a(t) = k_2 y'^2(t)$ with an open chute, with $k_1 = 1/150$, $k_2 = 4/150$. At what height does the chute open, how long does it take to reach the ground and what is the "impact velocity?" Hint: Mass $= 200/32 = 6.25$ slugs.

P8–4.–An old-fashioned spherical cannonball, or a low elevation rocket is fired with a given muzzle velocity $v(0) = v_0$ and elevation angle $\theta(0) = \theta_0$. It is desired to determine the trajectory with respect to a rectangular Cartesian coordinate system with horizontal x axis, vertical y axis, and origin at the launch point. The only forces acting on the projectile are its weight mg in the vertical direction, the rocket thrust $T(t)$ which is in the direction of the velocity ($T = 0$ in the case of a cannonball), the aerodynamic drag $D(t)$ which opposes the velocity, and the wind $W(t)$ which is assumed to be only in the x direction. The equations describing the motion of the projectile are given by (see for example Ortega and Poole (1981))

$$x' = v\cos\theta + W, \qquad y' = v\sin\theta,$$

and

$$\theta' = -\frac{g}{v}\cos\theta, \qquad mv' = T - D - mg\sin\theta - m'v.$$

For this problem take $D(t) = c\rho s v^2/2$, where $c = 0.2$ is the drag coefficient, $\rho = 1.29$ kg/meter is air density, $s = 0.25$ meter^2 is the projectile's cross sectional area, $g = 9.81$ meter/sec^2 is the acceleration of gravity, and $v_0 = 50$ meter/sec.

(a) Let $T = 0$, $m = 15$ kg, $m' = 0$, and $W(t) = 0$. For elevation angles $0.3 \le \theta_0 \le 1.5$ tabulate the downrange distance, impact velocity and flight time at one tenth radian intervals. Also print the amount of work as measured by the number of calls to subroutine F. From your table estimate the elevation angle to achieve the maximum range.

(b) Now let $W(t) = 10$ meter/sec, $1 \le t \le 2$ and repeat the calculations in (a). Since W only acts in the horizontal direction the flight times should not change but the impact point should increase by about 10 meters. Is this problem more difficult for SDRIV2? Why?

(c) The projectile is to be fired in gusty wind conditions. Repeat the calculations in (a) but with $W(t) = 10*\text{RNOR}()$, $1 \le t \le 2$ where RNOR() is a normally distributed random number with zero mean and unit variance (see Chapter 10). Why is this problem even more difficult for SDRIV2?

P8–5.–Consider a simple ecosystem consisting of rabbits that have an infinite food supply and foxes that prey upon the rabbits for their food. A classical mathematical model due to Volterra describes this system by a pair of nonlinear first-order equations:

$$dr/dt = 2r - \alpha rf, \qquad r(0) = r_0,$$
$$df/dt = -f + \alpha rf, \qquad f(0) = f_0,$$

where t is time, $r = r(t)$ is the number of rabbits, $f = f(t)$ is the number of foxes, and α is a positive constant. When $\alpha = 0$, the two populations do not interact, and the rabbits do what rabbits do best and the foxes die off from starvation. When $\alpha > 0$, the foxes encounter the rabbits with a probability which is proportional to the product of their numbers. Such an encounter results in a decrease in the number of rabbits and (for less obvious reasons) an increase in the number of foxes.

Investigate the behavior of this system for $\alpha = 0.01$ and various values of r_0 and f_0 ranging from 2 or 3 to several thousand. Draw graphs of interesting solutions. Include a phase plane plot. Since we are being rather vague about the units of measurement, there is no reason to restrict r and f to integer values.

(a) Compute the solution with $r_0 = 300$ and $f_0 = 150$. You should observe from the output that the system is periodic with a period close to five time units. In other words $r(0) \approx r(5)$ and $f(0) \approx f(5)$.

(b) Compute the solution with $r_0 = 15$ and $f_0 = 22$. You should find that the number of rabbits eventually drops below 1. This could be interpreted by saying that the rabbits become extinct. Find initial conditions which cause the foxes to become extinct. Find initial conditions with $r_0 = f_0$ which cause both species to become extinct.

(c) Is it possible for either component of the true solution to become negative? Is it possible for the numerical solution to become negative? What happens if it does? (In practice, the answers to the last two questions may depend on the value used for your error tolerance.)

(d) Many modifications of this simple model have been proposed to more accurately reflect what happens in nature. For example, the number of rabbits may be prevented from growing indefinitely by changing the first equation to

$$\frac{dr}{dt} = 2\left(1 - \frac{r}{R}\right) f - \alpha r f.$$

Now, even when $\alpha = 0$, the number of rabbits can never exceed R. Pick some reasonable value of R, and consider some of the above questions again. In particular, what happens to the periodicity of the solutions?

This model has been studied extensively by both mathematicians and biologists. Comparisons with actual observations of lynx and hare populations in the Hudson Bay area have even been made. Many interesting properties are described in the readable book by Haberman (1977).

P8–6.–The following differential equations describe the motion of a body in orbit about two much heavier bodies. An example would be an Apollo capsule in an earth-moon orbit. The coordinate system is a little tricky. The three bodies determine a plane in space and a two-dimensional cartesian coordinate in this plane. The origin is at the center of mass of the two heavy bodies, the x axis is the line through these two bodies, and the distance between them is taken as the unit. Thus, if μ is the ratio of the mass of the moon to that of the earth, then the moon and the earth are located at coordinates $(1 - \mu, 0)$ and $(-\mu, 0)$, respectively, and the coordinate system moves as the moon rotates about the earth. The third body, the Apollo, is assumed to have a mass which is negligible compared to the other two, and its position as a function of time is $(x(t), y(t))$. The equations are derived from Newton's law of motion and the inverse square law of gravitation. The first derivatives in the equation come from the rotating coordinate system and from a frictional term, which is assumed to be proportional to velocity with proportionality constant f:

$$x'' = 2y' + x - \frac{\tilde{\mu}(x + \mu)}{r_1^3} - \frac{\mu(x - \tilde{\mu})}{r_2^3} - fx',$$

$$y'' = -2x' + y - \frac{\tilde{\mu}y}{r_1^3} - \frac{\mu y}{r_2^3} - fy',$$

with

$$\mu = \frac{1}{82.45}, \quad \tilde{\mu} = 1 - \mu, \quad r_1^2 = (x + \mu)^2 + y^2, \quad r_2^2 = (x - \tilde{\mu})^2 + y^2.$$

Although a great deal is known about these equations, it is not possible to find closed-form solutions. One interesting class of problems involves the study of periodic solutions in the absence of friction. It is known that the initial conditions

$$x(0) = 1.2, \quad x'(0) = 0, \quad y(0) = 0, \quad y'(0) = -1.04935751$$

lead to a solution which is periodic with period $T = 6.19216933$, when $f = 0$. This means that the Apollo starts on the far side of the moon with an altitude of about 0.2 times the earth-moon distance and a certain initial velocity. The resulting orbit brings the Apollo in close to the earth, out in a big loop on the opposite side of the earth from the moon, back in close to the earth again, and finally back to its original velocity on the far side of the moon.

(a) Use SDRIV2 to compute the solution with the given initial conditions. Verify that the solution is periodic with the given period. How close does the Apollo come to the *surface* of the earth in this orbit? In the equation, distances are measured from the centers of the earth and moon. Assume that the moon is 238,000 miles from the earth and that the earth is a sphere with radius of 4000 miles. Note that the origin of the coordinate system is within this sphere but not at its center.

(b) When $f = 1$ with the same initial conditions as in (a), integrate from $0 \leq t \leq 5$. Plot the phase plane of the solution. That is, plot $x(t)$ versus $y(t)$. In this case, the Apollo is captured by the earth and eventually "crashes."

(c) When $f = 0.1$ repeat the computations of (b). By looking at the phase plane, can you guess what is happening? This will be easier to understand if you perform the integration for a longer time, say to $t = 8$.

P8–7.–A clever scientist was solving two equations and wanted to keep his own count of the number of times F was called. Rather than using COMMON he decided to enlarge the Y array. A portion of his program looked like

```
REAL Y(3), WORK(500), ...
EXTERNAL F
N=2
      .
      .
Y(3)=0     [Counter for F is initialized to zero]
CALL SDRIV2(N, T, Y, F, ... )
WRITE(*,*) 'Count = ', Y(3)
      .
      .
END
SUBROUTINE F(N, T, Y, YDOT)
     REAL Y(*), YDOT(*)
        .
        .
     Y(3)=Y(3)+1.
     RETURN
END
```

This worked, but when he tried the same thing with another popular solver it dramatically altered the stepsize strategy. Explain what feature of SDRIV2 permits this. Without looking at the source listing, explain how the two solvers must differ in the way they call F. Why might this be a difficult bug to locate?

P8–8.–A ball is dropped from a height $H_0 = 4$ meters. When it hits the ground it bounces. That is, its velocity changes discontinuously. A model describing the height as a function of time (ignoring air resistance) is given by the system $y_1' = y_2$, $y_2' = -g$, $y_1(0) = H_0, y_2(0) = 0$ as long as $y_1(t)$ is positive. When $y_1(t)$ becomes zero, say at $t = \tau$ the value of $y_2(\tau)$ must be replaced by $-ky_2(\tau)$, $0 < k < 1$.

(a) Use SDRIV2 to find the first ten bounce times as well as the height of the first ten bounces when $k = 0.7$. Generate a plot of $y(t)$ versus t. Hint: The only way to change a value of one of the solution components is to restart SDRIV2 with new initial conditions.

(b) Repeat (a) but alter the problem so that when y_1 becomes zero y_2 is replaced by $k|y_2|^{0.9}$.

P8–9.–In Problem P7–3, if the rod is initially at temperature T_∞ before the current is turned on, the temperature as a function of time satisfies

$$\frac{dT}{dt} = Q.$$

How long does it take for the rod to get within two percent of its steady state value? (Answer, about 0.58 sec.)

P8–10.–Drawing a Circle. The phase plane $(x(t), y(t))$ of the initial value problem

$$x' = -y, \quad y' = x \quad\quad 0 \le t \le 2\pi, \quad\quad x(0) = 1, \quad y(0) = 0$$

is a circle. Both Euler's method and the trapezoid rule are economical ways to generate points on the cirle. Furthermore, if the step h is a power of 2, most Fortran compilers will implement the apparent floating point multiplication by h as a shift.

(a) Compute the solution of this system by both of these methods and compare the effort and accuracy obtained. Also compare this approach to that given in problem P4–10.

(b) Integrate from $t = 0$ to $t = 2\pi$ with various values of h. Tabulate the error in the radius at $t = 2\pi$. Do your results confirm that Euler's method is first order and that trapezoid rule is second order?

P8–11.–In aerodynamics, one of the most important early applications is Blasius's equation,

$$2f''' + ff'' = 0,$$

which gives the incompressible velocity profile for a flat plate. The independent variable is usually denoted η. Two initial conditions for this problem are $f(0) = f'(0) = 0$. It is also known that $f'(\eta) \to 1$ as $\eta \to \infty$.

(a) Experiment with different values for the third initial condition, using `SDRIV2` to solve this problem. Then plot η versus $f'(\eta)$ and compare your results with standard published profiles, see for example Anderson (1984), page 531. (Answer: $f''(0) = 0.332$.) Caution: As f' approaches a constant, then $f'' \to 0$. If f'' gets small enough it can underflow. In some ways this signals success rather than failure but an abort on underflow is an ugly way for a program to terminate. Some compilers allow you to convert underflows to zero and that is an appropriate choice here.

(b) The momentum thickness for an incompressible flow is proportional to

$$\theta(u) = \int_0^u f' \cdot (1 - f') \, d\eta.$$

Find $\theta(5.0)$.

P8–12.–Techniques have been proposed for solving a system of nonlinear equations by thinking of the solution values as the solution of an ODE that has reached steady state. A good ODE solver is then used to "creep up" on the solution. You will sometimes see these described as **continuation methods**. While there are many technical difficulties this is often a practical approach because our ODE software is quite robust. Here is one typical example. We want to solve the nonlinear system

$$f_i(y_1, \ldots, y_n) = 0, \quad\quad i = 1, \ldots, n.$$

Let $y_1(t), \ldots, y_n(t)$ be unknown functions of t, $0 \leq t \leq 1$, and let $k_1, \ldots, k_n$ be arbitrary, but fixed, numbers. If the $y_i(t)$ were known and substituted into the left hand side of each of the equations above the result would be a function of t,

$$f_i(y_1(t), y_2(t), \ldots, y_n(t)).$$

We shall require that each of the resulting functions be a specific scalar multiple of $1 - t$,

$$f_i(y_1, y_2, \ldots, y_n) = f_i(k_1, k_2, \ldots, k_n)(1 - t).$$

Notice that the first term on the right hand side is a known number when the k_i's are given. If such functions y_i could be found then their values at $t = 1$ would be the solution to the nonlinear system. (Why?) To find y_i, differentiate the relations above with respect to t,

$$\frac{\partial f_1}{\partial y_1} y_1' + \cdots + \frac{\partial f_1}{\partial y_n} y_n' = -f_1(k_1, \ldots, k_n)$$

$$\vdots$$

$$\frac{\partial f_n}{\partial y_1} y_1' + \cdots + \frac{\partial f_n}{\partial y_n} y_n' = -f_n(k_1, \ldots, k_n),$$

or

$$Jy' = -f, \qquad 0 \leq t \leq 1, \qquad y_i(0) = k_i,$$

where J is the Jacobian of the system of equations. Use `SDRIV2` to solve the system given in problem P7-7, beginning with $k_1 = k_2 = k_3 = 0$. You will also need to use `SGEFS` although this is not efficient. Verify that your results agree with those in of problem P7-7.

P8–13.–A famous problem of nonlinear mechanics is known as the **inverted pendulum**. The pendulum is a stiff bar of length L which is supported at one end by a frictionless pin. The support pin is given a rapid up-and-down motion s by means of an electric motor, $s = A \sin \omega t$. An application of Newton's second law of motion yields the equation of motion

$$L\theta'' = (g - A\omega^2 \sin \omega t) \sin \theta,$$

where θ is the angular position of the bar (the bar is directly above the pin when $\theta = 0$), and $g = 386.09$ in/sec^2 is the acceleration due to gravity. (For small values of θ, $\sin \theta \approx \theta$, and this equation becomes the well-known Mathieu equation.) For $A = 0$ it is called the pendulum equation, which you may have seen in elementary physics courses,

$$\theta'' = \frac{g}{L} \sin \theta.$$

Even this equation is not solvable in terms of elementary functions. But it is known that when $A = 0$, and the pendulum is released from rest, i.e., $\theta'(0) = 0$, the period T of the pendulum is given by

$$T = 4\sqrt{\frac{L}{g}} K(\theta(0)/2),$$

where

$$K(u) = \int_0^{\pi/2} \frac{d\psi}{\sqrt{1 - \cos^2 u \sin^2 \psi}}$$

is the complete elliptic integral of the first kind.

The most interesting aspect of this problem is that there are regions in which the equation of motion is stable for initial values corresponding to an inverted configuration, and they have been observed experimentally. Write a program using SDRIV2 to compute the motion $\theta(t)$ when values of $L, A, \omega, \theta(0)$ and $\theta'(0)$ are input. Debug your program using the case $L = 10$ inches, $A = 0$ inches, $\omega = 0$ radians/second, $\theta(0) = \pi/3$ radians, and $\theta'(0) = 0$ radians/second by comparing the period computed by SDRIV2 with the value of T above. Use Q1DA to calculate K or look the value up in a table. Print θ and θ' for two or three oscillations of the pendulum.

When your program appears to be working satisfactorily, try the more interesting cases:

L	A	ω	$\theta(0)$	$\theta'(0)$
10	0.5	5.3	3.10	0
10	10.0	100.0	3.10	0
10	10.0	100.0	0.10	0
10	2.0	100.0	0.10	0
10	0.5	200.0	0.05	0

In each case plot the phase plane and then interpret your solutions physically.

P8–14.–The following laboratory chemical system has been used to study chemicals that are active in the atmospheric ozone depletion cycle. It involves the rapid formation of Chlorine atoms from Oxygen and Hydrogen atoms and molecular Chlorine.

$$
\begin{array}{lll}
Cl + H_2 & \xrightarrow{k_1} HCl + H & k_1 = 1.6 \times 10^{-14} \\
H + Cl_2 & \xrightarrow{k_2} HCl + Cl & k_2 = 2.0 \times 10^{-11} \\
H + O_2 & \xrightarrow{k_3} HO_2 & k_3 = 3.6 \times 10^{-13} \\
Cl + O_2 & \xrightarrow{k_4} ClO_2 & k_4 = 1.3 \times 10^{-14} \\
Cl + ClO_2 & \xrightarrow{k_5} Cl_2 + O_2 & k_5 = 1.4 \times 10^{-10}
\end{array}
$$

(a) Write the eight ordinary differential equations determined by these reactions.

(b) Use SDRIV2 to integrate the equations from $t = 0$ to $t = 10^{-4}$ seconds. Use MINT=2 and EPS=0.001. Take as initial conditions (in molecules per cubic centimeter) $[Cl] = 10^{14}$, $[Cl_2] = 3.25 \times 10^{16}$, $[H_2] = 1.62 \times 10^{18}$, $[O_2] = 4.84 \times 10^{18}$, and zero for the others. Print the solution at fifty equally spaced points. Also print the number of integration steps, IWORK(3), and the amount of computer time used. For some print points you may discover that no integration steps are required. Explain this. Do the concentrations approach steady state values? Hint: With such small concentrations, intermediate calculations in SDRIV2 may underflow. Set your compiler options to convert underflows to zero.

(c) Repeat (b) with MINT=1. This problem is mildly stiff. You should discover that for the first couple of print points the nonstiff mode is more efficient because the integrator has to resolve the most rapid changes in the concentrations. After that, the concentrations change slowly and the nonstiff mode restricts the step size.

P8–15.–This problem deals with the propagation of waves in a medium with variable speed of propagation. The particular setting is underwater propagation of sound, but the techniques are applicable to other situations. It involves a combination of data fitting, ordinary differential equation solving, and zero finding. Subroutines PCHEZ, SDRIV2 , and FZERO are used. The problem is derived from a paper by Moler and Solomon (1970).

The speed of sound in ocean water depends on pressure, temperature, and salinity, all of which vary with depth in fairly complicated ways. Let z denote depth in feet under the ocean surface (so that the positive z axis points down) and let $c(z)$ denote the speed of sound at depth z. We shall ignore the changes in sound speed observed in horizontal directions. It is possible to measure $c(z)$ at discrete values of z; typical results can be found in the table below. To obtain values of $c(z)$ between data points, we can use cubic spline interpolation. We shall also need to evaluate $c'(z) = dc(z)/dz$.

z (ft)	$c(z)$ (ft/sec)
0	5,042
500	4,995
1,000	4,948
1,500	4,887
2,000	4,868
2,500	4,863
3,000	4,865
3,500	4,869
4,000	4,875
5,000	4,875
6,000	4,887
7,000	4,905
8,000	4,918
9,000	4,933
10,000	4,949
11,000	4,973
12,000	4,991

Since the sound speed varies with depth, sound rays will travel in curved paths. The effect is a continuous version of the familiar refraction of light waves caused by the air-water interface in a fish bowl. The basic equation is a continuous version of Snell's law. Let x denote the horizontal (radial) distance in feet from a source of sound and $z(x)$ denote the depth of a particular ray at distance x. Let $\theta = \theta(x)$ denote the angle between a horizontal line and the tangent to the ray at x, i.e.,

$$\tan \theta = \frac{dz}{dx}.$$

Snell's law can be written

$$\frac{\cos \theta}{c(z)} = \text{constant} = A.$$

A fixed underwater point emits rays in all directions. Given a particular point and initial direction we would like to follow the ray path. Thus we know

$$z(0) = z_0, \qquad \frac{dz}{dx}(0) = \tan \theta_0$$

To use these we need a second order ODE. We can obtain obtain this by differentiating both equations above with respect to x and combining the results to obtain

$$\frac{d^2 z}{dx^2} = -\frac{c'(z)}{A^2 c(z)^3}, \qquad \text{where} \qquad A^2 = \left(\frac{\cos \theta_0}{c(z_0)}\right)^2.$$

(a) Use SDRIV2 to trace the ray beginning at $z_0 = 2000$ feet and $\theta_0 = 5.4$ degrees. (You should use PCHEZ and PCHEV to compute $c(z)$ and $c'(z)$ wherever they are needed.) Follow the ray for 24 nautical miles, printing its depth at intervals of 1 mile. Assume 1 nautical mile is 6076 feet and do not forget that the Fortran trigonometric functions expect arguments in radians. You should find that the depth at 24 miles is close to 3000 feet. Plot the ray path as a function of x.

Now suppose that a sound source at a depth of 2000 feet transmits to a receiver 24 miles away at a depth of 3000 feet. The above calculation shows that one of the rays from the source to the receiver leaves the source at an angle close to 5.4 degrees. Because of the nonlinearity of the equation there may be other rays leaving at different angles that reach the same receiver. Let $x_f = 24$ nautical miles. As θ_0 varies, $z(x_f)$ also varies. We are interested in finding values of θ_0 for which $z(x_f) = 3000$.

(b) Write a function subprogram, F(THETA), which traces the ray with initial angle THETA and returns the value $z(x_f) - 3000$. Print a table of values of this function for THETA in the range -10 to 10 degrees. Since this function is fairly expensive to evaluate, the increment used in this table will depend on the amount of computer time you have available and the efficiency of your program.

(c) Use FZERO with starting values obtained from part (b) to find rays which pass through (or as near as possible to) the receiver.

(d) Assume that the floor of the ocean is at $12,000$ feet and that the surface is at 0 feet. If you repeat (b) but allow THETA to be in the range -15 to 15 you should discover that some rays will reach the ocean floor or surface. Modify your program so that the integration terminates whenever a ray reaches the ocean floor. Additionally, modify the program so that if a ray reaches the surface it is reflected downward with negative angle. Generate a plot of the ray paths for the 31 rays.

P8–16.–Solve the following nonlinear boundary-value problem for $y(x)$ on the interval $0 \le x \le 1$:

$$y'' = y^2 - 1, \quad y(0) = 0, \ y(1) = 1.$$

(a) Suppose that we knew the value of $z = y'(0)$. Then we could apply routine SDRIV2 to the problem

$$y'' = y^2 - 1, \quad y(0) = 0, \ y'(0) = z.$$

For any particular value of z the output would be a corresponding value of $y_z(1)$. One way to solve the boundary-value problem, called the **shooting method**, is to solve the nonlinear equation

$$f(z) \equiv y_z(1) - 1 = 0.$$

Use the shooting method and FZERO to solve this boundary-value problem.

(b) Now try to solve the problem another way: Partition the interval into n equal subintervals:

$$0 = x_0 < x_1 < x_2 < \cdots < x_{n-1} < x_n = 1.$$

Replace the differential equation by a difference equation in $n - 1$ unknowns $y_1, y_2, \ldots, y_{n-1}$, where y_i approximates $y(x_i)$:

$$y_{i+1} - 2y_i + y_{i-1} = h^2(y_i^2 - 1), \quad i = 1, 2, \ldots, n - 1,$$

where $y_0 = 0$ and $y_n = 1$. Solve the nonlinear tridiagonal system using SNSQE for $n = 50$, say.

(c) For a third approach, try the following. Observe that $y'' = y^2 - 1$ can be written

$$\frac{dy}{dx}\left(\frac{(y')^2}{2} - \frac{y^3}{3} + y\right) = 0.$$

Thus

$$\frac{(y')^2}{2} - \frac{y^3}{3} + y = c$$

for some constant c. Since $y(0) = 0$, we have $y'(0) = \sqrt{2c}$. So if we can compute c, the boundary-value problem has been converted into an initial value problem. Integrating the above equation gives

$$x = \int_0^y \frac{dy}{\sqrt{2(c + y^3/3 - y)}}.$$

Now, use FZERO and Q1DA to find c by solving

$$1 = \int_0^1 \frac{dy}{\sqrt{2(c + y^3/3 - y)}},$$

and then use SDRIV2 to obtain the desired solution.

8.22 PROLOGUE: SDRIV2

```
      SUBROUTINE SDRIV2 (N,T,Y,F,TOUT,MSTATE,NROOT,EPS,EWT,MINT,WORK,
     8   LENW,IWORK,LENIW,G)
C***BEGIN PROLOGUE  SDRIV2
C***DATE WRITTEN   790601   (YYMMDD)
C***REVISION DATE  871105   (YYMMDD)
C***CATEGORY NO.   I1A2,I1A1B
C***KEYWORDS  ODE,STIFF,ORDINARY DIFFERENTIAL EQUATIONS,
C             INITIAL VALUE PROBLEMS,GEAR'S METHOD,
C             SINGLE PRECISION
C***AUTHOR  KAHANER, D. K., NATIONAL BUREAU OF STANDARDS,
C           SUTHERLAND, C. D., LOS ALAMOS NATIONAL LABORATORY
C***PURPOSE  The function of SDRIV2 is to solve N ordinary differential
C            equations of the form dY(I)/dT = F(Y(I),T), given the
C            initial conditions Y(I) = YI.  The program has options to
C            allow the solution of both stiff and non-stiff differential
C            equations.  SDRIV2 uses single precision arithmetic.
C***DESCRIPTION
C    From the book "Numerical Methods and Software"
C       by  D. Kahaner, C. Moler, S. Nash
C           Prentice Hall 1988
C
C  I.  ABSTRACT .................................................
C
C    The function of SDRIV2 is to solve N ordinary differential
C    equations of the form dY(I)/dT = F(Y(I),T), given the initial
C    conditions Y(I) = YI.  The program has options to allow the
C    solution of both stiff and non-stiff differential equations.
C    SDRIV2 is to be called once for each output point of T.
C
C  II.  PARAMETERS ..............................................
C
C    The user should use parameter names in the call sequence of SDRIV2
C    for those quantities whose value may be altered by SDRIV2.  The
C    parameters in the call sequence are:
C
C    N      = (Input) The number of differential equations.
C
C    T      = The independent variable.  On input for the first call, T
C             is the initial point.  On output, T is the point at which
C             the solution is given.
C
C    Y      = The vector of dependent variables.  Y is used as input on
C             the first call, to set the initial values.  On output, Y
C             is the computed solution vector.  This array Y is passed
```

```
C            in the call sequence of the user-provided routines F and
C            G.  Thus parameters required by F and G can be stored in
C            this array in components N+1 and above.  (Note: Changes
C            by the user to the first N components of this array will
C            take effect only after a restart, i.e., after setting
C            MSTATE to +1(-1).)
C
C     F      = A subroutine supplied by the user.  The name must be
C            declared EXTERNAL in the user's calling program.  This
C            subroutine is of the form:
C                  SUBROUTINE F (N, T, Y, YDOT)
C                  REAL Y(*), YDOT(*)
C                    .
C                    .
C                  YDOT(1) = ...
C                    .
C                    .
C                  YDOT(N) = ...
C                  END (Sample)
C            This computes YDOT = F(Y,T), the right hand side of the
C            differential equations.  Here Y is a vector of length at
C            least N.  The actual length of Y is determined by the
C            user's declaration in the program which calls SDRIV2.
C            Thus the dimensioning of Y in F, while required by FORTRAN
C            convention, does not actually allocate any storage.  When
C            this subroutine is called, the first N components of Y are
C            intermediate approximations to the solution components.
C            The user should not alter these values.  Here YDOT is a
C            vector of length N.  The user should only compute YDOT(I)
C            for I from 1 to N.  Normally a return from F passes
C            control back to  SDRIV2.  However, if the user would like
C            to abort the calculation, i.e., return control to the
C            program which calls SDRIV2, he should set N to zero.
C            SDRIV2 will signal this by returning a value of MSTATE
C            equal to +6(-6).  Altering the value of N in F has no
C            effect on the value of N in the call sequence of SDRIV2.
C
C     TOUT   = (Input) The point at which the solution is desired.
C
C     MSTATE = An integer describing the status of integration.  The user
C            must initialize MSTATE to +1 or -1.  If MSTATE is
C            positive, the routine will integrate past TOUT and
C            interpolate the solution.  This is the most efficient
C            mode.  If MSTATE is negative, the routine will adjust its
C            internal step to reach TOUT exactly (useful if a
C            singularity exists beyond TOUT.)  The meaning of the
C            magnitude of MSTATE:
C                  1  (Input) Means the first call to the routine.  This
```

```
C                   value must be set by the user.  On all subsequent
C                   calls the value of MSTATE should be tested by the
C                   user.  Unless SDRIV2 is to be reinitialized, only the
C                   sign of MSTATE may be changed by the user.  (As a
C                   convenience to the user who may wish to put out the
C                   initial conditions, SDRIV2 can be called with
C                   MSTATE=+1(-1), and TOUT=T.  In this case the program
C                   will return with MSTATE unchanged, i.e.,
C                   MSTATE=+1(-1).)
C                 2 (Output) Means a successful integration.  If a normal
C                   continuation is desired (i.e., a further integration
C                   in the same direction), simply advance TOUT and call
C                   again.  All other parameters are automatically set.
C                 3 (Output)(Unsuccessful) Means the integrator has taken
C                   1000 steps without reaching TOUT.  The user can
C                   continue the integration by simply calling SDRIV2
C                   again.  Other than an error in problem setup, the
C                   most likely cause for this condition is trying to
C                   integrate a stiff set of equations with the non-stiff
C                   integrator option. (See description of MINT below.)
C                 4 (Output)(Unsuccessful) Means too much accuracy has
C                   been requested.  EPS has been increased to a value
C                   the program estimates is appropriate.  The user can
C                   continue the integration by simply calling SDRIV2
C                   again.
C                 5 (Output) A root was found at a point less than TOUT.
C                   The user can continue the integration toward TOUT by
C                   simply calling SDRIV2 again.
C                 6 (Output)(Unsuccessful) N has been set to zero in
C                   SUBROUTINE F.
C                 7 (Output)(Unsuccessful) N has been set to zero in
C                   FUNCTION G.  See description of G below.
C
C     NROOT  = (Input) The number of equations whose roots are desired.
C                   If NROOT is zero, the root search is not active.  This
C                   option is useful for obtaining output at points which are
C                   not known in advance, but depend upon the solution, e.g.,
C                   when some solution component takes on a specified value.
C                   The root search is carried out using the user-written
C                   function G (see description of G below.)  SDRIV2 attempts
C                   to find the value of T at which one of the equations
C                   changes sign.  SDRIV2 can find at most one root per
C                   equation per internal integration step, and will then
C                   return the solution either at TOUT or at a root, whichever
C                   occurs first in the direction of integration.  The index
C                   of the equation whose root is being reported is stored in
C                   the sixth element of IWORK.
C                   NOTE: NROOT is never altered by this program.
```

```
C
C   EPS   = On input, the requested relative accuracy in all solution
C           components. EPS = 0 is allowed. On output, the adjusted
C           relative accuracy if the input value was too small. The
C           value of EPS should be set as large as is reasonable,
C           because the amount of work done by SDRIV2 increases as
C           EPS decreases.
C
C   EWT   = (Input) Problem zero, i.e., the smallest physically
C           meaningful value for the solution. This is used inter-
C           nally to compute an array YWT(I) = MAX(ABS(Y(I)), EWT).
C           One step error estimates divided by YWT(I) are kept less
C           than EPS. Setting EWT to zero provides pure relative
C           error control. However, setting EWT smaller than
C           necessary can adversely affect the running time.
C
C   MINT  = (Input) The integration method flag.
C           MINT = 1 Means the Adams methods, and is used for
C                    non-stiff problems.
C           MINT = 2 Means the stiff methods of Gear (i.e., the
C                    backward differentiation formulas), and is
C                    used for stiff problems.
C           MINT = 3 Means the program dynamically selects the
C                    Adams methods when the problem is non-stiff
C                    and the Gear methods when the problem is
C                    stiff.
C           MINT may not be changed without restarting, i.e., setting
C           the magnitude of MSTATE to 1.
C
C   WORK
C   LENW  = (Input)
C           WORK is an array of LENW real words used
C           internally for temporary storage. The user must allocate
C           space for this array in the calling program by a statement
C           such as
C                 REAL WORK(...)
C           The length of WORK should be at least
C           16*N + 2*NROOT + 204      if MINT is 1, or
C           N*N + 10*N + 2*NROOT + 204   if MINT is 2, or
C           N*N + 17*N + 2*NROOT + 204   if MINT is 3,
C           and LENW should be set to the value used. The contents of
C           WORK should not be disturbed between calls to SDRIV2.
C
C   IWORK
C   LENIW = (Input)
C           IWORK is an integer array of length LENIW used internally
C           for temporary storage. The user must allocate space for
C           this array in the calling program by a statement such as
```

```
C                INTEGER IWORK(...)
C           The length of IWORK should be at least
C           21    if MINT is 1, or
C           N+21  if MINT is 2 or 3,
C           and LENIW should be set to the value used. The contents
C           of IWORK should not be disturbed between calls to SDRIV2.
C
C    G     = A real FORTRAN function supplied by the user
C           if NROOT is not 0. In this case, the name must be
C           declared EXTERNAL in the user's calling program. G is
C           repeatedly called with different values of IROOT to
C           obtain the value of each of the NROOT equations for which
C           a root is desired. G is of the form:
C                REAL FUNCTION G (N, T, Y, IROOT)
C                REAL Y(*)
C                GO TO (10, ...), IROOT
C              10 G = ...
C                    .
C
C                    .
C                END (Sample)
C           Here, Y is a vector of length at least N, whose first N
C           components are the solution components at the point T.
C           The user should not alter these values. The actual length
C           of Y is determined by the user's declaration in the
C           program which calls SDRIV2. Thus the dimensioning of Y in
C           G, while required by FORTRAN convention, does not actually
C           allocate any storage. Normally a return from G passes
C           control back to SDRIV2. However, if the user would like
C           to abort the calculation, i.e., return control to the
C           program which calls SDRIV2, he should set N to zero.
C           SDRIV2 will signal this by returning a value of MSTATE
C           equal to +7(-7). In this case, the index of the equation
C           being evaluated is stored in the sixth element of IWORK.
C           Altering the value of N in G has no effect on the value of
C           N in the call sequence of SDRIV2.
C
C***LONG DESCRIPTION
C
C III. OTHER COMMUNICATION TO THE USER ............................
C
C      The first three elements of WORK and the first five elements of
C      IWORK will contain the following statistical data:
C      AVGH     The average step size used.
C      HUSED    The step size last used (successfully).
C      AVGORD   The average order used.
C      IMXERR   The index of the element of the solution vector that
C               contributed most to the last error test.
C      NQUSED   The order last used (successfully).
```

```
C           NSTEP    The number of steps taken since last initialization.
C           NFE      The number of evaluations of the right hand side.
C           NJE      The number of evaluations of the Jacobian matrix.
C
C***REFERENCES  GEAR, C. W., "NUMERICAL INITIAL VALUE PROBLEMS IN
C                   ORDINARY DIFFERENTIAL EQUATIONS", PRENTICE-HALL, 1971.
C***ROUTINES CALLED  SDRIV3,XERROR
C***END PROLOGUE  SDRIV2
```

9

Optimization and Nonlinear Least Squares

9.1 INTRODUCTION

The word **optimization** denotes either the minimization or maximization of a function. Optimization problems often arise directly in the context of finding the best design of something. As a simple example, suppose a soup company were designing a cylindrical soup can so as to minimize the amount of metal required. Let h be the height of the can, r the radius of the top of the can, and assume that the can must hold 16 ounces of cream of mushroom soup. Then the following optimization problem would be obtained

$$\text{minimize} \quad \text{Metal Area} = 2\pi r^2 + 2\pi r h,$$
$$\text{subject to} \quad \text{Volume} = 2\pi r^2 h = 16.$$

The solution of this problem is approximately $r = 1.08$, $h = 2.17$, with minimal Metal Area = 22.14. If you compare the solution to an actual soup can, you will probably find that the can shape is not optimal in terms of metal use. This is because other factors are used to determine the design, such as the design of the label, and the way the cans will pack on the shelves of the supermarket. These factors would lead to additional constraints in the definition of the problem.

More generally, an optimization problem seeks to determine the maximum or minimum (and the corresponding arguments) of a real-valued function $F(x_1, \ldots, x_n)$ of n real variables over a set S in an n-dimensional space. The function F often represents a goal or objective, and the set S represents the constraints on the problem. For the soup-can example above, let $x_1 = r = $ radius and $x_2 = h = $ height. Then the problem becomes

$$\text{minimize} \quad F(x_1, x_2) = 2\pi x_1^2 + 2\pi x_1 x_2,$$
$$\text{subject to} \quad x \in S = \left\{ (x_1, x_2) \mid 2\pi x_1^2 x_2 = 16 \right\}.$$

Sometimes the set S is the entire n-dimensional space; if so, the optimization problem is said to be **unconstrained**. Otherwise, the problem is **constrained** by whatever conditions define the set S. The set of constraints on the solution depends on the application being modelled. A simple but common constraint requires that the variables be non-negative, that is $S = \{\, x \mid x \geq 0 \,\}$; for example, it might not make sense to build -7 factories or to buy -12 washing machines. More generally, S could be defined by a set of functions satisfying equality or inequality conditions, e.g., $S = \{\, x \mid g_i(x) \geq 0,\ i = 1, \ldots, m \,\}$ where the g_i are prescribed functions of x. For the non-negativity constraints $x \geq 0$, we would have $g_i(x) = x_i$. For the soup-can example, we could use

$$S = \left\{\, (x_1, x_2) \mid 2\pi x_1^2 x_2 - 16 = 0 \,\right\}.$$

A point x that satisfies the constraints is said to be **feasible**.

Sometimes optimization problems arise indirectly as a means of solving some other problem. A data fitting problem with a nonlinear model (see Section 1 of Chapter 6) such as

$$b(t) \approx x_1 e^{x_2 t},$$

would give rise to the system

$$b(t_1) - x_1 e^{x_2 t_1} \approx 0,$$
$$b(t_2) - x_1 e^{x_2 t_2} \approx 0,$$
$$\vdots$$
$$b(t_m) - x_1 e^{x_2 t_m} \approx 0.$$

Ideally, we would like to satisfy all of the equations exactly, but in general this will not be possible. If instead a least-squares approach were used, the following optimization problem would be obtained

$$\min_{x_1, x_2} F(x_1, x_2) \equiv \sum_{j=1}^{m} [b(t_j) - x_1 e^{x_2 t_j}]^2.$$

This is called a **nonlinear least-squares problem**. This is discussed in more detail in Section 6. Other data fitting techniques, such as maximum likelihood estimation in statistics, can also lead to optimization problems.

An optimization problem can be either a minimization or a maximization. Techniques for minimizing are immediately translatable into techniques for maximizing, because minimizing $F(x)$ is the same as maximizing $-F(x)$. Although the two problems are equivalent, it will be convenient to concentrate on the minimization problem in this chapter.

Optimization problems are closely related to systems of nonlinear equations. If x^* minimizes $F(x)$ and there are no constraints, then the first derivatives of the objective function $F(x)$ will be zero; that is,

$$\frac{\partial}{\partial x_1} F(x_1, x_2, \ldots, x_n)|_{x=x^*} = 0,$$

$$\frac{\partial}{\partial x_2} F(x_1, x_2, \ldots, x_n)|_{x=x^*} = 0,$$

$$\vdots$$

$$\frac{\partial}{\partial x_n} F(x_1, x_2, \ldots, x_n)|_{x=x^*} = 0.$$

This is a system of nonlinear equations in the variables x.

If x is a point where the first derivatives of $F(x)$ are zero, it is called a **critical point**. Not all critical points minimize $F(x)$. In fact, a critical point can be a minimum, maximum, or (if it is not optimal in either sense) a **saddle point** of the function. Two other terms require definitions. We say that $F(x)$ has a **global minimum** at x^* if x^* is feasible, and if $F(x^*) \leq F(x)$ for all feasible points x. Similarly, $F(x)$ has a **local minimum** at x^* if x^* is feasible, and if $F(x^*) \leq F(x)$ for all feasible points x that are "near" x^*. Since the numerical techniques we will use are based on Taylor series, and hence allow us to examine the function $F(x)$ in a neighborhood of a particular point, it will be possible to recognize a local minimum, but difficult to recognize a global minimum.

It should not be too surprising that the techniques for solving nonlinear equations (see Chapter 7) are similar to the techniques for solving optimization problems. In fact, the techniques of Chapter 7 could be applied to the above system of nonlinear equations. Why then is there a separate chapter on optimization methods? There are three main reasons:

(1) An optimization method can guarantee convergence to a local minimum (a point where the first derivatives vanish), whereas a nonlinear equations method cannot (it might not find such a point, or it might find a maximum or saddle point of the function).

(2) An optimization method can provide savings in storage and arithmetic.

(3) Optimization problems frequently have constraints on the variables x which are harder to manage in an algorithm for nonlinear equations.

We shall confine our attention to finding *approximate local* minima or maxima for *unconstrained* problems. Justification for this remark follows.

It is easier to find local minima of a function than it is to find the absolute or global minimum over the entire domain. In part this is because the techniques of mathematical analysis, such as Taylor series expansions, provide information about how a function behaves near a given point, but do not give detailed information about the general behavior of a function over an entire region. Sometimes extra information about the problem (such as the result of a physical experiment) is available that can be used to determine approximately the location of the global minimum. Then local minimization techniques can be used to obtain a more precise solution. A few more general techniques for global optimization are available, though, and these are discussed in the references.

Constrained problems are considerably more difficult to solve than unconstrained problems, especially if the constraint functions are nonlinear, as they are in the soup-can example. However, since constrained problems are often solved using variants of unconstrained techniques, the material here would serve as a basis for studying constrained optimization methods. As an illustration, Section 9 of Chapter 6 discusses how problems with linear constraints can be converted to unconstrained problems.

Because of the close relationship between optimization problems and systems of nonlinear equations, many of the comments in Section 1 of Chapter 7 apply here. In particular, there is no way to minimize a general function in finite time, even to find a local minimum. There is no formula for the solution except in a few special cases. As a result, all the methods we derive will be iterative, and will only find approximate solutions to the problem.

From the point of view of computer arithmetic, however, there is a finite guaranteed way to determine the global minimum of a function, at least if there are upper and lower bounds on all the variables. To illustrate, consider minimizing a one-variable function $F(x)$ where x is constrained to satisfy $1 \leq x \leq 2$. On a computer, numbers have only finitely many digits; so if the relative machine precision is given by $\epsilon_{mach} = 10^{-16}$, then there are only about 10^{16} possible values of x to consider. If it were possible to evaluate $F(x)$ every 10^{-7} seconds, and if the search for the minimum were begun just after midnight on January 1, 1987, then the global minimum will have been determined just in time for dinner on Sunday, November 22, 2303. This is not very practical.

9.2 ONE-DIMENSIONAL OPTIMIZATION

We will begin with a discussion of techniques for functions of one variable. There are several reasons for this:

(1) One-dimensional problems are common.

(2) Especially efficient and reliable methods are available in one dimension.

(3) Many of the ideas used in higher-dimensional problems can be illustrated in one dimension.

(4) A major step in our n-dimensional algorithm will be the approximate solution of a one-dimensional problem.

We will describe two techniques for one-dimensional problems. The first uses derivative values to attain a high rate of convergence. The second uses only function values, and is reliable but somewhat slow.

9.2.1 Newton's Method

We consider minimization of a one-dimensional function $F(x)$, without any restrictions on x. We will assume that $F(x)$ has at least two continuous derivatives and that it is bounded below. Since we will be using derivatives within the numerical methods, the first assumption is necessary to make the methods practical. If the function is not bounded below, it may not have a local minumum, and so the problem may not make sense.

For a general function, there will be no formula for the point that minimizes $F(x)$. The idea behind Newton's method is to approximate $F(x)$ by a simpler function that we can minimize, and use the minimizer of the simpler function as the new estimate of the minimizer of $F(x)$. The process is then repeated from this new point. Since a linear function does not have a finite minimum, a quadratic function is the simplest that can be used. To form a quadratic approximation, let x_k be the current estimate of the solution x^*, and consider a Taylor series expansion of F about the point x_k:

$$F(x_k + p) = F(x_k) + pF'(x_k) + \tfrac{1}{2}p^2 F''(x_k) + \cdots.$$

The original minimization problem can be approximated using a Taylor series expansion

$$
\begin{aligned}
F(x^*) &= \min_x F(x) \\
&= \min_p F(x_k + p) \\
&= \min_p [F(x_k) + pF'(x_k) + \tfrac{1}{2}p^2 F''(x_k) + \cdots] \\
&\approx \min_p [F(x_k) + pF'(x_k) + \tfrac{1}{2}p^2 F''(x_k)].
\end{aligned}
$$

Now that the higher-order terms in the series have been ignored, this last problem is just a minimization of a quadratic function of p. We hope that minimizing this quadratic approximation will be an effective substitute for directly minimizing $F(x)$. To minimize the quadratic, take the derivative with respect to p and set it equal to zero giving

$$p = -F'(x_k)/F''(x_k).$$

Since p is an approximation to the step that would take us from x_k to the solution x^* of the original problem, then $x^* \approx x_k + p$, and the algorithm is defined by the formula

$$x_{k+1} = x_k + p = x_k - F'(x_k)/F''(x_k).$$

It is called **Newton's method**. This is exactly the same formula that would be obtained if the nonlinear equation

$$F'(x) = 0$$

were solved using the form of Newton's method described in Chapter 7.

Newton's method can also be described graphically. Let $q(x)$ be the quadratic which interpolates to $F(x)$ in the following way

$$q(x_k) = F(x_k),$$
$$q'(x_k) = F'(x_k),$$
$$q''(x_k) = F''(x_k)$$

(see Section 1 of Chapter 4). The point that minimizes $q(x)$ is exactly x_{k+1}, the Newton estimate derived in the preceding paragraph. This approach is illustrated in Figure 9.1.

Newton's Method

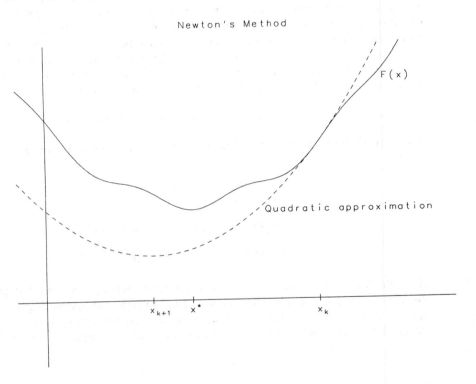

Figure 9.1   Newton's Method

The principle advantage of Newton's method is that it converges quickly, with a quadratic rate of convergence. That is, if x_k is sufficiently close to x^* and if $F''(x^*) > 0$, then

$$|x_{k+1} - x^*| \le \beta |x_k - x^*|^2$$

where β is a non-negative constant that depends on the function being minimized. Roughly speaking, this means that the number of accurate digits in x_k will double at every iteration. Newton's method is demonstrated in Example 9.1.

Example 9.1   Newton's Method.

As an illustration, Newton's method is applied to the function $F(x) = \sin x - \cos x$ with $x_0 = -0.5$. At the first iteration

$$F'(x_0) = \quad \cos x_0 + \sin x_0 = \quad \cos(-0.5) + \sin(-0.5) = 0.3982,$$

$$F''(x_0) = -\sin x_0 + \cos x_0 = -\sin(-0.5) + \cos(-0.5) = 1.3570,$$

$$p = -F'(x_0)/F''(x_0) = -0.3982/1.3570 = -0.2934,$$

$$x_{k+1} = x_k + p = -0.5 - 0.2934 = -0.7934.$$

The complete iteration is given in the Table below. As expected, the convergence is rapid, and the solution is obtained in only three iterations.

k	x_k	$|x_k - x^*|$
0	-0.5000000000000000	2.8×10^{-1}
1	-0.7934079930260234	8.0×10^{-3}
2	-0.7853979920965160	1.7×10^{-7}
3	-0.7853981633974483	1.3×10^{-17}

The assumption that $F''(x^*) > 0$ is not unusual. If $F''(x^*) < 0$, then x^* is a maximizer and not a minimizer of $F(x)$. If $F''(x^*) = 0$, then $F'(x)$ has a multiple zero at x^*. In this case, Newton's method will converge linearly, i.e., $|x_{k+1} - x^*| \approx c|x_k - x^*|$ for some constant $c \leq 1$.

Newton's method, when it works, is remarkably effective at minimizing a function. However, it is not guaranteed to converge. It can fail in several ways:

(1) Each iteration of Newton's method is based on approximating the function $F(x)$ by the first three terms of its Taylor series. If this is not a good approximation to the function, then there is no reason to believe that x_{k+1} will be closer to x^* than x_k is. This is illustrated in Figure 9.2.

(2) Although the step $p = -F'(x_k)/F''(x_k)$ is defined whenever $F''(x_k) \neq 0$, the quadratic approximation only has a minimum if $F''(x_k) > 0$. Unless this happens, Newton's method could converge to a maximum of $F(x)$ rather than a minimum. This is illustrated in Figure 9.3.

(3) A final disadvantage of Newton's method is that it requires the computation of derivatives of $F(x)$ in addition to just function values; this may be difficult or impossible in a complicated real-world problem.

These comments, troubling though they sound, do not mean that Newton's method is useless or impractical for real problems. With a few simple modifications it can be made into an effective general method, one guaranteeing progress toward the solution at every iteration. In fact, the most widely used and effective optimization methods are based on Newton's method.

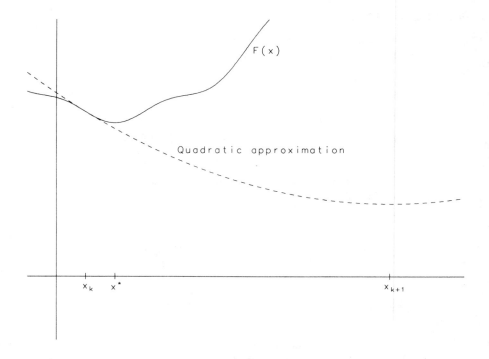

Figure 9.2   Failure of Newton's Method to Converge

Newton's method is modified in two ways to make it more reliable. First, the quadratic approximation must always have a finite minimum. If this is not true, then a different quadratic function must be used to approximate the function, one with a sensible (finite) minimum value. This is only a temporary modification to the method; at the next iteration we will attempt to use the standard Newton method again. Second, the function value must be improved at every iteration, that is, the new iterate x_{k+1} must satisfy $F(x_{k+1}) < F(x_k)$; this can be guaranteed by choosing $x_{k+1} = x_k + \alpha p$ for some $\alpha \in (0, 1)$ so as to approximately minimize $F(x_k + \alpha p)$ as a function of α. These ideas will be discussed in greater detail below, in the section on n-dimensional optimization. For many problems, those where $F''(x^*) > 0$, these modifications will only be necessary far away from the solution; near the solution x^* the method will work in the ideal manner indicated above.

The requirement for derivative calculations can be relaxed at the cost of extra function evaluations. As described in Example 2.8 of Chapter 2, the derivative $F'(x_k)$ can be approximated by the forward-difference formula

$$F'(x_k) \approx \frac{F(x_k + h) - F(x_k)}{h}$$

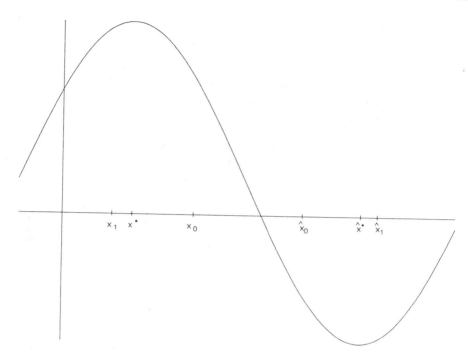

Figure 9.3   Failure of Newton's Method to Find a Minimum

for $h \approx \sqrt{\epsilon_{mach}}$. When $|F'(x_k)|$ is small relative to $|F(x_k)|$ the error in the forward-difference estimate can be large. In this case the subtraction $F(x_k + h) - F(x_k)$ can produce a large relative error. In minimization algorithms, since $F'(x^*) = 0$, this will happen as the algorithm converges. When $|F'(x_k)|$ is small a more accurate central-difference approximation should be used, such as

$$F'(x_k) \approx \frac{F(x_k + h) - F(x_k - h)}{2h},$$

which is accurate to $O(h^2)$ in exact arithmetic. As in Example 2.8, h is chosen to balance the approximation error and the rounding error, giving

$$h = \sqrt[3]{\frac{3|F(x_k)|\epsilon_{mach}}{|F'''(x_k)|}}.$$

If $F(x)$ is well-scaled, then it is possible to use the simpler formula $h = \sqrt[3]{\epsilon_{mach}}$.

The finite-difference intervals h were derived under the assumption that $F(x)$ could be computed to full working accuracy. If this is not true, then ϵ_{mach} should be replaced by the relative accuracy of the function values. For example, if $F(x)$ were accurate to 8 digits, then the relative accuracy would be $\frac{1}{2}10^{-7}$ and $h \approx 2.2 \times 10^{-4}$. Although we have only discussed approximations to $F'(x_k)$, similar approximations can be derived for the second derivative $F''(x_k)$, which is needed for the Newton formula.

If these approximations are used, then the derivative estimates will be less accurate than the function values. For example, the first derivative value will only have half as many accurate digits as $F(x_k)$. Away from the solution, this will have little effect on the behavior of the algorithm. However, it may limit the accuracy with which the solution can be computed.

Newton's method was derived by forming a quadratic approximation to $F(x)$ using function and derivative values at a point. Another iterative method based on quadratic approximations, this time using only function values, can be derived as follows: Start with three arbitrary real numbers x_1, x_2 and x_3. At the general stage one has x_{k-2}, x_{k-1}, and x_k. Let x_{k+1} be the minimizer of the quadratic function that interpolates $(x_i, F(x_i))$, $i = k-2, k-1, k$. Now discard the old point x_{k-2} and continue the iteration with x_{k-1}, x_k, and x_{k+1}. This algorithm is called **successive parabolic interpolation**. Near to the solution x^*, the iteration can be proved to converge with convergence rate approximately equal to $1.324\ldots$, provided $F''(x^*) > 0$. Note that this is the same condition that was imposed on Newton's method.

9.2.2 Golden-section Search

We now discuss more reliable but slower methods for one-dimensional minimization, ones that do not require the derivatives of $F(x)$. The methods in this subsection do not correspond to methods for solving nonlinear equations. Suppose that F is a real-valued function defined on $[0, 1]$. Suppose, moreover, that there is a unique value x^* such that $F(x^*)$ is the minimum of $F(x)$ on $[0, 1]$ and that $F(x)$ strictly decreases for $x \leq x^*$ and that $F(x)$ strictly increases for $x^* \leq x$. Such a function is called **unimodal**, and its graph takes one of the three forms shown in Figure 9.4. The restriction to the interval $[0, 1]$ is not essential, since other intervals can be considered if a change of variables is made; see Section 3 of Chapter 5. It is only important that some bound on the solution be provided.

Notice that a unimodal function need not be smooth or even continuous. For minimization purposes, though, it has a very useful property. For any two points x_1, x_2 in the interval satisfying $x_1 < x_2 \leq x^*$, we can conclude that $F(x_1) > F(x_2)$. Similarly, if $x^* \leq x_1 < x_2$, then $F(x_1) < F(x_2)$. Conversely, if $x_1 < x_2$ and $F(x_1) > F(x_2)$, then $x_1 \leq x^* \leq 1$, and if $F(x_1) < F(x_2)$, then $0 \leq x^* \leq x_2$. (Of course, if $F(x_1) = F(x_2)$, we have the extra information that $x_1 \leq x^* \leq x_2$, but we never use this.) These facts enable us to refine our estimate of the minimizer of $F(x)$ using function values alone. The problem is to find a set of points $x_1, x_2, \ldots, x_k$ and corresponding function values $F(x_1), \ldots, F(x_k)$, and then determine that the optimal value of F lies in the interval $x_{i-1} \leq x^* \leq x_{i+1}$ for some i. Such an interval is called the **interval of uncertainty**.

Suppose that we are allowed to evaluate the function k times, where $k > 1$ is given. How can we use these evaluations to locate x^* with the smallest possible interval of uncertainty?

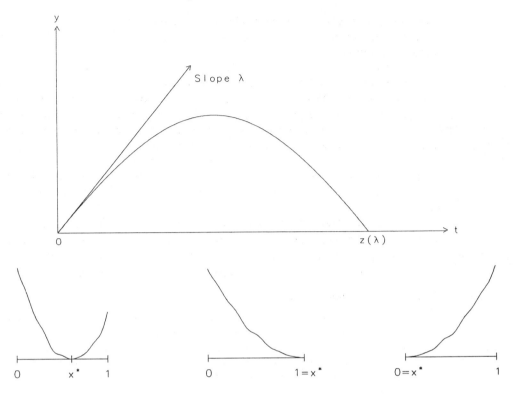

Figure 9.4   Unimodal Functions

The algorithm for choosing the x_i $(i = 1, \ldots, k)$ is called a **search plan**. If we know only that F is unimodal, what is the optimal strategy for finding x^*? An **optimal** search plan for a given number of function evaluations is one that results in the smallest interval of uncertainty. To derive such a plan, we shall assume that the function F will be evaluated sequentially, that is, at one point at a time.

A theory for choosing sequential search plans was started by J. Kiefer in the early 1950s. The algorithm for an optimal strategy will produce k test points sequentially, which we choose to number x_k, x_{k-1}, $\ldots$, x_2, x_1. We will first describe it in general and then apply it to an example. At the first stage two points x_k and x_{k-1} are chosen simultaneously, with $x_{k-1} < x_k$. If $F(x_{k-1}) \geq F(x_k)$, then the interval of uncertainty is reduced to $[x_{k-1}, 1]$, *and we already have the value of $F(x_k)$ to help further refine the interval of uncertainty.* On the other hand, if $F(x_{k-1}) \leq F(x_k)$, then the interval of uncertainty is reduced to $[0, x_k]$, and we know $F(x_{k-1})$ already. The key to the optimal choice of x_{k-1} and x_k is to be sure that the one test point inherited by the reduced interval of uncertainty is itself located at an optimal position for the continued search.

In order to describe the algorithm in more detail, and to be able to apply the algorithm on a general interval rather than on just $[0, 1]$, it is useful to define the **coordinate**

r **of a point** x **relative to an interval** $[a, b]$ as the number $r = (x - a)/(b - a)$. Thus, a has the coordinate 0 and b the coordinate 1, relative to the interval $[a, b]$, while the midpoint has the coordinate $\frac{1}{2}$.

Let $f_0 = f_1 = 1$, $f_2 = 2$, $f_3 = 3$, $f_4 = 5$, $f_5 = 8$, ..., $f_k = f_{k-1} + f_{k-2}$. The f_i are the famous Fibonacci numbers. The optimal strategy for the sequential search for a minimum is called a **Fibonacci search** because it is closely related to these numbers. In the optimal strategy one picks $x_{k-1} = f_{k-2}/f_k$ and $x_k = f_{k-1}/f_k$. Whichever interval $[0, x_k]$ or $[x_{k-1}, 1]$ becomes the reduced interval of uncertainty, the inherited point will be located at one of the following two coordinates relative to the new interval: f_{k-3}/f_{k-1} or f_{k-2}/f_{k-1}. Then x_{k-2} is selected to be the other one of the above two coordinates relative to the new interval. From $F(x_{k-2})$ and the function value inherited from the first interval, one can further reduce the interval of uncertainty and inherit one function value.

At the last stage, one has some interval of uncertainty $[a, b]$ with its midpoint $\frac{1}{2}$ as the inherited point. Then x_1 is selected to have relative coordinate $\frac{1}{2} + \epsilon$, where ϵ is some small positive number, and the final interval of uncertainty is either $[0, \frac{1}{2} + \epsilon]$ or $[\frac{1}{2}, 1]$ relative to $[a, b]$. This method is demonstrated in Example 9.2.

Example 9.2 Fibonacci Search.

Consider minimizing $F(x) = (x - .65)^2$ on the interval $[0, 1]$. We will use a Fibonacci search beginning with $f_7 = 21$. See Figure 9.5 and the table below. The value $\epsilon = .01/21$ was used. At iteration 1, the function is evaluated at the points $f_6/f_7 = 13/21$ and $f_5/f_7 = 8/21$, giving the function values .000958 and .0724, respectively. Since the function is unimodal, the minimum must lie in the interval $[8/21, 1]$. And so forth.

Iteration	Interval	(x_{k-1}, x_k) Relative	(x_{k-1}, x_k) Absolute	$(F(x_{k-1}), F(x_k))$
1	$[\ 0/21, 21/21]$	$(8/21, 13/21)$	$(8/21, 13/21)$	$(.0724, .000958)$
2	$[\ 8/21, 21/21]$	$(5/13, 8/13)$	$(13/21, 16/21)$	$(.000958, .0125)$
3	$[\ 8/21, 16/21]$	$(3/8, 5/8)$	$(11/21, 13/21)$	$(.0159, .000958)$
4	$[11/21, 16/21]$	$(2/5, 3/5)$	$(13/21, 14/21)$	$(.000958, .000278)$
5	$[13/21, 16/21]$	$(1/3, 2/3)$	$(14/21, 15/21)$	$(.000278, .000413)$
6	$[13/21, 15/21]$	$(1/2, 1/2 + \epsilon)$	$(14/21, 14.01/21)$	$(.000278, .000294)$

After the final iteration, the interval of uncertainty has been reduced to $[13/21, 14.01/21] \approx [.62, .67]$. The midpoint of this interval .645 is a good approximation to the solution .65. ∎

There is a disadvantage to Fibonacci search: The total number of function evaluations must be chosen in advance. On most problems, we would prefer to iterate until the function value or the point x is obtained to sufficient accuracy. A close variant of Fibonacci search called **golden section search** is a solution to this problem. To derive it, we examine Fibonacci search in more detail.

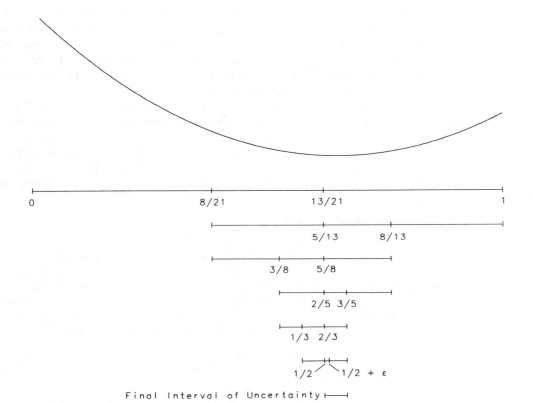

Figure 9.5 Fibonacci Search

At the first stage the length of the interval of uncertainty is reduced from 1 to f_{k-1}/f_k. At succeeding stages the intervals are reduced by the factors

$$\frac{f_{k-2}}{f_{k-1}}, \frac{f_{k-3}}{f_{k-2}}, \ldots, \frac{f_2}{f_3}, \frac{f_1}{f_2}(1+2\epsilon).$$

Thus the length of the final interval of uncertainty is $(1+2\epsilon)/f_k$. Ignoring ϵ, it can be shown that $1/f_k \approx r^k$ as $k \to \infty$, where

$$r = \frac{\sqrt{5}-1}{2} \approx 0.6180.$$

Thus, asymptotically for large k, each step of the Fibonacci search reduces the interval of uncertainty by the factor 0.6180. This should be compared with 0.5, the reduction of the interval of uncertainty of the bisection method for finding a zero of a function (see Section 2.1 of Chapter 7).

For large values of k, the locations of x_{k-1} and x_k are close to $1 - r \approx 0.3820$ and $r \approx 0.6180$, respectively, relative to the initial interval. If the search plan is started with these values, rather than those based on the Fibonacci numbers, the plan will be close to an optimum strategy. To see how to proceed further, suppose for example that $F(0.3820) < F(0.6180)$. The x^* is known to be in the interval $[0, 0.6180]$. Hence one wants to evaluate F at the points $(1 - r) * 0.6180 \approx 0.3820 * 0.6180$ and $r * .6180 \approx 0.6180 * 0.6180$, the points r and $1 - r$ relative to the reduced interval. But since $0.6180 * 0.6180 \approx 0.3820 \approx x_{k-1}$, F is already known there. Thus, again, only one evaluation of F is required at each stage of the iteration after the first, and each stage reduces the length of the interval of uncertainty by the factor 0.6180. In contrast with the Fibonacci search, one does not have to fix the number k before the search begins. For this reason, it is a more practical algorithm. Golden section search is illustrated in Example 9.3.

Example 9.3   Golden Section Search.

We again consider minimizing $F(x) = (x - .65)^2$ on the interval $[0, 1]$, but now using golden section instead of Fibonacci search. See the table below. At iteration 1, the function is evaluated at the points $r \approx 0.6180$ and $1 - r \approx 0.3820$, giving the function values 0.001022 and 0.071842, respectively. Since the function is unimodal, the minimum must lie in the interval $[a, b] = [0.3820, 1]$. At the next iteration, the function is evaluated at the points $y = (b - a)r + a \approx 0.7639$ and $x = (b - a)(1 - r) + a \approx 0.6180$; these are the coordinates of the points r and $1 - r$ with respect to the new interval. And so forth.

Iteration	Interval $[a, b]$	(x, y) Relative	(x, y) Absolute	$(F(x), F(y))$
1	[.0000, 1.0000]	(.3820, .6180)	(.3820, .6180)	(.071842, .001022)
2	[.3820, 1.0000]	(.3820, .6180)	(.6180, .7639)	(.001022, .012981)
3	[.3820, .7639]	(.3820, .6180)	(.5279, .6180)	(.014917, .001022)
4	[.5278, .7639]	(.3820, .6180)	(.6180, .6738)	(.001022, .000565)
5	[.6180, .7639]	(.3820, .6180)	(.6738, .7082)	(.000565, .003388)
6	[.6180, .7082]	(.3820, .6180)	(.6525, .6738)	(.000006, .000565)

After the final iteration, the interval of uncertainty has been reduced to $[0.6180, 0.7082]$. The midpoint of this interval 0.6631 can be taken as an approximation to the solution $.65$.

■

Because of the history of the number $r \approx 0.6180$, this method of finding x^* is called the golden-section search. The number $1/r = \phi = (1 + \sqrt{5})/2 = 1.6180\ldots$ is called the **golden ratio**. In Chapter 7, we noted that the rate of convergence for the secant method is equal to ϕ. The golden ratio ϕ is a fundamental ratio that appears in many different places. For an interesting discussion of ϕ see Gardner (1961).

The golden-section search is analogous to the method of bisection for finding a real zero of a function (see Chapter 7) in that it is guaranteed to work in the worst possible case, and the price of this safety is slowness of convergence, which is only linear. Golden-section search takes no advantage whatever of the possible smoothness of F.

9.3 **SUBROUTINE** FMIN

Brent (1973) has created a minimum finder which uses a combination of the golden-section search and successive parabolic interpolation. The algorithm is entirely analogous to the zero finder FZERO, which uses a combination of bisection and inverse quadratic interpolation. A Fortran translation of Brent's algorithm is described in this section. It is a function subprogram

$$\text{FUNCTION FMIN (A, B, F, TOL)}.$$

Here [A, B] is the initial interval on which the function F is defined. The value returned by FMIN is an estimate of the point x^* where $F(x)$ is minimized. TOL is an input parameter that, roughly speaking, gives the desired length of the final interval of uncertainty. Thus, if you set TOL to 10^{-3}, and if the result returned by FMIN has a magnitude of about 1, then the answer has about three significant digits. If you set TOL to 10^{-6} and the answer is (say) 0.00000385746, then this result probably has no significant digits, but if you interpret a result of this magnitude as zero, then you do not need any significant digits. FMIN uses the golden-section search, switching when possible to successive parabolic interpolation.

The subprograms FZERO and FMIN have similar parameters—an interval to be searched, a function to be evaluated, and a tolerance. Both subroutines attempt to reduce the length of the interval until it is less than the tolerance. However, there is an important difference between the two routines which affects the choice of the tolerance. Consider first the nonlinear equation problem. If $F(x^*) = 0$ and $F'(x^*) \neq 0$, then for small ϵ

$$F(x^* + \epsilon) = F(x^*) + \epsilon F'(x^*) + \epsilon^2 \frac{F''(x^*)}{2} + \cdots$$
$$\approx c\epsilon,$$

where $c = F'(x^*)$. Thus small changes in x^* cause proportionally small changes in $F(x^*)$. It is reasonable to choose as the tolerance for FZERO a value roughly the size of the errors in the function values. Often this can be as small as the machine roundoff level ϵ_{mach}.

Returning to the minimization problem, if we are seeking a minimum where $F'(x^*) = 0$ and $F''(x^*) \neq 0$, then the first-derivative term in the Taylor series above vanishes. For small ϵ

$$F(x^* + \epsilon) = F(x^*) + \epsilon^2 \frac{F''(x^*)}{2} + \cdots$$
$$\approx F(x^*) + c\epsilon^2,$$

where $c = F''(x^*)/2$. A change of order ϵ in x^* now causes a change of order ϵ^2 in $F(x^*)$. So it is unreasonable to choose as the tolerance for FMIN a value smaller than the square root of the error in the function values. In other words, simple zeros of a function can often be found to nearly *full* machine precision, but minimizers of a function may be found to only about *half* precision.

If TOL is the tolerance input to FZERO or FMIN, and ϵ_{mach} is the machine precision, then FZERO never evaluates the function at points closer together than

$$2 * \text{TOL} * (\text{ABS}(X) + 1.0),$$

while FMIN never evaluates the function at points closer together than

$$\text{SQRT}(\epsilon_{\text{mach}}) * \text{ABS}(X) + \text{TOL}/3.$$

Thus small values of TOL are much more likely to be overridden by FMIN than by FZERO.

Unimodal functions occur relatively infrequently in practice, and it can be difficult to verify if a given function is unimodal. If in doubt, several approaches are possible. Since many functions are unimodal in a neighborhood of a local minimum, FMIN is likely to be successful if the initial interval [A, B] provides a good estimate of the solution. Even if the initial interval is large, in many cases FMIN will find a local minimum of the function. A second idea is to use the routine UNCMIN described in Section 5; it is intended for n-dimensional minimization but can be applied when $n = 1$ if the function is sufficiently smooth.

If the function is not differentiable, and there is no estimate of the minimizer, then there is little theoretical basis for designing an algorithm. (This is because the value of the function at one point gives no information about the values of the function at other points.) In this case, one possibility is to evaluate the function at a sequence of points and pick the lowest value. For example, a program might pick some $a > 0$ and then evaluate $F(0)$, $F(a)$, $F(2a)$, $F(2^2a)$, $F(2^3a)$, etc., as long as the values continue to decrease, accepting as an approximate x^* the point $2^k a$ with least $F(x)$. Or, if $F(a) > F(0)$, the program evaluates $F(2^{-1}a)$, $F(2^{-2}a)$, ... until a value $F(2^{-r}a)$ is accepted as an approximate x^*. The reader can devise various ways of refining the x^* accepted by the above crude methods. This is not efficient; whenever possible the more sophisticated algorithms discussed in this chapter should be used.

For a sample program using FMIN, we use the same function as we used for FZERO, $F(x) = x^3 - 2x - 5$. Since $F'(x)$ is a quadratic, the extrema can be found analytically. Note that the value of TOL used here is the square root of the value used with FZERO.

The output is XSTAR $= 0.8168136$.

```
C SAMPLE PROGRAM FOR FMIN
C
      REAL A, B, XSTAR, TOL, FMIN
      EXTERNAL F
```

```
A      = 0.1
B      = 0.9
TOL    = 1.0E-5
XSTAR = FMIN (A, B, F, TOL)
WRITE (*,*) ' XSTAR =', XSTAR
STOP
END
C

REAL FUNCTION F(X)
REAL X
F = X*(X*X - 2.0) - 5.0
RETURN
END
```

9.4 OPTIMIZATION IN MANY DIMENSIONS

The local minimization of a function of n variables is so important that algorithms have been devised for it for over 140 years. Many of them are based on the following metaphor: Imagine the graph of a function as a landscape with hills and valleys. The bottoms of the valleys represent minima of the function. If we are not at a minimum, then a natural idea is to move downhill until we find one.

An old method, known now as the **method of steepest descent**, uses this idea. It was proposed by A. Cauchy in 1845. Let the vector $(x_1, \ldots, x_n)^T$ be denoted by x, and assume that the function $F(x)$ has continuous partial derivatives of several orders. Let $\nabla F = \nabla F(x)$ denote the gradient of F at x, the vector whose i-th component is $(\nabla F)_i(x) = \partial F / \partial x_i$. For fixed x and varying α, $x - \alpha \nabla F$ is a ray through x. It is known that $-\nabla F(x)$ is a "downhill" direction for the function $F(x)$ at the point x in the sense that for small enough positive α, the function value will decrease if we move a small distance along the direction $-\nabla F(x)$, i.e., it is guaranteed that $F(x - \alpha \nabla F) < F(x)$ for "small" $\alpha > 0$.

This approach may seem *ad hoc* since it suggests that any downhill direction will be satisfactory. However $-\nabla F(x)$ is the "best" downhill direction since the function value will decrease more rapidly along the ray $x - \alpha \nabla F$ than along any other ray starting at the point x. The direction $-\nabla F(x)$ is "best" in a slightly different sense also. When we derived Newton's method in Section 2.1, the objective function was approximated using three terms of its Taylor series to obtain the step. It can be shown that $-\nabla F(x)$ is the step obtained by minimizing the first *two* terms of the Taylor series. For these reasons, $-\nabla F(x)$ is called the direction of *steepest* descent for F at x.

Cauchy proposed searching the half line defined by $x - \alpha \nabla F$ ($0 < \alpha < \infty$) for a minimum value of F. This is a one-dimensional minimization of the type discussed in Section 2. Having found this minimum, one starts over and searches along the half line of steepest descent from the new x. Under weak hypotheses, this method will converge to a local minimum of F.

Theoretical analysis has predicted that Cauchy's method will converge extremely slowly in some cases, and experience has confirmed this in many practical cases, even for such modest values of n as 2, 3, or 4. The poor behavior of the steepest-descent method is illustrated in Example 9.4.

Example 9.4   The Steepest-Descent Method.

We will use steepest descent to solve

$$\text{minimize } F(x_1, x_2, x_3) = x_1^2 + 10x_2^2 + 100x_3^2$$

with starting guess $(1, 1, 1)$. The solution is $x^* = (0, 0, 0)^T$. The gradient of this function is given by

$$\nabla F(x) = (2x_1, 20x_2, 200x_3)^T.$$

At each iteration, the new point is determined by solving the one-dimensional problem

$$\min_{\alpha} F(x - \alpha \nabla F(x)) = (x_1 - 2\alpha x_1)^2 + 10(x_2 - 20\alpha x_2)^2 + 100(x_3 - 200\alpha x_3)^2.$$

This is a quadratic function in α, and can be minimized by taking the derivative and setting it equal to zero, giving

$$\alpha = \frac{x_1^2 + 10^2 x_2^2 + 10^4 x_3^2}{2(x_1^2 + 10^3 x_2^2 + 10^6 x_3^2)}.$$

(For general problems it will not be possible to find a formula for α.) The first few iterations of the algorithm are shown in the table below.

Iteration	x	$F(x)$
0	(1.0000, 1.0000, 1.0000)	111.0000
1	(0.9899, 0.8991, −0.0091)	9.0718
2	(0.8981, 0.0661, 0.0751)	1.4148
3	(0.8890, 0.0593, −0.0016)	0.8258
4	(0.7368, −0.0422, 0.0255)	0.6260
5	(0.7287, −0.0375, 0.0027)	0.5458
6	(0.6706, −0.0076, 0.0190)	0.4863

The convergence is slow. After another 180 iterations, reasonable results are obtained: $x = (1.07 \times 10^{-4}, 0, 3.01 \times 10^{-6})^T$ and $F(x) = 1.24 \times 10^{-9}$. ∎

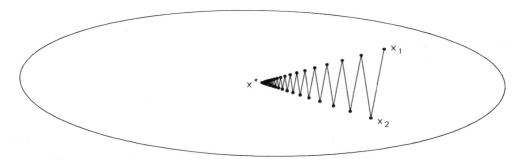

Figure 9.6   Steepest Descent Method

As a further illustration, the graph of a steepest-descent iteration for a two-dimension problem is given in Figure 9.6. The path to the minimum at the center proceeds very slowly by a route that gradually works down the valley, oscillating back and forth in the local gradient directions. There is great need to accelerate the convergence.

The big difficulty is that the local gradient does not even approximately point toward the solution x^*. We can improve on this by considering the n-dimensional analogue of Newton's method. As before, it is based on a quadratic approximation to the function $F(x)$ derived from a Taylor expansion about the point x. (It can also be derived using the techniques of Chapter 7 applied to the system of nonlinear equations $\nabla F(x) = 0$.)

We will write the Taylor series in matrix/vector form. In two dimensions, the second-order Taylor series approximation is

$$F(x_1 + p_1, x_2 + p_2) \approx F(x_1, x_2) + p_1 \frac{\partial F(x_1, x_2)}{\partial x_1} + p_2 \frac{\partial F(x_1, x_2)}{\partial x_2}$$

$$+ \frac{1}{2}\left(p_1^2 \frac{\partial^2 F(x_1, x_2)}{\partial x_1^2} + p_1 p_2 \frac{\partial^2 F(x_1, x_2)}{\partial x_1 \partial x_2} + p_2 p_1 \frac{\partial^2 F(x_1, x_2)}{\partial x_2 \partial x_1} + p_2^2 \frac{\partial^2 F(x_1, x_2)}{\partial x_2^2} \right).$$

Let $\nabla^2 F$ be the constant matrix of second partial derivatives of F at x^*—the so-called **Hessian matrix**:

$$\nabla^2 F_{i,j} = \frac{\partial^2 F}{\partial x_i \partial x_j}.$$

The Taylor series above may seem complicated, in part because of the number of terms. If the notations for the gradient and the Hessian matrix are used, however, the two-dimensional Taylor series closely resembles the Taylor series in one dimension. If we compare terms with the series above we see that

$$F(x + p) \approx F(x_1, x_2) + (\, p_1 \quad p_2 \,) \begin{pmatrix} \frac{\partial F(x_1, x_2)}{\partial x_1} \\ \frac{\partial F(x_1, x_2)}{\partial x_2} \end{pmatrix}$$

$$+ \frac{1}{2}(\, p_1 \quad p_2 \,) \begin{pmatrix} \frac{\partial^2 F(x_1, x_2)}{\partial x_1^2} & \frac{\partial^2 F(x_1, x_2)}{\partial x_1 \partial x_2} \\ \frac{\partial^2 F(x_1, x_2)}{\partial x_2 \partial x_1} & \frac{\partial^2 F(x_1, x_2)}{\partial x_2^2} \end{pmatrix} \begin{pmatrix} p_1 \\ p_2 \end{pmatrix}$$

$$= F(x) + p^T \nabla F(x) + \tfrac{1}{2} p^T \nabla^2 F(x) p.$$

Even though this result has only been derived in two dimensions, this final expression is also valid in n dimensions. (A similar Taylor series expansion was derived in Section 5 of Chapter 7.)

To apply this to Newton's method we write

$$F(x + p) \approx F(x) + p^T \nabla F(x) + \tfrac{1}{2} p^T \nabla^2 F(x) p \equiv F(x) + Q(p),$$

where the higher-order terms are considered as a quadratic function in p. To obtain the step p, we now minimize this quadratic as a function of p by forming its gradient with respect to p

$$\nabla_p Q(p) = \nabla_p (p^T \nabla F(x) + \tfrac{1}{2} p^T \nabla^2 F(x) p) = \nabla F(x) + \nabla^2 F(x) p$$

and setting it equal to zero

$$\nabla^2 F(x) p = -\nabla F(x).$$

This is a set of n linear equations in the n unknowns $p = (p_1, p_2, \ldots, p_n)^T$. These linear equations are called the **Newton equations**. Thus, $x_{k+1} = x_k + p = x_k - \nabla^2 F(x_k)^{-1} \nabla F(x_k)$. (As discussed in Chapter 3, the step p is computed by solving the system of equations and not by explicitly forming the inverse matrix.)

Newton's method in n dimensions has the same rapid convergence properties as in one dimension, namely it converges quadratically near the solution,

$$\|x_{k+1} - x^*\|_2 \le \beta \|x_k - x^*\|_2^2,$$

for some non-negative constant β depending on $F(x)$. As an illustration, see Example 9.5. This is not the same minimization problem as that considered in Example 9.4. Newton's method would solve that problem in one iteration, since it is a quadratic function (why?).

Example 9.5   Newton's Method.

Consider the problem

$$\text{minimize } F(x_1, x_2) = 100(x_2 - x_1^2)^2 + (1 - x_1)^2,$$

with initial guess $x = (-2.0, 5.0)^T$. The gradient and Hessian of this function are

$$\nabla F(x) = \begin{pmatrix} -400x_1(x_2 - x_1^2) + 2(x_1 - 1) \\ 200(x_2 - x_1^2) \end{pmatrix}, \qquad \nabla^2 F(x) = \begin{pmatrix} 1200x_1^2 - 400x_2 + 2 & -400x_1 \\ -400x_1 & 200 \end{pmatrix}.$$

Graphs of the contours of this function are given in Figure 9.7. They also show the results of the Newton and steepest-descent algorithms for this function. The results of applying Newton's method to this problem are given in more detail in the table below.

Iteration	x	$F(x)$
0	$(-2.000000, \quad 5.000000)$	$1.09 \times 10^{+2}$
1	$(-2.015075, \quad 4.060302)$	$9.09 \times 10^{+0}$
2	$(\ 0.868913, -7.562379)$	$6.92 \times 10^{+0}$
3	$(\ 0.868992, \quad 0.755147)$	1.70×10^{-2}
4	$(\ 0.999999, \quad 0.982837)$	2.93×10^{-2}
5	$(\ 0.999999, \quad 0.999999)$	1.58×10^{-14}
6	$(\ 1.000000, \quad 0.999999)$	2.51×10^{-26}
7	$(\ 1.000000, \quad 1.000000)$	0.0

As in one dimension, there are disadvantages to Newton's method. For example, it can fail to converge. To remedy this, it is typically incorporated into a general algorithm of the following form. The descent algorithm below is quite general; the steepest descent method could also be used inside it to produce a convergent method. It is mainly used to guarantee convergence; the properties of the descent direction computed in step 2 determine how rapidly the overall algorithm converges.

Descent Algorithm for Nonlinear Minimization
Given an initial guess x_0, set $k \leftarrow 0$.
1. Compute $F_k = F(x_k)$, $\nabla F_k = \nabla F(x_k)$, the function and gradient values. Test for convergence. If converged, stop.
2. Compute a descent direction p, i.e., a direction p such that $F(x_k + \epsilon p) < F_k$ for ϵ small. This is equivalent to requiring that $p^T \nabla F_k < 0$.
3. Line search: Find $\alpha > 0$ such that $F(x_k + \alpha p) < F_k$. Set $x_{k+1} \leftarrow x_k + \alpha p$, $k \leftarrow k+1$. Go to step 1.

By using appropriate techniques in each step, it is possible to guarantee convergence of such a minimization method to a local minimum for a large class of problems. Step 3 is an approximate one-dimensional minimization problem, and is solved using techniques similar to those described earlier. This step is necessary since, even when the Newton direction is satisfactory as a downhill direction, it may not have the correct magnitude or scaling if x_k is far from x^*.

*9.4.1 Modifications to Newton's Method

Ideally, when using Newton's method in Step 2 of the Descent Algorithm above, $p = -\nabla^2 F^{-1} \nabla F$. However, this may not be a descent direction. As a simple illustration, consider the one-dimensional example

$$F(x) = 2x^3 - 5x^2 + 4x$$

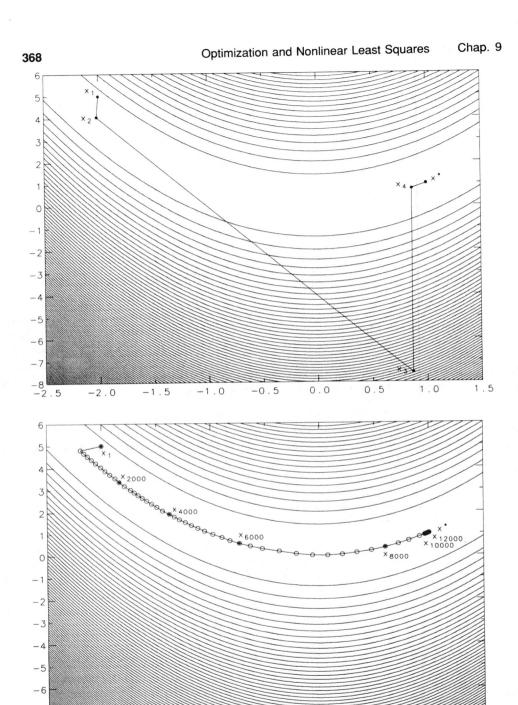

Figure 9.7   Contour Plots for Example 9.5

with initial guess $x_0 = 0$. Then $F(x_0) = 0$. We compute the Newton step using

$$p = -\nabla^2 F^{-1} \nabla F = -(12x_0 - 10)^{-1}(6x_0^2 - 10x_0 + 4) = 0.4.$$

Then $F(x_0 + p) = F(.4) = .9280 > F(x_0)$. In fact, it is straightforward to check that $F(x_0 + \alpha p) > F(x_0)$ for all $\alpha \in (0, 1]$. Thus p is not a descent direction.

It is sometimes possible to guarantee that Newton's method will produce a descent direction. Suppose that inverse Hessian matrix $\nabla^2 F^{-1}$ is positive definite, that is, it satisfies the condition $z^T \nabla^2 F^{-1} z > 0$ for all $z \neq 0$. In this case, the Newton direction is guaranteed to be a downhill direction since (via a Taylor series expansion with $\nabla F = \nabla F(x) =$ gradient of $F(x)$)

$$
\begin{aligned}
F(x + \epsilon p) &= F(x) + \nabla F^T(\epsilon p) + O(\epsilon^2) \\
&= F(x) + \epsilon \nabla F^T(-\nabla^2 F^{-1} \nabla F) + O(\epsilon^2) \\
&= F(x) - \epsilon \nabla F^T \nabla^2 F^{-1} \nabla F + O(\epsilon^2).
\end{aligned}
$$

Since $\nabla^2 F^{-1}$ is positive definite, $\nabla F^T \nabla^2 F^{-1} \nabla F > 0$ as long as $\nabla F \neq 0$. Thus if ϵ is small and $\nabla F \neq 0$, then $F(x + \epsilon p) < F(x)$, that is, p is a downhill direction. If $\nabla F = 0$, then x is a critical point, and further conditions involving second derivatives must be checked to determine if x minimizes the function.

Since few functions will have $\nabla^2 F^{-1}(x)$ positive definite for all x, some strategy is required to guarantee a descent direction. One successful idea, due to Gill and Murray (1974), is to replace G by a "nearby" positive-definite matrix $\bar{\nabla}^2 F$ as necessary, and to compute the step by solving $\bar{\nabla}^2 F p = -\nabla F$. This modified direction will be a downhill direction whenever $\nabla F \neq 0$. The modified Hessian is only used for one iteration. At the next iteration, the algorithm will try to use the true Hessian again, as long as it is positive definite.

Even with these modifications, there are still disadvantages to Newton's method. First, there is the need to compute the second-derivative matrix. As in one dimension, $\nabla^2 F$ can be approximated using finite differences of function or gradient values. In n dimensions, though, this would require n additional gradient evaluations or n^2 additional function evaluations, which would be expensive unless the function were especially easy to compute.

Another disadvantage of Newton's method is that computing the Newton step p requires the solution of a system of n linear equations at a cost of about $n^3/6$ arithmetic operations. For large n, this is a huge amount of work per iteration.

The steepest-descent method suffers from neither of these difficulties, but results in a great loss in performance. Instead, effective and practical methods can be developed by building an approximation to the Hessian as the function is minimized. The approximation is based on secant approximations to the Hessian, and is a generalization of the secant method described in Section 2.3 of Chapter 7. If we have $B_k \approx \nabla^2 F_k = \nabla^2 F(x_k)$, then the step at the k-th iteration will be defined by

$$B_k p = -\nabla F_k,$$

and so such a method is a compromise on Newton's method, where an approximate rather than an exact Hessian is used. This step p will be used within the general descent method above. After the line search obtains $x_{k+1} = x_k + \alpha p$, the approximate Hessian B_k will be updated to produce the new approximation B_{k+1}, using the values of x_{k+1} and $\nabla F(x_{k+1})$.

If $F(x)$ were a quadratic function, then it would satisfy

$$(\nabla^2 F)(x_{k+1} - x_k) = \nabla F_{k+1} - \nabla F_k.$$

The new Hessian approximation will be chosen so that

$$B_{k+1}(x_{k+1} - x_k) = \nabla F_{k+1} - \nabla F_k.$$

In one dimension, this determines B_{k+1} uniquely. In higher dimensions, further conditions are necessary to specify B_{k+1}. Typically, B_k will be updated by a simple, low-rank matrix, as discussed in Section 9.1 of Chapter 3.

What are the advantages of such an approach?

(1) Near the solution the method will still converge rapidly, at a superlinear (better than linear) rate.

(2) Only gradient values are used, and not second derivatives.

(3) It is possible to choose B_k to be positive definite, so that a descent direction is always obtained.

(4) Because B_k is only modified by a low-rank matrix, the work per iteration (excluding the cost of evaluating the function) can be reduced to $O(n^2)$ (see Section 9.1 of Chapter 3).

Methods of this type are called **quasi-Newton** methods. An early algorithm was devised by Davidon in 1959. There have been many improvements of quasi-Newton methods since Davidon's original paper. The reader is referred to a survey by Dennis and Moré (1974). Convergence theory for the methods is also discussed in this paper.

9.5 SUBROUTINE UNCMIN

The routine UNCMIN is designed for the unconstrained minimization of a real-valued function $F(x)$ of n variables. It can be used when $n = 1$, but in this case it is likely to be less efficient than subroutine FMIN. It is based on the general descent method outlined above and uses a quasi-Newton algorithm.

As with FMIN, the user must provide a routine to evaluate the function. The gradient values required by UNCMIN will be computed using finite differencing. In addition, work space must be provided for storage of the approximate Hessian matrix and some subsidiary vectors.

Also as with FMIN, if we are seeking a minimum where $\nabla F(x) = 0$, a change of order ϵ in x now causes a change of order ϵ^2 in $F(x)$. So it is unreasonable to choose as the tolerance for UNCMIN a value smaller than the square root of the error in the function values.

UNCMIN will find a local minimum of the function $F(x)$ if the function is bounded below. It is not possible to guarantee that a global minimum will be found. If $F(x)$ is not bounded below, but still has local minima, UNCMIN may still be effective if the initial guess is sufficiently good.

In the sample problem below, we are minimizing a function of $n = 10$ variables, known as the Generalized Rosenbrock Function. $F(x)$ is defined by

$$F(x_1, \ldots, x_n) = 1 + \sum_{i=2}^{n} \left[100 \left(x_i - x_{i-1}^2 \right)^2 + \left(1 - x_{i-1} \right)^2 \right].$$

The initial estimate of the minimizer is

$$\left(\frac{1}{11}, \frac{2}{11}, \frac{3}{11}, \ldots, \frac{10}{11} \right)^T,$$

and the optimal solution is

$$x^* = (1, 1, 1, \ldots, 1)^T$$

with optimal function value $F(x^*) = 1$.

```
C MAIN PROGRAM TO MINIMIZE A FUNCTION REPRESENTED BY ROUTINE CALCF
C
      PARAMETER (N = 10, LWORK = N*(N+10))
      REAL X0(N), X(N), F, WORK(LWORK)
      EXTERNAL CALCF
C
C SPECIFY INITIAL ESTIMATE OF THE SOLUTION
C
      DO 10 I = 1,N
         X0(I) = I / FLOAT(N+1)
10    CONTINUE
C
C MINIMIZE FUNCTION
C
      CALL UNCMIN (N, X0, CALCF, X, F, IERROR, WORK, LWORK)
C
C PRINT RESULTS
C
      IF (IERROR .NE. 0) WRITE (*,*) ' ERROR CODE =', IERROR
      WRITE (*,*) ' F(X*) =', F
      WRITE (*,*) ' X* ='
      WRITE (*,800) (X(I), I = 1,N)
C
      STOP
```

```
800    FORMAT (5F12.6)
       END
C
C OBJECTIVE FUNCTION
C
       SUBROUTINE CALCF (N, X, F)
       REAL X(N), F, T1, T2
C
       T1 = 0.0
       T2 = 0.0
       DO 10 I = 2,N
          T1 = T1 + (X(I)-X(I-1)**2)**2
          T2 = T2 + (1.0-X(I-1))**2
10     CONTINUE
       F = 1.0 + 100.0*T1 + T2
C
       RETURN
       END
```

For this particular problem, which is essentially a sum of squares, the results are strongly dependent on the precision of the computer's floating point arithmetic (see Section 5 of Chapter 2 and Section 3 of this chapter). For example, when this program is run on a CDC Cyber with $\epsilon_{mach} = 6 \times 10^{-15}$, the computed optimal solution agrees with x^* to better than six digits. The same program run on an IBM PC/AT ($\epsilon_{mach} = 1 \times 10^{-7}$) produces

```
UNCMIN WARNING - - INFO = 3: CANNOT FIND LOWER POINT
ERROR CODE =            3
F(X*) =     1.00116
X* =
  0.999745   0.999598   0.999370   0.998893   0.998021
  0.996177   0.992586   0.985264   0.970914   0.942591
```

In such a case it is often helpful to use double precision arithmetic. As a convenience a double precision version UNCMND, is provided on the disk at the rear of this book.

9.6 NONLINEAR DATA FITTING

One of the most common unconstrained minimization problems involves fitting a nonlinear model to data. If least-squares data fitting is appropriate (see Chapter 6), then this problem can be written in the form

$$\min_{x} F(x_1, \ldots, x_n) \equiv \sum_{j=1}^{m} [f_j(x_1, \ldots, x_n)]^2,$$

where each nonlinear function $f_j(x_1, \ldots, x_n)$ represents the residual for one data point.

Example 9.6   Nonlinear Data Fitting.

Consider the nonlinear least-squares problem based on the exponential model of Section 1

$$b_j \approx x_1 \exp(x_2 t_j),$$

given data $\{(t_j, b_j)\}_{j=1}^{m}$. The residual functions are

$$f_j(x_1, x_2) = b_j - x_1 \exp(x_2 t_j),$$

and the least-squares objective function is

$$F(x_1, x_2) \equiv \sum_{j=1}^{m} [b_j - x_1 \exp(x_2 t_j)]^2.$$

With the actual data $\{(t_j, b_j)\} = \{(0, 20), (1, 9), (2, 3), (3, 1)\}$ we obtain (via UNCMIN) the solution $x_1^* = 19.9900$, $x_2^* = -20.6230$. ∎

The nonlinear least-squares problem is an unconstrained minimization problem, and hence can be solved using the techniques discussed in this chapter. Listed below is the main program for the example above.

```
C MAIN PROGRAM FOR NONLINEAR LEAST-SQUARES DATA FITTING
C
       PARAMETER      (N = 2, LWORK = N*(N+10), MD = 4)
       REAL           X0(N), X(N), F, WORK(LWORK), T(MD), B(MD)
       COMMON /EXPDAT/ T, B, M
       EXTERNAL       CALCF
C
C DATA FOR DATA FITTING
C
       T(1) =   0.0
       T(2) =   1.0
       T(3) =   2.0
       T(4) =   3.0
       B(1) = 20.0
       B(2) =  9.0
       B(3) =  3.0
       B(4) =  1.0
C
C SPECIFY INITIAL ESTIMATE OF THE SOLUTION
C
       M    = 4
       X0(1) = 1.0
       X0(2) = 1.0
C
C MINIMIZE FUNCTION
C
```

```
       CALL UNCMIN (N, X0, CALCF, X, F, IERROR, WORK, LWORK)
C
C PRINT RESULTS
C
       IF (IERROR .NE. 0) WRITE (*,*) ' ERROR CODE =', IERROR
       WRITE (*,*) ' F (X*) =', F
       WRITE (*,*) ' X* =', (X(I), I = 1,N)
C
       STOP
       END
C
C OBJECTIVE FUNCTION
C
       SUBROUTINE CALCF (N, X, F)
       REAL        X(N), F, T(4), B(4)
       COMMON /EXPDAT/ T, B, M
C
       F = 0.0
       DO 10 J = 1,M
          F = F + (B(J) - X(1)*EXP(X(2)*T(J)))**2
10     CONTINUE
C
       RETURN
       END
```

There are several things to notice. First, it is not much different than the sample program in Section 5. Second, it is necessary to get the data points $\{(t_j, b_j)\}$ to the subroutine CALCF, but this cannot be done using the subroutine call statement. Normally the data points would be passed as parameters to the subroutine; but since CALCF is called by UNCMIN, the set of parameters to CALCF must not be modified. The alternatives are: either use a COMMON statement as is done here; or use DATA, PARAMETER, or assignment statements within subroutine CALCF.

This approach will work well for many problems, and it has the advantage of not requiring additional software. However, it is not the ideal approach to the nonlinear data-fitting problem. This formulation is analogous to the forming of the normal equations in a linear least-squares problem (see Section 3 of Chapter 6), and all the remarks about the normal equations apply here. The double precision version, UNCMND may be helpful in this case. More specialized algorithms for nonlinear data fitting avoid this problem; such software can be found, for example, in Minpack (see the Bibliography).

9.7 HISTORICAL PERSPECTIVE: SIR ISAAC NEWTON (1642–1727)

In 1543, Nicolas Copernicus initiated a revolution in scientific thought by announcing that the earth revolved around the sun. For many hundreds of years it had been thought that the earth was the center of the universe, with man as the pinnacle of creation, and

Copernicus' discovery represented a "fall from grace" in the religiously-inspired society of that time.

The discovery also began a period, continuing to the present day, in which scientific analysis is used to explain the world around us. Copernicus' work was continued by Galileo, who developed the telescope and invented the concept of inertia. Even though Galileo was working almost a century after Copernicus, his ideas were still radical, and Galileo was persecuted by the Vatican for his "heretical" studies.

Galileo died in 1642. Isaac Newton was born on Christmas Day in the same year. Except for disruptions caused by his father's death and his mother's remarriage, his childhood was unremarkable. Luckily, his intellectual interests were noticed, and he was sent to Cambridge University in 1661. Officially, university education was still based on Greek models and ideas, but the newer ideas of Copernicus and others were being discussed widely. Newton studied the work of the French philosopher René Descartes (after whom "Cartesian coordinates" are named) and the chemist Robert Boyle. Newton graduated in 1665 without much attention.

That same year, the university was closed because of the plague. For two years, Newton worked in isolation for fear of disease. It was during these two years that he made many of his major discoveries. He developed the calculus; he discovered through experiments with prisms that light was made up of colors; and he determined that the radial force on a planet decreases with the square of its distance from the sun, the basis for his subsequent law of universal gravitation.

When the university reopened in 1667, Newton was given a fellowship. He lectured there on optics, presenting his new ideas on the nature of light, ideas directly opposed to the then accepted theories of Descartes. The lectures apparently made little impression. It was only in 1671 when the Royal Society heard of his invention of a reflecting telescope, that Newton's work began to receive attention.

The reception of Newton's work was not completely favorable. Robert Hooke, a renowned scientist within the Royal Society, criticized Newton's ideas. Newton reacted irrationally, far out of proportion to the nature of the criticism. When Newton submitted later work on optics in 1675, Hooke accused him of plagiarism. Other criticisms so enraged Newton that in 1678 he apparently suffered a nervous breakdown, and he isolated himself for six years.

During this period Newton continued his studies of planetary motion, leading to the writing of *Philosophiae Naturalis Principia Mathematica*, considered the most important book in modern science. In it Newton states three laws of motion: (1) a body remains at rest unless acted upon by force, (2) force is proportional to mass times acceleration, (3) to every action there is an equal and opposite reaction. From these he goes on to derive the law of universal gravitation, that any two bodies attract each other with a force proportional to the product of their masses divided by the square of their distance.

Hooke again accused Newton of plagiarism. In response, Newton removed every reference to Hooke from his book, and delayed publication of his book *Opticks* until Hooke was dead. In spite of Hooke's accusations, the *Principia* brought Newton immense international fame. He was elected president of the Royal Society, he was knighted (the first scientist ever to be so honored), and he was made master of the mint, a position that brought with it a generous income.

In his later years, Newton spent many hours studying alchemy, expending as much effort as on his more respectable scientific pursuits. He was also occupied with religious studies. He rejected the notion of the Trinity, an opinion that could have resulted in his expulsion from the university had it been discovered. He was also active in the management of the mint. In the last years of his life he continued to edit his major works, retaining his anger for those who had criticized him, and presided over the Royal Society. He died in 1727.

9.8 FURTHER IDEAS

This chapter has only discussed unconstrained optimization. Constrained problems are usually considerably more difficult to solve than unconstrained problems. If the function F and all of the constraints g_i are linear functions, the problem is one of **linear programming**; it can be shown that the set S in this case is a convex polyhedron in n dimensions, and that the solution is a corner or **vertex** of this polyhedron. The usual method of solution is to search over the vertices by moving from one vertex to an adjacent vertex. The difficulties in linear programming are principally associated with solving problems with very large n which lead to sparse matrices. Such problems are rendered difficult by the combinatorial complexity of a general polyhedron in n dimensions. If either the function F or any of the constraints are nonlinear, the problem is one of **nonlinear programming**. Linear and nonlinear programming are beyond the scope of this book.

Methods suitable for constrained problems, both linear and nonlinear, are described in the book by Luenberger (1984); additional material on this topic can be found in the books by Orchard-Hays (1968), Dantzig (1963), or Gill et al. (1981).

It should also be noted that the techniques discussed here are really only suitable for smaller problems, where the number of variables n is not much bigger than 100, say. This is because they require matrix storage and manipulation, with associated costs that grow rapidly as the problem size increases. In some cases, sparse-matrix techniques can be used in combination with Newton's method to effectively solve such problems; but there are methods with lower storage and work-per-iteration costs that are more generally applicable. A survey of methods for large-scale problems can be found in Gill et al. (1981).

Subroutines for nonlinear optimization can be classified as to whether they handle unconstrained or constrained problems, whether the objective function is a sum of squares or a general nonlinear function, and whether or not evaluation of derivatives is required. Collections of these routines also usually include programs for the solution of simultaneous nonlinear equations.

A collection of optimization routines can be found in the software library developed by the Systems Optimization Laboratory at Stanford University. For large problems, an effective routine is MINOS (discussed in the paper by Murtagh and Saunders (1980)). Subroutines from the Minpack collection can be used for the solution of nonlinear equations and nonlinear data-fitting problems.

Methods for finding global optima are still being actively researched. For further information on available techniques, see the articles in the book by Boggs et al. (1985).

9.9 PROBLEMS

P9–1.–When the definition of erf(x) is extended to complex arguments, one of the terms involved is a real function of a real variable known as Dawson's integral,

$$D(x) = e^{-x^2} \int_0^x e^{t^2}\, dt.$$

Since $D(x)$ approaches 0 as x approaches either 0 or ∞ and $D(x)$ is positive in between, it must have a finite maximum. Use FMIN to find the maximum of $D(x)$.

P9–2.–Use FMIN to find the maximum of the function

$$F(x) = (\sin x)^6 \tan(1 - x)e^{30x}$$

on the interval $[0, 1]$.

P9–3.–Use UNCMIN to minimize

$$F(x_1, x_2) = \tfrac{1}{2}(x_1^2 + x_2^2)\exp(x_1^2 - x_2^2)$$

with initial guesses $(1, 1)^T$ and $(.1, .1)^T$.

P9–4.–For any $\lambda > 0$, let $z(\lambda)$ be the least positive zero of the function $y(t)$ that solves the ordinary differential equation problem

$$y''(t) + I_0(y) + \frac{t}{10} = 0,$$

$$y(0) = 0, \quad y'(0) = \lambda,$$

where $I_0(t)$ is the zero-order modified Bessel function (see the book by Abramowitz and Stegun (1965), pp. 374–375). A subroutine BESIO to compute this function is on the disk provided with the book. The graph of the solution $y(t)$ is approximately as shown here.

(a) Find λ_{max}, the unique value of λ such that $z(\lambda)$ is a maximum.

(b) Also find $z(\lambda_{max})$.

(c) Give a table of values of $y(t)$ for $t = 0(0.1)z(\lambda_{max})$ for the $y(t)$ which maximizes $z(\lambda)$, i.e., for $y(t)$ satisfying $y'(0) = \lambda_{max}$.

(d) Give a discussion of the accuracy of your results.

P9–5.–The following problem was proposed in 1696 by Johann Bernoulli and Gottfried Leibniz, with the idea that Newton would be able to solve it; he did, thereby creating **the calculus of variations**. Suppose we are given two points on a wall, call them P_1 and P_2, with P_1 higher than P_2. If a ball is dropped at P_1, what path should it follow so as to reach P_2 as rapidly as possible? If P_1 were directly above P_2, the path would be a straight line, but in general it is more complicated. It is possible to obtain an approximate solution to this problem as follows. Consider Figure P9.5. The solution path will be denoted by $y = p(t)$. If we knew the value of $p(t)$ at discrete times $t_0, t_1, \ldots$ then we could approximate the solution using interpolation. The true solution to this problem would satisfy

$$\min_{p(t)} \int_{P_1}^{P_2} \sqrt{\frac{1 + (p'(t))^2}{-p(t)}}\, dt.$$

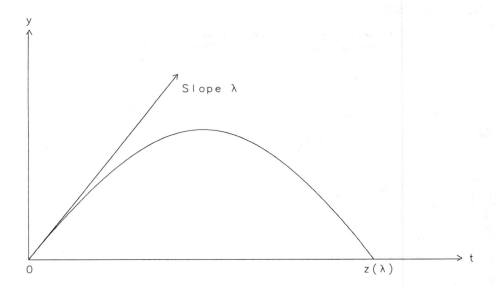

Figure P9.4

In order to approximate the solution, let $t_0, t_2, \ldots, t_{n+1}$ be evenly spaced points in the interval $[0, x_2]$, with $t_0 = 0$ and $t_{n+1} = x_2$. Let $p_i = p(t_i)$. Use UNCMIN to solve

$$\min_{p_1, \ldots, p_n, p_{n+1}} \int_0^{x_2} \sqrt{\frac{1+(g'(t))^2}{-g(t)}}\, dt,$$

where $g(t) = -p_{n+1}\sqrt{t} + h(t)$, and $h(t)$ is the Hermite cubic interpolant for the points $(t_0, 0 + p_{n+1}\sqrt{t_0})$, $(t_1, p_1 + p_{n+1}\sqrt{t_1}) \ldots$, $(t_n, p_n + p_{n+1}\sqrt{t_n})$, $(t_{n+1}, y_2 + p_{n+1}\sqrt{t_{n+1}})$ as computed by routine PCHEZ. The term $-p_{n+1}\sqrt{t}$ is included to model a derivative singularity at the point $(0, 0)$. Evaluate the integral using QK15; to evaluate the integrand you will have to use PCHEV. As an initial guess, use the straight-line segment connecting P_1 and P_2; take $n = 5$. Try your program on three problems, taking $P_1 = (0, 0)$ and $P_2 = (1, -5)$, $(1, -1)$, and $(5, -1)$. The problem is known as the "brachistochrone" ("least time") problem, and the true solution is a curve called a "cycloid."

P9–6.–(*Separable least squares*) Suppose m data points (t_i, y_i), $i = 1, \ldots, m$, are to be fitted in the least-squares sense by the following function of t:

$$b(t) = c_1 + c_2 t + c_3 t^2 + c_4 e^{\lambda t}.$$

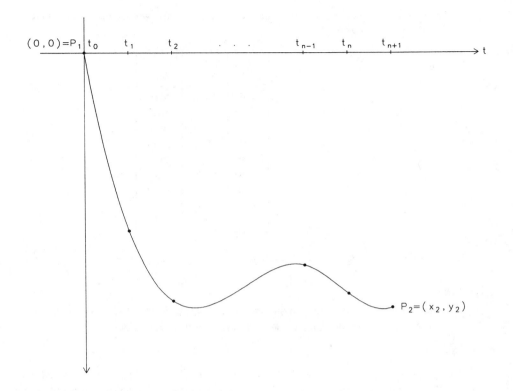

Figure P9.5

The functions involve five parameters: c_1, c_2, c_3, c_4, and λ. The four c's are involved linearly, but λ is involved in a nonlinear way.
Let $A(\lambda)$ be the $m \times 4$ matrix with elements

$$a_{i,j} = t_i^{j-1}, \; j = 1, 2, 3,$$
$$a_{i,4} = e^{\lambda t_i}.$$

Let b be the given m vector of data b_i, and let c be the unknown 4-vector of coefficients c_j. We then have the following optimization problem:

$$\min_{\lambda} \min_{c} \| A(\lambda)c - y \|_2^2 \, .$$

The *inner* minimum involving the four linear parameters can be found for any λ using SQRLS. The *outer* minimum involving the single nonlinear parameter λ can be found using FMIN. Note that SQRLS will be called by the function subprogram called by FMIN.

Here are two sets of data on which to carry out this technique. The first set should present no difficulty, but the second set leads to a degeneracy which can be detected by monitoring the value of the inner minimum and the rank of $A(\lambda)$. Find the c's and the λ which give a best fit.

Are the c's unique? Assume the data are accurate to only two decimal places, so that the y_i may have errors as large as 0.005.

t	y Set 1	y Set 2
0.00	20.00	20.00
0.25	51.58	24.13
0.50	68.73	26.50
0.75	75.46	27.13
1.00	74.36	26.00
1.25	67.09	23.13
1.50	54.73	18.50
1.75	37.98	12.13
2.00	17.28	4.00

P9–7.–Fit the census data from Problem P4–2 by

$$b(t) \approx c_1 + c_2(t - 1900) + c_3 e^{\lambda(t - 1900)}$$

with variable λ as described in the preceding problem. Predict the 1980 population.

P9–8.–Fit erf(t) by the same model as in Problem P6–1 (c) but with

$$z = \frac{1}{1 + \lambda t}.$$

The nonlinear parameter λ can be found using FMIN and the technique described in Problem P9–6.

P9–9.–Type I supernovae have been found to follow a specific pattern of luminosity (brightness). Beginning a few days after the maximum luminosity this pattern may be described by

$$L(t) = C_1 \exp(-t/\alpha_1) + C_2 \exp(-t/\alpha_2),$$

where t is the time in days after the maximum luminosity and $L(t)$ is the luminosity relative to the maximum luminosity. The table below gives the relative luminosity for the type I supernova SN1939A, measured in 1939. The peak luminosity occured at day 0.0, but all measurements taken before day 7.0 are omitted because the model above cannot account for the luminosity before and immediately after the maximum.

Day	Luminosity	Day	Luminosity	Day	Luminosity
7.0	0.6310	19.0	0.1318	57.0	0.05754
7.0	0.8318	20.9	0.1585	85.0	0.03631
14.8	0.2754	25.8	0.1096	109.0	0.02291
16.0	0.1445	26.8	0.1445	110.0	0.02291
16.9	0.2089	28.0	0.09120	141.0	0.01738
17.0	0.1585	53.0	0.06310	142.0	0.01585
18.8	0.1585	54.0	0.06918	168.0	0.009120

(a) Plot the data. You should notice two distinct regions, and thus two exponentials are required to provide an adequate fit.

(b) Use UNCMIN to fit the data to the model above. Plot the fit along with the data. Also plot the residuals. Do the residuals look random? Experience plays a role in choosing the starting values. It is known that the time constants, α_1 and α_2 are about 5.0 and 60.0 respectively. Try different starting values for C_1 and C_2. How sensitive is the resulting fit to these values?

P9–10.–Refer to Problem P6–8, involving a model of CO_2 in the atmosphere,

$$y(t) = B + d \exp(\alpha t) + \sum_{i=1}^{n} A_i \sin \left[\frac{2\pi}{(T/i)} (t + \phi_i) \right],$$

where t is in months, and the other parameters are unknown. Use the technique of P6–8 and UNCMIN to find these parameters for various values of n. Plot the fit and the residuals. Does this model do a good job fitting the data? [Note: The term $\exp(\alpha t)$ is sensitive to small changes in α. We suggest that you solve this problem in two stages. First fix $\alpha = 0.0037$ and estimate the other parameters, then use these estimates as an initial guess for the full model. Even so, there may be difficulties with overflow.]

P9–11.–Apply Newton's method to solve

$$\min F(x) = (x - 2)^4 - 9$$

with initial guess $x = 1$. Why is the convergence so slow?

9.10 PROLOGUES: FMIN AND UNCMIN

```
          REAL FUNCTION FMIN(AX,BX,F,TOL)
C***BEGIN PROLOGUE   FMIN
C***DATE WRITTEN    730101   (YYMMDD)
C***REVISION DATE   730101   (YYMMDD)
C***CATEGORY NO.   G1A2
C***KEYWORDS   ONE-DIMENSIONAL MINIMIZATION, UNIMODAL FUNCTION
C***AUTHOR   BRENT, R.
C***PURPOSE   An approximation to the point where F attains a minimum on
C             the interval (AX,BX) is determined as the value of the
C             function FMIN.
C***DESCRIPTION
C
C     From the book, "Numerical Methods and Software" by
C                D. Kahaner, C. Moler, S. Nash
C                Prentice Hall, 1988
C
C
C     The method used is a combination of golden section search and
C     successive parabolic interpolation.  Convergence is never much
C     slower than that for a Fibonacci search.  If F has a continuous
```

```
C        second derivative which is positive at the minimum (which is not
C        at AX or BX), then convergence is superlinear, and usually of the
C        order of about 1.324....
C
C        The function F is never evaluated at two points closer together
C        than EPS*ABS(FMIN) + (TOL/3), where EPS is approximately the
C        square root of the relative machine precision.  If F is a unimodal
C        function and the computed values of F are always unimodal when
C        separated by at least EPS*ABS(XSTAR) + (TOL/3), then FMIN
C        approximates the abcissa of the global minimum of F on the
C        interval AX,BX with an error less than 3*EPS*ABS(FMIN) + TOL.
C        If F is not unimodal, then FMIN may approximate a local, but
C        perhaps non-global, minimum to the same accuracy.
C
C        This function subprogram is a slightly modified version of the
C        ALGOL 60 procedure LOCALMIN given in Richard Brent, Algorithms for
C        Minimization Without Derivatives, Prentice-Hall, Inc. (1973).
C
C INPUT PARAMETERS
C
C  AX     (real)  left endpoint of initial interval
C  BX     (real) right endpoint of initial interval
C  F      Real function of the form REAL FUNCTION F(X) which evaluates
C         F(X)  for any  X in the interval  (AX,BX)
C         Must be declared EXTERNAL in calling routine.
C  TOL    (real) desired length of the interval of uncertainty of the
C         final result ( .ge. 0.0)
C
C
C OUTPUT PARAMETERS
C
C FMIN    abcissa approximating the minimizer of F
C AX      lower bound for minimizer
C BX      upper bound for minimizer
C
C***REFERENCES  RICHARD BRENT, ALGORITHMS FOR MINIMIZATION WITHOUT
C                 DERIVATIVES, PRENTICE-HALL, INC. (1973).
C***ROUTINES CALLED  (NONE)
C***END PROLOGUE  FMIN


        SUBROUTINE UNCMIN (N,X0,FCN,X,F,INFO,W,LW)
C***BEGIN PROLOGUE  UNCMIN
C***DATE WRITTEN  870923    (YYMMDD)
C***REVISION DATE  871222   (YYMMDD)
C***CATEGORY NO.  G1B1A1
C***KEYWORDS  UNCONSTRAINED MINIMIZATION
C***AUTHOR  NASH, S.G., (GEORGE MASON UNIVERSITY)
```

```
C***PURPOSE  UNCMIN minimizes a smooth nonlinear function of n variables.
C            A subroutine that computes the function value at any point
C            must be supplied, but derivative values are not required.
C            UNCMIN provides a simple interface to more flexible lower
C            level routines.  User has no control over options.
C
C***DESCRIPTION
C     From the book, "Numerical Methods and Software" by
C              D. Kahaner, C. Moler, S. Nash
C              Prentice Hall, 1988
C
C     This routine uses a quasi-Newton algorithm with line search
C     to minimize the function represented by the subroutine FCN.
C     At each iteration, the nonlinear function is approximated
C     by a quadratic function derived from a Taylor series.
C
C     The quadratic function is minimized to obtain a search direction,
C     and an approximate minimum of the nonlinear function along
C     the search direction is found using a line search.  The
C     algorithm computes an approximation to the second derivative
C     matrix of the nonlinear function using quasi-Newton techniques.
C
C     The UNCMIN package is quite general, and provides many options
C     for the user.  However, this subroutine is designed to be
C     easy to use, with few choices allowed.  For example:
C
C     1.  Only function values need be computed.  First derivative
C     values are obtained by finite-differencing.  This can be
C     very costly when the number of variables is large.
C
C     2.  It is assumed that the function values can be obtained
C     accurately (to an accuracy comparable to the precision of
C     the computer arithmetic).
C
C     3.  At most 150 iterations are allowed.
C
C     4.  It is assumed that the function values are well-scaled,
C     that is, that the optimal function value is not pathologically
C     large or small.
C
C     For more information, see the reference listed below.
C
C PARAMETERS
C - - - - - - - - - - - - - - - - - - - - - - - - - - - - - - - -
C N              -> INTEGER
C                   Dimension of problem
C X0(N)          -> REAL
C                   Initial estimate of minimum
C FCN            -> Name of routine to evaluate minimization function.
```

```
C                        Must be declared EXTERNAL in calling routine, and
C                        have calling sequence
C                            SUBROUTINE FCN(N, X, F)
C                        with N and X as here, F the computed function value.
C X(N)          <-  REAL
C                        Local minimum
C F             <-  REAL
C                        Function value at local minimum X
C INFO          <-  INTEGER
C                        Termination code
C                            INFO = 0:  Optimal solution found
C                            INFO = 1:  Terminated with gradient small,
C                                       X is probably optimal
C                            INFO = 2:  Terminated with stepsize small,
C                                       X is probably optimal
C                            INFO = 3: Lower point cannot be found,
C                                       X is probably optimal
C                            INFO = 4: Iteration limit (150) exceeded
C                            INFO = 5: Too many large steps,
C                                       function may be unbounded
C                            INFO = -1: Insufficient workspace
C W(LW)         --> REAL
C                    Workspace
C LW            --> INTEGER
C                    Size of workspace, at least N*(N+10)
C
C***REFERENCES R.B. SCHNABEL, J.E. KOONTZ, AND BE.E. WEISS, A MODULAR
C             SYSTEM OF ALGORITHMS FOR UNCONSTRAINED MINIMIZATION,
C             REPORT CU-CS-240-82, COMP. SCI. DEPT., UNIV. OF
C             COLORADO AT BOULDER, 1982.
C***ROUTINES CALLED OPTDRV, XERROR
C***END PROLOGUE UNCMIN
```

10

Simulation and Random Numbers

10.1 INTRODUCTION

When we use the term "random" in everyday life it is usually as a synonym for an unpredictable event, such as a lightning strike, the result of a coin toss, the income of the next customer in a store, or the time between customers. Actually none of these are really unpredictable; for example, the coin toss depends upon the initial orientation of the coin, force of the toss, height above ground, air resistance, etc., in a complicated way. It is easier to think of the result as random, especially if we are interested in averages rather than the next coin. "Nothing in Nature is random ... A thing appears random only through the incompleteness of our knowledge."[1] "Snow" on our televisions, and the lifespan of a lightbulb are random in that same sense. In these examples the long run or average properties are known but the specific events are uncertain. We might know that a hospital emergency room averages 200 patients on a Friday night but we do not know if Friday, December 13 will have 192 or 225 patients. We know that if we roll two die that the chance of snake eyes (a one on each die) is 1/36, but we do not know whether the next toss of the die will produce snake eyes, a seven or eleven, etc.

The knowledge about which outcomes are more likely than others is important in forecasting. For example, farmers, construction companies and airlines make decisions based on forecasts of the weather. Gamblers develop systems for trying to win at roulette, blackjack, and poker. City management decides how many fire companies to have, hospitals determine staffing for emergency rooms, and barbershop owners decide how many barbers need to be hired. If one barber can handle about 20 people per day

---

[1] Spinoza, *Ethics I*. Readers with knowledge of quantum mechanics might disagree with this conclusion.

and the shop expects about 60 customers should they hire 3 barbers? If 64 customers arrive in a day, will they grumble because the wait for a haircut is too long?

Generally, all of the items of interest in these situations can be written in the form of equations, usually of considerable complexity and rarely amenable to explicit solution. The usual approach for handling these situations is **simulation**. A typical simulation model contains statements in a computer language which look much like the actual situation being modelled. These include items as shown below:

```
    Barber Segment (barber is ready for a new customer):
        If at least one customer is waiting
            Then
                    Remove the customer who has been waiting longest from
                    the queue
*                   Determine the time until the haircut is completed
            Else
                    Mark the Barber as Idle
        End If
    End Barber Segment
    Customer Segment (a customer has just arrived):
*           Determine the time until the next customer arrives
            If at least one barber is Idle then
*               Select a barber
*               Determine the time until the haircut is completed
        Else
                Add the customer to the queue
        End If
    End Customer Segment
```

Typical Simulation Algorithm in Pseudocode

These statements are the basic logic in a "queuing" simulation, but many details such as scheduling events and collecting statistics need to be added. The pseudoprogram above can be converted into Fortran or any other language, but special purpose simulation languages such as GPSS (Schriber, 1974), Simscript (Russell, 1983), and SLAM (Pritsker, 1986), are also available. A good general reference on simulation is Bratley et al., (1987).

In the statements above notice that four are starred (∗). Everything except those four statements is deterministic. The starred statements represent probabilistic events. The problem here is to generate times and pick barbers in a method that resembles the real world situation. This means that the times between customer arrivals and the time to complete haircuts should look like some probabilistic process with, say, a known mean and known variance, see Section 2 of Chapter 6. Or if both barbers are idle, the younger barber should be chosen two-thirds of the time, etc. In order to do this, we need a method for obtaining times and choices where on average the real-world properties hold. This requires random numbers.

There are other problems which do not by themselves have a random element but where it is useful to take random samples for one purpose or another. For example, in United States presidential elections it is both unnecessary and too expensive to query every potential voter to obtain accurate predictions of the winning margins. Instead a rather small sample of a few thousand voters is carefully selected and conclusions are drawn about the general population. In a different context, we have seen in Chapter 5 how the Monte Carlo method can be used to estimate the value of an integral in any number of dimensions and we explained how this method was and continues to be used to design complicated nuclear reactors.

10.2 RANDOM NUMBERS

For applications on a computer it is necessary to produce a long sequence of numbers that behave as if they are random. A formal definition of randomness is difficult to obtain because it means defining the absence of a pattern. However, an intuitive definition is that a sequence of numbers is random if the simplest way to describe the sequence is to write it down. Thus

```
0101010101010101010101010101010101010101010101010101
```

can be specified in a short algorithm, whereas

```
0110110010000010101000010101001011111010100001010101
```

cannot.

The troubles with this definition are (1) it does not tell you how to construct a random sequence and (2) the definition is basically wrong, because it indicates that you can detect randomness by looking at a sequence. You cannot. If in tossing a coin we let zero signify "tail" and one signify "head," the probability that 52 tosses will produce exactly the first sequence is $(1/2)^{52}$; the probability of producing exactly the second sequence is the same, $(1/2)^{52}$. Yet we might reject the first sequence and would accept the second! What we seem to want is a sequence to pass statistical tests for randomness and to have a long interval between repetitions. This interval is called "the period," and plays an important role in the theory. We discuss some of these issues in Section 3.

There are two well known classes of methods for producing random numbers.

(**1**) *Methods based on real phenomena.* This includes such things as counting spacings between ticks of a Geiger counter, hiring somebody to flip coins, or drawing balls from an urn. These methods are slow and may not satisfy statistical tests of randomness. In any case the physical process, being random, cannot be reproduced and it is difficult to debug a computer program that uses it. Furthermore, for purposes such as Monte Carlo quadrature, theoretical results indicate that such random numbers would be less effective than the artificially generated sequences described in this chapter. Machines for generating random numbers are rarely used today in computations because they are too slow. However, drawing items from an urn is still in use for lotteries. Trust in the computer takes a long time to come.

(**2**) *Methods based on algorithms.* One can algorithmically compute the random numbers off-line and store them in a disk or tape file, to be consumed by a program as needed. Long before the invention of the computer and before the invention of Monte Carlo methods, there was a practical need for random numbers. A table of 40,000 random digits was published in 1927, based on census information. Later on, special purpose machines were used to generate random information, culminating in 1955 with the publication of *One Million Random Digits and 100,000 Normal Deviates* by the RAND Corporation. Such collections had value since the random information was reproducible, and hence calculations could be repeated and verified at a later date. These numbers were eventually made available on a computer tape and tended to be overused. One possible solution was to start using the table at a random location, but this brings us back to the problem of how to generate a random location in a file. Elaborate methods for entering a table of random numbers have been devised by statisticians in the past.

The current, commonest, method is to use an algorithm on-line with your program to generate numbers one after another. These are available when they are needed and are perfectly reproducible for program checkout.[2] Since such numbers are generated by an algorithm, they are deterministic, not random at all and are called **pseudorandom**. The program that produces the numbers is called a random number **generator**. We think of pseudorandom numbers as equivalent to random if "each term is unpredictable to the uninitiated and whose digits pass a certain number of tests traditional with statisticians."[3] However, there is only a finite amount of testing which we can do and the possibility exists that some unusual failure lies yet to be discovered. Furthermore, most algorithmic generators are **cyclic**, or **periodic**, that is, they eventually repeat. Nevertheless, this approach seems to be the most useful and will be the only method discussed in this book.

It should be evident that although the mathematical definition of randomness can be made precise, the practical definition (especially for an algorithmic generator) depends on the application. A generator for an arcade game need not be as good as one which will be used in a multi-dimensional quadrature. A number of criteria have been proposed for judging the quality of these generators. Generally, it will not be possible for a single generator to satisfy all of them. They include:

(a) *High quality*. The generator should pass all the statistical tests and have an extremely long period.

(b) *Efficiency*. Execution should be rapid and storage requirements minimal.

(c) *Repeatability*. Specifying the same starting conditions will generate the same sequence. The user should be able to restart the generator at any time, but explicit initialization is not necessary. A slight change in the starting procedure will result in a different random sequence.

(d) *Machine independence and portability*. The algorithm should work on different kinds of computers; in particular, no operation should cause the program to stop.

---

[2] In the future we may see a return to "table-look-up" generators—a "compact disk" could store about 125 million well tested random real numbers and might sell for a few dollars.

[3] Lehmer, 1951

The same sequence of random numbers should be produced on different computers by initializing the generator in exactly the same way.

(e) *Simplicity*. The algorithm should be easy to implement and use.

No generator can be successful in satisfying all of these criteria. For example, a simulation that runs on a supercomputer may require an exceptionally efficient generator. It may not be necessary for that generator to produce the same numbers, or even run at all, on another computer.

10.3 GENERATION OF UNIFORMLY DISTRIBUTED NUMBERS

The most often used random number is "uniformly distributed," either real or integer. A uniform random number generator will produce values uniformly on the interval $[0,1)$. By this we mean that the chances of a number being in a subinterval $0 \leq a < b \leq 1$, is equal to the length of the interval, i.e., $b - a$. Thus, if we generate say, 50,000 numbers, then about 25,000 will lie in the interval $[.4, .9)$.

For example, Figure 10.1 illustrates one particular experiment with the random number generator RND provided by IBM with the BASIC interpreter on their personal computer models. We divide $[0,1)$ into 10 subintervals, $[0, .1), [.1, .2), \ldots, [.9, 1.0)$, generate 50,000 random numbers, and count how many fall into each subinterval or "bin." There ought to be about 5,000. Of course, there will not be *exactly* 5,000 in each, but our intuition says that the fraction in each should not be very far from $1/10$th. Specific statistical tests, such as the χ^2 test, Kennedy et. al., (1980) can be used to confirm that for this particular example, the results are consistent with the generator being uniform. A figure such as this one, that shows in bar-chart form the number or fraction of items that fall into one of a set of bins is called a **histogram**.

If the uniform generator produces integers, then these are uniform, in the above sense, over the range of nonnegative representable integers. If the largest representable integer is denoted IMAX, see Section 2.1 of Chapter 2, then the generator will produce integers in the interval $[0, \text{IMAX}]$. An integer generator can also be used to generate real numbers, for example, by a final floating divide by IMAX, although this one operation can be as time consuming as the rest of the algorithm. Often a uniform integer is just as useful as a real, but as the range of integers varies greatly among computers, a program using random integers is likely to run differently on a new machine. At one time it was also universally true that integer arithmetic was much faster than floating point, and this encouraged the development of integer algorithms. But today this is computer dependent. Until the mid 1980s most generators operated with integers. Today this is changing and generators such as UNI that deal exclusively in floating point are the favorites.

The numbers coming from our generator must be independent as well as uniform. That means that there should not be any apparent pattern between successive numbers. If a sequence fails this, it is said to be "serially correlated." A famous example of such a generator was an early version of RND. Figure 10.2 uses the same 50,000 numbers as the previous figure, but represents them in a different way. As numbers are returned from the generator they are paired, $(x_1, x_2), (x_3, x_4), \ldots, (x_{49999}, x_{50000})$ and each pair is plotted

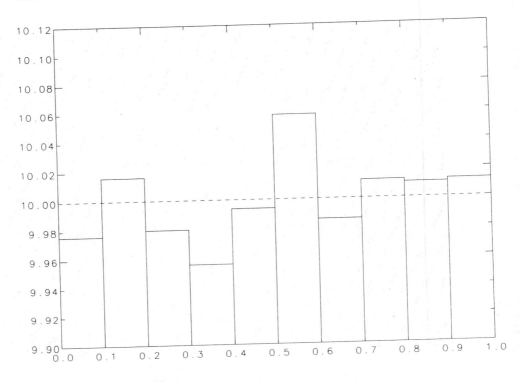

Figure 10.1   50,000 RND Calls in 10 Bins

as a point in the unit square. A clear striped pattern is apparent because the numbers are serially correlated. Imagine the effect of such a generator if we used it to estimate a two-dimensional integral over the unit square (see Section 11 of Chapter 5) and the integrand had an important peak in one of the unsampled regions. Marsaglia (1968) has proved that to a certain degree most generators of the type currently in use suffer correlations among successive numbers. The title of his article succinctly describes the situation: "Random numbers fall mainly in the planes," but for well designed generators the effect is far smaller than for others. This version of RND was sufficiently bad that IBM replaced it in later versions of their Basic. But we urge you not to use RND or any other built-in random number generator unless you have tested it on a model problem in your field of application with answers that are known to you.

The almost universally used method of generating pseudorandom numbers is to select a starting or **seed** value, usually an integer, and then to compute x_0 from the seed according to some rule. Then the $(k + 1)$-st number, x_{k+1}, is obtained from the preceding ones. Sometimes, only the last value is needed to generate the next. Initially the rules were chosen by intuition, to be as complicated, confusing and little understood as possible—for example, the "midsquare" method in which you square x_k and take x_{k+1} as a fixed number of digits from the middle of the result. But the lack of a theory

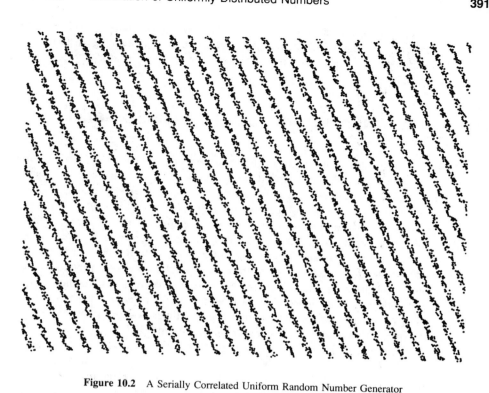

Figure 10.2   A Serially Correlated Uniform Random Number Generator

about such methods often proved disastrous. In the midsquare method x_k can turn out to be zero at unpredictable times and then all subsequent x_i are zero too. The method was proposed by von Neumann who was aware of its deficiencies. So quite early people changed to rules whose properties were as well understood as possible. Incidentally, the subscript notation x_k is only suggestive; random number generators never keep all the numbers they have generated, but only those few needed to produce the next one.

Another important characteristic of a random number generator is its **period**, p. This is the maximum length of the sequence $\{x_k\}$ before it begins to repeat. With either integer of floating point generators there are only a finite number M of possible values of x_k. If the rule for computing x_{k+1} only depends on x_k, once x_k duplicates any preceding x_i then the entire sequence repeats. Thus $p \leq M$. It might be much less. If x_{k+1} depends on x_k and x_{k-1} the maximum period is M^2 even though an individual x_k will appear more frequently. With the aid of number theory it is possible to choose a rule so that the period is known *a priori* to be as long, or nearly as long, as possible. The further use of number theory can to some extent predict the character of the sequence to give the user some degree of confidence that it will serve well enough to simulate a random sequence of numbers.

There is no way known to be sure that any particular generator will satisy all possible criteria for randomness. Lehmer's (1951) definition in Section 2 is heuristic. The early version of RND does not generate acceptable random numbers for many applications. Number theory can eliminate some bad generators, and can suggest good ones, but it is only by passing more and more stringent tests that a generator can get a reputation as being reliable.

10.4 APPLICATIONS OF RANDOM NUMBERS: BROWNIAN MOTION AND FRACTALS

In the early nineteenth century, the Scottish botanist Robert Brown examined a drop of water which had been trapped for millions of years in a piece of quartz. Under a microscope he discovered scores of tiny particles ceaselessly moving with completely irregular motion. He had observed such behavior in liquids before, but explained it on the basis that "vitality is retained [by the molecules of a plant] long after the [plant's] death," but the source and age of this droplet ruled out any such biological explanation. It is now known that this "Brownian motion" is due to the bombardment of suspended dust particles by molecules in the fluid in a random way—the path is randomized by random fluctuations in the velocities of colliding molecules. Interestingly, in 1905 while investigating atomic theory, Einstein wrote that "I discovered that there would have to be [observable] movement of suspended particles without knowing the observations concerning Brownian motion were already long familiar."

Fortunately, it is not necessary to look under a microscope to study Brownian motion. Using a random number generator it is possible to simulate the movement of such a particle accurately. Figure 10.3 shows such a computer-generated Brownian path, and you can generate a similar one by working Problem P10–3.

The simulated motion only begins to suggest the complexity of the true path. If we photographed a real particle and then increased the magnification by 1000, the effect of bombardment by smaller groups of molecules could be observed. A portion of the original trajectory which appeared to be straight would now look just as jagged and irregular as the entire initial path. B. B. Mandelbrot, (1982) of IBM has coined the term "fractals" for geometric objects which are self-similar at all magnifications, and Brownian motion was one of the first natural phenomena modelled this way. Of course "no real structure can be magnified repeatedly an infinite number of times and still look the same. The principle of self-similarity is nonetheless realized approximately in nature; in coastlines and riverbeds, in cloud formations and trees, in the turbulent flow of liquids, and in the hierarchical organization of living systems."[4]

One of the most interesting things about fractals is that random numbers are not always necessary in order to to generate them—they are deterministic. This is analogous to the completely reproducible methods that we use for generating pseudo random numbers on computers. We can illustrate this with one of the most famous examples. Let c and z_0 be two *complex* numbers,

$$c = p + iq, \qquad z_0 = x_0 + iy_0,$$

---

[4] H. Peitgen and P. Richter (1986)

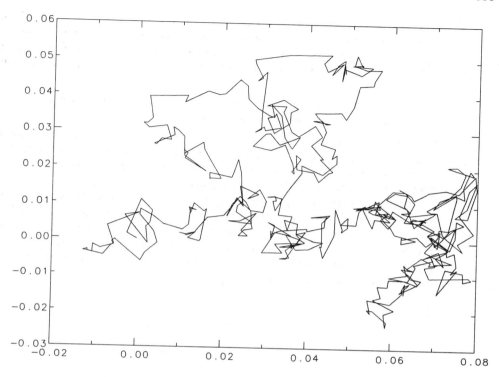

Figure 10.3 Brownian Motion

and consider the iteration

$$z_{n+1} = z_n^2 + c,$$

or its equivalent form in terms of the real and imaginary parts

$$x_{n+1} = x_n^2 - y_n^2 + p, \qquad y_{n+1} = 2x_n y_n + q.$$

For specific values of c and z_0 we can get different behavior. For example, if $c = 0$ and $|z_0| < 1$, then $z_n \to 0$, but if $|z_0| > 1$ then z_n diverges, or in the terminology of dynamics "converges to infinity." Complicated figures of extraordinary beauty can be produced from one of the following two algorithms.

(1) Fix c and "loop" over different values of z_0 in some regular region, for example the unit square. For each z_0 iterate enough times to decide if the sequence is converging to infinity or not. This is usually done by seeing if $x_n^2 + y_n^2 > R$ where R is a constant, such as $R = 100$ within a fixed number N of iterations, say 100. If the sequence converges to infinity in this sense the point $z_0 = (x_0, y_0)$ is displayed on a graphical device in a color, otherwise it is displayed as black. Usually the colors are keyed to the rapidity of convergence to infinity—hotter colors meaning that $x_n^2 + y_n^2 > R$ more quickly than cooler ones. The boundary separating the region of convergence to infinity from other points in the plane is called the Julia set.

(2) Fix $z_0 = 0$ and repeat (1) by looping over values of c in a square. The boundary of the region is called the Mandelbrot set.

In typical computer experiments the square in (1) and (2) is associated with the graphical display monitor. If your computer provides software to color point (I, J) on the monitor in color K, then programs for either algorithm are only a few lines but the results can be spectacular. See Problem P10–12.

Fractals, with some randomization, have also found their way into computer graphics and animation as a way to describe the outline of fictitious but exceedingly realistic mountain ranges and coastlines. Figure 10.4 shows a fractal coastline. To provide the necessary detail for this application, a random number generator must have very long period, be without measurable bias or correlation, and be as fast as possible. You can try it yourself in Problem P10–4.

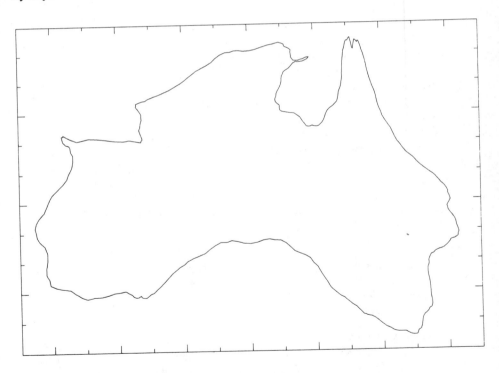

Figure 10.4　A Fractal Coastline

Usually, the algorithmic description of a fractal is concise, even if the resulting figure is quite complex. For example, the algorithm for the Julia set is given by specifying just a few numbers. Some work has been done to invert this relation. That is, given a complicated figure, for example a photographic image, can it be described as a fractal? An application of this would be to image compression. See Barnsley and Sloan (1988). A simple experiment for you to try is given in Problem P10–13.

10.5 CONGRUENTIAL AND FIBONACCI GENERATORS

One of the most common integer generators, the **congruential** generator was proposed by the mathematician, D. H. Lehmer. It uses

$$x_{k+1} = (ax_k + c)(\text{mod } M), \qquad x_0, a, c < M \leq \texttt{IMAX}$$

with x_0, a and c given integers. The mod function $\alpha \ (\text{mod } \beta)$, is defined as the remainder when α is divided by β. For example, 15(mod 4) = 3. The number M is a large integer. A simple physical experiment suggests why this generator might be useful: Mount a camera above a billiard table and set it to snap a picture at fixed intervals, perhaps every five seconds. A ball is hit and caroms about the table, without friction. The position of the ball every five seconds is perfectly well determined, but the succession of photographs give it the appearance of randomness, especially if the cushions are not shown. The mod function is analogous to these cushions.

Some choices of a and c are good, and some are not. For example, $a = c = 1$ would obviously be awful. Other, less obvious values can also lead to the type of serial correlation described in Section 3. Some values of a, c, and M produce acceptable random numbers. Commonly used values are $a = 69069$, $c = 0$, $M = 2^{32}$ or $M = 2^{31}$. Knuth (1969) summarizes the number theory necessary to pick a and c. A good implementation of this generator can produce almost M elements before it cycles. Its other statistical characteristics are also good. Congruential generators are fine for many simulations. Because they have been studied for years you will find them in most of the standard statistical packages. But bigger and faster computers allow us to tackle larger problems and these require longer cycles and even better statistical properties.

For producing floating point numbers on the interval $[0, 1)$ we prefer the **Fibonacci** generators. These are characterized by computing the new element x_{k+1} as the difference, sum, or product of two preceding elements. For example, the algorithm implemented in `UNI` (Section 6) is the Fibonacci generator

$$x_{k+1} = x_{k-17} - x_{k-5}.$$

The values in the sequence depend upon the initial seventeen x_i and whatever strategy we decide to adopt if $x_i < 0$. If this occurs, a standard approach is $x_i \leftarrow x_i + 1.0$; this will keep all the elements in the unit interval. Because x_{k+1} depends upon 17 preceding values, its theoretical period p, can be large, $p = (2^{17} - 1)2^n$, where n is the number of bits in the fractional part of x_i, i.e., the signed mantissa. For (IEEE) standard 32-bit floating point arithmetic, $n = 24$; thus p is about $2^{17} \cdot 2^{24} \approx 10^{12}$.

This generator is said to have **lags** 17 and 5. Fibonacci generators with lag 1 were studied and dismissed years ago because of a high degree of serial correlation. Recently though, the high-lag generators have been rediscovered. They are extremely efficient and have excellent statistical properites. Fibonacci generators are currently the first choice for large scale random number applications even though they require more storage than congruential generators.

If we use a Fibonacci generator requiring more than three or four past values it is usual to set these up in a small array which is thought of as joined at the ends into a circle. Two integer variables, I, J are used to point to the two elements which are to be subtracted. For example UNI uses a 17-element real array, U(1),...,U(17), with I and J initialized to 17 and 5. Then

```
UNI = U(I)-U(J)
IF (UNI .LT. 0.0) UNI = UNI+1.
U(I) = UNI
I = I-1
J = J-1
IF (I .EQ. 0) I = 17
IF (J .EQ. 0) J = 17
```

Such a generator produces numbers that pass almost all known tests for randomness, but it can still be improved. One improvement is to increase the lags. For example, on some supercomputers Fibonacci generators with lags 97 and 33 are in use. In addition it has been found that if two good generators are combined the result is better than either. In UNI the output u, of the Fibonacci generator is combined with v, the output of a floating point congruential generator, and UNI= $u - v$ or UNI= $u - v + 1.0$ as $u - v$ is positive or negative. The combination passes all known tests and has a period that is the product of the periods of each separate generator, or $(2^{17} - 1)2^{48} \approx 10^{19}$.

To start the sequence we need to generate 17 "random" values in [0,1). UNI does this as follows. Every element of the starting array is thought of as being represented in its binary form

$$s = \frac{s_1}{2} + \frac{s_2}{2^2} + \cdots + \frac{s_m}{2^m} \qquad s_i = 0 \text{ or } 1, \qquad m \leq n.$$

The initializing algorithm sets each bit, s_i randomly as 0 or 1, depending on whether the output of a simpler built-in integer generator is greater or less than zero. The only user input is a seed for this last generator.

If the parameter m in the expression for s is set equal to n (the number of bits in the floating point mantissa) then this generator produces different sequences when it is run with the same seed on computers with different word-lengths, say a Cray and an IBM personal computer. On each it has period $(2^{17} - 1)2^n$. If m is set to the smaller of the two mantissa lengths ($m = 24$), exactly the same sequences will be generated. In UNI $m = 24$ for portability but this is easy to change.

10.6 FUNCTION UNI

Using UNI requires one initializing evaluation of USTART(ISEED) with an integer seed. USTART returns a floating point echo of the seed. Occasionally a user will forget to initialize the generator, so the internal array U is preset to reasonable default values. Subsequently, each evaluation of UNI(), without an argument, returns a uniformly distributed number on $[0, 1)$. USTART is actually an "entry" into UNI, so there is only one routine rather than two. Null argument functions are supported in Fortran 77. Because the initializing is done by a different entry, UNI only has to do one thing, and the lack of an argument results in the most efficient linkage between UNI and the routine that evaluates it.

```
C Typical UNI usage:
C
      REAL     U,UNI,USTART,USEED
      INTEGER ISEED,I
C Set initial seed
      ISEED = 305
      USEED = USTART(ISEED)
C Ustart returns floating echo of iseed
      WRITE (*,*) ISEED, USEED
      DO 1 I = 1,1000
         U = UNI()
    1 CONTINUE
      WRITE (*,*) U
      END
C
C Produces the following output
C
C        305     305.000
C               0.157039
C
```

10.7 SAMPLING FROM OTHER DISTRIBUTIONS

Often the generation of a random number in $[0, 1)$ is merely the means to making some random decision. As an example, suppose in a simulation program it is desired to take some branch one time in ten, at random. One way to do this is to select a y uniformly on $[0, 1)$ and take the branch if $y < 0.1$. Another way is to select a random integer x on $[0, IMAX]$ and take the branch if $x < IMAX/10$. In the same way, any finite distribution with various weights can be easily produced from uniformly distributed random numbers.

Often it is necessary to generate a sequence of uniform random numbers from a given finite interval $[a, b]$. If u_i are generated on $[0, 1)$, say by UNI, then $(b - a)u_i + a$ is uniform on $[a, b]$.

Much more difficult is the computation of a random number from a non-uniform continuous distribution, for example a normal distribution. If $f(x)$ is the density function (in the case of the normal this is the familiar bell-curve) then the **cumulative distribution function** is defined to be

$$F(x) = \int_{-\infty}^{x} f(t)\, dt.$$

For the normal distribution $F(x)$ has an S-shaped curve, beginning at $F(x) = 0$ when $x \to -\infty$ and rising to $F(x) = 1$ as $x \to \infty$. A general way to sample from this kind of distribution is to find a uniform y on $[0, 1)$ and then find the unique x such that $F(x) = y$. See Figure 10.5.

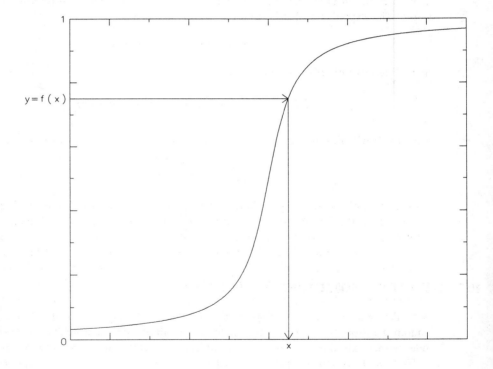

Figure 10.5   Sampling From a General Distribution

We can understand this as follows. Imagine sampling y's many times; every y we pick corresponds to a unique x. Selecting 50000 y's generates the same number of x's. Thus the fraction of selected y's between two fixed values, say y_1 and y_2, is exactly equal to the fraction between x_1 and x_2. But the former is equal to $|y_2 - y_1|$ if the y's are uniformly selected hence

$$\text{fraction in } [x_1, x_2] = \text{fraction in } [y_1, y_2] = y_2 - y_1 = F(x_2) - F(x_1)$$

$$= \int_{-\infty}^{x_2} f(t)\, dt - \int_{-\infty}^{x_1} f(t)\, dt = \int_{x_1}^{x_2} f(t)\, dt,$$

which says that if the y's are uniform the x's have a density given by $f(t)$.

This method is equivalent to computing the inverse function $F^{-1}(y)$. How to form this accurately and rapidly enough is the big problem. For some common distributions special and ingenious techniques have been developed. All these methods rely upon a fast and accurate uniform generator. See for example, the book by Kennedy and Gentle (1980), or by Rubinstein (1981) for a survey of methods for a great many different generators. As an example, consider the **exponential** distribution with parameter $\lambda > 0$. This distribution is often used in "queuing" problems to model interarrival and service times. The exponential distribution has density function

$$f(t) = \lambda \exp(-\lambda t), \qquad t > 0$$

and cumulative distribution function

$$F(x) = \int_0^x \lambda \exp(-\lambda t)\, dt = 1 - \exp(-\lambda x).$$

For this distribution it is easy to solve for $F^{-1}(y)$, i.e., to solve $F(x) = y$. This is

$$x = -\frac{\ln(1 - y)}{\lambda}.$$

To generate x's with this distribution it is only necessary to generate uniform y's on $[0, 1)$ and use the formula above. As a practical matter there are more efficient ways that do not require the slow evaluation of a logarithm for each random number.

The exponential distribution might be used in the barbershop simulation in Section 1. If the average time for a barber to complete a haircut is twenty minutes and he is able to give one hundred haircuts in a typical week we might model the distribution of times for the barber to complete a particular haircut as an exponential with parameter $\lambda = 1/20$. By using UNI to obtain uniform numbers and the formula above to get x's we can get a list of 100 cutting times. Figure 10.6 shows a histogram for these values. As you can see, most of the haircuts take eighteen minutes or less; one fastidious customer requires an hour and a half.

In contrast to the exponential distribution, for the normal distribution it is relatively difficult to compute $F^{-1}(y)$. You already know from the problems in preceding chapters that it is not possible to solve the equation $F^{-1}(y) = x$ explicitly. Straightforward numerical methods can be slow. An algorithm due to Marsaglia is to generate a pair of uniforms on $[-1, 1)$, say, u and v and compute $r^2 = u^2 + v^2$. If $r > 1$ discard u and v and regenerate two new uniforms. We lose 22% of our efficiency here. If $r \leq 1$

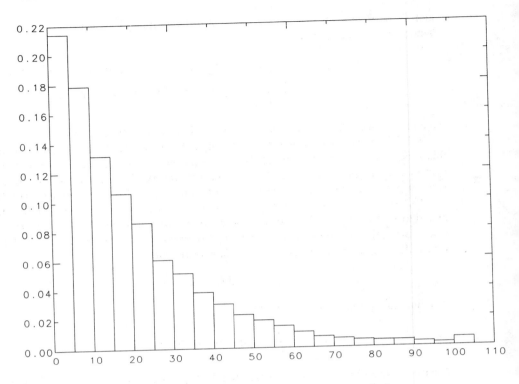

Figure 10.6   A Histogram From an Exponential Distribution

compute $w = \sqrt{-2(\ln r)/r}$. Then both $x = uw$ and $y = vw$ can be shown to be normally distributed with mean $\mu = 0$ and variance $\sigma^2 = 1$, denoted $N(\mu, \sigma^2)$. The slowest part is the calculation of the logarithm, but if maximum speed is not crucial this method is highly recommended because of its simplicity.

It is sometimes necessary to generate normal random numbers with mean and standard deviation different from $N(0, 1)$. In that case, if r_i is $N(0, 1)$, for example the output of RNOR below, then $r_i \sigma + \mu$ is $N(\mu, \sigma^2)$. Thus any normally distributed numbers can be obtained from RNOR.

10.8   FUNCTION RNOR

The function we include to generate standard normal $N(0, 1)$ numbers is based on the "ziggurat" method first described by Marsaglia (1984). It relies on an array of precomputed values which partition the normal distribution into regions of 1/32-nd of its area, and is very fast. Most evaluations of RNOR() only require generating one uniform and need no logarithms or square roots. Some evaluations can require several uniforms and other more complex computations, but these occur infrequently. To speed the computation further, the uniforms are generated with a UNI-like algorithm in-line, rather than by explicit evaluations of UNI (). Like UNI, RNOR requires an initializing evaluation of RSTART (ISEED) with an input seed, followed by evaluations of RNOR () without an argument.

```
C Typical RNOR usage:
      REAL      R,RNOR,RSTART,RSEED
      INTEGER ISEED,I
C Set initial seed
      ISEED = 305
      RSEED = RSTART(ISEED)
C Rstart returns floating echo of iseed
      WRITE (*,*) ISEED, RSEED
      DO 1 I = 1,3
         R = RNOR()
         WRITE (*,*) R
    1 CONTINUE
      END
C
C Produces the following output (can vary on different computers)
C
C        305      305.000
C             0.335785
C             0.230143
C             1.02517
```

Example 10.1   A Histogram from the Normal Distribution.

Using RNOR to produce 10, 000 values let us group them into thirty-two intervals $(-\infty, -3.0]$, $(-3.0, -2.8]$, $(-2.8, -2.6], \ldots, (2.8, 3.0]$, $(3.0, \infty)$ and then count the number and fraction in each subinterval. The program below shows how this is done. Plotting the fractions in "bar graph" form produces the histogram that is displayed in Figure 10.7. This figure also plots the "cumulative histogram," which is obtained from the former by summing the fractions from the left. The shape of a histogram is strongly dependent on the number of subintervals. If we had decided to use 5, 000 subintervals the histogram would look erratic because of many intervals with counts of zero. The cumulative histogram is less sensitive to the number of subintervals.

```
C HISTOGRAM FOR RNOR
C
      PARAMETER(NBINS=32,A=-3.0,B=3.0)
      INTEGER ISEED,I,J,H(NBINS),NR,INBIN
      REAL    R,RNOR,RSTART,RSEED,WIDTH
C
      ISEED = 305
      RSEED = RSTART(ISEED)
```

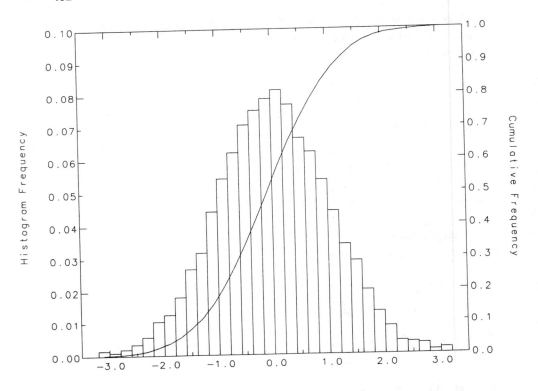

Figure 10.7   A Histogram of Numbers from RNOR and a Cumulative Histogram.

```
        WIDTH=(B-A)/(NBINS-2)
100     WRITE (*,*) 'EX 10.1: ENTER NUMBER OF NORMALS: '
        DO 200 I = 1,NBINS
            H(I)=0
200     CONTINUE
        READ (*,*) NR
        IF (NR .LE. 0) STOP
        DO 1 I = 1,NR
            R = RNOR()
            J = INBIN(A,B,NBINS,WIDTH,R)
            H(J) = H(J)+1
    1 CONTINUE
C
        WRITE (*,*) 'HISTOGRAM FOR RNOR: NUMBER IN BIN 1,...,32'
        WRITE (*,*) '  (-INFINITY,-3],(-3,-2.8],...,(2.8,3],(3,INFINITY)'
```

```
    WRITE (*,*) ' (VALUES ARE SLIGHTLY COMPUTER DEPENDENT)'
    WRITE (*,'(9I8)') (H(I),I=1,NBINS)
    WRITE (*,*)
    GO TO 100
    END
C
    INTEGER FUNCTION INBIN (XMIN,XMAX,NBINS,WIDTH,DATA)
C
C This function takes a real value in DATA, and finds
C the correct bin for it.  Vales below XMIN come back
C in 1.  Values above XMAX come back in NBINS.
C
    INTEGER NBINS
    REAL    XMIN,XMAX,WIDTH,DATA
    IF (DATA .LT. XMIN) THEN
        INBIN = 1
    ELSE IF (DATA .GE. XMAX) THEN
        INBIN = NBINS
    ELSE
        INBIN = 2 + (DATA-XMIN)/WIDTH
    ENDIF
    RETURN
    END    ■
```

*10.9 EXAMPLE: RADIATION SHIELDING AND REACTOR CRITICALITY

The earliest large scale uses of the Monte Carlo method occured in the study of uncharged particles moving through a medium. (Uncharged particles move in straight lines between collisions.) At any given time each neutron has a position, direction of motion, and energy. If it does not collide with an atom of the medium it will continue to travel in the same direction with the same energy. However at each point it has a probability of colliding; while traversing a length δs it has probability $\sigma_c \delta s$ of colliding. The quantity σ_c is called the "cross section," and it usually depends upon both the energy and the medium. Experiments and theory are used to determine cross sections, and extensive tables are needed to accurately describe a complicated medium. Normally though, at one energy the cross section is constant in each material but changes abruptly on passing from one region to the next. Reactors are composed of sections of many different cross sections, i.e., uranium rods immersed in water with pipes, walls, lead shields, etc. It is usual to assume that the distance traveled by the particle between one collision and the next is given by an exponential distribution with $\lambda = \sigma_c$. If the path between the current and next collision does not intersect a region boundary the particle is allowed to collide, otherwise it only proceeds as far as the boundary. If this is the outside then the particle is said to have escaped. (In programs of this type the two most time consuming portions are interpolating in tables to determine the cross section at a given point and energy, and solving the equations to decide whether a particle path intersects a region boundary.) If the particle collides, it can be absorbed into the medium and its life ends. In that case the energy is converted to heat; too much heat and the system will "melt down." Otherwise the particle might be scattered, and leave the point of collision with a new direction and new energy, or it can fission, i.e., several other particles leave the point of collision with different energies. Each of these can occur with certain probabilities.

There are two problems of interest.

(1) What fraction of the neutrons will exit and what is their average energy? This is the "shielding" problem.

(2) Are the total number of neutrons growing, i.e., are those generated due to fission exceeding those exiting? This is the "criticality" problem.

It is well beyond the scope of this text to go into these in more detail, but problem P10–11 will give you a sense of what reactor designers must deal with. Two widely used references are the books by Carter (1975) and Bell and Glasstone (1970).

10.10   PROBLEMS

P10–1.–If x is a random variable drawn from the normal distribution $N(0, 1)$ and z is some constant, then the probability that x is less than z, denoted by $P[x < z]$, can be shown to be

$$P[x < z] = \frac{1}{\sqrt{2\pi}} \int_{-\infty}^{z} \exp(-t^2/2)\, dt = \frac{1}{2}\left(1 + \text{erf}\frac{z}{\sqrt{2}}\right).$$

Write a subroutine HIST(VAL,H,N,HX) that will help you to form a histogram. That is, the input array HX should contain N increasing abscissas, and whenever the input VAL is between HX(I-1) and HX(I), HIST should increment H(I-1) by one, I=2, . . . ,N. Let M be a large integer, say 10,000. Generate M random variables x_i from $N(0, 1)$ using RNOR and use HIST to form a histogram of thirty-two subintervals as in Example 10.1. Then let z take on the values from -3.0 to 3.0 in steps of 0.2. For each z, print out and compare two quantities: (number of $x_i < z$)/M and $\frac{1}{2}(1 + \text{erf}(z/\sqrt{2}))$. Obtain the values of erf using any of the techniques described in the preceding chapters of this text. Try the experiment several times with different seed values for RNOR and compare the results. How much agreement do you get?

P10–2.–Use the uniform generator UNI to produce 10,000 numbers.

(a) Use the HIST subroutine from problem P10–1 to count the number which fall into ten equal size "bins" on [0,1]. Look up the χ^2 test in an applied statistics book, and use it to determine if the results appear to be uniform.

(b) Generate 5,000 pairs from the numbers you produced in (a). Using graphics software on your computer plot these pairs as points in the unit square. Do you see any pattern?

P10–3.–Use RNOR to generate two sequences $N(0, s)$, of normal random numbers, a_i and b_i. Let $x_1 = y_1 = 0$, $x_i = x_{i-1} + a_i$, $y_i = y_{i-1} + b_i$. Plot each point (x_i, y_i) and connect it to the preceding point with a straight line in order to generate a "Brownian" motion. Experiment with different values of s to keep the plot from straying more than one unit from the origin.

P10–4.–A map of Australia was overlaid on a grid and the following measurements were made of 13 points on the coastline: $(0, 0)$, $(20, 7.4)$, $(38.7, -4.8)$, $(47, 4)$, $(50, 13.5)$, $(41.5, 27)$, $(36, 40.5)$, $(32, 30)$, $(25, 34.5)$, $(26.8, 39)$, $(9, 34.5)$, $(2, 25.5)$, $(-7.5, 20)$. If an accurate rendition of the coast is required, for example, for topographic purposes, then many more points are needed. But for some applications, such as animated films, it is sufficient that the coast look like some country, but not any particular one. The basic idea is that when viewed from far away 13 points connected by straight lines are enough to represent the coast, but as the viewer comes closer there should be more detail, i.e., more points. Consider the following technique that generates a "fractal" coastline containing these points as well as an arbitrary number of others which are obtained by a randomization technique. Thus you can determine as many points as necessary depending on the viewing distance. The only criteria is that the image "look right" even if it is inaccurate. We use the algorithm of P4–11,

$$\mathbf{p}_{i+1/2} = (1/2 + w)(\mathbf{p}_i + \mathbf{p}_{i+1}) - w(\mathbf{p}_{i-1} + \mathbf{p}_{i+2}),$$

to generate a new point between $\mathbf{p}_i$ and $\mathbf{p}_{i+1}$, except that we choose w not as a constant but rather as a normally distributed variable with mean μ and variance σ^2, $N(\mu, \sigma^2)$. Experiment with different values of μ and σ. Reasonable values are $0 \leq \mu \leq 0.25$, $\sigma \approx \mu^2$, although $\mu \leq 0.333$ might be interesting too. Plot some of your coastlines. (See Fournier et al. (1982).)

P10–5.–Generating points inside a circle.

(a) Use UNI to generate $N = 10,000$ pairs of points X, Y uniformly on the interval $[-1, 1]$. Accept (X, Y) as a point in the unit circle if $X * X + Y * Y < 1$, otherwise reject the pair and generate two new points. What is the efficiency of this method? That is, what fraction of the generated pairs are actually inside the circle? If you can, plot the points in the circle. Do they appear uniform?

(b) Devise an analagous method for generating points inside the unit sphere. What is the efficiency of the method?

(c) Modify the method in (a) as follows: select X uniform on $[-1, 1]$ using UNI, then select Y (using UNI) uniform on $[-P, P]$, where $P = \sqrt{1 - X^2}$. This method is 100 percent efficient but (X, Y) is *not* uniform in the circle. Demonstrate this by plotting the points. Can you provide an explanation?

(c) In part (a), if the points are outside the circle do not reject them automatically. Instead view the unit circle as within the unit square. At each corner of the square we can just fit a quarter circle of radius $\sqrt{2} - 1$. If a pair (X, Y) is inside one of the quarter circles convert the pair to a point in the unit circle by translation and scaling. Revise your program using this algorithm. Now what percent of the points are accepted? Which of the two algorithms do you suppose is faster at generating 10,000 points in the circle?

P10–6.–Using UNI and the method described in Section 7, generate a histogram of 1000 points from an exponential distribution with parameter $\lambda = 1$. Compare the area in each section of the histogram to the exact area under the density function.

P10–7.–Two companies of archers face each other on either side of a ridge; "red" company has 100 archers and "white" has 45. Each minute one archer shoots an arrow. The arrow comes from red or white according to the fraction of archers left. So red shoots the first arrow $100/(100+45) \times 100$ percent of the time.

(a) If both companies always hit an opponent then red will almost always annihilate white. What is the average number of archers that remain in the red company when white has been exterminated?

(b) The commander of white company realizes the need to make it more difficult for red to succeed, and hence repositions the company so that red is only successful at scoring a hit a fraction of the time, say $0 \le p \le 1$. Approximately what value must p be for white to annihilate red, on the average, half the time. Closed form solutions to combat problems of this type can be obtained with "Lanchester" equations.

P10–8.–The Parking Problem. Cars that are one unit long try to park randomly along a street of length $N = 10$ units. A car can "park" if there is at least one half unit of space on either side of its center. In other words, no extra manuevering room is needed. What is the average number of cars that can park? To get an average you will have to run this problem repeatedly. (Answer: $0.7459N$. This is an unsolved problem in two dimensions, where the cars are squares of side one in a lot that is of size $N \times N$.)

P10–9.–A Simple Monkey Test.

(a) A monkey sitting at a typewriter generates letters at random (no numbers, spaces, punctuation, and only upper case). It is known that the average number of keystrokes before a specific three letter word appears is 26^3. Using UNI generate some sequences of "letters" and count the number before CAT appears. Does this agree with prediction? Try some other words, such as SEX, DOG, WOW, ETC. (Expect substantial variability in the count. If you have enough computer time, average the results over a few dozen tries.)

(b) If you have access to another generator, for example, if one is provided with your Fortran compiler, try it and compare the results.

P10–10.–An accelerator generates a beam of H^- atoms (i.e., hydrogen atoms weakly bonded with an extra electron). We would like these particles to hit a target 1000 kilometers away. There is a small random component to the angle of the particles as they leave the accelerator. If the path to the target were a straight line, it would be expected that both the x and y coordinate of the particle impact point would be normal with known means and standard deviations.

(a) Generate a plot of the distribution of particles on the target plane if the x and y coordinates are each $N(0,1)$. Do this by calling RNOR 5000 times and plotting the coordinates of each pair.

Unfortunately, the path of a charged particle is altered by the earth's magnetic field. To eliminate this difficulty the particles are sent through a low pressure chamber containing neutral oxygen or nitrogen molecules which strip the electron from H^- to produce a neutral particle beam which then will travel in a line. However, electron loss occurs at the expense of an angular deflection from the beam line. In this problem we assume that a particle's position on target plane x_1, x_2 depends on two factors—its direction as it leaves the accelerator and its scattering due to collision with the neutralizing gas,

$$x_1 = y_1 + z_1, \qquad x_2 = y_2 + z_2,$$

where (y_1, y_2) represents the effect of the accelerator, i.e., they are each $N(0,1)$, and (z_1, z_2) represents the additional (and independent) effect due to collision. (A standard deviation of 1.0 at the target corresponds to the assumption that the standard deviation at any given distance d meters from the accelerator is $10^{-6}d$ and that the target plane is 1000 km away.) With respect to z_1 and z_2, scientists claim that the gas deflects the particles uniformly with respect to angle and with a radius r that has a cumulative distribution function

$$F(r) = 1 - \frac{1}{\sqrt{r^2 + 1}}, \qquad r > 0.$$

Thus z_1 and z_2 can be generated from

$$z_1 = r\cos(\theta), \qquad z_2 = r\sin(\theta)$$

where θ is uniform on $[0, 2\pi)$. (This is sometimes called the Lorenz scattering distribution.)

(b) Generate a plot showing the distribution of points on the target plane when both the acceleration deflection and gas deflection are taken into account. Generate r's by using the inverse function F^{-1} as described in Section 7.

(c) Generate a histogram of the distribution of points on the target plane as a function of distance from the center.

P10–11.–This problem will give you an opportunity to think a little about the issues confronting a reactor designer. An infinite homogeneous slab is T centimeters thick. On one side is a point source of neutrons. The task is to analyze the neutrons that leave the source. We make the following assumptions.

(1) The source only emits neutrons into the slab. If μ is the cosine of the angle between the the normal to the wall and the neutron we assume that μ is uniform on $[0, 1]$. Thus neutrons are more likely to be injected along the normal than perpendicular to it. This defines the neutron direction on a cone from the source. The position around the edge of the cone is given by the "azimuthal" angle ϕ, which we assume to be uniform on $[0, 2\pi]$.

(2) The emitted neutrons have energy E, $E_{min} = 0.001 \leq E \leq E_{max} = 2.5$MeV (millions of electron volts). The energy is distributed according to the density function $f(E) = c/\sqrt{E}$. That means that $\int f(E)\,dE = 1$. Thus low energy neutrons are more likely.

(3) A neutron travels a distance D centimeters before it is absorbed, exits or is scattered. There are no fissions. D has an exponential distribution with parameter σ, the cross section. D is called the **distance to collision**.

(4) After updating the neutron's position we look to see if it has left the slab. If it has not, then ten percent of the time we allow the neutron to be absorbed and its path ends. The remaining ninety percent of the time it is scattered.

(5) Scattering is "isotropic," i.e., the neutron has no preferred direction.

(6) The scattered neutron usually has less energy. Its new energy is uniformly distributed in $[0.3E, E]$.

   For a given slab thickness, the task is to write a program to simulate a large number of emitted neutrons and accumulate some statistics. In particular to find the percentages of neutrons that are absorbed, transmitted (if the origin is at the source and the positive x direction is into the slab, a neutron is transmitted whenever its x coordinate is greater than T), or reflected ($x < 0$). In addition it is required to find the average energy of the absorbed, reflected and transmitted particles and their standard deviations.

(a) Take the cross section to be a constant, $\sigma = 2.0$. Use the progam in the file REACTOR.FOR to help you. The program is on a disk at the end of this book. (Because of the symmetry this is really a one dimensional problem. But reactor calculations are almost always three dimensional and REACTOR.FOR is written in this way.)

(b) A more realistic, but still simplified, cross section is given by the Fortran function CROSS below. Plot this cross section as a function of energy over the interval $[E_{min}, E_{max}]$. Use this function in your program. Approximately how thick does the slab have to be to allow no more than about one percent of the neutrons to be transmitted? Use at least $1,000$ particles to decide. A realistic simulation will need at least $100,000$ particles.

```
      REAL FUNCTION CROSS (E)
C RETURNS CROSS SECTION (FICTIONAL) FOR ENERGY IN RANGE [EMIN,EMAX]
      REAL E,S,ABS,SIN,Y,EXP
      S = ABS( SIN(100.*(EXP(E)-1.0)) + SIN(18.81*(EXP(E)-1.0)) )
      Y = MAX(0.02,S)
      CROSS = 10.0*EXP(-0.1/Y)
      RETURN
      END
```

P10–12.–Fractals.

(a) Write a program to display the Julia set for the iteration

$$z_{n+1} = z_n^2 + c,$$

where $c = 0.27334 + 0.00742i$. Sweep over points in the region $-1 \leq \text{Real}(z_0) \leq 1$, $-1.3 \leq \text{Imag}(z_0) \leq 1.3$. The amount of computation depends upon the choice of R and more importantly N. For points near the boundary of the set, N has to be large before you can decide if the iteration is going to converge. Nevertheless, $R = 100$, $N = 100$ will result in an interesting display.

(b) For the same iteration as in (a) generate the Mandelbrot set. Sweep over values of c in the region $-2.25 \leq \text{Real}(c) \leq 0.75$, $-1.5 \leq \text{Imag}(c) \leq 1.5$.

P10–13.–An iterated function system.

An **affine** transformation is a rule that associates arbitrary points in the plane (x, y) and (x', y') by the relation

$$x' = ax + by + e, \ \ y' = cx + dy + f,$$

where a, b, c, d, e, and f are numbers that define the transformation. An iterated function system is a set of n affine transformations and frequencies $p_i, i = 1, \ldots, n$. The transformations are applied iteratively, with each transformation being used $p_i \times 100$ percent of the time, at random. One starts with $(x, y) = (0, 0)$, generates a random number, picks a transformation based on the random number, computes the next (x, y) by using this transformation, then repeats the process, say, 2500 times. In other words

$$x' = a_k x + b_k y + e_k, \ \ y' = c_k x + d_k y + f_k,$$

where k is the index of the transformation that has been selected. Write a program to implement this algorithm. Your program should read the transformation parameters from a data file; that is, the values of n, a_i, b_i, e_i, c_i, d_i, f_i, and p_i, $i = 1, \ldots, n$. When your program has been checked out, run the transformations

k	a	b	c	d	e	f	p
1	0.5	0.0	0.0	0.5	0.0	0.0	0.33
2	0.5	0.0	0.0	0.5	1.0	0.0	0.33
3	0.5	0.0	0.0	0.5	0.5	0.5	0.34

which generates a "Sierpinski" triangle, and

k	a	b	c	d	e	f	p
1	0.00	0.00	0.00	0.16	0.0	0.00	0.01
2	0.20	−0.26	0.23	0.22	0.0	1.60	0.07
3	−0.15	0.28	0.26	0.24	0.0	0.44	0.07
4	0.85	0.04	−0.04	0.85	0.0	1.60	0.85

which generates a "fern." Hints: The largest and smallest x and y that will be generated by the transformations cannot be determined in advance. Either experiment with different values or save all the (x, y) pairs until the end and then determine an appropriate range. Also, the calculation takes a few iterations to settle down, so do not plot the first ten pairs.

P10–14.–It is often required that a random sample of k people be drawn without replacement from a population of n individuals. A component of a typical algorithm for drawing one such sample would be as follows:

```
(a) Number the individuals from 1,2,...,n
(b) Set i=1 and j=0
(c) If j=k stop. The sample is complete.  Otherwise go to d.
(d) Call a uniform (0,1) random number generator
(e) Select the ith individual with probability (k-j)/(n-i+1)
(f) If the ith individual was selected then add 1 to j
(g) Add 1 to i
(h) Go to c
```

The key criteria for a random sample is that every possible sample has the exact same chance of being drawn. Using the algorithm above generate 1000 samples of size 2 from a population of 6 people. There are 15 different possibilities. Do each of the different samples occur with about the same frequency? Are persons 1 and 2 as likely to be together in a sample as persons 1 and 6? Would this statement still be true if you used the old version of RND as the random number generator? Why?

P10–15.–Consider successive flips of a fair coin. Find the average number of flips until
  (a) HHTT appears,
  (b) HTHT appears,
  (c) HHHH appears.

(Hint: they are not the same.) This problem can be solved analytically, but can be approached easily with simulation. See the book by Ross (1983), page 231, for the analytic solution.

P10–16.–An intoxicated waiter in a restaurant takes food orders from a party of n people. Each person orders a different dish from the menu. The waiter gets the correct dishes but, forgetting who gave each order, gives out the food at random. What is the probability that no person gets the dish he or she ordered? To simulate this problem consider an array x with n elements. The problem is then to randomly assign one number from 1 to n to each element, where no number can be used more than once. (This rearrangement is called a permutation and is also related to sampling without replacement.) After you have created the permutation consider person i to have gotten the order if $x_i = i$. Try this problem for $n = 5, 10, 20$ and 50. As n increases, the answer approaches a famous constant. Find that constant.

10.11 PROLOGUES: UNI AND RNOR

```
      REAL FUNCTION UNI()
C***BEGIN PROLOGUE   UNI
C***DATE WRITTEN    810915 (YYMMDD)
C***REVISION DATE   871210 (YYMMDD)
C***CATEGORY NO.  L6A21
C***KEYWORDS  RANDOM NUMBERS, UNIFORM RANDOM NUMBERS
C***AUTHOR    KAHANER, DAVID, SCIENTIFIC COMPUTING DIVISION, NBS
C             MARSAGLIA, GEORGE, SUPERCOMPUTER RES. INST., FLORIDA ST. U.
C
C***PURPOSE  THIS ROUTINE GENERATES REAL (SINGLE PRECISION) UNIFORM
C            RANDOM NUMBERS ON [0,1)
C***DESCRIPTION
C        Computes real (single precision) uniform numbers on [0,1).
C        From the book, "Numerical Methods and Software" by
C              D. Kahaner, C. Moler, S. Nash
C              Prentice Hall, 1988
C
C     USAGE:
C            To initialize the generator
C                 USEED = USTART(ISEED)
C            where: ISEED is any NONZERO integer
C                   will return floating point value of ISEED.
C
C            Subsequently
C                 U = UNI()
C            will return a real uniform on [0,1)
C
C            One initialization is necessary, but any number of evaluation
C            of UNI in any order, are allowed.
C
```

```
C           Note: Depending upon the value of K (see below), the output
C                     of UNI may differ from one machine to another.
C
C           Typical usage:
C
C               REAL U,UNI,USTART,USEED
C               INTEGER ISEED
C                 Set seed
C               ISEED = 305
C               USEED = USTART(ISEED)
C               DO 1 I = 1,1000
C                   U = UNI()
C             1 CONTINUE
C                   NOTE: If K=24 (the default, see below) the output value of
C                           U will be 0.1570390462475...
C               WRITE(*,*) U
C               END
C
C       NOTE ON PORTABILITY: Users can choose to run UNI in its default
C               mode (requiring NO user action) which will generate the same
C               sequence of numbers on any computer supporting floating point
C               numbers with at least 24 bit mantissas, or in a mode that
C               will generate numbers with a longer period on computers with
C               larger mantissas.
C       TO EXERCISE THIS OPTION:  B E F O R E  invoking USTART insert
C               the instruction        UBITS = UNIB(K)        K >= 24
C               where K is the number of bits in the mantissa of your
C               floating point word (K=48 for Cray, Cyber 205). UNIB returns
C               the floating point value of K that it actually used.
C                   K input as .LE. 24, then UBITS=24.
C                   K input as .GT. 24, then UBITS=FLOAT(K)
C               If K>24 the sequence of numbers generated by UNI may differ
C               from one computer to another.
C
C***REFERENCES  MARSAGLIA G., "COMMENTS ON THE PERFECT UNIFORM RANDOM
C                   NUMBER GENERATOR", UNPUBLISHED NOTES, WASH S. U.
C***ROUTINES CALLED   (NONE)
C***END PROLOGUE UNI


        REAL FUNCTION RNOR()
C***BEGIN PROLOGUE   RNOR
C***DATE WRITTEN    810915 (YYMMDD)
C***REVISION DATE   870419 (YYMMDD)
C***CATEGORY NO.   L6A14
C***KEYWORDS   RANDOM NUMBERS, NORMAL DEVIATES
C***AUTHOR     KAHANER, DAVID, SCIENTIFIC COMPUTING DIVISION, NBS
```

```
C               MARSAGLIA, GEORGE, SUPERCOMPUTER RES. INST., FLORIDA ST. U.
C
C***PURPOSE  GENERATES NORMAL RANDOM NUMBERS, WITH MEAN ZERO AND
C                UNIT STANDARD DEVIATION, OFTEN DENOTED N(0,1).
C***DESCRIPTION
C
C          RNOR generates normal random numbers with zero mean and
C          unit standard deviation, often denoted N(0,1).
C              From the book, "Numerical Methods and Software" by
C                    D. Kahaner, C. Moler, S. Nash
C                    Prentice Hall, 1988
C    Use
C        First time...
C                    Z = RSTART(ISEED)
C                    Here ISEED is any  n o n - z e r o  integer.
C                    This causes initialization of the program.
C                    RSTART returns a real (single precision) echo of ISEED.
C
C        Subsequent times...
C                    Z = RNOR()
C                    Causes the next real (single precision) random number
C                         to be returned as Z.
C
C.................................................................
C                    Typical usage
C
C                    REAL RSTART,RNOR,Z
C                    INTEGER ISEED,I
C                    ISEED = 305
C                    Z = RSTART(ISEED)
C                    DO 1 I = 1,10
C                       Z = RNOR()
C                       WRITE(*,*) Z
C                  1 CONTINUE
C                    END
C
C
C***REFERENCES  MARSAGLIA & TSANG, "A FAST, EASILY IMPLEMENTED
C                METHOD FOR SAMPLING FROM DECREASING OR
C                SYMMETRIC UNIMODAL DENSITY FUNCTIONS", TO BE
C                PUBLISHED IN SIAM J SISC 1983.
C***ROUTINES CALLED  (NONE)
C***END PROLOGUE  RNOR
```

11

Trigonometric Approximation and the Fast Fourier Transform

11.1 INTRODUCTION

Nature imposes a cyclical character on many physical phenomena. Daily and seasonal fluctuations and the repetitive structure in materials such as crystals suggest the importance of periodic functions in explaining the world around us. The essence of Fourier analysis is that it is possible to represent complicated cyclical structure by suitable combinations of simpler cyclical functions, such as sines and cosines. This "transforms" the original information into another, potentially more useful form. Once this has been done our perspectives change and we can get new insights. You are already familiar with at least one such transform, the logarithm, and appreciate its significance. If you have read about the "power" or "frequency response" of your stereo amplifier, or purchased a "graphical equalizer" for your automobile you have been thinking in the style of Fourier analysis. We will see several examples of this approach in this chapter, and in Section 9 we use it to analyze a weather pattern called "el Niño."

The tools we will discuss are the the Fourier series and Fourier transform. These are masterpieces of applied mathematics which were first introduced by Daniel Bernoulli in the 1750s while studying the vibrations of a string. Almost 60 years later Jean B. Fourier gave a similar, although somewhat more formal development during his study of heat conduction. Fourier series and transforms are used in almost every branch of science, including electrical circuits, mechanical and vibrational systems, optics, acoustics and heat flow. A good general reference is the book by Weaver (1983). Fourier analysis has also been used in the analysis of **time series**, a sequence of numerical values arranged in a natural order. Usually, each observation is associated with a particular instant of time, such as the total weekly sales of a particular stock, and this provides the ordering. An elementary discussion is given in the book by Bloomfield (1976).

413

We will discuss three major topics which are heavily interrelated.

(1) The Fourier integral transform which converts a real function into a pair of real functions, or converts one complex function into another.

(2) The Fourier series which converts a periodic function to a sequence of Fourier series coefficients.

(3) The discrete Fourier transform which converts one sequence into another. The computational tool implementing this is the fast Fourier transform, for which a good reference is the book by Brigham (1974).

This chapter draws upon material from several preceding sections. We make use of the trapezoid rule from Chapter 5 to approximate integrals. Truncated Fourier series satisfy both a least squares condition (Chapter 6) and also interpolate (Chapter 4). The concepts we discuss can be subtle and require advanced mathematical techniques to prove. We will omit most proofs and rely on heuristics or diagrams for justification. We also assume that the reader has had some elementary exposure to complex arithmetic and the ideas of Fourier analysis.

11.2 FOURIER INTEGRAL TRANSFORM, DISCRETE FOURIER TRANSFORM, AND FOURIER SERIES

A function $t(x)$ is **periodic** with period $p > 0$ whenever $t(x + p) = t(x)$. If $t(x)$ has period p, then $t(x)$ also has period $2p, 3p$, etc., but we reserve the term for the *smallest* period of the function. Note that $T(x) \equiv t(fx)$ has period p/f because $T(x + p/f) = t(f[x + p/f]) = t(fx + p) = t(fx) = T(x)$. Thus we can create functions with any period we like. It is usually easy to decide if a function is periodic. The assumption of periodicity carries over to data but it is more difficult to determine. If the data you are analyzing are obviously *not* periodic, for example the CO_2 data in problem P6–4, an important first step is to determine (by eye, least-squares fitting, etc.) the nonperiodic trend and subtract it from the data.

The fundamental idea in this chapter is that a function or data can be expanded in terms of certain trigonometric functions. In these expansions the coefficients of the expansion play a central role and we will study their properties. This is different from, say, a Taylor Series expansion where the coefficients are often not of independent interest. We begin by defining the Fourier integral transform, the discrete Fourier transform and the Fourier series. Each of them has special properties, but they are also interrelated as we will illustrate in the following sections. This chapter only deals with real functions and real data, although we will use complex variables where necessary. We begin by stating (without proof) the fundamental results we will need.

(1) A real function $g(x)$ can be written as a difference of two semi-infinite integrals

$$g(x) = \int_0^\infty A(\omega)\cos(2\pi\omega x)\,d\omega - \int_0^\infty B(\omega)\sin(2\pi\omega x)\,d\omega,$$

with functions A and B defined by

$$A(\omega) = 2 \int_{-\infty}^{\infty} g(x) \cos(2\pi\omega x)\, dx, \quad \text{and} \quad B(\omega) = 2 \int_{-\infty}^{\infty} g(x) \sin(2\pi\omega x)\, dx, \quad \omega \geq 0.$$

A and B are called the **Fourier integral cosine (sine) transforms** of g, and g is called the **inverse Fourier integral transform** of A and B.

Example 11.1   Fourier Integral Transform for Runge's Function on $(-\infty, \infty)$.

Runge's function $R(x) = 1/(1 + 25x^2)$ is an even function, so $B(\omega) = 0$. The Fourier cosine integral transform can be shown to be

$$A(\omega) = 2 \int_{-\infty}^{\infty} \frac{\cos 2\pi\omega x}{1 + 25x^2}\, dx = \frac{2\pi}{5} \exp(-2\pi\omega/5).$$

Then, from (1)

$$\frac{1}{1 + 25x^2} = \frac{2\pi}{5} \int_{0}^{\infty} \exp(-2\pi\omega/5) \cos(2\pi\omega x)\, dx.$$

In some texts the function $1/(1 + ax^2)$ is called a **Cauchy distribution.** ■

(2) If $g_0, g_1, \ldots, g_{N-1}$ are arbitrary real numbers, they can be expressed as a finite series that we write, formally, as

$$g_j = a_0 + \sum_{k=1}^{N/2} a_k \cos\left(kj\frac{2\pi}{N}\right) + \sum_{k=1}^{N/2} b_k \sin\left(kj\frac{2\pi}{N}\right), \quad j = 0, \ldots, N-1.$$

This requires some clarification. First, the upper summation limit $N/2$ should be written $[N/2]$ the largest integer in $N/2$, but we will use the briefer notation. Thus $N/2$ is taken to be $(N-1)/2$ if N is odd. Second, the $b_{N/2}$ term is omitted if N is even. The numbers a_k and b_k are defined by

$$a_0 = \frac{1}{N} \sum_{j=0}^{N-1} g_j, \quad a_{N/2} = \frac{1}{N} \sum_{j=0}^{N-1} g_j \cos(j\pi) = \frac{1}{N} \sum_{j=0}^{N-1} (-1)^j g_j,$$

and

$$a_k = \frac{2}{N} \sum_{j=0}^{N-1} g_j \cos\left(jk\frac{2\pi}{N}\right), \quad b_k = \frac{2}{N} \sum_{j=0}^{N-1} g_j \sin\left(jk\frac{2\pi}{N}\right), \quad 1 \leq k < N/2.$$

The a's and b's are called the **finite, or discrete, Fourier cosine (sine) transform** of the g's and the latter are called the **inverse finite (discrete) Fourier transform** of the a's and b's. The discrete Fourier transform is often abbreviated DFT.

(3) If $g(x)$ is a periodic function on the interval $[a, b]$ then $g(x)$ can be written as an infinite series of sines and cosines,

$$g(x) = \frac{A_0}{2} + \sum_{n=1}^{\infty} A_n \cos\left(n\frac{2\pi x}{b-a}\right) + \sum_{n=1}^{\infty} B_n \sin\left(n\frac{2\pi x}{b-a}\right),$$

where the **Fourier coefficients**, A_n and B_n are given by

$$A_n = \frac{2}{b-a} \int_a^b g(x) \cos\left(n\frac{2\pi x}{b-a}\right) dx, \quad B_n = \frac{2}{b-a} \int_a^b g(x) \sin\left(n\frac{2\pi x}{b-a}\right) dx.$$

The function $\sin 2\pi\mu x$ repeats every $1/\mu$ radians and is said to have **radial, or angular frequency** $2\pi\mu$, and **circular frequency** μ. If x represents time in seconds $\sin 2\pi\mu x$ has radial frequency $2\pi\mu$ radians per second and circular frequency μ cycles per second. The period $1/\mu$ of $\sin 2\pi\mu x$ is the reciprocal of its circular frequency. If x is in spatial units, e.g., millimeters, the period is called the **wavelength** and the radial frequency is called the **wavenumber**. When you see the term **frequency** used alone, it can refer to either, but we will use it to mean circular frequency.

Each of the three representations above for g can be interpreted as a "sum" over frequencies.

(1) For the Fourier integral transform ω is a "frequency" variable and the integrals defining g at a point x are a "sum" over all possible frequencies.

(2) For the discrete Fourier transform it is convenient to think of the g_i as values that have been sampled every Δ time units. The quantity $T \equiv N\Delta$ is called the **record length** or **fundamental period**. The **fundamental circular frequency** is defined as $1/T$. In the discrete Fourier transform an expression like $kj2\pi/N$ can be written as $j\Delta k2\pi/T$ so at the point $j\Delta$ the sum defining g_j is over frequencies that are integer multiples of the fundamental frequency. Each frequency differs from the next by the fundamental frequency. The highest multiple of the fundamental is $N/2$, corresponding to a highest frequency of $N/2 \cdot 1/T = 1/(2\Delta)$, called the **Nyquist frequency** after the engineer Harry Nyquist (1889–1976) who described the concept while working on problems related to the telegraph. Frequencies greater than this cannot be determined by the data. Notice that a_0 is the average of the g's, and is called the **DC component**. The discrete Fourier series states that at the points $k = 0, \ldots, N-1$, g can be written as a DC component and a finite sum of terms with different frequencies.

(3) For the Fourier series the sum is also over integer multiples of the fundamental frequency, $1/(b-a)$. Each frequency differs from the next by this amount. The $A_0/2$ term is the average value of $g(x)$ on $[a, b]$. The Fourier series states that on $[a, b]$, $g(x)$ can be written as a DC component and an infinite sum of terms of different frequencies. The frequencies are not arbitrary, but in the ratio $1 : 2 : 3 : 4 : \cdots$. *Not every periodic function is a sinusoid.* But every periodic function can be written as a series of sinusoids of frequencies that are integral multiples of the fundamental frequency.

For each of these cases there is an equivalence between g and its transforms, or between g and its Fourier coefficients; one can be obtained from the other. There are many transform pairs in mathematics; you may have seen some such as the Laplace, Hankel, Z, etc., in other places. Whether we work with g, or its transforms, or its Fourier series, is a matter of convenience and depends upon the particular application at hand.

In some applications the integral transforms are the key mathematical tools. In optics, the effect of a lens can be studied with Fourier transforms. An interferometer is a precise optical measuring device; it is necessary to compute the Fourier transform of its output to analyze the measurements. In other fields such as digital signal processing the quantities that are studied are discrete, so integrals are replaced by infinite sums. But regardless of the origin, in practice what is likely to be available is a set of numbers representing experimentally determined "samples." This will naturally be a finite set. The engineer is continually asking "What inferences can I draw about the original problem from the finite samples?"

Normally a scientist will compute the discrete Fourier transform of the data. There are exciting and innovative techniques for this, of which the fast Fourier transform is the most important, but for applications this is only a first step. From this it is possible to obtain estimates of either the Fourier series or the Fourier integral transform of the original function. The interval between the sample points is determined by the instrumentation but the engineer must understand its implications. It is beyond the scope of this book to develop these applications in detail. So we will content ourselves with a few specific examples. The main point will be to illustrate how to use the subroutines and how the various transforms are related.

11.3   ENERGY AND POWER

When we speak about an explosion or about the output of a stereo amplifier, we often use the terms energy or power. These have intuitive as well as technical meanings. Fourier transforms and series can be used to make these terms precise, and this is the topic of the current section.

The **power at frequency** ω is defined as

$$P(\omega) = \frac{1}{2}\left(A(\omega)^2 + B(\omega)^2\right).$$

$P(\omega)$ is also the **energy** per unit frequency at frequency ω. For a discrete sequence the power at frequency k is

$$P_0 = Na_0^2, \quad P_k = \frac{N}{2}(a_k^2 + b_k^2), \ 0 < k \le N/2, \quad \text{but } P_{N/2} = Na_{N/2}^2, \text{ when } N \text{ is even,}$$

and for periodic functions the power at frequency k is

$$P_0 = \frac{b-a}{2}A_0^2, \qquad P_k = \frac{b-a}{2}(A_k^2 + B_k^2) \qquad k > 0.$$

In each case, the **total energy** E is the integral (or sum) of the power over all the frequencies.

If x in $g(x)$ represents time, $[g(x)]^2$ is the power at time x and also the energy per unit time at x. The total energy is the integral (or sum) of the power over all time. The relationship in energy between time and frequency is given by the **conservation of energy** which states that the total energy can be computed either from the Fourier coefficients or from the original function. Mathematically,

$$E = \int_{-\infty}^{\infty} [g(x)]^2 \, dx = \int_0^{\infty} P(\omega) \, d\omega,$$

$$\text{or} \quad E = \sum_{k=0}^{N-1} g_k^2 = \sum_{k=0}^{N/2} P_k,$$

$$\text{or} \quad E = \int_a^b [g(x)]^2 \, dx = \sum_{k=0}^{\infty} P_k,$$

depending on which transform or series we are dealing with. We interpret these as saying that the energy in g is spread among all the frequencies in the representation.

Often a plot is made of power versus frequency. This is called the **power spectrum** in the case of Fourier integral transform, or **periodogram** in the case of the discrete Fourier transform or Fourier series. For the discrete Fourier transform, power or energy is plotted against frequency k/T which then varies from 0 to $1/(2\Delta)$. In many applications the $k = 0$ term is not plotted. Another common convention is to invert the labelling on the horizontal axis, without otherwise altering the periodogram. The rightmost abscissa value then becomes 2Δ, the leftmost one becomes ∞, and abscissas are read as cycle lengths. In other words sinusoids with longer periods contribute to the left portion of the periodogram.

Example 11.2   Power Spectrum for Runge's Function.

The Fourier integral transform for Runge's function was computed in Example 11.1. In Figure 11.1 we have plotted its power spectrum.

We can also verify conservation of energy for this function. The integral of $R^2(x)$ can be shown to be

$$E = \int_{-\infty}^{\infty} \left(\frac{1}{1 + 25x^2} \right)^2 \, dx = \frac{\pi}{10},$$

and directly,

$$\frac{1}{2} \int_0^{\infty} [A(\omega)]^2 \, d\omega = \frac{2\pi^2}{25} \int_0^{\infty} \exp(-4\pi\omega/5) \, d\omega = \frac{\pi}{10}. \quad \blacksquare$$

In problem P6–8 we explained the importance of measuring the carbon dioxide concentration in the atmosphere and stated that the data can be fit with a growing exponential plus a few sinusoidal terms. To analyze it further in the context of this chapter we need to remove all but the sinusoids, that is subtract the quantity $B + d\exp(\alpha t)$, $t = 0, 1, \ldots, 215$

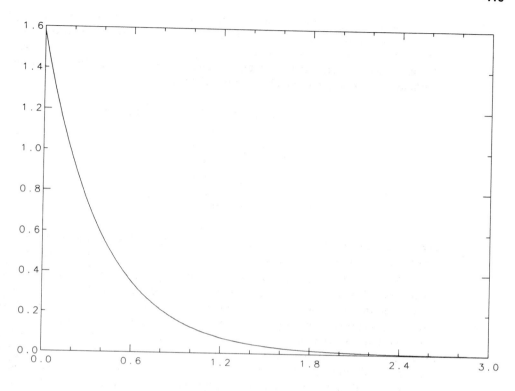

Figure 11.1   Power Spectrum for Runge's Function.

from each data value, where $B \approx 300.81$, $\alpha \approx 0.0037$, and $d \approx 14.18$. (Recall that there are 216 data points.) The result is plotted at the top of Figure 11.2. After the subtraction we can compute the discrete Fourier transform of the data. From this we can compute the periodogram which is plotted in the lower half of the figure. There are two distinct peaks at frequencies 18/216 and 36/216, and perhaps another at the far left. The large peaks correspond to frequencies of eighteen and thirty-six times the fundamental, 1/216, or cycle lengths of twelve months and six months respectively. The largest peak is associated with the annual (twelve month) cycle that we can see clearly on the top plot. The second peak corresponds to a semiannual (six month) cycle. It is worth trying to understand why this should have been expected. The strong annual variation in the data cannot be represented as a pure sinusoid with frequency 1/12, but can be represented as a sum of sinusoids with frequencies 1/12, 2/12, 3/12, 4/12, etc. These correspond to cycles of twelve, six, four, three months, etc. The six month peak is real; there is also a peak at three months but it is too small to be significant.

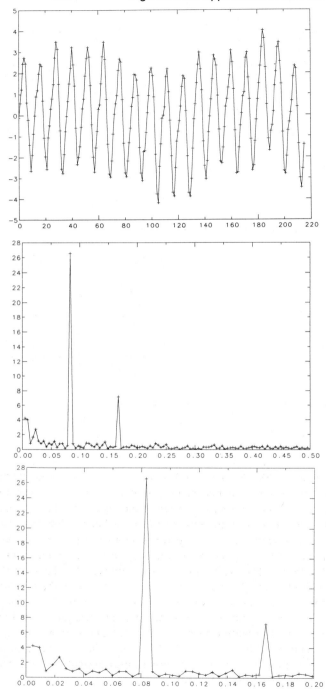

Figure 11.2 Carbon Dioxide in the Atmosphere (trend removed) and its Periodogram.

The bottom figure expands the vertical scale so that the peaks near the origin are easier to see. The one on the far left of the periodogram corresponds to a much longer cycle of 100–110 months that is barely detectable in the data. The smaller one just to its right corresponds to a period of about 40–44 months. These cycles are real, although more careful analysis is required to decide that they are not just statistical fluctuations. See for example the paper by Rust et al. (1979). The purpose of the example is only to indicate that the periodogram is a useful concept; we leave further discussion to the references.

11.4 HISTORICAL PERSPECTIVE: FOURIER (1786–1830)

On 21 December 1807, in one of the most memorable sessions of the French Academy, Jean Baptiste Joseph Fourier, a 21-year-old mathematician and engineer announced a thesis which began a new chapter in the history of mathematics. "Fourier claimed that an arbitrary function, defined in a finite interval by an arbitrary and capricious graph, can always be resolved into a sum of pure sine and cosine functions. The academicians, including the great analyst Lagrange, felt that this was entirely incredible. After all, any superposition of such functions could never give anything other than an infinitely differentiable function called 'analytic,' very far from the arbitrary function claimed by Fourier. Of course subsequent investigations demonstrated that Fourier's claim was entirely justified, although he himself was not able to provide the exact proofs, because he did not have the tools required for the operation with infinite series."[1] Further developments have involved the greatest mathematical minds including Dirichlet, Fejer, Lagrange, Riemann, Lebesgue, Gauss and Borel. But Fourier's method, which was developed in connection with his investigations into the theory of heat conduction, and in particular the form of the "Fourier Integral", had no predecessors.

Fourier's name is known and respected today, but neither his professional nor personal life were free from hardships. Born in a small village in south France, he was orphaned at age nine when his father, a tailor, died; he was then placed in the town's military school. His growing interest in mathematics was coupled with an active involvement in local affairs: during the French Revolution he was arrested briefly for the courageous defense of the victims of the Terror. An early teaching career was interrupted as his administrative abilities became known, in 1798 he was selected to join Napoleon's Egyptian campaign. He became secretary of the Egyptian Institute and held other diplomatic posts. He proposed a grand compendium of all the treasures which had been discovered during the Egyptian campaign; this was the first complete list of its kind ever published. On return to France he was given a prefecture near Grenoble and in 1808 Napoleon conferred a barony on him, but by the end of Napoleon's reign Fourier was forced to resign his position. Louis XVIII at first opposed his nomination to the Academy of Sciences because of his association with Napoleon but relented in 1817. From then until his death he was an active contributor to scientific thought. Throughout his career,

---

[1] C. Lanczos, *Discourse on Fourier Series*, Oliver and Boyd, 1966.

Fourier won the loyality of younger friends by his unselfish support and encouragement; most older colleagues were impressed with his achievements. A major exception was Poisson from whom there was continual controversy, criticism and enmity.

The last few years of Fourier's life were spent increasingly in confinement due to an illness (possibly myxedema) contracted during his stay in Egypt. But even during this time he continued publishing in mechanics, heat transfer, theory of equations and statistics. Various memorials have been made in his honor, including the renaming of his secondary school, in Auxerre, to Lycee Fourier. Interestingly, one of his most lasting accomplishments may be among his least known. Champollion, a student of his from the Rhone, excited about the Egyptian discoveries that Fourier was cataloging, eventually was able to translate the "Rosetta Stone." This is credited as the major breakthrough in understanding ancient Egyptian hieroglyphics.

11.5 PRACTICAL COMPUTATION OF FOURIER COEFFICIENTS; THE DISCRETE FOURIER TRANSFORM

To compute a Fourier series for a given function g we must first evaluate the integrals which determine the coefficients. It is rarely possible to compute these in closed form and they must be approximated using numerical quadrature. In this section we show that the discrete Fourier transform is such an approximation.

A truncated Fourier series $T_n(x)$, including terms through A_n and B_n, is called a **trigonometric polynomial** of degree n. For example, a trigonometric polynomial of degree one is

$$T_1(x) = \frac{A_0}{2} + A_1 \cos 2\pi f x + B_1 \sin 2\pi f x.$$

Subsequently, we will distinguish between a Fourier series with exact or approximate coefficients by using lower case t_j for the latter.

Example 11.3   Fourier Coefficients and Approximations.

Runge's function does not have a Fourier series because it is not periodic. We can define a new function that is periodic and agrees with $R(x)$ on $[-1, 1]$. It is called the **periodic protraction** of $R(x)$ to $[-1, 1]$. Let us compute its Fourier coefficients and compare the truncated series T_n with Runge's function on $[-1, 1]$.

Both Runge's function and its protraction are even, so $B_j = 0$. We take $[a, b] = [-1, 1]$; the fundamental frequency f is equal to $1/2$. The Fourier coefficients are given by

$$A_j = \int_{-1}^{1} \frac{\cos(j\pi x)}{1 + 25x^2}\, dx = 2\int_{0}^{1} \frac{\cos(j\pi x)}{1 + 25x^2}\, dx, \qquad j = 0, 1, \ldots.$$

For modest values of j the integrals on the right can be accurately computed by Q1DA from Chapter 5. These values are listed in the third column of the table below labelled Q1DA, and are accurate to all the digits printed. The Fourier coefficient A_j can be associated with its index j or with its frequency $\omega_j = j \cdot f = j/2$. This frequency is listed in the second column.

j	ω_j	A_j Q1DA	Trap. Rule $N = 100$ $\Delta = .02$	Trap. Rule $N = 50$ $\Delta = .04$	Trap. Rule $N = 20$ $\Delta = .1$	Trap. Rule $N = 9$ $\Delta = .2222$
0	0	5.49360×10^{-1}	5.49355×10^{-1}	5.49341×10^{-1}	5.49242×10^{-1}	5.44373×10^{-1}
1	1/2	3.44106×10^{-1}	3.44111×10^{-1}	3.44126×10^{-1}	3.44235×10^{-1}	3.39447×10^{-1}
2	2/2	1.75749×10^{-1}	1.75745×10^{-1}	1.75730×10^{-1}	1.75632×10^{-1}	1.66760×10^{-1}
3	3/2	9.69064×10^{-2}	9.69114×10^{-2}	9.69263×10^{-2}	9.70499×10^{-2}	8.28955×10^{-2}
4	4/2	5.00138×10^{-2}	5.00089×10^{-2}	4.99939×10^{-2}	4.99077×10^{-2}	2.18574×10^{-2}
5	5/2	2.77280×10^{-2}	2.77330×10^{-2}	2.77481×10^{-2}	2.79186×10^{-2}	-2.18574×10^{-2}
6	6/2	1.40807×10^{-2}	1.40757×10^{-2}	1.40604×10^{-2}	1.40275×10^{-2}	-8.28955×10^{-2}
7	7/2	8.02722×10^{-3}	8.03220×10^{-3}	8.04772×10^{-3}	8.36474×10^{-3}	-1.66760×10^{-1}
8	8/2	3.89227×10^{-3}	3.88728×10^{-3}	3.87152×10^{-3}	4.05238×10^{-3}	-3.39447×10^{-1}
9	9/2	2.38204×10^{-3}	2.38705×10^{-3}	2.40309×10^{-3}	3.20095×10^{-3}	-5.44373×10^{-1}
10	10/2	1.02504×10^{-3}	1.02001×10^{-3}	1.00367×10^{-3}	1.98009×10^{-3}	-3.39447×10^{-1}

Using these numbers as the Fourier coefficients we have plotted $T_n(x), 0 \leq x \leq 1$ for $n = 0, 1, 2, 4$, and 8, on the right-hand side of Figure 11.3. Note that T_8 is hardly distinguishable from Runge's function. Finally as a verification of the conservation of energy, we have computed (using Q1DA)

$$E^2 = \int_{-1}^{1} \left(\frac{1}{1 + 25x^2} \right)^2 dx = 0.31314169\ldots,$$

which ought to be equal to

$$= (b - a) \left(\left(\frac{A_0}{2} \right)^2 + \frac{1}{2} \sum_{k=1}^{\infty} A_k^2 + B_k^2 \right) \approx 2 \left(\left(\frac{A_0}{2} \right)^2 + \frac{1}{2} \sum_{k=1}^{10} A_k^2 \right) = 0.3131410\ldots,$$

computed from the values in the third column of the table.

Figure 11.1 plots the periodogram associated with the protraction so you can compare it with the power spectrum for Runge's function on $(-\infty, \infty)$. ∎

There are other ways of computing the Fourier coefficients which may be easier than using Q1DA. The integrand for each of the coefficients is a periodic function on $[a, b]$. In Section 8.1 of Chapter 5 we pointed out that the trapezoid rule is particularly effective when applied to such functions and thus it seems a likely candidate here. While it is possible to consider the approximation of each coefficient as an independent problem, a more direct approach is to apply the same trapezoid rule (the same number of evaluation

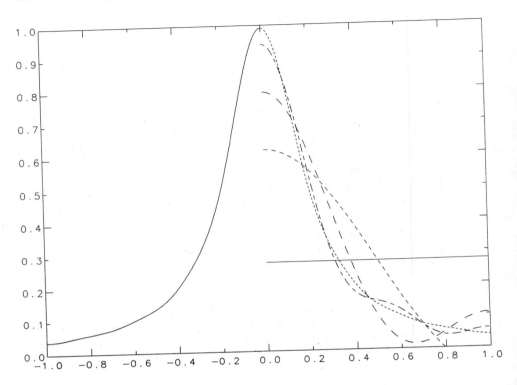

Figure 11.3 Approximations to Runge's Function. Left: t_8 on $[-1,0]$. Right: T_0, T_1, T_2, T_4, T_8 on $[0,1]$

points) to all of the coefficients that we want to approximate. If we label the trapezoid points $a = x_0, x_1, \ldots, x_N = b$ and let $2\pi f = 1/(b-a)$, we have

$$
\begin{aligned}
A_j &= \frac{2}{b-a} \int_a^b g(x) \cos{(j2\pi f x)} \; dx \\
&\approx \frac{2}{b-a} \cdot \frac{b-a}{N} \left[\frac{g(x_0) \cos(j2\pi f x_0)}{2} + g(x_1) \cos(j2\pi f x_1) + \cdots \right. \\
&\qquad\qquad \left. + g(x_{N-1}) \cos(j2\pi f x_{N-1}) + \frac{g(x_N) \cos(j2\pi f x_N)}{2} \right] \\
&= \frac{2}{N} \sum_{k=0}^{N-1} g(x_k) \cos(j2\pi f x_k) = \frac{2}{N} \sum_{k=0}^{N-1} g(x_k) \cos{\left(j2\pi f[x_0 + k(b-a)/N] \right)}.
\end{aligned}
$$

Notice that the sum ends at $N-1$ because both g and the cosine are periodic over $[a, b]$. By simplifying the cosine this becomes (note $g_k = g(x_k)$)

$$A_j \approx \frac{2}{N} \cos\left(\frac{2\pi j x_0}{b-a}\right) \sum_{k=0}^{N-1} g_k \cos(2\pi k j/N) - \frac{2}{N} \sin\left(\frac{2\pi j x_0}{b-a}\right) \sum_{k=0}^{N-1} g_k \sin(2\pi k j/N).$$

The important point is that the expression above contains the j-th components of the discrete cosine and sine Fourier transforms of the numbers $g_0, \ldots, g_{N-1}$,

$$A_0 \approx 2a_0, \qquad A_j \approx a_j \cos\left(\frac{2\pi j x_0}{b-a}\right) - b_j \sin\left(\frac{2\pi j x_0}{b-a}\right), \qquad 0 < j \le N/2.$$

In an analagous way the trapezoid rule approximation to B_j is given by

$$B_j \approx a_j \sin\left(\frac{2\pi j x_0}{b-a}\right) + b_j \cos\left(\frac{2\pi j x_0}{b-a}\right).$$

In fact, if $x_0 = 0$

$$A_0 \approx 2a_0, \qquad A_j \approx a_j, \qquad B_j \approx b_j, \qquad 0 < j \le N/2.$$

Thus we can view the discrete Fourier transform as a way to approximate the exact Fourier coefficients of a periodic function. The fact that $a_n \approx A_n$ only when $x_0 = 0$ is natural. Since $g(x)$ is assumed to be periodic the integrals can be over any interval of length $b - a$. Nevertheless the Fourier series coefficients, A_j, B_j and A_j', B_j' of respectively the two functions $g(x)$ and $g(x + x_0)$ are not identical, but are related by

$$A_j' = A_j \cos\left(\frac{2\pi j x_0}{b-a}\right) - B_j \sin\left(\frac{2\pi j x_0}{b-a}\right),$$

etc. The formulas for the discrete Fourier transform approximations to A_j and B_j reflect this shift of origin.

Example 11.4   Fourier Coefficient Approximations for Runge's Function, Continued.

Return to the table in the preceding example and examine columns four, five, and six. These values have been computed by using the trapezoid rule with N points, $N = 20$, 50, and 100. Notice that the discrete Fourier transform produces excellent approximations to the values of A_j. ■

When we approximate the Fourier coefficients by quadrature, their values depend upon the number of evaluation points. For the trapezoid rule this is N. Given N, how many Fourier coefficients can we reasonably approximate? We have already remarked that the most direct approach is to use the same number of trapezoid points for all the coefficients we want. With N points, approximations for the first $N/2$ A_j's and B_j's are given by the discrete Fourier transform. What happens if we attempt to compute additional coefficients with the same N points? The answer is that we get "worthless" numbers. To illustrate, look at the last ($N = 9$) column in the table above. It shows the values obtained by using $N = 9$. The formula in Section 2 tells us that $A_0, \ldots, A_4$ ought to be approximated by the first five values in the last column. From the table we see the first *four* values are reasonable approximations, but subsequent ones are not. In fact, the sixth through eleventh numbers are exactly the negatives of the first set. A discussion of the fifth (2.18574×10^{-2}) is delayed until Section 7.2.

Why does this happen? To explain it we must look again at the trapezoid rule, displayed below for $j = 5$

$$A_5 \approx \frac{2}{N} \left[\frac{g(x_0) \cos(5 \cdot 2\pi f x_0)}{2} + g(x_1) \cos(5 \cdot 2\pi f x_1) + \cdots + \frac{g(x_N) \cos(5 \cdot 2\pi f x_N)}{2} \right].$$

When we try to compute A_j for $j > N/2$ each of the terms in the trapezoid sum is exactly the negative of a term from an earlier trapezoid sum for a value of $j' < N/2$. For $N = 9$ and $j = 5$ the sum above is equal to

$$A_5 \approx -\frac{2}{N} \left[\frac{g(x_0) \cos(4 \cdot 2\pi f x_0)}{2} + g(x_1) \cos(4 \cdot 2\pi f x_1) + \cdots + \frac{g(x_N) \cos(4 \cdot 2\pi f x_N)}{2} \right]$$

$$= -a_4.$$

The equality occurs because

$$\cos\big(5\pi[x_0 + 2k/9]\big) = -\cos\big(4\pi[x_0 + 2k/9]\big), \qquad k = 0, \ldots, 9.$$

You can check this, using trigonometric identities. The integrands are not the same, but they agree at the mesh points. In words, the higher frequency sinusoid $\cos(5 \cdot 2\pi f x)$ generates the same values at the meshpoints (except for sign) as the lower frequency sinusoid $\cos(4 \cdot 2\pi f x)$. This is called **aliasing**, a term promolgated by the statistician John W. Tukey. In Figure 11.4 we have plotted $\cos(4\pi x)$ and $-\cos(5\pi x)$ on $[-1, 1]$. Also shown are the ten equally spaced points on the interval at which they agree. Aliasing refers, generally, to the fact that at equally spaced points the values of a sine (or cosine) function *could* have come from many other sine functions of higher frequency. This fact is used by an instrument called a stroboscope which appears to "slow down" rapidly rotating or oscillating machinery. For another example be sure to read the latter part of Section 7.2.

From this discussion you see that using the trapezoid rule, i.e., the discrete Fourier transform with N points, permits us to compute Fourier coefficients up to the Nyquist frequency $1/(2\Delta)$, or period 2Δ.

If we increase the number of points N we can approximate additional Fourier coefficients, *but the highest frequency only goes up if the spacing between the points goes down*. This is the situation in the table of Example 11.3. If we have an analog telephone signal which we sample at a rate of 5,000 times a second, we can only get information about frequencies in the signal through 2,500 cycles per second, *no matter how long we sample at this rate*. In most practical problems there is a range of frequencies in which we are interested, and the sampling rate is much more important than the number of Fourier coefficients we can approximate.

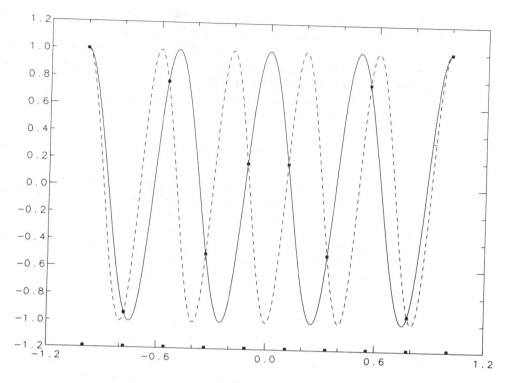

Figure 11.4   The Functions $\cos 4\pi x$ and $-\cos 5\pi x$ Agree at 10 Points on $[-1, 1]$.

11.6 **SUBROUTINES** EZFFTF **AND** EZFFTB

Because of the many applications of the discrete Fourier transform, almost every computing facility has one or more packages available for their calculation. The **fast Fourier transform** (FFT) is an algorithm which allows us to perform these operations very rapidly. (Section 12 briefly describes how this algorithm works.) Some vendors even provide special purpose hardware so that FFTs and related operations can be done quickly, in "real" time. The routines given here are part of the package by Swarztrauber (1975). The three subroutines are

EZFFTF   Computes the discrete sine and cosine Fourier transform given N real data values.

EZFFTB   Computes the N real data values given their discrete sine and cosine Fourier transform. This is the inverse or "back" transform.

EZFFTI   Initializing routine which must be called once before either EZFFTF or EZFFTB can be used with a new value of N.

The package from which these subroutines have been taken also includes routines for dealing with complex valued data (see Section 11.1), routines which are less "EZ" to use but more flexible, and routines which can handle data which are to be expanded in terms of only sines or cosines. Often, packages for FFTs require that N be a power of 2; these routines have no such restriction, but work most efficiently when N is a product of many small numbers.

We summarize the relevant formulas of Section 5 here.

(1) If the data values can be assumed to begin with abscissa $x_0 = 0$, then the discrete Fourier transform from EZFFTF approximates the Fourier series coefficients of the periodic function which generated the data.

(2) Otherwise, if $x_0 \neq 0$, the output arrays must be adjusted. To more clearly show the correspondence between the output and the data we begin our indexing at 1, rather than 0. Thus we assume that we are given N equally spaced real data values R(1), R(2), ..., R(N), with Δ as the spacing between the abscissas. The first and last data values are not assumed to be equal but the period of the underlying function is $N\Delta$. Thus $f = 1/(N\Delta)$. The output of EZFFTF is the number AZERO and the two arrays A(J) and B(J), J = 1,...,N/2, but B(N/2) = 0 if N is even. From these we can form a trigonometric polynomial of degree $N/2$ which approximates the truncated Fourier series for the function generating the data,

$$t_{N/2}(x) = \text{AZERO} + \sum_{j=1}^{N/2} \Big(A(J)\cos(j2\pi f x_0) - B(J)\sin(j2\pi f x_0) \Big) \cos j2\pi f x$$
$$+ \Big(A(J)\sin(j2\pi f x_0) + B(J)\cos(j2\pi f x_0) \Big) \sin j2\pi f x.$$

(3) The numbers A(J) and B(J) correspond to frequency J times the fundamental frequency, e.g. $J \times 1/(N\Delta)$.

The program below illustrates the use of EZFFTF. We will compute the approximate Fourier coefficients for Runge's function on $[-1, 1]$ using 16 and 17 sample points. Thus we will compute approximations to $A_0, \ldots, A_8$, to $B_1, \ldots, B_7$ and also to B_8 if $N = 17$. The program also evaluates the approximate Fourier series at 101 points between 0 and 1. It is this program for various N which we used to generate the numbers in the table of Example 11.3 as well as the graph in Figure 11.3.

```
C Using the real discrete Fourier transform, find the approximate
C Fourier coefficients to Runge's function on [-1,1] with N=16 and
C N=17.
C
      PARAMETER (MCOEF=17)
      REAL     A(MCOEF/2),B(MCOEF/2),R(MCOEF),WSAVE(3*MCOEF+15)
      REAL     DFTA(MCOEF/2),DFTB(MCOEF/2),C(MCOEF/2),S(MCOEF/2)
C
C Arithmetic statement function for Runge's function.
      RUNGE(X) = 1.0/(1+25.0*X*X)
C
      X0 = -1.0
```

```
        PI = ASIN(1.0)*2.0
C
        DO 10 N = MCOEF-1,MCOEF
            CALL EZFFTI (N,WSAVE)
C Function assumed to be periodic on [-1,1], of length 2.
            DEL = 2.0/N
            F = 2.0*PI/(N*DEL)
            DO 1 J = 1,N
C First sample point at -1, last at 1-DEL
                XJ = (-1.0) + (J-1)*DEL
                R(J) = RUNGE(XJ)
C Compute sines and cosines to adjust output of EZFFTF to give
C approximate Fourier coefficients.
            IF (J .LE. N/2) THEN
                C(J) = COS(J*F*X0)
                S(J) = SIN(J*F*X0)
            END IF
    1       CONTINUE
            CALL EZFFTF (N,R,AZERO,A,B,WSAVE)
C
C As a convenience this loop can go to N/2. If N is even last B is
C zero.
            DO 11 J = 1,N/2
                DFTA(J) = A(J)*C(J) - B(J)*S(J)
                DFTB(J) = A(J)*S(J) + B(J)*C(J)
    11      CONTINUE
            WRITE (*,*) ' RESULTS FOR N= ' ,N, ' AZERO = ',AZERO
            WRITE (*,*) '   J              DFTA(J)                 DFTB(J) '
            DO 12 J = 1,N/2
                WRITE(*,*) J, DFTA(J), DFTB(J)
    12      CONTINUE
            M = 101
C
C Evaluate interpolant at 101 points on [-1,1]
            WRITE (*,*) ' RESULTS FOR N= ',N
            DO 20 K = 1,M
                X = -1.0 + 2.0*(K-1.0)/(M-1.0)
                TN = AZERO
                DO 19 J = 1,N/2
                    TN = TN + DFTA(J)*COS(J*F*X) + DFTB(J)*SIN(J*F*X)
    19          CONTINUE
                ER = TN - RUNGE(X)
                WRITE (*,*) X,TN,ER
    20      CONTINUE
C
            WRITE (*,*)
C
    10  CONTINUE
```

```
        STOP
        END
C
C PRODUCES THE FOLLOWING RESULTS (THE 101 EVALUATIONS ARE OMITTED)
C
C   RESULTS FOR N=     16 AZERO =      0.274611
C   J               DFTA(J)              DFTB(J)
C         1     0.344365        -0.100161E-07
C         2     0.175654        -0.307123E-07
C         3     0.972947E-01    -0.203292E-07
C         4     0.501320E-01    -0.175307E-07
C         5     0.285903E-01    -0.176853E-07
C         6     0.149957E-01    -0.786582E-08
C         7     0.105194E-01     0.327562E-08
C         8     0.383778E-02    -0.268407E-08
C
C
C   RESULTS FOR N=     17 AZERO =      0.274581
C   J               DFTA(J)              DFTB(J)
C         1     0.344243        -0.350539E-07
C         2     0.175520        -0.364332E-07
C         3     0.969902E-01    -0.295525E-07
C         4     0.496441E-01    -0.205552E-07
C         5     0.275968E-01    -0.149523E-07
C         6     0.132331E-01    -0.103949E-07
C         7     0.709945E-02    -0.110340E-07
C         8     0.141324E-02    -0.116999E-07
C
```

11.7   TRUNCATED FOURIER SERIES AS AN APPROXIMATION

In some applications the Fourier series is most important, for others it is the Fourier coefficients. Truncating an infinite Fourier series leads to error, and approximating the coefficients with a discrete Fourier transform is an additional problem. We have already seen in Example 11.3 that the truncated series with exact coefficients can do a good job representing the underlying function. This section explains how these truncated series approximate the original $g(x)$, which is a periodic function on a fixed interval $[a, b]$.

11.7.1 Using Exact Fourier Coefficients

What do we mean when we say that the infinite Fourier series "equals" $g(x)$? That is, on $[a, b]$ the truncated series comes closer and closer to $g(x)$ as we take more and more terms, but for any particular x may never be exactly equal to $g(x)$, i.e., the truncated series converges to $g(x)$. For engineering problems the convergence is at all values of x at which $g(x)$ is "smooth." The series converges slowly where g changes rapidly.

Furthermore, in contrast to some other series approximations, Fourier series can often be differentiated (repeatedly) to obtain series approximations to derivatives.

T_n approaches g as n increases. In addition, for each fixed n, T_n has another important property. If $T_n(x)$ is an arbitrary trigonometric polynomial including terms through $\cos(n2\pi fx)$ and $\sin(n2\pi fx)$, then any selection of coefficients other than the "official" Fourier coefficients in T_n gives a worse approximation to g, *in a certain specific sense.* In this sense (described below) the truncated Fourier series is the best approximation to $g(x)$. The precise meaning of "best" is as follows. A measure of the average difference between g and T_n is

$$\int_a^b [g(x) - T_n(x)]^2 \, dx,$$

the **integral least square** difference. This difference, which depends upon the values of the coefficients A_i and B_i, is minimized when these are given as the Fourier coefficients of g. We interpret this to mean that on the average, truncated Fourier series are good approximations to smooth periodic functions, how good depends upon g. A rule of thumb is "the smoother—the smaller," i.e., the smoother the function g the more quickly its Fourier coefficients approach zero and the better an approximation T_n is likely to be. In particular this means that if $g(x)$ is smooth at $x = a$ and at $x = b$ the convergence is improved. For example the function $g(x) = \exp(-x^2)$ on $[a, b] = [-5, 5]$ joins smoothly to its periodic protraction. On the other hand on $[a, b] = [0, 5]$ g has a jump at each endpoint when we make it periodic. The Fourier series on the first interval converges much more quickly than on the second interval.

11.7.2 Using Approximate Fourier Coefficients

Let $T_n(x)$ be the truncated Fourier series with exact coefficients and let $t_n(x)$ be the similar series with coefficients given by the discrete Fourier transform. This requires that we use $N = 2n + 1$ points. Two properties of these trigonometric polynomials are obvious:

(1) Given T_n, to compute T_{n+1} only two new coefficients are needed, A_{n+1} and B_{n+1}, but both are integrals. For t_{n+1} two more data points must be given and all the coefficients need to be recomputed $(a_0, \ldots, a_{n+1}, b_1, \ldots, b_{n+1})$, but these are sums.

(2) There is no reason to expect t_n to satisfy the same integral least squares condition that T_n does. Intuition suggests that t_n may not be nearly as good an approximation to g as T_n. The left half of Figure 11.3 plots t_8, with the coefficients calculated by the 17-point trapezoid rule. It appears to be every bit as good as the "exact" truncated Fourier series. Remarkably, series computed in this way are known to converge just as rapidly to g as the ordinary Fourier series. We will try to explain why in this section.

We might replace the integral least squares condition with a discrete one,

$$\sum_{k=0}^{N-1} [g(x_k) - t_n(x_k)]^2,$$

where the integral has been replaced by a sum over the evaluation points. It is possible to show that the minimum of this sum is obtained when the coefficients of t_n are given by the discrete Fourier transform. So in this sense t_n is also a best approximation. In fact the minimum is zero, because

$$g(x_k) = t_n(x_k), \qquad k = 0, \ldots, N-1.$$

In the terminology of Chapter 4 the function t_n interpolates $g(x)$ at the equally spaced values x_k. The function T_n with exact coefficients also satisfies an interpolation property, but not at such a predictable set of points.

We can see this clearly in the case $x_0 = 0$ by writing

$$T_{N/2}(x) = \frac{A_0}{2} + \sum_{j=1}^{N/2} A_j \cos(2\pi j x/(b-a)) + B_j \sin(2\pi j x/(b-a)).$$

Now if $x_0 = 0$

$$t_{N/2}(x) = a_0 + \sum_{j=1}^{N/2} a_j \cos(2\pi j x/(b-a)) + b_j \sin(2\pi j x/(b-a))$$

$$= a_0 + \sum_{j=1}^{N/2} a_j \cos(2\pi j x/(N\Delta)) + b_j \sin(2\pi j x/(N\Delta)).$$

If we set $x = x_k = k\Delta$ then

$$t_{N/2}(x_k) = g_k = g(x_k),$$

because of the formula for the discrete Fourier transform in Section 2, item (2). This shows that $t_{N/2}$ interpolates $g(x)$ at x_k.

Recall that interpolation at increasing numbers of points is not always a good idea. For polynomials, adding another point often makes the interpolation worse, but not so for trigonometric interpolants. Suppose we consider a truncated series $t_m(x)$, with $m < n$ terms, but we still measure the quality of our approximation by the discrete least squares difference over the original $N = 2n + 1$ points. This is now a traditional least squares problem in the sense of Chapter 6, with model functions which are sines and cosines. Remarkably the solution to this problem, that is, the coefficients of t_m, are exactly $a_0, a_1, \ldots, a_m$, and $b_1, b_2, \ldots, b_m$, the same as the first m coefficients in t_n. This occurs because the normal equations for this problem—usually not recommended—are a diagonal system and can be solved immediately. Of course, t_m does not interpolate unless $m = n$, but because it is a least squares solution we feel that it ought to give reasonable approximations. Thus interpolation and least squares work together.

The idea of conservation of energy, Section 3, can also be applied here to help explain why t_n is a good approximation to $g(x)$. We know that

$$\frac{b-a}{N} \sum_{k=0}^{N-1} g_k^2 \approx \int_a^b [g(x)]^2 \, dx.$$

If you examine the formulas for conservation of energy in Section 3 you will see that the sum above equals

$$(b-a)\left(a_0^2 + \frac{1}{2} \sum_{k=1}^{N/2} (a_k^2 + b_k^2) \right),$$

while the integral on the right equals

$$(b-a)\left(\left(\frac{A_0}{2} \right)^2 + \frac{1}{2} \sum_{k=1}^{\infty} A_k^2 + B_k^2 \right).$$

Since these are approximately equal, the total energy from all the Fourier frequencies must spill into the approximate coefficients somehow. For $x_0 = 0$ the following formula can be shown to hold,

$$a_s = A_s + A_{N+s} + A_{N-s} + A_{2N+s} + A_{2N-s} + \cdots,$$
$$b_s = B_s + B_{N+s} - B_{N-s} + B_{2N+s} - B_{2N-s} + \cdots, \quad s \le N/2.$$

This is an analytic statement of the concept of aliasing—the approximate Fourier coefficient is altered by the exact Fourier coefficients corresponding to higher frequencies. We say that the higher frequencies $(N+s)f$, $(N-s)f$, etc., have been aliased into frequency sf. The difference between the two coefficients A_s and a_s arises from the presence of higher harmonics which are simply ignored in an ordinary truncated Fourier series T_n while t_n is cognizant of them. The sampling process, which is effectively what we do to compute the discrete Fourier transform, converts a high frequency oscillation to a low frequency oscillation, and any Fourier coefficient with index larger than $N/2$ will show its influence on the coefficient of a certain lower frequency.

We can rephrase this in a somewhat more practical way. The Fourier series is an infinite series. If we truncate it after N terms without adjusting the coefficients, information about the function g is lost. The approximate coefficients allow some of that information to be retained in a simple and explicit way as given in the formula above. Generally, when we take a finite discrete sequence of observations of a continuous function, information is lost. Partly this occurs because the discrete sequence is of finite extent. It is an advantage of the trigonometric functions that the loss of information due to discretization is manifest in the form of aliasing.

We can also use this formula to complete our explanation for the difference that we saw in Example 11.3 between $A_4 = 5.00138 \cdot 10^{-2}$ and $a_4 = 2.18574 \cdot 10^{-2}$ when

$N = 9$. The formula above must be adjusted to reflect the fact that $x_0 = -1$. In that case it can be shown that these two table entries are related by

$$a_4 = 2.18574 \cdot 10^{-2} = A_4 - A_{N-4} - A_{N+4} + \cdots \approx$$
$$A_4 - A_5 = 5.00138 \cdot 10^{-2} - 2.77280 \cdot 10^{-2} = 2.22858 \cdot 10^{-2}.$$

In other words the first missing frequency, $5/2$, aliases strongly into the frequency $4/2$. The lower frequencies are also affected but not as much. For example, the constant term estimated by the trapezoid rule is

$$a_0 = 5.44373 \cdot 10^{-1} = A_0 - 2A_N - 2A_{2N} - \cdots \approx$$
$$A_0 - 2A_9 = 5.49360 \cdot 10^{-1} - 2 \cdot 2.38204 \cdot 10^{-3} = 5.44596 \cdot 10^{-1}.$$

If the periodic function $g(x)$ has Fourier coefficients of large amplitudes beyond index $N/2$, then there will be a marked difference between A_s and a_s. In that case the truncated Fourier series will not be a close approximation to g whether we use exact or approximate coefficients. But if N is large enough that the tail of the series is small, then the high harmonics must have small and quickly diminishing amplitudes, and the difference between A_s and a_s will likewise be small.

An example of aliasing is given in Figure 11.5a–d. Figure 11.5a shows 36 sample values from the function $\sin(\pi x/4)$, $x = 0, \ldots, 35$. The sine curve is drawn to show how the samples were obtained. This function has frequency $1/8$. Using these data values and EZFFTF the discrete Fourier transform is computed and its periodogram is plotted to the right of the data. The record length is $N\Delta = 36$, thus the fundamental frequency is $1/36$ and the Nyquist frequency is $1/(2\Delta) = 1/2$. Notice the horizontal axis on the periodogram, which is in multiples of the fundamental frequency up to the Nyquist frequency, $1/2$. The frequency associated with the function $\sin(\pi x/4)$ is $1/8$. This is *less* than $1/2$ so the periodogram shows a large magnitude for those values of frequency near $1/8$. Since $1/8$ is not exactly represented on the periodogram, contributions to frequencies around $1/8$ are large. Figure 11.5b shows the same computation, but with 12 data points, sampled with $\Delta = 3$. In this case the Nyquist frequency is $1/6$, still larger than the frequency of the data, so once again the periodogram has large components near $1/8$. In Figure 11.5c the sampling interval is $\Delta = 6$ and $N = 6$. Now the Nyquist frequency $1/12$ is *less* than the frequency of the data. The high frequency data is aliased into a lower frequency, i.e., the same data values can be generated by a lower frequency sinusoid and that function, $-\sin(2\pi x/24)$, is displayed along with the data. Consequently, the periodogram shows a large magnitude at the two frequencies ($1/36$ and $2/36$) surrounding $1/24$. In Figure 11.5d $\Delta = 9$, $N = 4$ and the Nyquist frequency is $1/18$, less than the frequency of the data. Once again the data can be generated by a lower frequency sinusoid, $\sin(2\pi x/72)$ that is shown with the data. The periodogram has a large component at $1/36$, the frequency nearest $1/72$.

In summary:

(1) The discrete Fourier transform approximates the Fourier series coefficients. With N points $a_s \approx A_s$, $b_s \approx B_s$, $s \le N/2$.

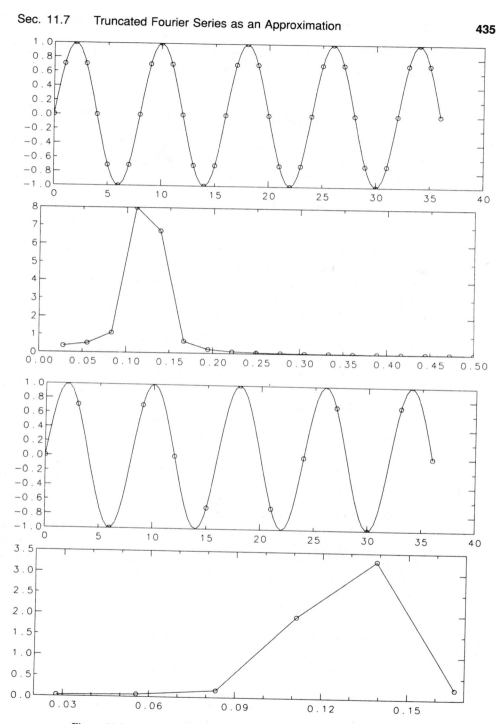

Figure 11.5   Aliasing a High Frequency into a Lower Frequency (a) and (b).

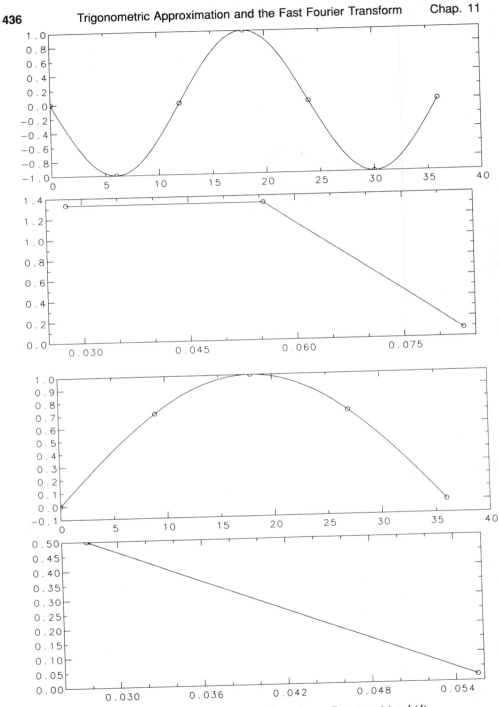

Figure 11.5   Aliasing a High Frequency into a Lower Frequency (c) and (d).

(2) The coefficients of the approximate Fourier series t_n are affected by *all* the Fourier coefficients. As a consequence t_n often provides more information than T_n.

(3) The approximate Fourier series t_n interpolates at $2n + 1$ equally spaced points, T_n need not.

(4) The exact and approximate Fourier series T_n and t_n are the best approximation to $g(x)$ when measured by an integral and a sum respectively.

A concrete illustration of case (2) occurs when we see a film of a moving wagon whose spokes appear to slow, stop, or even reverse. The wheel rotation can be described by a high frequency sinusoid, i.e., placing a paint spot at one point on the wheel allows us to express the position of the spot as a function of time as $g(x) = A_j \cos(j2\pi fx)$ for some large j. The Fourier series for g only has one nonzero term. The truncated Fourier series $T_n(x)$ (with exact coefficients) is identically zero until $n \geq j$. The movie camera samples $g(x)$ at about 1/24-th of a second intervals. By Nyquist's result we cannot approximate any Fourier coefficient which corresponds to a frequency higher than twelve cycles per second. But because of aliasing the sampling alters the approximate Fourier coefficients for much lower frequencies. The approximate truncated Fourier series contains some of this high frequency information. Because we "see" wheel rotation, presumably our brain is also taking these sample values and forming the same approximate truncated Fourier series even if it gives surprising values between the data points.

11.8 RELATIONSHIPS BETWEEN FOURIER TRANSFORMS AND FOURIER SERIES

Section 7 showed the relation between the discrete Fourier transform and Fourier series. We would now like to show the relationship between Fourier series and Fourier integral transform. The important idea here is that the Fourier integral transform is a Fourier series for a function with a very large (infinite) interval of periodicity. We will illustrate this by reconsidering Example 11.3, and studying the effect of altering the period of the Fourier series by expanding the interval $[a, b]$ from $[-1, 1]$ to $[-2, 2]$. The period of the function has now doubled. The frequencies in the Fourier series change and the spacing between them is halved because the fundamental frequency has been halved from $1/2$ to $1/4$. In a sense the new Fourier series has "twice" as many terms as the old one. The first two columns of the table below essentially duplicate the second and third columns of the table in Example 11.3. By expanding the interval our new table of Fourier coefficients will have contributions at all the frequencies listed including the ones which are blank in column two.

How shall we define the function on the larger interval, that is for $-2 \leq x \leq -1$ and $1 \leq x \leq 2$? One way is to take it to agree with Runge's function, another way is to define it to be zero there. The second case is particularly simple because we can relate the new Fourier coefficients to the old ones when they both are associated with the same frequency. For example, if A_5 is the old coefficient associated with frequency $5 \cdot 1/2$, then the new coefficient with index 10 corresponds to the same frequency, $10 \cdot 1/4$. The formula for this coefficient (remember that g is zero over the larger interval) is

$$\frac{2}{4} \int_{-2}^{2} \cos(10 \cdot 2\pi x/4) g(x)\, dx = \frac{1}{2} \int_{-1}^{1} \cos(5x\pi) g(x)\, dx = \frac{A_5}{2},$$

i.e., it is exactly half as large. So that our table entries can be compared we will scale each entry by the frequency interval over which it applies, i.e. divide column two by π, and divide column three by $\pi/2$. Hence, both tabular values at this frequency are identical. Of course, at the intermediate frequencies we do not have anything to compare with. Column three gives these values, computed using Q1DA. The highest frequency we've listed, 4, is eight times the fundamental for the interval $[-1, 1]$ and sixteen times the fundamental for the interval $[-2, 2]$. Notice that the spacing between the frequencies will not change if we displayed more terms.

If we define g to agree with Runge's function on the larger interval we cannot expect to get exact agreement at the common frequencies, but the numbers should not differ too much. These values are shown in the fourth column.

COMPARISON OF FOURIER COEFFICIENTS ON TWO INTERVALS.

ω	$1/\pi \times$Coeff. on $[-1, 1]$	$2/\pi \times$Coeff. on $[-2, 2]$ zero off $[-1, 1]$	$2/\pi \times$Coeff. on $[-2, 2]$ $R(x)$ on $[-2, 2]$
0/4	1.7487×10^{-1}	1.7487×10^{-1}	1.8731×10^{-1}
1/4		1.5395×10^{-1}	1.4755×10^{-1}
2/4	1.0953×10^{-1}	1.0953×10^{-1}	1.0618×10^{-1}
3/4		7.3552×10^{-2}	7.8184×10^{-2}
4/4	5.5943×10^{-2}	5.5943×10^{-2}	5.6774×10^{-2}
5/4		4.4470×10^{-2}	4.1673×10^{-2}
6/4	3.0846×10^{-2}	3.0846×10^{-2}	3.0299×10^{-2}
7/4		2.0043×10^{-2}	2.2231×10^{-2}
8/4	1.5920×10^{-2}	1.5920×10^{-2}	1.6162×10^{-2}
9/4		1.3521×10^{-2}	1.1864×10^{-2}
10/4	8.8261×10^{-3}	8.8261×10^{-3}	8.6178×10^{-3}
11/4		4.9203×10^{-3}	6.3334×10^{-3}
12/4	4.4820×10^{-3}	4.4820×10^{-3}	4.5934×10^{-3}
13/4		4.5517×10^{-3}	3.3826×10^{-3}
14/4	2.5551×10^{-3}	2.5552×10^{-3}	2.4470×10^{-3}
15/4		7.6741×10^{-4}	1.8078×10^{-3}
16/4	1.2389×10^{-3}	1.2389×10^{-3}	1.3024×10^{-3}

Generalizing this idea further we may let $[a, b] \to (-\infty, \infty)$. Each new Fourier series contains more terms at finer frequencies. In the limit the fundamental frequency goes to zero and the Fourier series, which is a sum over multiples of the fundamental, becomes the Fourier integral over all positive frequencies. The integrals over $[a, b]$ defining the Fourier coefficients become integrals from $(-\infty, \infty)$; the index k denoting the k-th Fourier coefficient is replaced by a continuous variable, ω. If the function is zero off $[a, b]$ then we have seen that the *scaled* Fourier coefficients at the frequencies $k \cdot 1/(b - a)$ are equal in all the subsequent Fourier series, and therefore are equal in the limit function, which is the Fourier integral transform, i.e.,

$$A_k \cdot (b - a) = A\left(\frac{k}{b - a}\right), \quad \text{and} \quad B_k \cdot (b - a) = B\left(\frac{k}{b - a}\right).$$

We can thus think of the Fourier coefficients as sampled values of the Fourier integral transform of the function which is zero off $[a, b]$. To restate this, if a function goes to zero at both ends of $[a, b]$, the Fourier series coefficients and the value of the Fourier transform corresponding to the same frequency will be equal within a factor of $b - a$. In that case the discrete Fourier transform a_k, b_k will approximate the Fourier transform, $a_k(b - a) \approx A(k/(b - a))$, $b_k(b - a) \approx B(k/(b - a))$. A good example of such a function is $\exp(-x^2)$ on $[a, b] = [-5, 5]$.

*11.8.1 Functions That Do Not Have Fourier Transforms or Fourier Series

Not every function can be written as a Fourier integral, and in Section 2 we ignored the question of what functions have Fourier integral transforms. The most elementary view is to realize that the formulas for energy involve the infinite integral of $g(x)^2$, $A(\omega)^2$ and $B(\omega)^2$, so these functions must go to zero for large $|x|$ or $|\omega|$ in order for the integrals to exist. This is impossible for a periodic function like $g(x) = \sin x$, or even for a constant function. They do not have Fourier integral transforms in the normal sense. But the concept is so useful that it has been generalized to cover these cases too.

For example, what would you expect as the Fourier integral transform of $\sin 2\pi \alpha x$? Intuitively, we think of the Fourier integral transform as a sum over all possible frequencies. The contribution at frequency α is one, and is zero at other frequencies. Another way to see this is to examine the value of the integrals defining $A(\omega)$ and $B(\omega)$ on a large finite interval and then allow the interval length to become infinite. Because $\sin 2\pi \alpha x$ is an odd function $A(\omega)$ will be be zero. For $B(\omega)$ we can calculate

$$B(\omega) \approx 2 \int_{-U}^{U} \sin(2\pi \alpha s) \sin(2\pi \omega s)\, ds = \frac{\sin(2\pi U[\alpha - \omega])}{\pi[\alpha - \omega]} - \frac{\sin(2\pi U[\alpha + \omega])}{\pi[\alpha + \omega]}.$$

This is an odd function of ω, has peaks at at $\omega \approx \pm \alpha$ of height about $\pm U$, and decays for large ω. As $U \to \infty$ we get a "function" which is zero everywhere except at $\pm \alpha$ where it goes to $\pm \infty$. We usually describe this by saying that the Fourier integral transform of $\sin(2\pi \alpha x)$ is a pair of δ functions at $\pm \alpha$. In a similar way any linear combination of trigonometric functions has a Fourier integral transform which is the same linear combination of δ functions.

Analogously not every function has a Fourier series. Runge's function $R(x)$ is not periodic. The Fourier series we wrote for it in Example 11.3 converges to $R(x)$ on $[-1, 1]$ but nowhere else. We are usually interested in functions over a finite domain and do not care what happens outside this domain. We can deal with this in three related ways.

(1) Assume that the function we are studying is identically zero or goes to zero very rapidly outside the domain we are interested in. This function has a Fourier integral transform but not a Fourier series.

(2) Assume that the function is periodic. This function has a Fourier series, but not a normal Fourier integral transform.

(3) Assume that the function is defined from $-\infty$ to ∞ but the values available to us are its product with another function that is zero off the interval of interest and one within it. This is called **windowing** and the zero/one function is called the **window**.

A comparison of these three approaches is beyond the scope of this text. Interested readers are referred to the book by Weaver (1983).

There are two kinds of functions that play an important role in understanding Fourier transforms and series, these are **time limited** and **band limited**. A time limited function is nonzero only on a finite interval. (Depending on the units of the variable these functions are sometimes called space limited.) A time limited function cannot have a Fourier series because it is not periodic. A band limited function is one whose Fourier integral transforms $A(\omega)$ and $B(\omega)$ are nonzero only on a finite interval, e.g., $A(\omega) = B(\omega) = 0$, if $|\omega| > \Omega > 0$. In that case $g(x)$ is said to have **bandwidth** 2Ω. There are two fundamental theoretical results about such functions. For details we again refer the reader to Weaver (1983).

(1) The Heisenberg uncertainty principle: It is not possible for a function to be both time limited *and* band limited. If a finite interval exists beyond which g is zero, then no such finite interval exists for its transform, and conversely. In introductory physics courses, this is sometimes presented by stating, that it is not possible to determine both the position and the momentum of an atomic particle with exact accuracy.

(2) The Sampling Theorem: If we know the values of a band limited function at all the points $k/(2\Omega), k = -\infty, \ldots, \infty$ then we can (theoretically) determine its values for *all* x. The sampling rate of $1/(2\Omega)$ represents two samples per cycle of the highest frequency present. As mentioned earlier, it is called the **Nyquist rate** and the sequence of samples obtained using this rate, the **Nyquist samples**.

In most cases we do not have any control over the sampling rate. But (2) says that if our samples are 2Ω apart, i.e., at a rate of $1/(2\Omega)$, we can fully reproduce any function whose highest frequency is Ω. This is a crucial idea. A simple application occurs in our telephone system. Suppose the phone company wants to transmit voice signals up to a frequency of 2,500 cycles/sec. According to the Sampling theorem they can sample the incoming signal at 5,000 points/sec and exactly reproduce the signal. This means that the entire analog signal does not have to be transmitted from sender to receiver, but only 5,000 samples per second of it. Perhaps this seems like a great deal of data, but it actually leaves the transmission line with nothing to do for most of each second. Of course, the same line can then be utilized for another call.

If our samples are 2Ω apart we cannot infer anything about frequencies higher than Ω. If we try anyway we will get spurious results. We saw this behavior before, in the last column of the table in Example 11.3.

11.9   LEAST SQUARES APPLICATIONS: EL NIÑO

Recall from Section 7 that Fourier series deal with functions on an interval. Once the coefficients have been estimated by the trapezoid rule the series only relies on a discrete set of data. The truncated sum $t_n(x)$ with coefficients computed by the $2n + 1$ point trapezoid rule interpolates the original data. As you have seen, in some ways it is more useful than the "ordinary" series and you should avoid thinking of it as an approximation but rather as an entity in its own right. For example, we explained in Section 7.2 that if we drop off some of the high frequency terms, then t_m, $m < n$, is the best least squares approximation to the data g_k, $k = 0, \ldots, N - 1$, using the first m Fourier sines and cosines.

As a typical application we consider "el Niño." Scientists have observed that in the southern Pacific the prevailing winds are from East to West. This causes the surface water to move in the same direction, resulting in an upwelling of lower level, colder water on the West coast of South America. The colder water is richer in nutrients and allows large numbers of marine organisms to thrive on the continental shelf off these coasts. The coastal weather is also affected. The effect is not constant but varies during the year and from year to year in a more or less regular cycle called **el Niño**. As food production and weather prediction are of great economic importance scientists try to analyze these cycles. One way is by use of the "Southern Oscillation Index," the difference in atmospheric pressure between Easter Island and Darwin Australia, measured at sea level at the same moment. There is one data point per month representing an average of a number of values, and it is thought of as occuring at the middle of the month. Figure 11.6a shows this index for the 14 year period 1962–1975. There are 168 points plotted at the half integers (representing mid-months) between 0 and 168.

Let us consider various "fits" to this data. On Figure 11.6a we have also plotted the trigonometric interpolant, t_n, $n = 84$, whose coefficients have been computed using EZFFTF. The function t_n has then been evaluated at enough extra points to generate the plot shown there. You can see that t_n interpolates and is also satisfactory between the data points. Interpolating 168 points with a single algebraic polynomial would be hopeless. The remaining portions of this figure plot t_m for $m = 42, 21, 15$, and 10 terms. In each case we begin with the trigonometric interpolant t_n and discard more of the high frequency terms. These are also the least squares fits with the given number of terms. Figure 11.6d, corresponding to 15 terms, is perfectly reasonable, and we can even see the major fluctuations which are due to annual variation.

Figure 11.6e shows t_{10}. Why does it give such a bad fit? The annual variation in the data leads to a t_n which has one especially large coefficient, corresponding to a sinusoid oscillating once per year. You can see this by examining the periodogram, shown in Figure 11.7a. We see a large peak at $j = 14$. Each j point on the periodogram corresponds to a sinusoid which has j cycles within the 168 month data. Since 14 cycles

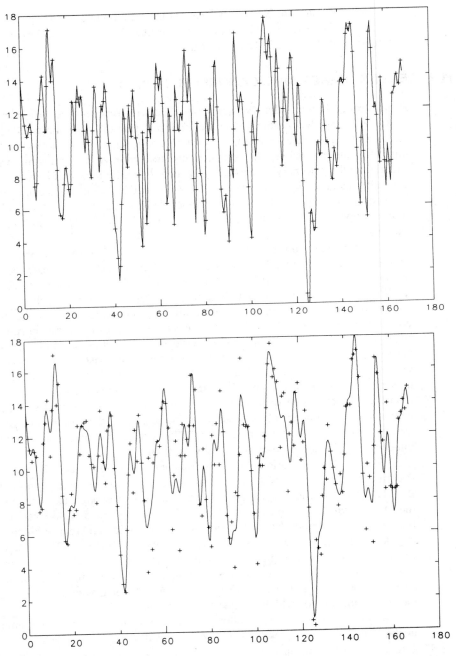

Figure 11.6 The Southern Oscillation Index and Trigonometric Approximations (a) and (b).

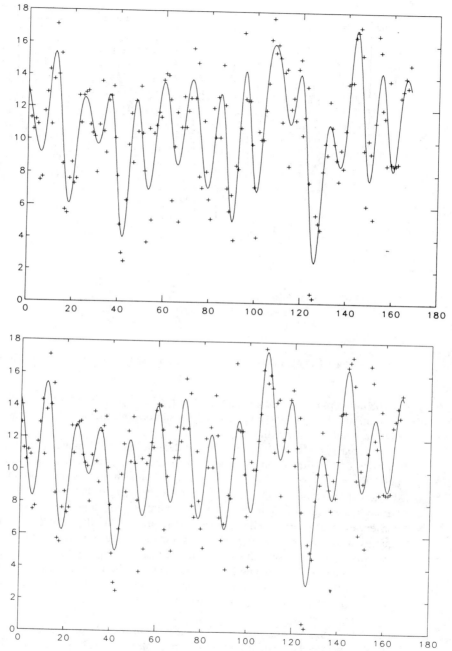

Figure 11.6   The Southern Oscillation Index and Trigonometric Approximations (c) and (d).

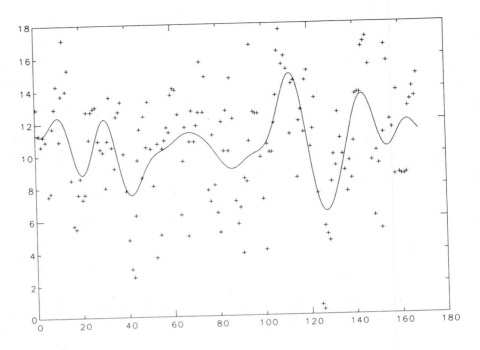

Figure 11.6 The Southern Oscillation Index and Trigonmetric Approximations (e).

in 168 corresponds to one per 12 months, this large peak is associated with regular annual fluctuations. If the trigonometric polynomial does not include this term we cannot expect to do a good job fitting the data.

What happens if the cycles in the data do not correspond to a frequency that is a multiple of the fundamental? After all, nature does not always operate with the same units we do. The ensemble of frequencies chosen by the periodogram may bracket the actual frequency in the data by such a wide margin that they fail to capture the full variation associated with it. We can guess what would happen by considering the discrete conservation of energy. The sum of the squares of the data must equal the sum of the squares of the Fourier coefficients. The power of the missing frequency leaks into the other frequencies. Mostly this occurs in the nearby frequencies but to some extent all are affected. Figure 11.7b is another periodogram of the el Niño, but we have deliberately set the 14-th term to zero so that this large component does not swamp the others. You can see that the terms associated with the fourth, fifth and the sixth components are large, corresponding to cycles of 42, 33.4 and 28 months respectively. Other analyses of this data suggest that the largest peak is from a cycle of about 44 months which has leaked over into the nearest Fourier components. The other peaks have been studied too and more accurate estimates made for them. One conclusion was that these sinusoids added together in 1982 and that the California mudslides which washed away million dollar homes might have been predicted. Maybe next time around they will be. For interesting articles on El Niño see the paper by Rasmusson (1985) or the survey papers in the Journal Oceanus, Volume 27 (1984).

11.10 THE FAST FOURIER TRANSFORM

By looking at the formulas in Section 2 you should verify that even if the numbers $\cos(jk2\pi/N)$ and $\sin(jk2\pi/N)$ are given, computing all the a_k and b_k requires about N^2 multiplications and a similar number of additions. We saw in Chapter 3 that solving a linear system of N equations takes about $N^3/3$ operations with Gaussian elimination, so N^2 does not seem like a large number. But the applications in which the two algorithms are applied are quite different. Many situations require the computation of discrete Fourier transforms with thousands, or even millions of points. It is rare to want to solve a system of linear equations that large, unless the matrix has special structure which can be utilized. If N is large it can take hours to compute the discrete Fourier transform by its definition even on a supercomputer. The fast Fourier transform (FFT) allows us to perform these same calculations with only about $N \log_2 N$ operations. For example, if $N = 1024$ the ratio of the explicit formulas to the FFT is $N^2/(N \cdot \log_2 N) = 102.4$. Thus the FFT requires about two orders of magnitude less effort. Depending upon the details of the programs the actual ratio of times can be better or worse.

Example 11.5 Computing the discrete Fourier transform directly and by the FFT.

To illustrate the differences between computing time for the discrete Fourier transform and the FFT, the subroutine below is a straightforward implementation of the discrete sine and cosine transform formulas in Section 2. It is one of the *least* efficient techniques because it does not utilize the FFT algorithm and also requires that the same sines and cosines be computed repeatedly. Figure 11.8 shows times for EZFFTF and DIRECT for various values of N. Notice that EZFFTF is always faster than DIRECT, even when N is prime. This occurs because DIRECT recomputes all the sines and cosines. If N is the product of many small numbers EZFFTF is much faster that DIRECT. (There is a certain amount of "jitter" in these curves, due to small inaccuracies in the timing routine.)

```
        SUBROUTINE DIRECT (N,DATA,AZERO,A,B)
C Direct use of definitions to compute real DFT
C No simplifications...SLOW
        REAL DATA(0:*),A(*),B(*)
        AZERO = 0.0
        DO 1 J = 0,N-1
            AZERO = AZERO+DATA(J)
    1 CONTINUE
        AZERO = AZERO/N
        TPN = 2*ASIN(1.0)*2./N
        DO 20 K = 1,N/2
```

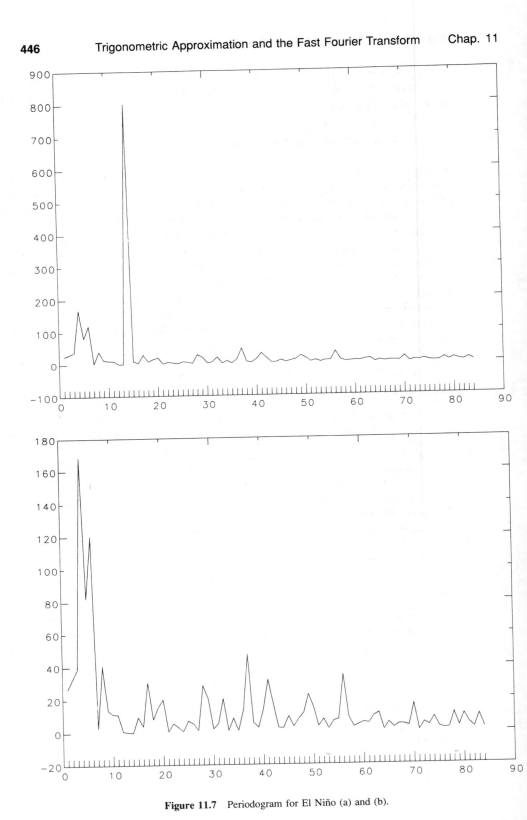

Figure 11.7   Periodogram for El Niño (a) and (b).

```
        A(K) = 0
        B(K) = 0
        DO 10 J = 0,N-1
            A(K) = A(K) + DATA(J)*COS(J*K*TPN)
            B(K) = B(K) + DATA(J)*SIN(J*K*TPN)
10      CONTINUE
        IF (K .NE. N/2) THEN
            A(K) = A(K)*(2./N)
            B(K) = B(K)*(2./N)
        ELSE
            A(K) = A(K)*(1./N)
            B(K) = B(K)*(1./N)
        END IF
20  CONTINUE
    RETURN
    END      ■
```

The essential ingredient of the FFT algorithm is that an N-point discrete Fourier transform can be performed by doing two $N/2$-point discrete Fourier transforms on slight modifications of the original data. Although this seems more complicated it actually takes less work than the direct calculation. Of course, each $N/2$-point calculations can be broken further into two $N/4$-point calculations, etc. In Section 6 we mentioned that some programs require that N be a power of two, and now you can see the reason for this. If N is prime then there is no way to reduce the amount of work other than by throwing away some data or adding (fictitious) zero values. This is sometimes called **zero padding** and you should be alert that some FFT routines do this automatically, thus altering the expected results.

There are several different ways to give a more detailed presentation of the FFT. These include but are not limited to the following.

(1) Divide and conquer (see the book by Aho et al (1974)).

(2) Matrix factorization (Kahaner (1970)).

(3) Polynomial evaluation (Kahaner (1978)).

(4) Butterfly (Brigham (1974)).

For some applications it is not necessary to know the details of the algorithm, but important to be able interpret its output. On the other hand if you are interested in implementing an FFT on a new computer the details are essential, since the different perspectives can lead to distinct data structures and programming details. In this text we will not discuss these issues and refer the reader to the references.

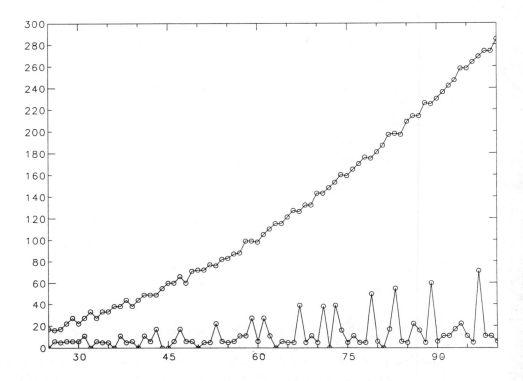

Figure 11.8   Time for EZFFTF and Direct Discrete Fourier Transform

*11.11   COMPLEX REPRESENTATION

Section 2 gave the Fourier integral sine and cosine transform, the discrete Fourier transform and the Fourier series of real functions and real data. These can be generalized to complex functions and complex data values. This has certain advantages.

(1) By using complex variable notation it is possible to rewrite more compactly most of the formulas from the precedings sections.

(2) Complex notation is used almost exclusively when two dimensional transforms are discussed, for example in image processing. It is also used extensively in digital signal processing.

(3) Many scientists feel that the complex plane is the "natural" setting for describing Fourier ideas.

Depending on your area of specialization you may be able to stick with sines and cosines, but to study the references or other advanced material you will have to master the complex formulation. This Section provides the essential background.

Recall Euler's formula: for arbitrary numbers f and t,

$$e^{i2\pi ft} = \cos 2\pi ft + i\sin 2\pi ft, \qquad i = \sqrt{-1}.$$

(1) If $g(t)$ is a complex function, its complex Fourier integral transform is denoted by $G(\omega)$. The functions g and G are coupled by

$$G(\omega) = \int_{-\infty}^{\infty} g(t)e^{-2\pi i\omega t}\, dt, \quad g(t) = \int_{-\infty}^{\infty} G(\omega)e^{2\pi i\omega t}\, d\omega.$$

Notice that $G(\omega)$ is defined for both positive and negative values of ω. If g is a *real* function it has both a complex Fourier integral transform $G(\omega)$ and (from Section 2) a cosine and sine Fourier integral transform $A(\omega), B(\omega)$. These are related. If we define $A(\omega)$ and $B(\omega)$ for negative ω by $A(-\omega) = A(\omega), B(-\omega) = -B(\omega)$, then it can be shown that $G(\omega) = (A(\omega) - iB(\omega))/2$. For real functions the complex transform can be confusing because it represents g as a "sum" over both positive and negative frequencies. The power at frequency $|\omega|$ is $|G(\omega)|^2 + |G(-\omega)|^2 = (A(\omega)^2 + B(\omega)^2)/2$. It is rarely important to distinguish between positive and negative frequencies and they are often spoken of as the same frequency.

(2) If g_j are a set of N complex numbers, its complex discrete Fourier transform is another set G_j. The numbers g_j and G_j are coupled by

$$G_j = \frac{1}{N}\sum_{k=0}^{N-1} g_k \exp\left(-ijk\frac{2\pi}{N}\right), \quad g_j = \sum_{k=0}^{N-1} G_k \exp\left(ikj\frac{2\pi}{N}\right), \quad j = 0, 1, \ldots, N-1.$$

Notice that G_j is defined for frequencies from 0 through $N - 1$. If g_j is a set of real numbers it has both a complex discrete Fourier transform and (from Section 2) a discrete cosine and sine Fourier transform $a_k, b_k,\ k = 0, \ldots, N/2$. What is the relation between these? It can be shown that

$$G_0 = a_0, \qquad G_k = (a_k - ib_k)/2, \quad 1 \leq k \leq N/2,$$

except if N is even the $1/2$ is omitted in $G_{N/2}$. The remaining G_k's are complex conjugates of the first half and are usually ignored. Specifically,

$$G_{N-k} = (a_k + ib_k)/2, \qquad 1 \leq k \leq N/2.$$

The sequences g_j and G_j are said to be a discrete Fourier transform pair. The distinction between odd/even N does not occur in the complex formulation, so these formulas are much easier to remember. For real data g_j the complex discrete Fourier transform can be confusing because it seems to generate frequency components twice as high as we expect. But we see from the formula above that there is no new information in these values. As an illustration let's compute the a's, b's, and G's for the sequence $g_0 = 1$, $g_1 = g_2 = g_3 = 0$. Using the formulas in Section 2 we have

$$a_0 = \frac{1}{4}, \quad a_1 = \frac{2}{4}, \quad a_2 = \frac{1}{4}, \quad b_1 = b_2 = 0.$$

From the summation formula above $G_i = 1/4$ for all i. Now

$$G_0 = a_0, \quad G_1 = \frac{a_1 - ib_1}{2}, \quad G_2 = a_2 - ib_2,$$

from the formula $G_k = (a_k - ib_k)/2$ for the first half of the G's. From the formula $G_{N-k} = (a_k + ib_k)/2$ for the second half

$$G_3 = \frac{a_1 + ib_1}{2}, \quad G_2 = a_2 + ib_2.$$

Since N is even, b_2 is automatically zero and we get again that $G_i = 1/4$.

(3) If g is a complex periodic function on $[a, b]$ it has a complex Fourier series,

$$g(t) = \sum_{j=-\infty}^{\infty} C_j e^{ij2\pi ft} \qquad C_j = \frac{1}{b-a} \int_a^b g(t) \exp(-ij2\pi ft)\, dt,$$

where $f = 1/(b - a)$. Notice that this also represents g as a sum over positive and negative frequencies. If g is a real periodic function it has a real Fourier series with coefficients A_j and B_j which are related to the C_j. Define A_j and B_j for $j < 0$ by

$$A_{-j} = A_j, \qquad B_{-j} = -B_j, \; j > 0.$$

Then it can be shown that

$$C_0 = \frac{A_0}{2}, \qquad C_j = \frac{1}{2}(A_j - iB_j), \; j \neq 0.$$

Using complex formalism does not add new information. Some texts start with the complex representation because it is more concise and makes proofs and algebraic manipulation much easier. However, the negative subscripts occurring in the complex form of the Fourier series, as well as the negative frequencies in the Fourier integral transform often cause confusion. The complex form is essential in two-dimensional applications as we illustrate in Section 12. Programs for computing the discrete Fourier transform either produce real a's and b's from real g's, or complex G's from arbitrary (real or complex) g's. In this chapter EZFFTF is an example of the former, while CFFTF is an example of the latter.

There is an alternative form of (2) that is also common in applications. We note again that in (1) and (3) frequencies can be negative. It is possible to define the discrete Fourier transform in (2) in a similar way

$$G'_j = \frac{1}{N} \sum_{k=0}^{N-1} g_k \exp\left(-ijk\frac{2\pi}{N}\right), j = -N/2, \ldots, 0, 1, \ldots, M,$$

where $M = \begin{cases} N/2, & \text{if } N \text{ odd}, \\ N/2 - 1, & \text{if } N \text{ even}. \end{cases}$

By studying this formula it is apparent that $G_0 = G'_0$, $G_1 = G'_1, \ldots,$ and $G_{N/2} = G'_{N/2}$. For G'_{-1}, G'_{-2}, etc., these are equal to the second half of the unprimed Gs. For example, $G'_{-1} = G_{N-1}$ since $\exp(-i(-1)k(2\pi/N)) = \exp(-i(N-1)k(2\pi/N))$. Similarly $G'_{-2} = G_{N-2}$, etc. It is then possible to write

$$g_j = \sum_{k=M}^{N/2} G'_k \exp\left(ijk\frac{2\pi}{N}\right), \qquad j = 0, \ldots, N-1.$$

Why is this form useful? We mentioned above that if the data g_j are real it is usual to ignore the second half of the G array. If we forget this and plot the magnitude of all the coefficients (the periodogram) the plot will usually decrease to its midpoint and then increase. In fact for real data the plot is symmetric about the midpoint. If the data are complex the plot has the same general shape but is no longer symmetric. It would be easier to interpret such a plot if the peak were at the center rather than at both ends. The representation above does that because the zero frequency is at the center, i.e., $-N/2, \ldots, -1, 0, 1, \ldots, M$. In practice if N is even there is a little trick that is used to avoid shuffling around the output array G of a complex FFT. The original input data g_k are often altered before they are passed to the complex FFT by multiplying them by $(-1)^k$. Since $(-1)^k = \cos(k\pi) = \exp(ik\pi)$, what we actually calculate is

$$\frac{1}{N} \sum_{k=0}^{N-1} (-1)^k g_k \exp\left(-ijk\frac{2\pi}{N}\right) = \frac{1}{N} \sum_{k=0}^{N-1} g_k \exp\left(-ik\frac{2\pi}{N}(j - N/2)\right).$$

When $j = N/2$ this is G'_0, when $j = 1 + N/2$ this is G'_1, when $j = (N/2) - 1$ it is G'_{-1}, etc. So the middle values in the output array correspond to the low frequencies. We will see an example of this in Section 12.1 when we examine an image processing application.

*11.11.1 Subroutines CFFTF and CFFTB

The three subroutines, CFFTF, CFFTB, and CFFTI are the complex analogues of the real routines in Section 6. In the preceding Section we explained that the complex discrete Fourier transform is always used for multi-dimensional applications. However, because it takes complex input g_i, CFFTF is easy to use for a real one-dimensional transform by setting the imaginary parts of the input array equal to zero. The complex output, contains the a's and b's as the real and imaginary parts of its first $N/2$ components. The output values are scaled by N in CFFTF and overwrite the input. The remaining components contain complex conjugates of these numbers. Just as in the real case, if the first abscissa value is $t_0 = 0$ the output of CFFTF approximates the complex Fourier series coefficients of the periodic function that generated the data. Otherwise the output array must be scaled by multiplying its j-th component by $\exp(-ij2\pi f t_0), j = 1, \ldots, N/2$. If the array of transformed data is indexed beginning with zero, then C(J) corresponds to frequency J times the fundamental frequency, e.g., $J \times 1/(N\Delta)$. In applications where maximum efficiency is not essential you will often see the complex transform used.

In the program below we illustrate the use of CFFTF to compute the complex discrete Fourier transform of Runge's function. You should compare this program with the similar one in Section 6 using EZFFTF.

```
C Using complex discrete Fourier transform, find the approximate Fourier
C coefficients to Runge's function on [-1,1] with N=16 and N=17.
C
      REAL    WSAVE(150)
      COMPLEX COEFF(0:16), SQTM1
C
C Note 0 subscript, which makes indexing easier, allowed in Fortran 77.
C
C Arithmetic statement function for Runge's function.
      RUNGE(X) = 1.0/(1+25.0*X*X)
C
      X0 = -1.0
      PI = ASIN(1.0)*2.0
      SQTM1 = CMPLX(0.0,-1.0)
C
      DO 10 N = 16,17
         CALL CFFTI (N,WSAVE)
C Function assumed to be periodic on [-1,1], of length 2.
         DEL = 2.0/N
         F = 2.0*PI/(N*DEL)
         DO 1 J = 0,N-1
C First sample point at -1, last at 1-DEL
            XJ = (-1.0) + J*DEL
            COEFF(J) = CMPLX(RUNGE(XJ),0.0)
    1    CONTINUE
         CALL CFFTF (N,COEFF,WSAVE)
C Returned coefficients must be divided by N for correct normalization.
C
C Note repetition after N/2 in original coefficients. Scaling because
C X0 not at origin destroys this to some extent.
C
         WRITE (*,*) ' RESULTS FOR N = ' ,N
         WRITE (*,*) ' CZERO = ',COEFF(0)/N*2
```

```
          WRITE (*,*)
    *'  J              OUTPUT FROM CFFTF,              SCALED COEFFICIENTS'
          DO 11 J = 1,N-1
               WRITE (*,'(I5,2E15.6,5X,2E15.6)')
    *                J, COEFF(J), EXP(-SQTM1*J*F*X0) * COEFF(J)/N *2
 11       CONTINUE
          WRITE (*,*)
 10   CONTINUE
      STOP
      END
C
C PRODUCES THE FOLLOWING OUTPUT
C
C    RESULTS FOR N =              16
C    CZERO =  (0.549222,0.000000)
C    J              OUTPUT FROM CFFTF,              SCALED COEFFICIENTS
C    1   -0.275492E+01    0.520041E-07       0.344365E+00   -0.366058E-07
C    2    0.140523E+01    0.000000E+00       0.175654E+00   -0.307123E-07
C    3   -0.778358E+00    0.730062E-08       0.972947E-01   -0.264299E-07
C    4    0.401056E+00    0.000000E+00       0.501320E-01   -0.175307E-07
C    5   -0.228722E+00   -0.730062E-08       0.285903E-01   -0.115846E-07
C    6    0.119966E+00    0.000000E+00       0.149957E-01   -0.786582E-08
C    7   -0.841548E-01   -0.730062E-08       0.105194E-01   -0.552484E-08
C    8    0.614045E-01    0.000000E+00       0.767556E-02   -0.536815E-08
C    9   -0.841548E-01   -0.374029E-07       0.105194E-01   -0.360132E-08
C   10    0.119966E+00    0.000000E+00       0.149957E-01   -0.131097E-07
C   11   -0.228722E+00    0.730062E-08       0.285903E-01   -0.284064E-07
C   12    0.401056E+00    0.000000E+00       0.501320E-01   -0.525921E-07
C   13   -0.778358E+00   -0.730062E-08       0.972947E-01   -0.109662E-06
C   14    0.140523E+01    0.000000E+00       0.175654E+00   -0.214986E-06
C   15   -0.275492E+01   -0.730062E-08       0.344365E+00   -0.450667E-06
C
C    RESULTS FOR N =              17
C    CZERO =  (0.549161,0.000000)
C    J              OUTPUT FROM CFFTF,              SCALED COEFFICIENTS
C    1   -0.292606E+01   -0.421539E-07       0.344243E+00   -0.251354E-07
C    2    0.149192E+01    0.488273E-07       0.175520E+00   -0.249445E-07
C    3   -0.824417E+00   -0.349775E-07       0.969903E-01   -0.213225E-07
C    4    0.421975E+00    0.271580E-07       0.496441E-01   -0.141651E-07
C    5   -0.234573E+00   -0.245598E-07       0.275968E-01   -0.917357E-08
C    6    0.112482E+00    0.293560E-07       0.132331E-01   -0.348762E-08
C    7   -0.603455E-01   -0.568597E-07       0.709948E-02    0.234479E-08
C    8    0.120123E-01    0.910478E-07       0.141321E-02    0.972312E-08
C    9    0.120123E-01   -0.910478E-07      -0.141321E-02    0.118234E-07
C   10   -0.603455E-01    0.568597E-07      -0.709948E-02    0.128959E-07
C   11    0.112482E+00   -0.293560E-07      -0.132331E-01    0.161793E-07
C   12   -0.234573E+00    0.245598E-07      -0.275968E-01    0.318405E-07
C   13    0.421975E+00   -0.271580E-07      -0.496441E-01    0.596154E-07
```

C	14	$-0.824417E+00$	$0.349775E-07$	$-0.969903E-01$	$0.122823E-06$
C	15	$0.149192E+01$	$-0.488273E-07$	$-0.175520E+00$	$0.235911E-06$
C	16	$-0.292606E+01$	$0.421539E-07$	$-0.344243E+00$	$0.486474E-06$

A more efficient way of getting the real transform from CFFTF is to place the zeroth, second, fourth, sixth etc., data values into the real components of g_i and the first, third, fifth, etc., into the imaginary components. This only requires using a complex array of length $N/2$. The transform of this array can be unscrambled to give the transform of the original data, see Brigham (1974), page 169.

*11.12 TWO-DIMENSIONAL TRANSFORMS

Image processing programs often make heavy use of FFTs. A two-dimensional photographic image is represented as an $N \times N$ matrix of positive numbers, g_{ij}, $i, j = 0, \ldots, N - 1$, where each number is the average light intensity (grey scale value) of a small rectangular region (pixel) on the image. For specific applications we may want to improve the quality of the image by sharpening or smoothing it, or extract some important feature such as the location of an edge, etc. All of the concepts in Section 2 can be generalized to two or more dimensions. In this case, the complex representation makes the description much easier, and two-dimensional programs use the complex transform.

The Fourier integral transform pair of a function $g(x, y)$ is given by

$$G(\omega, \mu) = \int_{-\infty}^{\infty} \int_{-\infty}^{\infty} g(x, y) \exp(-2\pi i[\omega x + \mu y]) \, dx \, dy,$$

and

$$g(x, y) = \int_{-\infty}^{\infty} \int_{-\infty}^{\infty} G(\omega, \mu) \exp(2\pi i[\omega x + \mu y]) \, d\omega \, d\mu.$$

The discrete Fourier transform pair of the $M \times N$ array of numbers g_{mn} is

$$G_{uv} = \frac{1}{NM} \sum_{m=0}^{M-1} \sum_{n=0}^{N-1} g_{mn} \exp\left(-2\pi i \left[\frac{mu}{M} + \frac{nv}{N}\right]\right), \qquad u = 0, \ldots, M - 1,$$
$$v = 0, \ldots, N - 1,$$

and

$$g_{mn} = \sum_{u=0}^{M-1} \sum_{v=0}^{N-1} G_{uv} \exp\left(2\pi i \left[\frac{mu}{M} + \frac{nv}{N}\right]\right), \qquad m = 0, \ldots, M - 1, \quad n = 0, \ldots, N - 1.$$

Just as in the one-dimensional case, the magnitude $|G_{uv}|$ is usually interpreted as the "amount" of the input at frequency (u, v), or at frequency u times the fundamental frequency in x and v times the fundamental frequency in y. In image processing, x

and y are usually distances and (u, v) is called **spatial frequency**. Note that $\omega x + \mu y = \sqrt{\omega^2 + \mu^2}(x \cos \theta + y \sin \theta)$, where $\tan \theta = \mu/\omega$. Since $x' \equiv x \cos \theta + y \sin \theta$ is a rotation by θ the expression $\cos(2\pi[\omega x + \mu y])$ has frequency $\sqrt{\omega^2 + \mu^2}$ in the x' direction and is constant in the perpendicular direction. Hence we often interpret the expansion for g by saying that any g is a complex linear combination of rotated sine and cosine "bar patterns." See Rosenfeld and Kak (1982), page 18.

One of the most important facts about two-dimensional transforms is that they can be computed using programs such as CFFTF for one-dimensional transforms. If we rewrite the formula for G_{uv} as

$$G_{uv} = \frac{1}{M} \sum_{m=0}^{M-1} \left[\frac{1}{N} \sum_{n=0}^{N-1} g_{mn} \exp\left(-inv\frac{2\pi}{N}\right) \right] \exp\left(-imu\frac{2\pi}{M}\right),$$

the quantity in the square brackets above is the one-dimensional discrete Fourier transform of the m-th row of the array g_{mn}. Thus computing the two-dimensional discrete Fourier transform of g_{mn} can be broken into the following steps.

(1) Compute the one-dimensional discrete Fourier transform of each *row* of the array g, and overwrite the row with the transform.

(2) Then compute the one-dimensional discrete Fourier transform of each *column* of the array g, and overwrite the column with the transform.

*11.12.1 Subroutine CFFT2D

This subroutine computes the two-dimensional discrete Fourier transform (or its inverse) of a square two-dimensional complex array. The example below illustrates its use on a 64 × 64 array of positive numbers that represent the light intensity of an image at the point (I, J). The image intensity is zero except for a small square at the center that has intensity one. The left part of figure 11.9 shows this image. The center figure displays the magnitude of the discrete Fourier components just as they are returned from CFFT2D. Notice the peaks at the four corners corresponding to low frequencies. This is typical and illustrates the remark made at the end of Section 11. Most people would prefer to study the rightmost figure. To generate that, we repeated the calculation but first altered the data by multiplying G_{ij} by $(-1)^{i+j}$. The particular type of plot that we have use in this figure is called **axonometric** and is described more fully in Problem P11–6.

In the following program we illustrate CFFT2D by using it to compute the discrete Fourier transform of a 64 × 64 digitized image. We also compute the magnitude of its Fourier coefficients, and then use CFFT2D to transform back to the original data.

```
C EXAMPLE 11.8:  Plot image and transform of 8 by 8 unit source
C in 64 by 64 (otherwise zero) array.
C
      PARAMETER (N=64)
      REAL      A(N,N),A2(N,N)
```

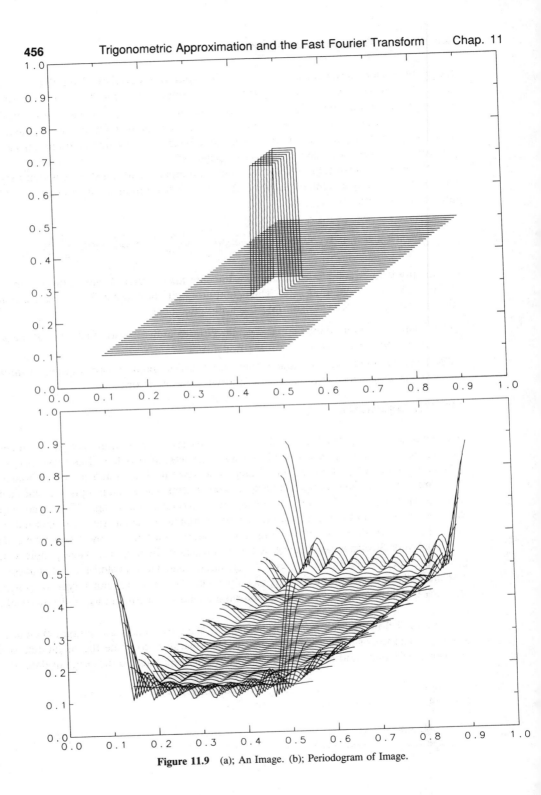

Figure 11.9   (a); An Image. (b); Periodogram of Image.

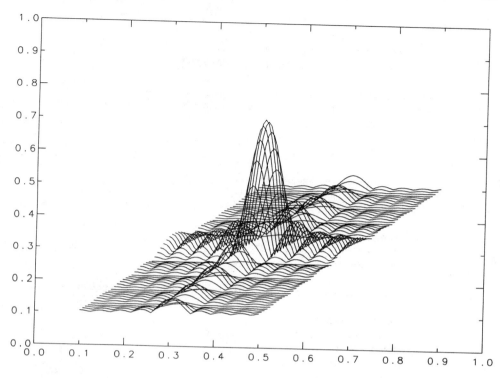

Figure 11.9 (c); Scaled Periodogram.

```
      COMPLEX    IMAGE(N,N),IMAGE2(N,N),W(3*N+8)
      LOGICAL    FORWD
C
      LDA = N
C Set up data. IMAGE is original. IMAGE2 is scaled by (-1)**(I+J)
C to place Fourier coefficients in the correct place for viewing.
      DO 1 I = 1,N
        DO 1 J = 1,N
          A(I,J) = 0.
          IF (I.GE.(N/2-4) .AND. I.LE.(N/2+4) .AND. J.GE.(N/2-4)
     *        .AND. J.LE.(N/2+4)) A(I,J) = 1.0
          IMAGE(I,J) = CMPLX(A(I,J),0.0)
          IMAGE2(I,J) = IMAGE(I,J)*(-1)**(I-1+J-1)
    1 CONTINUE
C
C Plot image  with axonometric plotting routine.
      CALL AXNPLT (A,LDA,N)
C
C Compute Fourier transform of IMAGE and IMAGE2
```

```
C
      FORWD = .TRUE.
      CALL CFFT2D (N,IMAGE,LDA,W,FORWD)
      CALL CFFT2D (N,IMAGE2,LDA,W,FORWD)
C
C Compute magnitude of components of transforms.
C Actual transforms are unscaled and need to be divided by N*N
C to be correct.
C
      DO 2 I = 1,N
        DO 2 J = 1,N
          A(I,J) = ABS(IMAGE(I,J))
          A2(I,J) = ABS(IMAGE2(I,J))
    2 CONTINUE
C
C PLOT MODULOUS OF TRANSFORM
      CALL AXNPLT (A,LDA,N)
      CALL AXNPLT (A2,LDA,N)
C
C Compute inverse transform of image and see if it agrees with original.
C
      FORWD = .FALSE.
      CALL CFFT2D (N,IMAGE,LDA,W,FORWD)
      ERMAX = 0.0
      DO 3 I = 1,N
        DO 3 J = 1,N
          DATA = 0.0
          IF (I.GE.(N/2-4) .AND. I.LE.(N/2+4) .AND. J.GE.(N/2-4)
     *        .AND. J.LE.(N/2+4)) DATA = 1.0
          ERR = ABS( DATA-ABS(IMAGE(I,J))/(N*N) )
          IF (ERR .GT. ERMAX) ERMAX = ERR
    3 CONTINUE
      WRITE (*,*) 'MAXIMUM ERROR IS ',ERMAX
C
      END
```

*11.13 CONVOLUTION AND CORRELATION

The convolution of two functions is basic to almost every field of science and engineering. It is called the superposition principle or Duhamel integral in mechanics, the impulse response integral in systems theory, and the point spread or smearing function in optics. The convolution of two functions is usually thought of as an averaging or smoothing operation. In practical measurement problems, if a scientist believes that each of the measured values is an average then the measurements can often be viewed as a convolution. An example is a photodigitizer in which the output (ostensibly the light intensity at a point) is really an average (convolution) of several nearby values.

If $f(x)$ and $g(x)$ are functions defined on $(-\infty, \infty)$, the **convolution**, $h(x)$ of f and g, written $f * g$ is the function defined by

$$h(x) = f * g \equiv \lim_{T \to \infty} \left[\int_{-T}^{T} f(t)g(x-t)\, dt \right].$$

Another integral of great importance is the correlation of two functions. If $f(x)$ and $g(x)$ are functions defined on $(-\infty, \infty)$, the **cross covariance** of f and g, written $f \circ g$, is the function

$$z(x) = f \circ g \equiv \lim_{T \to \infty} \left[\int_{-T}^{T} \left(f(t) - \overline{f} \right) \left(g(x+t) - \overline{g} \right) dt \right]$$

where

$$\overline{f} = \lim_{T \to \infty} \left[\frac{1}{2T} \int_{-T}^{T} f(t)\, dt \right], \qquad \overline{g} = \lim_{T \to \infty} \left[\frac{1}{2T} \int_{-T}^{T} g(t)\, dt \right]$$

are the **mean** values of f and g. If $f = g$, $z/z(0)$ is called the **autocorrelation** function and often written $\rho(x)$. The division by $z(0)$ amounts to a normalization so that $-1 \leq \rho(x) \leq 1$.

It is possible to interpret convolution and correlation graphically, and this is done in any of the standard references, such as Weaver (1983), or Brigham (1974).

The most important property about convolution is that the Fourier transform of $f * g$ is the product of the Fourier transform of f and the Fourier transform of g. Conversely, the Fourier transform of a product $f \cdot g$ is the convolution of the Fourier transform of f and the Fourier transform of g. Brigham (1974), page 58, says that this property is "possibly the most important and powerful tool in modern scientific analysis;" it is certainly one of the most heavily used. For correlation there is an analogous result, the Fourier transform of $f \circ g$ is the product of the Fourier transform of f times the complex conjugate of the Fourier transform of g.

If f and g have zero mean, $\overline{f} = \overline{g} = 0$, the formulas for convolution and cross covariance only differ by a sign, and the correlation only differs by a scale factor. You might think that these distinctions are minor. Actually convolution and correlation have different interpretations. As we mentioned above the convolution is often thought of as an average. The importance of this observation is that it is sometimes possible to "undo" the averaging by using Fourier transforms. The autocorrelation measures how different g is from a translate of itself. This is used in statistical tests and to find hidden cycles in data. See Example 11.6.

Some care is necessary to apply these definitions to finite length discrete sequences. If we replace the integrals by sums, the limits on the summation indices are not obvious. A standard approach is to extend the sequences g_i and h_i, $i = 0, \ldots, N-1$, to make them periodic, so that all subscripts, positive and negative, make sense. In image processing the period is left as N, but in signal processing and time series analysis the sequences are often first extended to length $2N$ by appending N zeros, making the period $2N$. Once we have periodic sequences, it is possible to define discrete correlation and convolution. Both conventions are common, but in what follows we have elected to use the second.

If f_i and g_i are numbers $i = 0, \ldots, N-1$ (extended to $2N$ by appending zeros) the **discrete cross covariance** of f and g is the set of numbers

$$z_i = \sum_{j=0}^{N-1} \left(f_i - \overline{f}\right)\left(g_{i+j} - \overline{g}\right) = \sum_{j=0}^{N-i-1} \left(f_i - \overline{f}\right)\left(g_{i+j} - \overline{g}\right), \qquad i = 0, \ldots, N-1,$$

where

$$\overline{f} = \frac{1}{N} \sum_{i=0}^{N-1} f_i, \qquad \overline{g} = \frac{1}{N} \sum_{i=0}^{N-1} g_i.$$

If $f_i = g_i$ the **discrete autocorrelation** of f is the set of numbers z_i/z_0, $i = 0, \ldots, N-1$. If the numbers are also thought of as periodic of period $2N$, the **discrete convolution** (also called the circular or cyclic convolution) of f and g is the set of numbers h_i, $i = 0, \ldots, N-1$, defined by

$$h_i = \sum_{j=0}^{N-1} f_i g_{i-j}, \qquad i = 0, \ldots, N-1.$$

The discrete Fourier transform of the discrete convolution of two sets of numbers f_i and g_i is the product of the discrete Fourier transform of f_i and the discrete Fourier transform of g_i, just as in the continuous case. There is a similar relation for the discrete autocorrelation. The discrete Fourier transform of $f \circ g$ is the product of the discrete Fourier transform of f times the complex conjugate of the discrete Fourier transform of g. As an application, the autocorrelation of a sequence g_i can be computed rapidly by using the convolution theorem—convolving g with itself as follows.

(1) Subtract the mean so that the sequence of length N has mean zero. Extend g to be of length $2N$ by appending N zeros.

(2) Compute the discrete Fourier transform of the $2N$-point sequence g.

(3) Compute the square of the magnitude of the transform components. This amounts to multiplying the transform by its conjugate. If we use the complex transform G_j this means $|G_j|^2$, $j < 2N$. If we use the real transform a_j, b_j, this means $(a_j^2 + b_j^2)/2$, $j < N$, and twice that for $j = N$. In the real case set the remaining N components to zero.

(4) Compute the inverse transform of this $2N$-point sequence. Throw away the second half of the result.

Example 11.6   Autocorrelation of El Niño Data by Direct and FFT Methods.

To illustrate how to compute a discrete autocorrelation we consider the el Niño data from Section 9. Let us compute the discrete autocorrelation of this data directly, that is by the formula above, and also by using the FFT.

```
C Find autocorrelation to el Nino data using direct and FFT methods.
C
      INTEGER N
      PARAMETER(N=168)
      REAL EL(0:2*N-1),WSAVE(4*(2*N)+15),ACOV(0:N-1),A(2*N),B(2*N),
     *          ACOVR(0:2*N-1)
      COMPLEX CEL(0:2*N-1),CORR(0:2*N-1)
      LOGICAL EX
C
C needs N locations for el, 2N for acov, cel, corr
C and 4*(2N)+15 for wsave in complex case. N has maximum of 168.
      INQUIRE (FILE='ELNINO.DAT',EXIST=EX,ERR=1000)
      IF(.NOT.EX)GOTO 1000
      OPEN (UNIT=8,FILE='ELNINO.DAT',ERR=1000)
C
C Read data, find mean.
      SUM = 0.0
      DO 100 I = 0,N-1
         READ (8,*) EL(I)
         SUM = SUM + EL(I)
  100 CONTINUE
      DO 101 I = 0,N-1
         EL(I) = EL(I) - SUM/N
C Subtract mean, and add N zeros for either complex or real FFT usage
         CEL(I) = CMPLX(EL(I),0.0)
         EL(I+N) = 0.0
         CEL(I+N) = 0.0
  101 CONTINUE
C
C- - - - - - - - - - - - - - - - - - - - - - - - - - - - -
C
C Direct calculation. Only sum as far as there is data.
C Simple, but slow.
```

```
C
      DO 110 J = 0,N-1
         ACOV(J) = 0.0
         DO 110 M = 0,N-1-J
            ACOV(J) = ACOV(J)+ EL(M) * EL(M+J)
  110 CONTINUE
C
C Write, scaled correlation
      WRITE (*,*)
    * 'EX 11.6: AUTOCORRELATION (DIRECT)'
      WRITE (*,'(5E14.6)') ( (ACOV(I)/ACOV(0)), I = 0,N-1)
C- - - - - - - - - - - - - - - - - - - - - - - - - - - - - -
C
C FFT approach (complex).
C    Compute FFT of data of length 2N.
C    Compute square of magnitude of transform components and place
C       in complex array as real parts.
C    Compute inverse transform, throwing away second half and
C       imaginary parts (which are zero), and multiply by length of
C       sequence, 2N.
      CALL CFFTI (2*N,WSAVE)
      CALL CFFTF (2*N,CEL,WSAVE)
C CFFTF returns unscaled transforms. Actual transforms are output
C divided by (2N).
      DO 120 I = 0,2*N-1
         CORR(I) = ABS(CEL(I) / (2*N)) **2
  120 CONTINUE
C Since we compute transform times its conjugate, must divide by
C (2N) for each, i.e., (2N)**2.
      CALL CFFTB (2*N,CORR,WSAVE)
C
      DO 121 I = 0,N-1
         ACOV(I) = REAL(CORR(I))*(2*N)
  121 CONTINUE
C Autocovariance is inverse transform times sequence length, 2N.
C Normally, all the scaling would be done  only once
C by dividing by 2N. We've broken it up for exposition.
      WRITE (*,*)
```

```
      *   'EX 11.6: AUTOCORRELATION (COMPLEX FFT)'
         WRITE (*,'(5E14.6)') ( (ACOV(I)/ACOV(0) ), I = 0,N-1)
C
C- - - - - - - - - - - - - - - - - - - - - - - - - - - - - - - -
C
C FFT approach (real).
C    Compute FFT of data of length 2N.
C    EZFFTF produces correctly scaled A's and B's so no extra scaling
C       needed to get transform.
C    Compute array of square of each frequency component and place
C       in cosine array (A's) to be back transformed. Set B's to 0.
C    There are N A's, and N B's.
C    Note that care must be taken to compute magnitude correctly,
C       0.5*(A(I)**2+B(I)**2) for I < N, twice that for I=N.
C    Compute back transform throwing away its second half.
C
         CALL EZFFTI (2*N,WSAVE)
         CALL EZFFTF (2*N,EL,AZERO,A,B,WSAVE)
         AZERO = AZERO*AZERO
C
         DO 150 I = 1,N
            IF(I.NE.N) THEN
               A(I) = (A(I)**2 + B(I)**2) / 2.0
            ELSE
               A(I) = (A(I)**2 + B(I)**2)
            ENDIF
            B(I) = 0.0
  150    CONTINUE
         CALL EZFFTB (2*N,ACOVR,AZERO,A,B,WSAVE)
         WRITE (*,*)
      *  'EX 11.6: AUTOCORRELATION (REAL FFT)'
         WRITE (*,'(5E14.6)') ( (ACOVR(I)/ACOVR(0) ), I = 0,N-1)
C
         STOP
 1000 WRITE (*,*) 'CANNOT FIND THE DATA FILE: ELNINO.DAT '
      END
```

Figure 11.10 shows this autocorrelation. Notice the peaks which occur every twelve points. We infer from this that data values at one point are strongly correlated to data at the same month but one or more years later. You can also see some evidence of a cycle with a longer period.

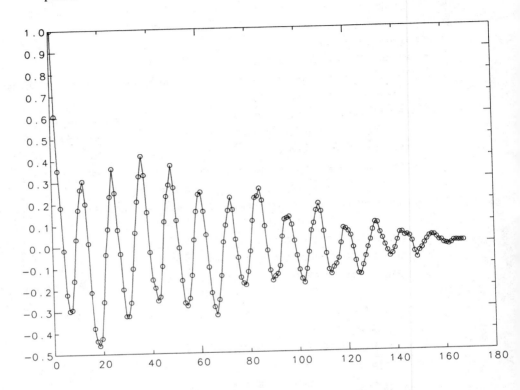

Figure 11.10   Autocorrelation for El Niño

As a final example in this section we describe, in simplified form, a typical convolution application. The problem is to sharpen a fuzzy, badly focused image on a photographic plate. The concept of a point source is convenient, that is a bright point on a dark background. If an arbitrary light source could be considered as the sum of point sources of different intensities then knowing the image of a point source could be used to determine the image resulting from the more general source. The output of a point source is called the **point spread function**, or **impulse response**. This can be thought of as a measure of the degradation of the source caused by whatever physical apparatus the light passes through before it hits the plate, because it describes the blurred output whenever a sharp point is input. In a standard model of this process we assume

(1) that the effect of the point source is the same, except for translation, wherever it is placed on the source plane (this is called the **shift invariant** property),

(2) that the process is linear, i.e., that the effect of adding sources has the same additive effect on the image.

Then the image can be shown to be given by the convolution of the source and the point spread function.

To say this in another way, let p be a point on the plate and $\psi_2(p)$ be the observed image light intensity. The source, which is separated from the image, has intensity $\psi_1(p)$ at a corresponding point. In a perfect world $\psi_2(p) = \psi_1(p)$, but dusty air and lenses that separate the source and plate destroy this. If we assume that the image intensity at p is given by a weighted average of source intensity values, then this can be represented as an integral

$$\psi_2(p) = \int k(p, q)\psi_1(q)\, dq.$$

That is, the effect on the image at p of the source near q is $k(p, q)\psi_1(q)\, dq$. If we further assume that the blurring occurs at all points on the source in the same way, then what matters is only the distance between p and q, and $k(p, q) = k(p - q)$. If the lens extent is infinite the formula for ψ_2 becomes a two-dimensional convolution. We would like to unblur the image. This is equivalent to finding ψ_1. Often we know something about k or can make an approximation to it. After all, the point spread function k is an averaging operation and it is likely that $k(p - q)$ is large when $p = q$ and that it decreases rapidly to zero as p and q become further apart. Using capital letters for Fourier transforms, we have from the previous paragraph that

$$\psi_2 = k * \psi_1 \qquad \text{implies} \qquad \Psi_2 = K \cdot \Psi_1.$$

Thus to find ψ_1 we divide the Fourier transform of ψ_1 by the Fourier transform of k and take the inverse transform of the quotient.

Actual photographic images can rarely be described by such a simple model. For example, if H is zero at even one point the formulas break down, and in any case there is no provision for the inclusion of the "noise" which always occurs. Nevertheless it does provide an opportunity to illustrate the fast Fourier transform.

11.14   HISTORICAL PERSPECTIVE: THE FAST FOURIER TRANSFORM

Even in precomputer days, Fourier analysis was an essential part of some fields of study, but performing a 32-point discrete Fourier transform required $32^2 = 1024$ major operations, a substantial computation. The earliest documented description of a rapid algorithm is due to Gauss, in Latin, in 1866. In 1903, Runge described a method which used the symmetries of the sine and cosine functions to shorten the computation, and in a later book by Koenig and Runge it was shown how one could use the periodicity of the sine-cosine functions to obtain a $2N$-point discrete Fourier transform from two N-point transforms with only slightly more than N operations. If the data are of length N and N is a power of 2, the data can be split into $\log_2 N$ subsets and the Koenig-Runge algorithm could be used to compute the discrete Fourier transform in $\log_2 N$

doublings, which would lead to a total effort of about $N \log_2 N$, rather than N^2. The use of symmetries only reduces the proportionality factor, while the successive doublings lead to the $\log_2 N$ factor. This distinction was not important for the small values of N that were considered in the days of Runge and Koenig and was apparently overlooked even though the technique was published by well known authors. In 1942 Danielson and Lanczos described the method in a more generalized way, which explicitly mentioned the $N \log_2 N$ operations. But doing the computation by hand was still a staggering task. In 1948, Thomas reported spending three months doing a transform using an office tabulating machine at IBM.

With the advent of the digital computer, one would have expected that the fast algorithm would be put to immediate and widespread use, but the methods were generally unknown, and discrete Fourier transforms remained prohibitively expensive into the 1960s. In the early 1960s Richard Garwin was studying solid helium and required some Fourier transforms. He approached a colleague, John Tukey, and asked for an efficient means of performing these computations. Tukey supplied him with the essence of the $N \log_2 N$ algorithm, but Garwin needed some programming help and went to IBM Research in Yorktown Heights to have the algorithm programmed. John Cooley was new at the computing center and was doing some of his own research. Since he seemed the only one with nothing important to do, he was given the problem to work on. Cooley thought it looked interesting, but that his own work was more important; it took some prodding to get him to work on it in his spare time. Cooley ultimately gave the program to Garwin and thought that he would be able to go back to doing real work. But Garwin saw a wide range of applications for the program and pointed these out to many of his colleagues. Cooley began receiving requests for the program and ultimately for a paper describing what he had done. A joint paper with Tukey was published in 1965 and this is generally credited as being the starting point for modern usage of the FFT. Like many good ideas, we can find snips of it in the work of a number of scientists, but it took the right combination of circumstances to bring it forward to widespread attention. There is no dispute, though, about the importance of the FFT. Many knowledgeable people feel it is the single most important contribution to computing since the advent of the stored programming concept. With respect to Gauss' work, which has largely been ignored, Cooley advises young scientists who aspire to recognition, "Do not publish papers in neo-classic Latin."

11.15 PROBLEMS

P11–1.–Compute 32 equally spaced values of erf(x) on $[0, 3]$. Use EZFFTF to find $t_n(x)$, for $n = 4, 8$, and 16. Plot these functions. How well do they approximate erf? Repeat the computation after modifying the data by subtracting erf$(3)\frac{x}{3}$. This will make the first and last data values equal to zero.

P11–2.–Using EZFFTF generate the periodogram that was displayed in Figure 11.2. If the values of B, d, and α that you computed in Chapter 9 are different use them instead. Are the locations of the peaks in your periodogram different from ours?

P11–3.–Using EZFFTF perform the computations that led to Figure 11.5.

P11–4.–Using RNOR from Chapter 10, generate 256 values $f_i = \sin(i/10) + \eta_i$, $i = 1, \ldots, 256$, with $\eta_i = N(0, 0.1)$. Plot this data.

(a) Using EZFFTF compute and plot the sine and cosine coefficients. Also compute and plot the periodogram.

(b) Set some of the high frequency coefficients to zero and then use EZFFTB to transform the data back. Plot the result.

P11–5.–Understanding the output of FFT routines.

(a) Write a program to sample the function $f(t) = \cos(2\pi t)$ at N equally spaced points $t_i = i\Delta$, $i = 0, \ldots, N - 1$. Using the sample values call EZFFTF and print out the values of the computed Fourier coefficients for various values of N and Δ. Interpret the numbers that are printed in terms of the Fourier series for $f(t)$. For a typical value of Δ, say $\Delta = 1/8$, compare the results you get for $N = 8$ and $N = 9$. In the latter case the first and last data value are equal.

(b) Repeat (a) using CFFTF. For the case $N = 8$, $\Delta = 1/8$, be sure that you can explain which of the complex coefficients should be zero, and how to scale the nonzero ones. Repeat the calculation after you have altered the data values by multiplying the i-th by $(-1)^i$. Print the absolute value of the DFT coefficients and explain the results that you get.

(c) Repeat (a) by sampling at the points $t_i = 0.25 + i\Delta$. Explain why your results are different from those in (a).

(d) Compute the discrete Fourier transform of an impulse. That is, use EZFFTF to find the DFT of the array $(1, 0, 0, \ldots, 0)$. What kind of a function $f(t)$ would result in such an array? Verify conservation of energy. Can you interpret the results?

(e) The function $f(x) = \exp(-x^2)$ has Fourier transform $A(\omega) = 2\sqrt{\pi} f(\pi\omega)$, $B(\omega) = 0$. In other words the transform is unchanged except for scaling. Sample $f(x)$ at $N = 4$, 8, 16, and 32 equally spaced points on $[a, b] = [-5, 5]$, that is at $a + i\Delta$, $i = 0, \ldots, N - 1$, $\Delta = (b - a)/N$. Use EZFFTF to compute the discrete Fourier transform of the sequence. Compare the output values to the exact transform. Do not forget to account for the fact that $a \neq 0$. What is the maximum error?

(f) Repeat (e) but use CFFTF. Do not center the frequencies.

(g) Repeat (f) but center the frequencies. That is, multiply the i-th data value by $(-1)^i$ before calling CFFTF.

P11–6.–Light propagation.

(a) Two parallel planes are separated by a distance z. An optical disturbance ψ_1 of coherent radiation (laser light, for example) in plane 1 propagates or **diffracts** to plane 2 and results in an optical disturbance ψ_2 on the second plane. All detectors, including the human eye, respond to the modulus squared of the disturbance. In other words, we see $|\psi|^2$ rather than ψ. It can be shown that if z is large enough ψ_2 is proportional to the Fourier transform of ψ_1, a property that is known as Fraunhoffer or far-field diffraction. If the disturbance on plane 1 consists of a square light source of unit intensity then the diffraction pattern will look like Figure 11.9. Compute the diffraction pattern of a circular light source of unit intensity. Plot its square root as in Figure 11.9 using an axonometric plot. It is easy to generate such a plot for a two-dimensional array A of nonnegative elements if you have an N-point plotting routine PLOT(N, X, Y). Let X0 be an array of consecutive integers. Add an increment $i\Delta$ to the i-th column (or row) of A and place that in Y. Add the same increment to X0 and place that in X. Then plot the two arrays.

(b) Let K=(N-1)/2 and R=((I-K)**2+(J-K)**2)/S**2. If P(I,J) is the array representing the disturbance on plane 1, define P(I,J)=EXP(-R). This is called a two dimensional "Gaussian" source. Take N=64 and let S range between 1 and 10. Compute the far-field diffraction of this source. You should discover that the diffraction looks just like the input to within a scale factor. For what value of S is the source most nearly equal to its image?

(c) A more accurate description of the the disturbance on plane 2 that does not assume the planes are far apart is the Fresnel (near-field) diffraction equation. For a light source such as in (a) the near-field is proportional to the Fourier transform of the product of $\psi_1(x,y)$ and $\exp(2\pi i r/z)$ where r is the distance from the center of the plane to (x,y). Find the near-field diffraction for $z = 0.01, 0.1, 1.0, 10$ and 100. Plot the results, and verify that the diffraction approaches that of (a) when z is large.

P11–7.–This problem illustrates the effect of removing high frequency components on an image. A 64×64 image P(I,J) is composed of two small squares of unit intensity on a black background. That is P(I,J) = 0, except P = 1 when $30 \le I \le 34$ and $30 \le J \le 34$, or when $31 \le I \le 33$ and $2 \le J \le 4$.

(a) Generate an axonometric plot of the image.

(b) Using CFFTF compute the DFT of the image, scaled so the zero frequency is at the center. Then set to zero any frequency components outside a radius $R > 0$ of the center, perform the inverse transform, and replot the image. Write your program so R can be varied. As R decreases the small square will become more difficult to "see." For what value of R does it disappear?

P11–8.–Fourier transform of a derivative.

(a) Using the formula for the complex Fourier transform show that the Fourier transform of $f'(t)$ is given by $2\pi i \omega F(\omega)$, where $F(\omega)$ is the Fourier transform of $f(t)$. You can assume that $f(t)$ goes to zero as t approaches either plus or minus infinity. Is this a reasonable assumption? There is a similar result about Fourier series: if $f(t)$ has complex Fourier series coefficients C_j, then $f'(t)$ has Fourier series coefficients $2\pi i j f C_j$, assuming both series converge.

(b) Start with values of the function $f(t) = \exp(-t^2)$ at N points in the interval $[a, b] = [-5, 5]$, $a + i\Delta$, $i = 0,\ldots,N-1$, $\Delta = (b-a)/N$ and compute the discrete Fourier transform using CFFTF. Multiply the components of the transform by $2\pi i \omega$ and then take the inverse transform of the result. Plot your results against the exact derivative. Compute the maximum error as a function of N, for $N = 8, 16, 32$, and 64. Hints: (1) Before transforming, multiply the k-th function value by $(-1)^{k-1}$, $k = 1, \cdots, N$ to center the frequencies. (2) For this problem to work, it is important to associate the k-th component of the transformed array with its correct frequency. As k varies from 1 to N let $j = (k-1) - N/2$ (see Section 11). Then the frequency ω associated with the k-th component of the transform is $\omega = j/(b-a)$. (3) After computing the inverse transform, again multiply the output by $(-1)^{k-1}$, and do not forget the final division by N.

P11–9.–Computing an indefinite integral. If $g(x)$ has the complex Fourier series

$$g(x) = \sum_{-\infty}^{\infty} C_n \exp(2\pi i n x/(b-a)),$$

show that (if the series converges) the indefinite integral of g has Fourier series

$$g_I(x) \equiv \int_a^x g(t)\,dt = C_0(x-a) + \sum_{n=-\infty,n\neq0}^{\infty} C_n \frac{\exp(2\pi i n x/(b-a))}{2\pi i n/(b-a)}.$$

Suppose $C_0(b-a) = \int_a^b g(t)\,dt$ is known. Here is an algorithm for estimating $g_I(x)$. Evaluate $g(x_i)$, $i = 1, \ldots, N$, $x_i = a + (i-1)\Delta$, $\Delta = (b-a)/N$ and compute the discrete Fourier transform of these numbers. Divide the transform component associated with frequency $n/(b-a)$ by $2\pi i n/(b-a)$, except for the zero frequency component, which is set to zero. Compute the inverse discrete Fourier transform of the resulting sequence. Then the i-th component of the result plus $C_0(x_i-a)$ approximates $g_I(x_i)$. Implement this algorithm on $g(x) = \exp(-x^2)$, $[a,b] = [-5,5]$ for $N = 8$, 16, and 32 and compare it to the trapezoid rule with the same values. Repeat with $g(x) = 1/(1+25x^2)$. For which function are the results better? Is this expected?

P11–10.–Let $g(x) = \exp(-x)$, for $x \geq 0$, and 0 for $x < 0$.

(a) Find $A(\omega)$ and $B(\omega)$. Hint: Compute $G(\omega)$.

(b) Verify that

$$\int_0^\infty (A^2(\omega) + B^2(\omega))\,d\omega = 2\int_{-\infty}^\infty g^2(x)\,dx.$$

(c) Find and plot the power spectrum of $g(x)$.

(d) Sample $g(x)$ at $N = 64$ equally spaced points on $[0,5]$ and use EZFFTF to calculate A(K) and B(K). Calculate and plot the periodogram on the same plot as in (c).

(e) Using the coefficients from (d) calculate and plot $t_{N/2}(x)$ and compare it to $g(x)$.

P11–11.–The file ELNINO.DAT on a disk at the back of this book contains the 168 data points discussed in Section 9.

(a) Plot the data. Use EZFFTF to compute A_k and B_k. Plot the periodogram.

(b) Using CFFTF compute and plot the autocorrelation for this data.

(c) Compute and plot $t_m(x)$, $m = 84, 42, 21$, and 15.

P11–12.–For centuries people have noted that the face of the sun is not constant or uniform in appearance, but that darker regions appear at random locations on a cyclical basis. In 1848 Rudolf Wolfer proposed a rule that combined the number and size of these dark spots into a single number. Using archival records astronomers have applied Wolfer's rule and determined sunspot numbers back to the year 1700. Today these are measured by many observatories and the worldwide distribution of the data is coordinated by the Swiss Federal Observatory on a daily, monthly, and yearly basis. Sunspot activity is cyclical and variation in the Wolfer numbers has been correlated with weather and other terrestrial phenomena of economic significance; this accounts for the continuing interest in them. The file SUNSPOT.DAT that is on a disk at the back of this text, contains the average Wolfer sunspot number for each year from 1700 to 1987. Write a program to read these numbers and plot them. Then use EZFFTF on this data. The numbers do not have any noticeable trend that needs to be removed. Plot the power as a function of frequency, omitting the $\omega = 0$ term. Is there a single frequency that dominates? What is the period of the dominant cycle? Between now and the turn of the next century, in what years will sunspot activity be a maximum?

P11–13.–According to Section 10, when N is prime the time for EZFFTF should be $O(N^2)$. Figure 11.8 shows that the time required for EZFFTF when N is prime is greater than for other values of N, but it is unclear if the trend is quadratic. The table below gives a more extensive set of timings for EZFFTF when N is prime.

N	Time	N	Time	N	Time	N	Time
41	22	89	104	149	285	199	506
43	28	97	121	151	291	211	571
47	27	101	132	157	319	223	631
53	33	103	137	163	335	227	659
59	44	107	148	167	357	229	670
61	49	109	149	173	385	233	692
67	55	113	165	179	407	239	725
71	66	127	208	181	418	241	742
73	72	131	220	191	466	251	802
79	83	137	242	193	472		
83	88	139	247	197	495		

Using subroutine SQRLS, fit the timing values through $N = 151$ to the model $a + bt + ct^2$. Evaluate the resulting model at the remaining values of N. Is it a good fit?

P11–14.–This simple problem illustrates that the highest frequency one can analyze in data is one half the sampling frequency. A wheel rotates in a dark room at a rate of 1 rpm. Flash pictures are taken at $t = 0$ and every Δ seconds thereafter. A dark spot on the wheel appears at the 12 o'clock position in the first picture and at other positions later. If $\Delta = 15$ what is the apparent rate of rotation of the wheel? What if $\Delta = 30, 45, 60$?

11.16 PROLOGUES: EZFFTF, EZFFTB, CFFTF, CFFTB AND CFFT2D

```
        SUBROUTINE EZFFTF(N,R,AZERO,A,B,WSAVE)
C***BEGIN PROLOGUE  EZFFTF
C***DATE WRITTEN   790601    (YYMMDD)
C***REVISION DATE  830401    (YYMMDD)
C***CATEGORY NO.   J1A1
C***KEYWORDS  FOURIER TRANSFORM
C***AUTHOR  SWARZTRAUBER, P. N., (NCAR)
C***PURPOSE  A simplified real, periodic, forward transform
C***DESCRIPTION
C  Subroutine EZFFTF computes the Fourier coefficients of a real
C  perodic sequence (Fourier analysis).  The transform is defined
C  below at Output Parameters AZERO, A and B.  EZFFTF is a simplified
C  but slower version of RFFTF.
C
C    From the book "Numerical Methods and Software"
C        by  D. Kahaner, C. Moler, S. Nash
C            Prentice Hall 1988
C
C  Input Parameters
C  N       the length of the array R to be transformed.  The method
C          is most efficient when N is the product of small primes.
C  R       a real array of length N which contains the sequence
```

```
C              to be transformed.  R is not destroyed.
C  WSAVE      a work array which must be dimensioned at least 3*N+15
C              in the program that calls EZFFTF.  The WSAVE array must be
C              initialized by calling subroutine EZFFTI(N,WSAVE), and a
C              different WSAVE array must be used for each different
C              value of N.  This initialization does not have to be
C              repeated so long as N remains unchanged.  Thus subsequent
C              transforms can be obtained faster than the first.
C              The same WSAVE array can be used by EZFFTF and EZFFTB.
C
C  Output Parameters
C  AZERO      the sum from I=1 to I=N of R(I)/N
C  A,B        for N even B(N/2)=0. and A(N/2) is the sum from I=1 to
C              I=N of (-1)**(I-1)*R(I)/N
C              for N even define KMAX=N/2-1
C              for N odd  define KMAX=(N-1)/2
C              then for  k=1,...,KMAX
C                  A(K) equals the sum from I=1 to I=N of
C                      2./N*R(I)*COS(K*(I-1)*2*PI/N)
C                  B(K) equals the sum from I=1 to I=N of
C                      2./N*R(I)*SIN(K*(I-1)*2*PI/N)
C***REFERENCES  (NONE)
C***ROUTINES CALLED  RFFTF
C***END PROLOGUE  EZFFTF
C
        SUBROUTINE EZFFTB(N,R,AZERO,A,B,WSAVE)
C***BEGIN PROLOGUE  EZFFTB
C***DATE WRITTEN   790601   (YYMMDD)
C***REVISION DATE  830401   (YYMMDD)
C***CATEGORY NO.  J1A1
C***KEYWORDS  FOURIER TRANSFORM
C***AUTHOR  SWARZTRAUBER, P. N., (NCAR)
C***PURPOSE  A Simplified real, periodic, backward transform
C***DESCRIPTION
C  Subroutine EZFFTB computes a real perodic sequence from its
C  Fourier coefficients (Fourier synthesis).  The transform is
C  defined below at Output Parameter R.  EZFFTB is a simplified
C  but slower version of RFFTB.
C
C    From the book "Numerical Methods and Software"
C        by  D. Kahaner, C. Moler, S. Nash
C            Prentice Hall 1988
C
C  Input Parameters
C  N          the length of the output array R.  The method is most
C              efficient when N is the product of small primes.
C  AZERO      the constant Fourier coefficient
C  A,B        arrays which contain the remaining Fourier coefficients.
```

```
C               These arrays are not destroyed.
C               The length of these arrays depends on whether N is even or
C               odd.
C               If N is even, N/2    locations are required.
C               If N is odd, (N-1)/2 locations are required
C  WSAVE        a work array which must be dimensioned at least 3*N+15
C               in the program that calls EZFFTB.  The WSAVE array must be
C               initialized by calling subroutine EZFFTI(N,WSAVE), and a
C               different WSAVE array must be used for each different
C               value of N.  This initialization does not have to be
C               repeated so long as N remains unchanged.  Thus subsequent
C               transforms can be obtained faster than the first.
C               The same WSAVE array can be used by EZFFTF and EZFFTB.
C
C  Output Parameters
C  R            if N is even, define KMAX=N/2
C               if N is odd,  define KMAX=(N-1)/2
C               Then for I=1,...,N
C                    R(I)=AZERO plus the sum from K=1 to K=KMAX of
C                    A(K)*COS(K*(I-1)*2*PI/N)+B(K)*SIN(K*(I-1)*2*PI/N)
C ******************** Complex Notation ************************
C               For J=1,...,N
C               R(J) equals the sum from K=-KMAX to K=KMAX of
C                    C(K)*EXP(I*K*(J-1)*2*PI/N)
C               where
C                    C(K)  = .5*CMPLX(A(K),-B(K))    for K=1,...,KMAX
C                    C(-K) = CONJG(C(K))
C                    C(0)  = AZERO          and I=SQRT(-1)
C ************** Amplitude - Phase Notation **********************
C               For I=1,...,N
C               R(I) equals AZERO plus the sum from K=1 to K=KMAX of
C                    ALPHA(K)*COS(K*(I-1)*2*PI/N+BETA(K))
C               where
C                    ALPHA(K) = SQRT(A(K)*A(K)+B(K)*B(K))
C                    COS(BETA(K))=A(K)/ALPHA(K)
C                    SIN(BETA(K))=-B(K)/ALPHA(K)
C***REFERENCES   (NONE)
C***ROUTINES CALLED   RFFTB
C***END PROLOGUE   EZFFTB
        SUBROUTINE CFFTF(N,C,WSAVE)
C***BEGIN PROLOGUE   CFFTF
C***DATE WRITTEN    790601    (YYMMDD)
C***REVISION DATE   870721    (YYMMDD)
C***CATEGORY NO.    J1A2
C***KEYWORDS   FOURIER TRANSFORM
C***AUTHOR   SWARZTRAUBER, P. N., (NCAR)
C***PURPOSE   Forward transform of a complex, periodic sequence.
C***DESCRIPTION
```

```
C
C     From the book "Numerical Methods and Software"
C        by  D. Kahaner, C. Moler, S. Nash
C           Prentice Hall 1988
C
C Subroutine CFFTF computes the forward complex discrete Fourier
C transform (the Fourier analysis).  Equivalently, CFFTF computes
C the Fourier coefficients of a complex periodic sequence.
C
C The transform is not normalized.  To obtain a normalized transform
C the output must be divided by N.  Otherwise a call of CFFTF
C followed by a call of CFFTB will multiply the sequence by N.
C
C The array WSAVE which is used by subroutine CFFTF must be
C initialized by calling subroutine CFFTI(N,WSAVE).
C
C Input Parameters:
C     N          length of the complex sequence C.  The method is
C                more efficient when N is the product of small primes.
C
C     C          COMPLEX array of length N which contains the sequence.
C                This array is destroyed on output.
C
C     WSAVE      REAL work array which must be dimensioned at least 4*N+15
C                in the program that calls CFFTF.  The WSAVE array must be
C                initialized by calling subroutine CFFTI(N,WSAVE), and a
C                different WSAVE array must be used for each different
C                value of N.  This initialization does not have to be
C                repeated so long as N remains unchanged.  Thus subsequent
C                transforms can be obtained faster than the first.
C                The same WSAVE array can be used by CFFTF and CFFTB.
C
C Output Parameters:
C     C          for J=1,...,N
C                   C(J)=the sum from K=1,...,N of
C                        C(K)*EXP(-I*(J-1)*(K-1)*2*PI/N)     where I=SQRT(-1)
C
C     WSAVE      contains initialization calculations which must not be
C                destroyed between calls of subroutine CFFTF or CFFTB
C***REFERENCES  (NONE)
C***ROUTINES CALLED  CFFTF1
C***END PROLOGUE  CFFTF
      SUBROUTINE CFFTB(N,C,WSAVE)
C***BEGIN PROLOGUE  CFFTB
C***DATE WRITTEN   790601   (YYMMDD)
C***REVISION DATE  870721   (YYMMDD)
C***CATEGORY NO.   J1A2
C***KEYWORDS   FOURIER TRANSFORM
```

```
C***AUTHOR  SWARZTRAUBER, P. N., (NCAR)
C***PURPOSE  Unnormalized inverse of CFFTF.
C***DESCRIPTION
C
C    From the book "Numerical Methods and Software"
C         by  D. Kahaner, C. Moler, S. Nash
C              Prentice Hall 1988
C
C  Subroutine CFFTB computes the backward complex discrete Fourier
C  transform (the Fourier synthesis).  Equivalently, CFFTB computes
C  a complex periodic sequence from its Fourier coefficients.
C
C  A call of CFFTF followed by a call of CFFTB will multiply the
C  sequence by N.
C
C  The array WSAVE which is used by subroutine CFFTB must be
C  initialized by calling subroutine CFFTI(N,WSAVE).
C
C  Input Parameters:
C     N       length of the complex sequence C.  The method is
C             more efficient when N is the product of small primes.
C
C     C       COMPLEX array of length N which contains the sequence.
C             This array is destroyed on output.
C
C     WSAVE   REAL work array which must be dimensioned at least 4*N+15
C             in the program that calls CFFTB.  The WSAVE array must be
C             initialized by calling subroutine CFFTI(N,WSAVE), and a
C             different WSAVE array must be used for each different
C             value of N.  This initialization does not have to be
C             repeated so long as N remains unchanged.  Thus subsequent
C             transforms can be obtained faster than the first.
C             The same WSAVE array can be used by CFFTF and CFFTB.
C
C  Output Parameters:
C     C       For J=1,...,N
C                   C(J)=the sum from K=1,...,N of
C                       C(K)*EXP(I*(J-1)*(K-1)*2*PI/N)    where I=SQRT(-1)
C
C     WSAVE   contains initialization calculations which must not be
C             destroyed between calls of subroutine CFFTF or CFFTB
C***REFERENCES  (NONE)
C***ROUTINES CALLED  CFFTB1
C***END PROLOGUE  CFFTB
      SUBROUTINE CFFT2D(N,F,LDF,W,FORWD)
C***BEGIN PROLOGUE  CFFT2D
C***DATE WRITTEN   870811   (YYMMDD)
C***REVISION DATE  870811   (YYMMDD)
```

```
C***CATEGORY NO.   J1B
C***KEYWORDS  TWO DIMENSIONAL FOURIER TRANSFORM, FFT
C***AUTHOR  KAHANER, DAVID K., (NBS)
C***PURPOSE  Two dimensional complex fast Fourier transform.
C***DESCRIPTION
C       Two dimensional fast Fourier transform, forward or backward
C          of complex N*N matrix F.
C
C    From the book "Numerical Methods and Software"
C        by  D. Kahaner, C. Moler, S. Nash
C            Prentice Hall 1988
C
C    Input:
C      N: (INTEGER)        Number of rows and columns in the matrix F to be
C                            transformed. You must set N > 0, NOT CHECKED.
C      F: (COMPLEX)        Array of N*N complex values to be transformed.
C                            This array is overwritten on output.
C      LDF: (INTEGER)      Leading (first) dimension of the complex
C                            in the subroutine that calls CFFT2D. For
C                            example, if you declare F by either
C                                 COMPLEX F(0:15,0:20)  OR  F(16,21) then
C                            set LDF=16. You must have LDF >= N, NOT CHECKED.
C      W: (COMPLEX)        Array for internal use as work storage.
C                            Must be dimensioned by calling program to be
C                            at least 6N+15 REAL words or 3N+8 COMPLEX words.
C      FORWD: (LOGICAL)    Direction of transform. Set to .TRUE. for
C                            forward transform, set to .FALSE. for backward
C                            transform.
C
C
C    Output:
C      F: (COMPLEX)            Forward or reverse transformed input matrix.
C                              Output is unscaled, that is, a call to CFFT2D
C                              with FORWD=.TRUE. followed by a call to CFFT2D
C                              with FORWD=.FALSE. returns original data
C                              multiplied by N*N.
C
C
C    Remark:
C      For some applications it is desirable to have the transform scaled so
C          the center of the N by N frequency square corresponds to zero
C          frequency. The user can do this replacing the original input data
C          F(I,J) by F(I,J)*(-1.)**(I+J),   I,J =0,...,N-1.
C
C***REFERENCES  (NONE)
C***ROUTINES CALLED CFFTI, CFFTF, CFFTB
C***END PROLOGUE  CFFT2D
```

Bibliography

SOFTWARE

The following software collections have been discussed earlier in the book.

1. The ACM collected algorithms—available from the ACM Algorithms Distribution Service, IMSL Inc., Sixth Floor, NBC Building, 7500 Bellaire Blvd., Houston TX 77036.

2. Itpack—see the paper by Kincaid et al. (1982) referenced below.

3. Linpack—available from National Energy Software Center, Argonne National Laboratory, Argonne IL 60439.

4. Minpack—available from National Energy Software Center, Argonne National Laboratory, Argonne IL 60439.

5. Eispack—available from National Energy Software Center, Argonne National Laboratory, Argonne IL 60439.

6. Quadpack—available from Prof. R. Piessens, Computer Science Department, Univ of Leuven, Celestijnenlaan 200, A3030 Heverlee, Belgium.

7. Hompack—described in the paper by Watson et al. (1987) referenced below.

8. PORT—available from AT&T Bell Laboratories, 600 Mountain Ave., Murray Hill NJ 07074.

9. SLATEC—described in paper by Buzbee (1984) referenced below.

10. Sparspack—described in the paper by George et al. (1980) referenced below.

11. Yale sparse matrix package—available from Prof S.C. Eisenstat, Department of Computer Science, Yale University, New Haven CT 06520.

476

BOOKS AND PAPERS

[AbrS65] M. Abramowitz and I.A. Stegun, *Handbook of Mathematical Functions,* Dover, New York, 1965.

[AhHU74] A. Aho, J. Hopcroft, and J. Ullman, *The Design and Analysis of Computer Algorithms,* Addison Wesley, Reading, Massachusetts, 1974.

[Aike85] R. Aiken (editor), *Stiff Computation,* Oxford University Press, Oxford, England, 1985.

[Ande84] J. Anderson, *Fundamentals of Aerodynamics,* McGraw-Hill, New York, 1984.

[AndP75] H.C. Andrews, C.L. Patterson, *Outer Product Expansions and Their Uses in Digital Image Processing,* Amer. Math. Monthly 82 (1975) pp. 1–13.

[Anon85] Anonymous, *IEEE Standard for Binary Floating Point Arithmetic,* ANSI/IEEE Standard 754–1985, Inst. Electrical and Electronics Engineers, Inc., New York (1985).

[Back79] J. Backus, *The History of Fortran I, II, and III,* Annals of the History of Computing 1 (1979) pp. 21–37.

[BarR] R. Barnhill and A. Riesenfeld, *Computer Aided Geometric Design,* Academic Press, New York, 1974.

[BarS88] M. Barnsley and A. Sloan, *A Better Way to Compress Images,* Byte 13 (1988) pp. 215–223.

[BaBB] R. Bartels, J. Beatty and B. Barsky, *An Introduction to Splines for use in Computer Graphics and Geometric Modeling,* Morgan Kaufmann, Los Altos, California, 1987.

[Bell75] E.T. Bell, *Men of Mathematics,* Simon and Schuster, New York, 1975.

[BelG70] G. Bell and S. Glasstone, *Nuclear Reactor Theory,* Van Nostrand Reinhold, New York, 1970.

[BeKW81] D.A. Belsley, E. Kuh, and R. Welsch, *Regression Diagnostics: Identifying Influential Data and Sources of Collinearity,* Wiley, New York, 1981.

[Bloo76] P. Bloomfield, *Fourier Analaysis of Time Series: An Introduction,* Wiley-Interscience, New York, 1976.

[Bogg85] P. Boggs, R.H. Byrd, and R.B. Schnabel, *Numerical Optimization 1984,* SIAM, Philadelphia, Pennsylvania, 1985.

[BoHKS84] R.F. Boisvert, S.E. Howe, and D.K. Kahaner, *The Guide to Available Mathematical Software,* National Technical Information Service Report PB 84-171305, Springfield, Virginia (1984).

[Boor78] C. de Boor, *A Practical Guide to Splines,* Springer, Berlin, 1978.

[BrFS87] P. Bratley, B. L. Fox, and L. Schrage, *A Guide to Simulation,* Springer, Berlin, 1987.

[Bren72] R.P. Brent, *Algorithms for Minimization Without Derivatives,* Prentice-Hall, Englewood Cliffs, NJ, 1972.

[Brig74] O. Brigham, *The Fast Fourier Transform,* Prentice Hall, Englewood Cliffs, New Jersey, 1974.

[Buzb84] B.L. BUZBEE, *The SLATEC Common Mathematical Library,* in *Sources and Development of Mathematical Software,* W.R. Cowell (editor), Prentice-Hall, Englewood Cliffs, New Jersey (1984).

[ByrH87] G. BYRNE and A. HINDMARSH, *Stiff ODE Solvers: A Review of Current and Coming Attractions,* J. Computational Physics 70 (1987) pp. 1–62.

[CarC75] L.L. CARTER and E.D. CASHWELL, *Particle-Transport with the Monte Carlo Method,* ERDA Critical Review Series, TID-26607, National Technical Information Service, Springfield, Virginia (1975).

[ChaP77] S. CHATTERJEE and B. PRICE, *Regression Analysis by Example,* Wiley, New York, 1977.

[Clin74] A. CLINE, *Scalar-and-Planar-Valued Curve Fitting Using Splines Under Tension,* Communications of ACM 17 (1974) pp. 218–220.

[CoGO76] P. CONCUS, G.H. GOLUB, and D.P. O'LEARY, *A generalized conjugate gradient method for the numerical solution of elliptic partial differential equations,* in *Sparse Matrix Computations,* J.R. Bunch and D.J. Rose (editors), Academic Press, New York (1976).

[Dant63] G.B. DANTZIG, *Linear Programming and Extensions,* Princeton University Press, Princeton, NJ, 1963.

[Davi63] P. DAVIS, *Interpolation and Approximation,* Blaisdell, New York, 1963.

[DavR84] P. DAVIS and P. RABINOWITZ, *Methods of Numerical Integration, 2nd Ed.,* Academic Press, New York, 1984.

[Dekk69] T.J. DEKKER, *Finding a zero by means of successive linear interpolation,* in *Constructive aspects of the fundamental theorem of algebra,* B. Dejon and P. Henrici (editors), Wiley-Interscience, New York (1969).

[DenM74] J.E. DENNIS Jr. and J. MORÉ, *Quasi-Newton methods, motivation and theory,* SIAM Review 19 (1974) pp. 46–89.

[DenS83] J.E. DENNIS Jr. and R.B. SCHNABEL, *Numerical Methods for Unconstrained Optimization and Nonlinear Equations,* Prentice-Hall, Englewood Cliffs, NJ, 1983.

[DonR84] E. de DONCKER and I. ROBINSON, *An Algorithm for Automatic Integration Over a Triangle Using Nonlinear Extrapolation,* ACM Transactions on Mathematical Software 10 (1984) pp. 1–16.

[DMBS79] J.J. DONGARRA, C.B. MOLER, J.R. BUNCH, and G.W. STEWART, *LINPACK Users' Guide,* SIAM, Philadelphia, 1979.

[DraS81] N.R. DRAPER and H. SMITH, *Applied Regression Analysis,* Wiley, New York, 1981.

[DyGL87] N. DYN, J. GREGORY, D. LEVIN, *A 4-Point Interpolatory Subdivision Scheme For Curve Design,* The Weizmann Institute of Science, Department of Applied Mathematics, Rehovot, Israel 76100 (1987).

[FoMM77] G. FORSYTHE, M. MALCOLM, and C. MOLER, *Computer Methods for Mathematical Computations,* Prentice-Hall, Englewood Cliffs, New Jersey, 1977.

[ForM67] G.E. FORSYTHE and C.B. MOLER, *Computer Solution of Linear Algebraic Systems,* Prentice-Hall, Englewood Cliffs, NJ, 1967.

[FoFC82] A. FOURNIER, D. FUSSELL, L. CARPENTER, *Computer Rendering of Stochastic Models,* Comm of ACM 25 (1982) pp. 371–384.

[FoHS78] P.A. Fox, A.D. Hall, and N.L. Schryer, *Framework for a portable library*, ACM Transactions on Mathematical Software 4 (1978) pp. 177–188.

[FriC80] F.N. Fritsch and R.E. Carlson, *Monotone Piecewise Cubic Interpolation*, SIAM J. Num. Anal. 17 (1980) pp. 238–246.

[Gaff87] P. Gaffney, *When Things Go Wrong ...*, IBM Bergen Scientific Centre Report BSC87/1 (1987).

[Gard61] M. Gardner, *Mathematical Puzzles and Diversions*, Simon and Schuster, New York, 1961.

[GeoL81] A. George and J.W. Liu, *Computer Solution of Large Sparse Positive Definite Systems*, Prentice-Hall, Englewood Cliffs, NJ, 1981.

[GeLN80] A. George, J.W. Liu, and E. Ng, *User Guide for SPARSPACK: Waterloo Sparse Linear Equations Package*, Report CS–78–30 (revised 1980), Computers Science Dept., University of Waterloo, Waterloo, Canada (1980).

[GGMS74] P.E. Gill, G.H. Golub, W. Murray, and M.A. Saunders, *Methods for modifying matrix factorizations*, Mathematics of Computation 28 (1974) pp. 505–535.

[GilM74] P.E. Gill and W. Murray, *Newton-type methods for unconstrained and linearly constrained optimization*, Mathematical Programming 28 (1974) pp. 311–350.

[GiMW81] P.E. Gill, W. Murray, and M.H. Wright, *Practical Optimization*, Academic Press, New York, 1981.

[GolV83] G.H. Golub and C.F. Van Loan, *Matrix Computations*, The Johns Hopkins University Press, Baltimore, 1983.

[Habe70] S. Haber, *Numerical Evaluation of Multiple Integrals*, SIAM Review 12 (1970) pp. 481–526.

[Habe77] R. Haberman, *Mathematical Models*, Prentice-Hall, Englewood Cliffs, NJ, 1977.

[Hamm62] R. Hamming, *Numerical Methods for Scientists and Engineers*, McGraw-Hill, New York, 1962.

[Hans69] E.R. Hansen, *Topics in Interval Analysis*, Oxford University Press, London, 1969.

[Hind80] A. Hindmarsh, *LSODE and LSODEI, Two Initial Value Ordinary Differential Equation Solvers*, ACM SIGNUM Newsletter 15 (1980) pp. 10–11.

[JenT70] M.A. Jenkins and J.F. Traub, *A three-stage algorithm for real polynomials using quadratic iteration*, SIAM Journal on Numerical Analysis 7 (1970) pp. 545–566.

[JonK83] R.E. Jones and D.K. Kahaner, *XERROR, the SLATEC Error-handling Package*, Software–Practice and Experience 13 (1983) pp. 251–257.

[Joyc71] D. Joyce, *Survey of Extrapolation Processes in Numerical Analysis*, SIAM Rev. 13 (1971) pp. 435–490.

[Kaha70] D. Kahaner, *Matrix Description of the Fast Fourier Transform*, IEEE Transactions on Audio and Electroacoustics AU–18 (1970) pp. 442–450.

[Kaha78] D. Kahaner, *The Fast Fourier Transform by Polynomial Evaluation*, ZAMP 29 (1978) pp. 387–394.

[KenG80] W. J. Kennedy and J. E. Gentle, *Statistical Computing*, Marcel Dekker, New York, 1988.

[KRYG82] D. Kincaid, G. Respess, D. Young, and R. Grimes, *Itpack 2C: A Fortran package for solving large sparse linear systems by adaptive accelerated iterative methods*, ACM Transactions on Mathematical Software 8 (1982) pp. 302–322.

[Knut69] D.E. Knuth, *The Art of Computer Programming: Volume 2, Seminumerical Algorithms*, Addison-Wesley, Reading, MA, 1969.

[Krog70] F. Krogh, *VODQ/SVDQ/DVDQ, Variable Order Integrators for the Numerical Solution of Ordinary Differential Equations*, Tech Brief NPO-11643, Jet Propulsion Laboratory, California Institute of Technology, Pasadena California (1970).

[Lamb73] J. Lambert, *Computational Methods in Ordinary Differential Equations*, John Wiley, New York, 1973.

[Lanc66] C. Lanczos, *Discourse on Fourier Series*, Oliver and Boyd, Edinburgh, 1966.

[LapS71] L. Lapidus and J. Seinfeld, *Numerical Solution of Ordinary Differential Equations*, Academic Press, New York, 1971.

[Laur78] D. Laurie, *Automatic Numerical Integration Over a Triangle*, Technical Report, National Research Center for Mathematical Sciences of the CSIR, P.O. Box 395, Pretoria, South Africa (1978).

[Laws79] C.L. Lawson et al, *Basic Linear Algebra Subprograms for Fortran Usage*, ACM Trans Math Software 5 (1979) pp. 308–323.

[LawH74] C.L. Lawson and R.J. Hanson, *Solving Least Squares Problems*, Prentice-Hall, Englewood Cliffs, NJ, 1974.

[Lehm51] D. Lehmer, *Mathematical Models in Large-scale Computing Units*, in *Proceedings of the Second Symposium on Large Scale Digital Computing Machinery*, Harvard University Press, Cambridge, Massachusetts, pp. 141–146 (1951).

[Luen84] D.G. Luenberger, *Introduction to Linear and Nonlinear Programming*, Addison-Wesley, Reading, Massachusetts, 1984.

[Lyne83] J. Lyness, *AUG2—Integration Over a Triangle*, Argonne National Laboratory Report, ANL/MCS–TM–13 (1983).

[Mand82] B. Mandelbrot, *The Fractal Geometry of Nature*, Freeman, New York, 1982.

[Mars68] G. Marsaglia, *Random numbers fall mainly in the planes*, Proc. Nat. Acad. Sci. 61 (1968) pp. 25–28.

[Mars84] G. Marsaglia, *A Fast, Easily Implemented Method for Sampling from Decreasing of Symmetric Unimodal Density Functions*, SIAM J. Sci. Stat. Comp. 5 (1984) pp. 349–359.

[Moor79] R.E. Moore, *Methods and Applications of Interval Analysis*, SIAM, Philadelphia, 1979.

[MoGH80] J.J. Moré, B.S. Garbow, and K.E. Hillstrom, *User Guide for MINPACK–1*, Report ANL–80–74, Argonne National Laboratory, Argonne, Illinois (1980).

[MolS70] C.B. Moler, L.P. Solomon, *Use of Splines and Numerical Integration in Geometrical Acoustics*, J. Acoustical Soc. Amer. 48 (1970) pp. 739–744.

[Mose71] J. Moses, *Symbolic Integration: The Stormy Decade*, Communications of ACM 14 (1971) pp. 548–560.

[MurI82] K. Murota and M. Iri, *Parameter Tuning and Repeated Application of the IMT Type Transformation in Numerical Quadrature*, Numer. Math. 38 (1982) pp. 347–363.

[MurS78] B.A. Murtagh and M. Saunders, *Large-scale linearly constrained optimization,* Mathematical Programming 14 (1978) pp. 41–72.

[Orch68] W. Orchard-Hays, *Advanced Linear-programming Computing Techniques,* McGraw-Hill, New York, 1968.

[OrtP81] J. Ortega and W. Poole, *An Introduction to Numerical Methods for Differential Equations,* Pitman, Marshfield, MA, 1981.

[OrtR70] J.M. Ortega and W.C. Rheinboldt, *Iterative Solution of Nonlinear Equations in Several Variables,* Academic Press, New York, 1970.

[Osbo66] M.R. Osborne, *On Nordsieck's Method for the Numerical Solution of Ordinary Differential Equations,* BIT 6 (1966) pp. 51–57.

[PeiR86] H. Peitgen and P. Richter, *The Beauty of Fractals,* Springer, Berlin, 1986.

[Pies83] R. Piessens, et al., *QUADPACK: A Subroutine Package for Automatic Integration,* Springer, Berlin, 1983.

[Pren75] P. Prenter, *Splines and Variational Methods,* Wiley, New York, 1975.

[Prit86] A. Pritsker, *Introduction to Simulation and SLAM II,* Wiley, New York, 1986.

[Rals65] A. Ralston, *A First Course in Numerical Analysis,* McGraw-Hill, New York, 1965.

[Rasm85] E. Rasmusson, *El Niño and Variations in Climate,* American Scientist 73 (1985) pp. 168–177.

[RosK82] A. Rosenfeld and A. Kak, *Digital Picture Processing,* Academic Press, New York, 1982.

[Ross83] S. Ross, *Stochastic Processes,* Wiley, New York, 1983.

[Roy85] M.R. Roy, *A History of Computing Technology,* Prentice-Hall, Englewood Cliffs, NJ, 1985.

[Rubi81] R. Rubinstein, *Simulation and the Monte Carlo Method,* John Wiley, New York, 1981.

[Russ83] E. Russell, *Building Simulation Models with Simscript II.5,* Consolidated Analysis Centers Inc., New York, 1983.

[RuRM79] B.W. Rust, R.M. Rotty, G. Maryland, *Inferences Drawn From Atmospheric CO_2 Data,* J. Geophysical Research 84 (1979) pp. 3115–3122.

[ScKW82] R.B. Schnabel, J.E. Koontz, and B.E. Weiss, *A modular system of algorithms for unconstrained minimization,* Report CU–CS–240–82, Computer Science Department, University of Colorado, Boulder, Colorado (1982).

[Schr74] T. Schriber, *Simulation Using GPSS,* Wiley, New York, 1974.

[Schu73] M. Schultz, *Spline Analysis,* Prentice-Hall, Englewood Cliffs, New Jersey, 1973.

[ShaB84] L. Shampine and C. Baca, *Error Estimators for Stiff Differential Equations,* J. Computational and Applied Mathematics 11 (1984) pp. 197–208.

[ShaG75] L. Shampine and M. Gordon, *Computer Solution of Ordinary Differential Equations: the Initial Value Problem,* Freeman, San Francisco, 1975.

[Skee79] R.D. Skeel, *Equivalent Forms of Multistep Formulas,* Math. Comp. 33 (1979) pp. 1229–1250.

[SBGI74] B.T. Smith, J.M. Boyle, B.S. Garbow, Y. Ikebe, V.C. Klema, and C.B. Moler, *Matrix Eigensystem Routines—EISPACK Guide,* Springer, Berline, 1974.

[Ster74] P. STERBENZ, *Floating Point Computation*, Prentice-Hall, Englewood Cliffs, New Jersey, 1974.

[Stro72] A. STROUD, *Approximate Calculation of Multiple Integrals*, Prentice-Hall, Englewood Cliffs, New Jersey, 1972.

[Swar75] P. SWARZTRAUBER, *Efficient Subprograms for the Solution of Elliptic Partial Differential Equations*, Report TN/LA–109, National Center for Atmospheric Research, Boulder, Colorado (1975).

[Timo56] S. TIMOSHENKO, *Strength of Materials, Part II*, Van Nostrand, Princeton, NJ, 1956.

[Varg62] R.S. VARGA, *Matrix Iterative Analysis*, Prentice-Hall, Englewood Cliffs, NJ, 1962.

[Wats86] L.T. WATSON, *Numerical linear algebra aspects of globally convergent homotopy methods*, SIAM Review 28 (1986) pp. 529–545.

[WaBM87] L.T. WATSON, S.C. BILLUPS, and A.P. MORGAN, *HOMPACK: A suite of codes for globally convergent homotopy algorithms*, ACM Transactions on Mathematical Software 13 (1987) pp. 281–310.

[Weav83] H. J. WEAVER, *Applications of Discrete and Continuous Fourier Analysis*, Wiley-Interscience, New York, 1983.

[Wend66] B. WENDROFF, *Theoretical Numerical Analysis*, Academic Press, New York, 1966.

[WhiR24] E. WHITTAKER and G. ROBINSON, *The Calculus of Observations*, Blackie and Son, London, 1924.

[Wilk63] J.H. WILKINSON, *Rounding Errors in Algebraic Processes*, Prentice-Hall, Englewood Cliffs, NJ, 1963.

[Wilk67] J.H. WILKINSON, *Two algorithms based on successive linear interpolation*, Tech. Report STAN–CS–67–60, Computer Science Department, Stanford University, Stanford, California (1967).

INDEX